Fuzzy Decision Making and Soft Computing Applications

Fuzzy Decision Making and Soft Computing Applications

Editors

Giuseppe De Pietro
Marco Pota

MDPI • Basel • Beijing • Wuhan • Barcelona • Belgrade • Manchester • Tokyo • Cluj • Tianjin

Editors
Giuseppe De Pietro
National Research Council of Italy (CNR) - Institute for High Performance Computing & Networking (ICAR)
Italy

Marco Pota
National Research Council of Italy (CNR) - Institute for High Performance Computing & Networking (ICAR)
Italy

Editorial Office
MDPI
St. Alban-Anlage 66
4052 Basel, Switzerland

This is a reprint of articles from the Special Issue published online in the open access journal *Applied System Innovation* (ISSN 2571-5577) (available at: https://www.mdpi.com/journal/asi/special_issues/fdmaking?authAll=true).

For citation purposes, cite each article independently as indicated on the article page online and as indicated below:

LastName, A.A.; LastName, B.B.; LastName, C.C. Article Title. *Journal Name* **Year**, *Volume Number*, Page Range.

ISBN 978-3-0365-4929-3 (Hbk)
ISBN 978-3-0365-4930-9 (PDF)

Contents

applied system innovation

MDPI

Editorial

Special Issue "Fuzzy Decision Making and Soft Computing Applications"

Giuseppe De Pietro and Marco Pota *

Institute for High Performance Computing and Networking–National Research Council of Italy (ICAR-CNR), 80131 Naples, Italy; giuseppe.depietro@icar.cnr.it
* Correspondence: marco.pota@icar.cnr.it

Research on fuzzy logic [1] and soft computing for decision making has a long history. In many fields of application, rule-based fuzzy systems have been employed [2–4] for their unique properties in solving modelling problems. In particular, decision-making systems often deal with uncertain data. Moreover, in some fields of application, such as differential diagnosis in medicine, a meaningful confidence measure is required to be associated with the classification result in order to show all possible outcomes with the relative likelihood. Finally, semantically meaningful systems are often required, providing clear and logical interpretation of the inference process, in order to encapsulate them in interactive frameworks of cognitive systems, or to enable validation by domain experts. These issues can be accomplished, on the one hand, by encoding uncertain numerical data in terms of interpretable linguistic variables [5]. On the other hand, fuzzy rules show a clear and logical justification for each conclusion [6,7]. Finally, if desired, fuzzy systems allow classification results to be presented associated with a confidence measure, such as the probability of different classes [8].

The remarkable progress made by these approaches in various fields underlines their benefits and is stimulating further research. In particular, despite the remarkable successes in different tasks, research on these approaches is a field of increasing interest [9], with regard to theoretical aspects, which are being deepened [10–12], as well as aspects regarding procedures for learning fuzzy systems optimizing accuracy and/or interpretability, or for solving mathematical tasks using fuzzy numbers and soft computing [13–18]. Moreover, these approaches are prone to easily and proficiently be employed in different new fields of application [19–21].

This Special Issue collects original research articles discussing cutting-edge work as well as perspectives on future directions in the whole range of theoretical and practical aspects in this research area. In particular, there are 12 contributions selected for this Special Issue, representing progresses in the following areas specifically addressed.

1. **Theory of fuzzy systems and soft computing.** The authors of [10] consider causal graphs and propose a procedure to explicitly understand underlying assumptions, the kind of data and methodology needed to understand a given relationship, and how to develop explicit assumptions with clear alternatives in order to apply a process of probabilistic elimination. In [11], the authors unambiguously define the relations "greater than", "equal to", and their combination, in the space of all ordered fuzzy numbers, to solve optimization tasks. Moreover, in [12], a problem of "acceptance of an optimal solution" is presented in the form of a vector problem of mathematical programming. The theory of vector optimization is proposed as a mathematical apparatus for the acceptance of optimal solutions of such a class of problems, and the analysis and problem definition of decision making under the conditions of certainty and uncertainty are presented.
2. **Learning procedures.** In [13], the authors propose two types ("infimum type" and "supremum type") of dual double fuzzy semi-metric, as well as different types of

Citation: De Pietro, G.; Pota, M. Special Issue "Fuzzy Decision Making and Soft Computing Applications". *Appl. Syst. Innov.* **2022**, *5*, 54. https://doi.org/10.3390/asi5030054

Received: 1 June 2022
Accepted: 8 June 2022
Published: 10 June 2022

triangle inequalities, which are used to investigate the convergence. Another contribution [14] proposes an adaptive-network-based fuzzy inference system (ANFIS) model for the accurate estimation of signal propagation. Results on benchmark data show that the proposed model outperforms nondeterministic models in terms of accuracy and presents flexibility, ease of use, robustness, generalization capability, and an alleviated training process for propagation prediction in complex scenarios. Some authors contributed with three different papers. In the first [15], to solve the Cauchy problem, three fuzzy numerical methods, based on the combination of fuzzy transform with one-step, two-step, and three-step numerical methods, are introduced. The error analysis of the new fuzzy methods is discussed, showing more accurate results compared with other existing methods. The other two papers by the same authors [16,17] report parts I and II of the same voluminous work, where new approximation methods for solving systems of ordinary differential equations (SODEs) using fuzzy transform are introduced and discussed. In particular, different modified numerical schemes and new representations of basic functions are proposed, the error analysis of the new approximation methods and the properties of the uniform fuzzy partition are examined, and numerical examples showing improved accuracy are presented. A further work [18] assesses how three shaking procedures affect the performance of a metaheuristic General Variable Neighbourhood Search algorithm. The different schemes were applied on benchmark instances of the Traveling Salesman Problem to examine the potential advantage of any of the three metaheuristic schemes, showing similarities and differences among different methods.

3. **Decision-making applications employing fuzzy logic and soft computing.** Contributions in this field show a variety of possible applications. In the context of the characterization of basmati rice product value using an image-based grading process, the authors in [19] propose a model for quality grade testing and identification, using a novel digital-image-processing- and knowledge-based ANFIS. This approach provides capabilities to simulate the behaviour of an expert in the characterization of rice grains using their physical properties, and compared to other machine learning techniques, its results are promising in terms of classification accuracy and efficiency. In the field of electron beam (EB) measurements, the author of [20] presents a novel method of restoring the EB measurements that are degraded by linear motion blur. The author's approach is based on a fuzzy inference system and a Wiener inverse filter, providing autonomy, reliability, flexibility, and real-time execution, in restoring highly degraded signals without requiring exact knowledge of EB probe size, and a demonstration is given by comparing ground truth signals with restorations. Finally, in [21], the motion control of mobile robots in a cluttered environment with obstacles is considered. In particular, to control the motion of a mobile robot using an eye gaze coordinate as inputs to the system, the paper presents an intelligent vision-based gaze guided robot control, utilizing an overhead camera, an eye-tracking device, a differential drive mobile robot, vision, and an interval-type-2 fuzzy inference tool. Experiments and simulation results indicate that the system can successfully perform operator intention, modulating speed and direction accordingly.

Funding: This research received no external funding.

Conflicts of Interest: The authors declare no conflict of interest.

References

1. Zadeh, L. Fuzzy sets. *Inf. Control* **1965**, *8*, 338–353. [CrossRef]
2. Pota, M.; Scalco, E.; Sanguineti, G.; Cattaneo, G.M.; Esposito, M.; Rizzo, G. Early classification of parotid glands shrinkage in radiotherapy patients: A comparative study. *Biosyst. Eng.* **2015**, *138*, 77–89. [CrossRef]
3. Pota, M.; Esposito, M.; De Pietro, G. Learning to rank answers to closed-domain questions by using fuzzy logic. In Proceedings of the 2017 IEEE International Conference on Fuzzy Systems (FUZZ-IEEE), Naples, Italy, 9–12 July 2017; pp. 1–6. [CrossRef]

4. Pota, M.; Scalco, E.; Sanguineti, G.; Farneti, A.; Cattaneo, G.M.; Rizzo, G.; Esposito, M. Early prediction of radiotherapy-induced parotid shrinkage and toxicity based on CT radiomics and fuzzy classification. *Artif. Intell. Med.* **2017**, *81*, 41–53. [CrossRef] [PubMed]
5. Pota, M.; Esposito, M.; De Pietro, G. Likelihood-fuzzy analysis: From data, through statistics, to interpretable fuzzy classifiers. *Int. J. Approx. Reason.* **2018**, *93*, 88–102. [CrossRef]
6. Pota, M.; Esposito, M.; De Pietro, G. Interpretability indexes for Fuzzy classification in cognitive systems. In Proceedings of the 2016 IEEE International Conference on Fuzzy Systems (FUZZ-IEEE), Vancouver, BC, Canada, 24–29 July 2016; pp. 24–31. [CrossRef]
7. Pota, M.; Esposito, M.; De Pietro, G. Best fuzzy partitions to build interpretable DSSs for classification in medicine. In Proceedings of the 2013 International Conference on Hybrid Artificial Intelligence Systems, Salamanca, Spain, 11–13 September 2013; Springer: Berlin/Heidelberg, Germany, 2013; pp. 558–567. [CrossRef]
8. Pota, M.; Esposito, M.; De Pietro, G. Designing rule-based fuzzy systems for classification in medicine. *Knowl.-Based Syst.* **2017**, *124*, 105–132. [CrossRef]
9. Yazdanbakhsh, O.; Dick, S. A systematic review of complex fuzzy sets and logic. *Fuzzy Sets Syst.* **2018**, *338*, 1–22. [CrossRef]
10. Levine, E.; Butler, J.S. Causal Graphs and Concept-Mapping Assumptions. *Appl. Syst. Innov.* **2018**, *1*, 25. [CrossRef]
11. Piasecki, K. Relation "Greater than or Equal to" between Ordered Fuzzy Numbers. *Appl. Syst. Innov.* **2019**, *2*, 26. [CrossRef]
12. Mashunin, Y.K. Mathematical Apparatus of Optimal Decision-Making Based on Vector Optimization. *Appl. Syst. Innov.* **2019**, *2*, 32. [CrossRef]
13. Wu, H.-C. Using Dual Double Fuzzy Semi-Metric to Study the Convergence. *Appl. Syst. Innov.* **2019**, *2*, 13. [CrossRef]
14. Hosseinzadeh, S.; Larijani, H.; Curtis, K.; Wixted, A. An Adaptive Neuro-Fuzzy Propagation Model for LoRaWAN. *Appl. Syst. Innov.* **2019**, *2*, 10. [CrossRef]
15. ALKasasbeh, H.; Perfilieva, I.; Ahmad, M.Z.; Yahya, Z.R. New Fuzzy Numerical Methods for Solving Cauchy Problems. *Appl. Syst. Innov.* **2018**, *1*, 15. [CrossRef]
16. ALKasasbeh, H.; Perfilieva, I.; Ahmad, M.Z.; Yahya, Z.R. New Approximation Methods Based on Fuzzy Transform for Solving SODEs: I. *Appl. Syst. Innov.* **2018**, *1*, 29. [CrossRef]
17. ALKasasbeh, H.; Perfilieva, I.; Ahmad, M.Z.; Yahya, Z.R. New Approximation Methods Based on Fuzzy Transform for Solving SODEs: II. *Appl. Syst. Innov.* **2018**, *1*, 30. [CrossRef]
18. Papalitsas, C.; Karakostas, P.; Andronikos, T. A Performance Study of the Impact of Different Perturbation Methods on the Efficiency of GVNS for Solving TSP. *Appl. Syst. Innov.* **2019**, *2*, 31. [CrossRef]
19. Mandal, D. Adaptive Neuro-Fuzzy Inference System Based Grading of Basmati Rice Grains Using Image Processing Technique. *Appl. Syst. Innov.* **2018**, *1*, 19. [CrossRef]
20. Hosseinzadeh, S. A Fuzzy Inference System for Unsupervised Deblurring of Motion Blur in Electron Beam Calibration. *Appl. Syst. Innov.* **2018**, *1*, 48. [CrossRef]
21. Dirik, M.; Castillo, O.; Kocamaz, A.F. Gaze-Guided Control of an Autonomous Mobile Robot Using Type-2 Fuzzy Logic. *Appl. Syst. Innov.* **2019**, *2*, 14. [CrossRef]

applied system innovation

MDPI

Article

Causal Graphs and Concept-Mapping Assumptions

Eli Levine [1,*] **and J. S. Butler** [2]

1 User-Supplied Information, University at Buffalo, 12 Capen Hall, Buffalo, New York, NY 14260-1660, USA
2 Martin School, Economics, University of Kentucky, Lexington, KY 40506, USA; j.s.butler@uky.edu
* Correspondence: elilevin@buffalo.edu

Received: 11 May 2018; Accepted: 20 July 2018; Published: 24 July 2018

Abstract: Determining what constitutes a causal relationship between two or more concepts, and how to infer causation, are fundamental concepts in statistics and all the sciences. Causation becomes especially difficult in the social sciences where there is a myriad of different factors that are not always easily observed or measured that directly or indirectly influence the dynamic relationships between independent variables and dependent variables. This paper proposes a procedure for helping researchers explicitly understand what their underlying assumptions are, what kind of data and methodology are needed to understand a given relationship, and how to develop explicit assumptions with clear alternatives, such that researchers can then apply a process of probabilistic elimination. The procedure borrows from Pearl's concept of "causal diagrams" and concept mapping to create a repeatable, step-by-step process for systematically researching complex relationships and, more generally, complex systems. The significance of this methodology is that it can help researchers determine what is more probably accurate and what is less probably accurate in a comprehensive fashion for complex phenomena. This can help resolve many of our current and future political and policy debates by eliminating that which has no evidence in support of it, and that which has evidence against it, from the pool of what can be permitted in research and debates. By defining and streamlining a process for inferring truth in a way that is graspable by human cognition, we can begin to have more productive and effective discussions around political and policy questions.

Keywords: causality; statistics; concept-mapping; causal graph

1. Introduction

Causal inference is a key goal for understanding the relationships among phenomena in the real world that researchers are attempting to study [1] This becomes a challenging task when possible causal phenomena are numerous, highly interrelated, complex, and complicated to study with validity [2,3]. As things currently stand, there is no clear method for either promoting correct facts and high quality and honestly treated evidence, or for eliminating incorrect facts and inferences of poor quality, or dishonestly treated evidence from the pool of knowledge that is acceptable in policy debates. This paper proposes a possible method to clarify researchers' intentions and work, determine what data are necessary to collect, guide the selection of the methodology of treating the evidence, and produce possible counterfactual arguments that can be tested to establish a greater probability that correct inferences are drawn from the data. The hope of this paper is to clarify what is more probably true from what is less probably true and to streamline the pool of evidence that is permissible in policy and political debates. High quality and honestly treated evidence gains precedence over, and is promoted in discussions and debates, at the expense of poor quality and dishonestly treated evidence.

2. Literature Review

"Causality" is defined as "the relationship between something that happens or exists and the thing that causes it" [4]. Determining causal relations among variables is a challenging and much studied topic [1,5–10]. Much of the literature on causal relations comes from the medical field of epidemiology [11] and is used to infer causal relationships in disease diagnoses and treatment effects of medical regimens [12]. Causality is also a much studied subject in the social sciences. Its inference is typically derived from a statistical method or technique or qualitative analysis [6,13–17]. Testing for Granger Causality, which is a statistical concept where variables that cause effect variables contain information that predicts effect variables within them, has used "path diagrams" in the literature [18]. This paper specifically draws upon the concept of the "causal graph" described by Pearl [19] as the basis of this methodology. The causal graph is used alongside "concept mapping" in order to tease out the underlying assumptions about the nature and relationships among the variables in question. Casual graphing and concept mapping promote better understandings of the researchers' assumptions, and they develop alternative counterfactual cases with different causal graphs. Causal graphing also could help design research to test the factual and causal validity of the causal graph and, by extension, the researchers' concept map [20]. In summary, the researchers and other stakeholders may make different concept maps and causal graphs according to existing methodologies. The difference with this proposed method is that it actively seeks to remove all or parts of concept maps and causal graphs to infer what is more probably true in the real world itself.

A "causal graph is a directed graph that describes the variable dependencies" [21]. Causal graphs were first developed in the fields of mathematics, computer science, machine learning, and statistics [1,22,23] but have since evolved to the study of complex phenomena, such as epidemiology [9] and planning [21,24,25]. While causal graphs are not new tools in several academic fields and have been used in statistical analyses for developing causal relations after the data collection, it does not appear that they have been widely used by researchers to sketch assumptions and hypotheses before the data has been collected.

The ideas expressed here are not new in the field of economics. One of the first two Nobel Laureates in Economics in 1969, Jan Tinbergen, collected all proposed macroeconomic models in the late 1930s and built models of the business cycle with a similar technique ([26] pp. 101–130). Tinbergen "explained his model building as an iterative process involving both hypotheses and statistical estimation" ([26], p. 103). Morgan (1990) points out that "Despite their usefulness, few copied his graphical methods" ([26], footnote 9, p. 111).

While Tinbergen's methods are similar to the concept of causal graph modeling that is described here, they are not quite the same. Tinbergen was aiming to understand economies and processes in economies, not to infer causal relations among different social, economic, ecological variables, and factual conditions. Indeed, the method that is described in this paper is more applicable in meta-analyses of existing studies and guiding the direction of future research, not as the centerpiece of individual topical studies. The intention behind this method is to understand what is true and what is not as true, and to provide a quantitative method for deriving those truths and assessing the quality of the evidence behind them.

Another process that is similar to this one is known as "group model building" [27–29]. Group model building is a process that was created by system dynamics researchers to facilitate diverse stakeholders sharing information across different fields. This is done to solve problems that are common to these stakeholders by unifying, standardizing, and connecting the information that is presented by and for the stakeholders in question [27,29]. While this is a useful technique for helping groups understand problems from many different angles, it is not a generalized way of inferring causality and truth. Creating and testing different causal graphs with the evidence that is available is a separate process that aims to produce general knowledge of empirically inferred reality. The goal with causal graph analyses is to produce a coherent and accurate map of a given concept or problem that is more probably true than competing alternative maps. It is the process of weeding out models that are not supported by evidence, more than it is just the production of different models.

Most people have implicit assumptions about how the world works, in addition to possible desires about how they would like the world to work [15,30–32]. One method for determining the underlying assumptions that are implicit in a research project is to map them out through a process known as "cognitive mapping" [33–37]. Cognitive mapping has been used to understand the implicit assumptions and decisions made by policymakers in the past [38]. Cognitive maps give rise to different concept maps, which then are used to produce different causal graphs. It is logically plausible that the creation of causal graphs in causal modeling is produced by the conceptual maps people make by the same cognitive maps used by researchers, policymakers, and stakeholders. Indeed, cognitive mapping is implicit in some research concerning Bayesian networks, which map out the probabilities that a set of causal conditions relates to a set of observed variables [39–43]. It also has been linked to modeling ecological systems by researchers [37].

The hypothesis that underpins this perspective is that implicit cognitive maps of researchers, policymakers, and stakeholders alike result in the production of different conceptual maps of the world. The interplay between cognitive maps and conceptual maps gives rise to different causal graphs being produced through the different perceived factual "nodes" (points) and connecting "edges" (relations or connections between factual nodes) of the researcher, policymaker, and stakeholder. This is different from existing methods for making the goals of researchers, policymakers, and stakeholders explicit, such as the Logic Model. Logic models display the connection between different inputs and activities with different outputs and outcomes [44], in that this method is more free-form and allows the cognitive maps and implicit biases to be made more readily apparent instead of confining the maps to a preset form. Different assumptions, perspectives, levels, and degrees of awareness in the cognition of researchers, policymakers, and stakeholders result in the perception of different "facts", different interpretations of those "facts", and different edges among the "facts". This could be done implicitly and subconsciously by the researcher and policymaker, but it also is hypothetically possible for it to be done deliberately through conscious choice and selection of facts, interpretations of facts, and edges among facts [10,32,45,46].

A hypothetical example of this is between people who identify as "conservatives" and "liberals" looking at the same situation facing their shared nation. The conservative may claim that the moral integrity of the society is eroding as time passes, while a progressive may have a different outlook on change and difference in a society from one time to another. The evidence suggests that people on both sides will look for, perceive, and interpret the situation differently in mutually exclusive way. For example, the environment cannot both be and not be affected by humans' economic activities, and it cannot both be and not be significant for human survival. Different problems are identified, different choices and preferences are made, and different actions are seen as more or less acceptable because of those differences between the general psychological phenotypes. The obvious problem with relying on these subconscious assumptions and biases alone is that the individual person who is making the policy decisions may not accurately understand, represent, or interpret the meaning of the world. Without an accurate map of how the world works, policymakers are less able to make the best possible choices for the people living in the society and for their own benefit as policymakers making decisions that affect the world they live in. One can think back to the times before navigational and weather/oceanographic sensory technology had advanced to the point where ships could orient themselves accurately on Earth. Without the production of these technologies, which aid navigation and the ability to detect and predict conditions around the ships, sailors' lives were easily lost on the tempestuous oceans, and valuable cargo was lost and destroyed in transit around the world more often than now. The analogy could be carried over to the fate and condition of nations and human societies.

3. The Methodology

The goal of this paper is not to advocate a singular methodology or tool for studying complex phenomena in our universe. Rather, the goal is to propose a new tool that can be used to help determine the appropriate tool(s) for studying complex phenomena, and to at least partially overcome the deficits of human cognition and perception in research and decision-making. By making the assumptions explicit rather than implicit in research and policy designs, we can get a firmer grip on what healthy priorities are, and how to achieve them. Below is an example of a theoretical causal graph (Figure 1).

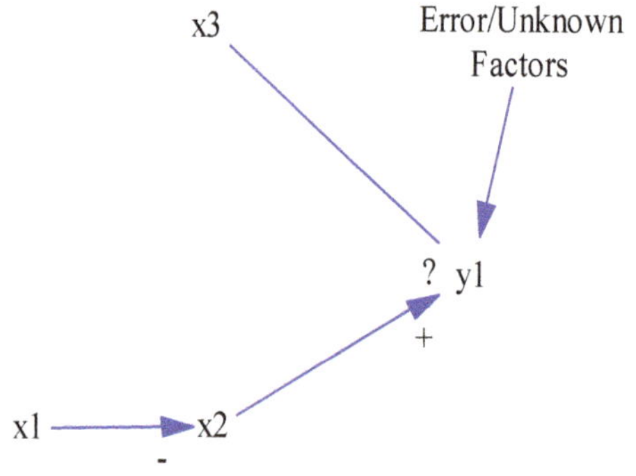

Figure 1. An example of a theoretical causal graph.

where x factors cause y effect when brought together in this combination. We see that x1 causes x2 which, when combined with x3, produces y1 effect. The plus represents x2 having a positive feedback effect on y effects (more of x2 leads to more of y1), while x3 has an unknown effect on y1. Notice the error/unknown factors variable to account for anything else the model misses.

The methodology is simple to describe and works as follows:

1. Draw out the causal graph as the researcher perceives it to be. This is the conceptual mapping stage, since all causal graphs are ultimately concept maps. Nodes or points in the graph are facts or conditions, edges between the nodes are interactions and associations among the facts. The researcher should be free to base this initial step on their own working knowledge, the existing literature on the subject in question, and any applicable theory;
2. Consult with other researchers, policymakers, and stakeholders to develop alternative facts and conditions and alternative ways for them to interact with each other through the interaction edges in the graph;
3. Design research projects to test the validity of the factual nodes and interaction edges that are produced from Steps 1 and 2:
   a It is important to note that this paper is agnostic about the specifics of the designs of the research, so long as it is logically valid and testable;
   b This is where any number of qualitative and quantitative methods can be used;
   c It is also a good idea to use multiple methods on the same factual node or interaction edge to increase the probability of validity. That is often called robustness in research;
4. Out of the population of causal graphs that were created, assign equal probabilities that each one is valid based on the total number of causal graphs that are explored.
   a The probability of the population of causal graphs can never truly equal 1 for complete validity because there is always an unknown quantity of potential error present in the population of models, i.e., the unknown unknowns;
   b The probabilities can be explicitly Bayesian, empirical Bayesian, or based on flat priors;
5. Consider the quality and source of the evidence that is presented. If quality evidence for a particular edge or node is present, then that adds to the probability that that edge or node is true at the expense of other edges and nodes. If there is evidence against a node or edge, it subtracts from the probability that that edge or node is true without necessarily affecting alternative

edges and nodes. Poorer quality evidence has less of an effect, or no effect on the probability of demonstrating truth;

6. Alter the probability of validity for each of the graphs as evidence becomes apparent through new research. This can be based on Bayesian updating or frequentist testing;
7. Repeat Steps 1 through 6 using a variety of techniques to examine each node and each edge in the causal graph.

It is important to again note that this procedure is agnostic about the specific research techniques that are used to infer causality or the truthfulness of factual nodes. Notice how the factual validity for each of the variables (the nodes in the causal graph) and causal edges (the links in the causal graph) are not necessarily known, and are rather hypothesized to exist based on past evidence and the circumspection of the researcher. From this model, we can derive various other models to test for and identify possible methods for gathering and examining the data. We can see other possible models below (Figures 2 and 3).

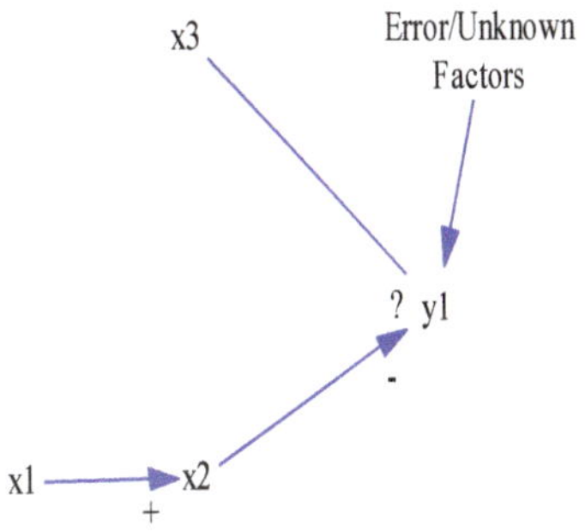

Figure 2. An alternative graph to Figure 1.

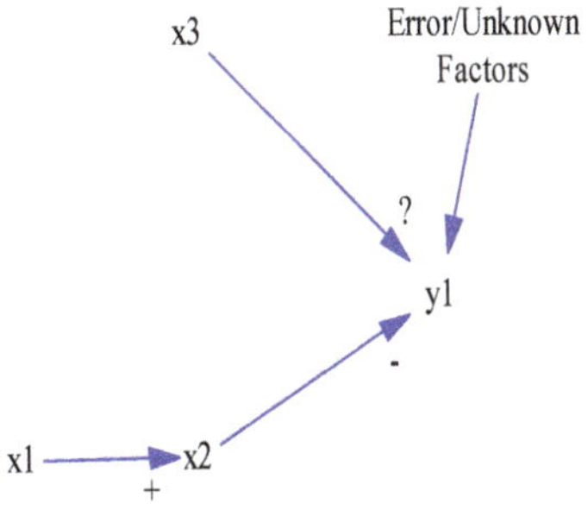

Figure 3. An alternative graph to Figure 2.

Notice how parts of the graphs in Figures 2 and 3 changed from Figure 1, representing different and mutually exclusive hypothetical models that may or may not be more accurate than the original hypothesis.

These assumptions (that are different from the original causal graph) each then have their own theoretical and observational bases and their own interpretations of what is present and happening in the real world outside of the researcher's perspective and assumptions. With this technique, it is also possible to model unknown or hypothesized interactions and facts, such as the question mark between variables x_3 and y_1. Other models can be constructed using all of the possibilities. For simplicity's sake, most of these options in the research design space have been left out. However, if the researcher(s) are able to get the largest possible collection of causal graphs together while staying relevant to the topic(s) at hand, the larger design space should provide a rich environment for testing the factual assumptions and interactions among the variables. Researchers can then work together across disciplines to design experiments and determine which data to collect and how in order to "shave away" at the hypothesis space of the research topic. The surviving causal graphs, which withstand the scrutiny of the researchers' efforts, can be said to be more probably true and valid than the other causal graphs that have aspects that are not valid or which have little to no evidence in support of them. These surviving causal graphs correspond to Bayesian posteriors or unrejected frequentist hypotheses, in that they are the end products of analyses.

Figure 4 is an example of a causal graph produced to clarify questions about education policy and the factors that link in to create academic, social, behavioral, and personal success in students. Using various data sets and methods of estimation, the most likely causal pathways could be found. Some researchers will add double-headed causal arrows and reversed arrows.

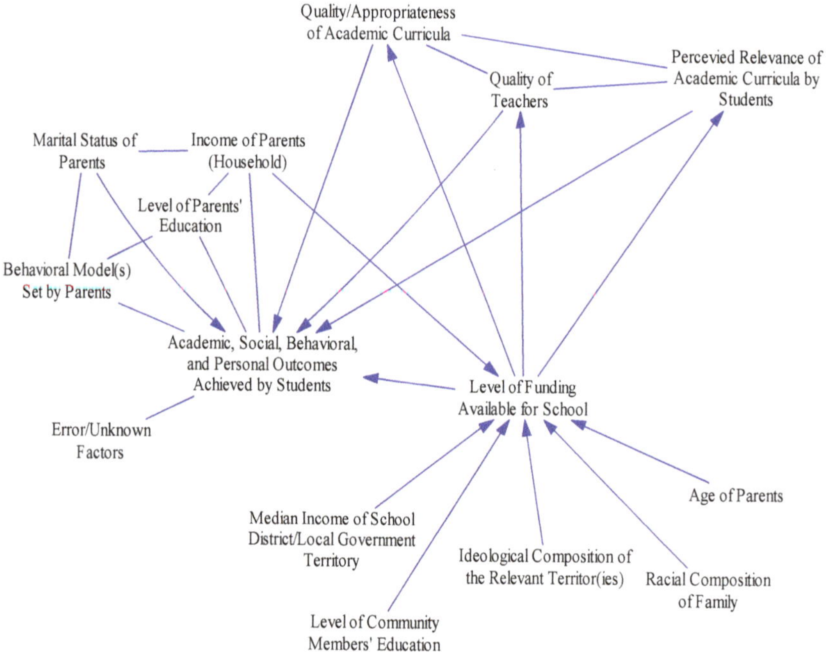

Figure 4. An example causal graph for hypotheses concerning outcomes in education.

There are two ideas that can be deconstructed from taking this holistic approach to education and educational success. The interrelated subject areas, such as the defined pedagogy, territorial demographics, the political environment, and parental/familial conditions that the child grows up in can be broken out from the causal graph into their own interrelated clusters as part of the larger graph that contains the whole. This would enable collaboration among experts in these various fields to create a more accurate model of the whole picture of how children develop, learn, and grow into adults, which can then give us a more accurate map for helping policymakers be better able to see where and how they might intervene in the given subject area. The second idea is that the whole causal graph is malleable to the perspective of the researcher in question, and alternatives for hypothesis testing can easily be developed by simply going through the graphical representation of the problem(s) at stake to find other possibilities and alternatives. Time stamps can be added to refine the temporal relationships among the variables.

4. Implications of this Method for Policy Research

The implications of this method for conducting social science research would allow policy researchers, policymakers, and stakeholders to better understand not only their own implicit, subconscious biases and explicit conscious biases, but to move beyond those biases in order to perceive and study our complex social and environmental worlds accurately. Communication of divergent beliefs and models would be easier. It is feasible that policymakers, and the researchers and stakeholders as well, will be able to move beyond disagreements over what may be just different cognitive maps in order that better choices may be made faster, with less debate, and with greater efficacy than would ordinarily happen without using this methodology explicitly to understand, design, and analyze situations and conditions in our social and environmental worlds. At least, it would be clearer what issues must be resolved and models estimated.

The methodology can also be used as a technique for holding policymakers and researchers more accountable for their assumptions and their chosen research techniques. Even though the method itself is agnostic about the methods that are used in research, there are practices for testing validity and causality that are more or less effective than others. By explicitly drawing the causal graph, it is easier to tell whether more or less appropriate methods are being used to test the nodes and causal edges of the graph. By explicitly stating the implicit and explicit biases of the individual through the process of mapping out their factual and causal assumptions, human societies and organizations that adopt this method for making choices and understanding the world may be able to more effectively understand political opponents' concerns and perspectives, as well as to be more effectively able to challenge those perspectives and opinions that are not based in evidence both behind closed doors in negotiations and in front of the fora of the general public. Assumptions totally lacking empirical verification would stand out.

The most significant benefit of using this methodology is that mutually exclusive opinions on facts and relations can be more clearly examined. Most of the common controversies in policy debates stem from competing, mutually exclusive ideas on how the world works, and how it ought to work for human well-being and survival. From whether to have public sector involvement in the economy, to the necessity of protecting the environment, the different attitudes, biases, and opinions cannot all be called of equal value for ensuring human health and well-being. Causal graphs can be used to sort through those differences in policy preferences and opinions to deliver a clearer picture of common reality and what is needed for human societies at given times. Those opinions that are supported by quality evidence can then take priority over those that simply are not, or have evidence against their empirical validity.

5. Caveats to this Method

The most glaring problem with this methodology is that it does not give instructions on how to collect data, what data to collect, or how to treat the data when they are collected. It may help inform research decisions, but it does not give explicit instructions on what to use or when to use it. This leaves the design of the experiments still open to possible researcher bias and the usual difficulties with inferring causality with researchers who have underlying assumptions and cognitive biases that they consciously or subconsciously prefer over other models and methods. Through explicitly stating the researcher's hypothesis space and cognitive bias, measures of robustness can be developed for causal models to see if researchers are truly ruling out other possibilities or whether they are honestly adhering to sound da identification, collection, and interpreting methods. Ignoring logical possibilities would be much more difficult.

Another caveat to this research methodology is the possibility for aspects of the causal graphs to change stochastically during the development of the models and throughout the experiments and analysis of the data. That is, the structure can change. A policymaker may be in the middle of developing a causal graph which is presently accurate, but may have dynamic aspects to it which can change in the near to distant future as aspects of our social world (such as technology and our understanding of the world itself) change. In addition to these probable knowledge based changes, there may also be some aspects of our social world which change due to aesthetic preferences or changes in relative perspective and attitude. This further complicates the development of these causal graphs, as the aspects and perspectives of them may change in ways that do not track neatly with the development and production of our knowledge and awareness. What may be in fashion and perceived of as desirable now may not be viewed as such in the near to distant future, thus altering the perspectives on the causal graphs that are developed today or rendering them potentially useless for achieving the goals of the society in the future. Thus, the dynamic and evolving nature of consciousness and preferences over time may influence the development of these causal graphs, if not affect the actual graphs themselves in the content of their facts, interpretations of facts, and interactions among the variables. In response, more basic social factors related to group dynamics can be added to the models, such as fundamental psychological processing common to humans. Change itself can become a part of the model. As different edges and nodes can change over time, and their changing nature can theoretically be observed, the changes and their effects can be noted and tested. This gives the resulting models significantly greater empirical validity, and thus enriches our understanding of common reality to the fullest possible extent that we can achieve.

A third possible problem with this methodology is that there is no method for keeping the model parsimonious and simple. While this may not be a problem when working with large and complex topics, it can be said that it is feasible that the models that researchers may make could become too unwieldy for practical use. A simple method for resolving this while not abandoning the potential complexity in a subject is for the researcher to narrow their focus initially to a given factual node or a specific interaction, and then to grow the model from there, limiting it to the practical relevance of the research in question. The researcher in question, or other researchers can then expand the web of knowledge in other directions at future times.

6. Conclusions

This paper presents a new tool for researchers and policymakers alike for understanding complex and interconnected topics of interest and importance to human society as a whole. Through explicitly stating the assumptions behind the subject, researchers and policymakers can then develop counterfactual alternative graphs for the subjects of their interest and research, identify data that is relevant to the subject, develop methods for collecting and analyzing the data, and then systematically shave away at factual assumptions and hypothetical interactions for which there is little to no quality evidence. Through this deductive process of elimination, researchers and policymakers alike can eliminate graphs for which all or parts do not have evidence, and thus, be left with a pool of possibilities that shrinks in size and increases in the chances of being probably accurate representations of reality itself.

It is possible that some specific aspects of the graphs may change over time with peoples' attitudes, preferences, and perspectives. However, it is assumed explicitly in this paper that the underlying method of creating causal graphs with fact nodes and interaction edges can be valid throughout time, space, and context, even if the specific models change over time. The process of shaving away at conceptual maps with this method can produce a more robust, accurate, and complete representation of reality that the human mind can comprehend and use for other purposes. By doing so, we can begin to constructively resolve key policy and political debates as they arise with this common process of gathering, analyzing, and evaluating evidence from our common reality. The political debates may be based ultimately in values and opinions. However, not all opinions and values are of equal value for human society's health and well-being. This proposed method hopes to help resolve these debates for that which is factually true and healthful, at the expense of those opinions that are not true, and are very likely unhealthful for humans in general.

Author Contributions: E.L. conceived and designed the methodology and its uses; E.L. wrote the paper and developed the graphical figures. J.S.B. helped edit the first draft.

Funding: This research received no external funding.

Acknowledgments: Special thanks to J.S. Butler for his assisting with the editing and advice to add Tinbergen's work into this paper.

Conflicts of Interest: The authors declare no conflict of interest.

References

1. Holland, P.W. Statistics and causal inference. *J. Am. Stat. Assoc.* **1986**, *81*, 945–960. [CrossRef]
2. Dennard, L.F.; Richardson, K.A.; Morcol, G. *Complexity and Policy Analysis: Tools and Concepts for Designing Robust Policies in a Complex World*; ISCE Pub: Goodyear, AZ, USA, 2008.
3. Sayama, H. *Introduction to the Modeling and Analysis of Complex Systems*; Open SUNY Textbooks; Milne Library: Geneseo, NY, USA, 2015.
4. Merriam-Webster Online Dictionary. Available online: http://www.merriam-webster.com/dictionary/causality (accessed on 2 May 2016).
5. Bennett, A.; Elman, C. Complex causal relations and case study methods: The example of path dependence. *Polit. Anal.* **2006**, *14*, 250–267. [CrossRef]
6. Blalock, H.M. (Ed.) *Causal Models in the Social Sciences*; Transaction Publishers: New York, NY, USA, 1985.
7. Higgins, E.T.; Kruglanski, A.W. (Eds.) Motivated social cognition: Principles of the interface. In *Social Psychology: Handbook of Basic Principles*; Guildford Press: New York, NY, USA, 1996; pp. 493–520, ISBN 9781572301009.
8. Holland, P.W. Causal inference, path analysis and recursive structural equations models. *ETS Res. Rep. Ser.* **1988**, *1988*, i-50. [CrossRef]
9. Robins, J.M.; Hernán, M.Á.; Brumback, B. Marginal Structural Models and Causal Inference in Epidemiology. *Epidemiology* **2000**, *11*, 550–560. [CrossRef] [PubMed]
10. Helmert, M.; Richter, S. Fast downward-making use of causal dependencies in the problem representation. In Proceedings of the International Planning Competition, Hosted at the 14th International Conference on Automated Planning and Scheduling (IPC4, ICAPS 2004), Whistler, BC, Canada, 3–7 June 2004; pp. 41–43.
11. Galea, S.; Riddle, M.; Kaplan, G.A. Causal thinking and complex system approaches in epidemiology. *Int. J. Epidemiol.* **2010**, *39*, 97–106. [CrossRef] [PubMed]
12. Plowright, R.K.; Sokolow, S.H.; Gorman, M.E.; Daszak, P.; Foley, J.E. Causal inference in disease ecology: Investigating ecological drivers of disease emergence. *Front. Ecol. Environ.* **2008**, *6*, 420–429. [CrossRef]
13. Granger, C.W. Testing for causality: A personal viewpoint. *J. Econ. Dyn. Control* **1980**, *2*, 329–352. [CrossRef]
14. Granger, C.W. Some recent development in a concept of causality. *J. Econom.* **1988**, *39*, 199–211. [CrossRef]
15. King, G.; Keohane, R.O.; Verba, S. *Designing Social Inquiry: Scientific Inference in Qualitative Research*; Princeton University Press: Princeton, NJ, USA, 1994.
16. Pierce, D.A.; Haugh, L.D. Causality in temporal systems: Characterization and a survey. *J. Econom.* **1977**, *5*, 265–293. [CrossRef]
17. Sobel, M.E. Causal inference in the social and behavioral sciences. In *Handbook of Statistical Modeling for the Social and Behavioral Sciences*; Springer: New York, NY, USA, 1995; pp. 1–38. ISBN 978-0306448058.
18. Eichler, M. Granger causality and path diagrams for multivariate time series. *J. Econom.* **2007**, *137*, 334–353. [CrossRef]
19. Pearl, J. Causal Inference in Statistics: An overview. *Stat. Surv.* **2009**, *3*, 96–146. [CrossRef]
20. Trochim, W.M. An introduction to concept mapping for planning and evaluation. *Eval. Progr. Plan.* **1989**, *12*, 1–16. [CrossRef]
21. Chen, H.; Giménez, O. Causal graphs and structurally restricted planning. *J. Comput. Syst. Sci.* **2010**, *76*, 579–592. [CrossRef]
22. Kiiveri, H.; Speed, T.P.; Carlin, J.B. Recursive causal models. *J. Aust. Math. Soc.* **1984**, *36*, 30–52. [CrossRef]
23. Kosko, B. Fuzzy cognitive maps. *Int. J. Man-Mach. Stud.* **1986**, *24*, 65–75. [CrossRef]
24. Helmert, M. A Planning Heuristic Based on Causal Graph Analysis. In Proceedings of the 14th International Conference on Automated Planning and Scheduling, Whistler, BC, Canada, 3–7 June 2004; pp. 161–170.

25. Helmert, M.; Geffner, H. Unifying the Causal Graph and Additive Heuristics. In Proceedings of the 18th International Conference on Automated Planning and Scheduling, Sydney, Australia, 14–18 September 2008; pp. 140–147.
26. Morgan, M.S. *The History of Econometric Ideas*; Cambridge University Press: Cambridge, UK, 1990.
27. Berard, C. Group Model Building Using System Dynamics: An Analysis of Methodological Frameworks. *J. Bus. Res.* **2010**, *8*, 13–24.
28. Hovmand, P. *Community Based System Dynamics*; Springer: New York, NY, USA, 2014.
29. Vennix, J.A.M.; Akkermans, H.A.; Rouwette, E.A.J.A. Group model-building to facilitate organizational change: An exploratory study. *Syst. Dyn. Rev.* **1996**, *12*, 39–58. [CrossRef]
30. Balcetis, E. Where the Motivation Resides and Self-Deception Hides: How Motivated Cognition Accomplishes Self-Deception. *Soc. Personal. Psychol. Compass* **2008**, *2*, 361–381. [CrossRef]
31. Baumeister, R.F. Self-regulation and ego threat: Motivated cognition, self deception, and destructive goal setting. In *The Psychology of Action: Linking Cognition and Motivation to Behavior*; Gollwitzer, P.M., Bargh, J.A., Eds.; Guilford Press: New York, NY, USA, 1996; pp. 27–47.
32. Jost, J.T.; Glaser, J.; Kruglanski, A.W.; Sulloway, F.J. Political conservatism as motivated social cognition. *Psychol. Bull.* **2003**, *129*, 339–375. [CrossRef] [PubMed]
33. Chown, E.; Kaplan, S.; Kortenkamp, D. Prototypes, location, and associative networks (PLAN): Towards a unified theory of cognitive mapping. *Cogn. Sci.* **1995**, *19*, 1–51. [CrossRef]
34. Eden, C. On the nature of cognitive maps. *J. Manag. Stud.* **1992**, *29*, 261–265. [CrossRef]
35. Ennis, R.H. Identifying implicit assumptions. *Synthese* **1982**, *51*, 61–86. [CrossRef]
36. Kitchin, R.; Freundschuh, S. *Cognitive Mapping: Past, Present, and Future*; Routledge: London, UK, 2000.
37. Özesmi, U.; Özesmi, S.L. Ecological models based on people's knowledge: A multi-step fuzzy cognitive mapping approach. *Ecol. Model.* **2004**, *176*, 43–64. [CrossRef]
38. Axelrod, R.M. *Structure of Decision: The Cognitive Maps of Political Elites*; Princeton University Press: Princeton, NJ, USA, 1976.
39. Nadkarni, S.; Shenoy, P.P. A Bayesian network approach to making inferences in causal maps. *Eur. J. Oper. Res.* **2001**, *128*, 479–498. [CrossRef]
40. Nadkarni, S.; Shenoy, P.P. A causal mapping approach to constructing Bayesian networks. *Dec. Support Syst.* **2004**, *38*, 259–281. [CrossRef]
41. Siau, K.; Tan, X. Improving the quality of conceptual modeling using cognitive mapping techniques. *Data Knowl. Eng.* **2005**, *55*, 343–365. [CrossRef]
42. Swan, J. Using cognitive mapping in management research: Decisions about technical innovation. *Br. J. Manag.* **1997**, *8*, 183–198. [CrossRef]
43. Korver, M.; Lucas, P.J. Converting a rule-based expert system into a belief network. *Med. Inform.* **1993**, *18*, 219–241. [CrossRef]
44. McCawley, P.F. *The Logic Model for Program Planning and Evaluation*; University of Idaho Extension: Moscow, ID, USA, 2010; CLS 1097; pp. 1–5. Available online: https://www.researchgate.net/publication/237568681_The_Logic_Model_for_Program_Planning_and_Evaluation (accessed on 11 June 2018).
45. Jost, J.T.; Amodio, D.M. Political ideology as motivated social cognition: Behavioral and neuroscientific evidence. *Motiv. Emot.* **2012**, *36*, 55–64. [CrossRef]
46. Jost, J.T.; Glaser, J.; Kruglanski, A.W.; Sulloway, F.J. Exceptions that prove the rule—Using a theory of motivated social cognition to account for ideological incongruities and political anomalies: Reply to Greenberg and Jonas (2003). *Psychol. Bull.* **2003**, *129*, 383–393. [CrossRef]

applied system innovation

MDPI

Article

Relation "Greater than or Equal to" between Ordered Fuzzy Numbers

Krzysztof Piasecki

Department of Investment and Real Estate, Poznań University of Economics and Business, Niepodległości 10, 61-875 Poznań, Poland; krzysztof.piasecki@ue.poznan.pl

Received: 13 July 2019; Accepted: 1 August 2019; Published: 3 August 2019

Abstract: The ordered fuzzy number (OFN) is determined as an ordered pair of fuzzy number (FN) and its orientation. FN is widely interpreted as imprecise number approximating real number. We interpret any OFN as an imprecise number equipped with additional information about the location of the approximated number. This additional information is given as orientation of OFN. The main goal of this paper is to determine the relation "greater than or equal to" on the space of all OFNs. This relation is unambiguously defined as an extension of analogous relations on the space of all FN. All properties of the introduced relation are investigated on the basis of the revised OFNs' theory. It is shown here that this relation is a fuzzy one. The relations "greater than" and "equal to" also are considered. It is proven that the introduced relations are independent on the orientation of the compared OFNs. This result makes it easier to solve optimization tasks using OFNs.

Keywords: ordered fuzzy number; fuzzy relation; preorder; strict order; equivalence relation

1. Introduction

The concept of ordered fuzzy number (OFN) was intuitively introduced by Kosiński [1–4] as an extension of the notion of fuzzy number (FN) which is widely interpreted as imprecise approximation of real number. OFNs' usefulness follows from the fact that it is interpreted as FN with additional information about the location of the approximated number. Kosiński [1–4] has determined arithmetic for OFNs as an extension of results obtained by Goetschel and Voxman [5] for FNs. For formal reasons, the Kosiński' theory was revised [6] in such a way that revised OFN definition fully corresponds to the intuitive Kosiński's definition of OFN. OFNs are always defined without use of any ordering relation between FNs. Knowing this fact makes it easier to read the section on ordering relationship between OFNs. This paper is linked to the revised OFNs' theory.

In decision analysis, economics and finance, OFNs are frequently employed to evaluate the alternatives in modelling a real-world problem [7–19]. On the other hand, the OFN theory has an important disadvantage. This disadvantage is due to the lack of formal mathematical models associated with OFNs. Therefore, an important goal of further formal research should be to fill these theoretical gaps.

If any alternatives are evaluated by OFNs then their ranking leads to OFNs' arrangement which is pre-given as an ordering relation "greater than or equal to" between OFNs.

Since the notion of OFN is interpreted as an extension of the notion of FN, any formal model of order between OFNs should be consistent with the fixed ordering relation between FNs. Unlike in the case of real numbers, FNs have no natural order. A straightforward approach to the ordering of FNs is to convert each compared FN into a real number. Any procedure of this conversion is called a "defuzzification method" [20]. Representative examples of FNs' arrangement using different defuzzification methods are presented in [20–56]. Each individual defuzzification method, however, pays attention to a special aspect of an FN. As a consequence, each approach suffers from some defects that only one real number is associated with each FN. Freeling [57] pointed out that "by reducing the whole of our analysis to a single number, we are losing much of the information. We have purposely been keeping throughout our calculations".

Kosiński and Sztyma [58] introduced defuzzification methods for OFNs. Some applications of OFN arrangement using defuzzification methods are presented in [8,16,18,19]. On the other side, in [17], it is shown that the use of defuzzification methods has a significant impact on the ordering of OFNs. In an extreme case, the use of defuzzification procedures can totally blur the true picture of arrangement of OFNs. It can lead to results deviating from real ranking of decision alternatives, which will increase the hazard of making a wrong decision. For this reasons, OFNs arrangement should be described by a fuzzy relation which compares OFNs pairwise. In this way, we can compare OFNs without losing information about the imprecision and orientation of evaluated OFNs. This approach is more realistic.

For FNs, fuzzy order relations can be defined in two ways. First of all, fuzzy order of FNs can be determined using α-cuts. Representative examples of FNs' arrangement using α-cuts are presented in [59–61]. At present, the α-cuts theory dedicated to OFNs is unknown. Therefore, in the current moment, any formal models of ordering with use of α-cuts cannot be extended to the case of OFNs. Moreover, Orlovsky [62] defined fuzzy order of FN applying the Zadeh's Extension Principle [63–65]. This method does not raise any objections.

Therefore, the main goal of presented work is to define such fuzzy order relation between OFN's which is consistent with fuzzy order introduced by Orlovsky. Setting such a relationship is needed to build each quantitative model based on comparison of OFNs. In general, the relation $\widetilde{GE}$ can be applied in any such quantitative model of the real world that a comparison of imprecise numbers is used. The tentative approach to this subject was presented in [66]. Obtained in this way fuzzy order of OFNs is applied in [12,17]. The results presented here are the final generalized version of such fuzzy ordering OFNs that it fulfils assumed condition.

The paper is organised in the following way. Section 2 presents considered models of imprecise quantity. Section 2.1 describes the basic concepts of FNs and arithmetic operations on FNs. The revised notion of OFN and arithmetic operations on OFNs are presented in Section 2.2. It is pointed out here that OFNs are always defined without use of any ordering relation between FNs. In Section 2.3, the disorientation map is introduced. Moreover, some differences between FNs and OFNs are explained here. In Section 3 the author proves that some simple properties are fulfilled by Orlovsky's fuzzy order of FN. In Section 4 the author introduced such relation "greater than or equal to" between OFNs which is consistent with Orlovsky's fuzzy order. Section 5 contains some basic problems linked with ordering of OFNs. In Section 6, all theoretical considerations are illustrated by case study devoted to the subject of investment decisions. Finally, Section 7 contains the final remarks.

2. Imprecise Quantities—Considered Models

Objects of any considerations may be given as elements of a predefined space $\mathbb{X}$. The basic tool for imprecise classification of these elements is the notion of fuzzy set introduced by Zadeh [67]. Any fuzzy set $\mathcal{A}$ is unambiguously determined by means of its membership function $\mu_A \in [0,1]^{\mathbb{X}}$, as follows

$$\mathcal{A} = \{(x, \mu_A(x)); x \in \mathbb{X}\}. \quad (1)$$

In all our considerations we use the multivalued logic determined by Łukasiewicz [68]. The truth value of the sentence $\mathcal{P}$ will be denoted by the symbol $tv(\mathcal{P})$. From the point-view of multi-valued logic, the value $\mu_A(x)$ is interpreted as the truth value $tv(``x \in \mathcal{A}")$. By the symbol $\mathcal{F}(\mathbb{X})$ we denote the family of all fuzzy sets in the space $\mathbb{X}$. Any fuzzy set $\mathcal{A} \in \mathcal{F}(\mathbb{X})$ may be described using the following notions:

For each $\alpha \in]0,1]$, the $\alpha-$ cuts $[\mathcal{A}]_\alpha$ determined as follows

$$[\mathcal{A}]_\alpha = \{x \in \mathbb{X} : \mu_A(x) \geq \alpha\}; \tag{2}$$

The support closure $[\mathcal{A}]_{0+}$ given in the following way

$$[\mathcal{A}]_{0+} = \lim_{\alpha \to 0^+} [\mathcal{A}]_\alpha. \tag{3}$$

An imprecise quantity is a family of real numbers belongs to it in a different degree. In this section, the fuzzy set notion is applied for describing imprecise quantities.

2.1. Fuzzy Numbers—Some Basic Notions

A commonly used model of an imprecise number is FN, defined as a fuzzy set in real line $\mathbb{R}$. The most general definition of FN is given as follows:

The most general definition of fuzzy number is given as follows:

Definition 1 [69]. *The fuzzy number (FN) is such a fuzzy subset $\mathcal{L} \in \mathcal{F}(\mathbb{R})$ with bounded support closure $[\mathcal{L}]_{0+}$ that it is represented by its upper semi-continuous membership function $\mu_L \in [0;1]^{\mathbb{R}}$ satisfying the conditions:*

$$\exists_{x \in \mathbb{R}} \ \mu_L(x) = 1 \tag{4}$$

$$\forall_{(x,y,z) \in \mathbb{R}^3} \ x \leq y \leq z \implies \mu_L(y) \geq \min\{\mu_L(x); \mu_L(z)\}. \tag{5}$$

The set of all FN we denote by the symbol $\mathbb{F}$. Let us consider any arithmetic operation $*$ defined on $\mathbb{R}$. The symbol $\circledast$ denotes an extension of arithmetic operation $*$ to $\mathbb{F}$. In [70], arithmetic operations on FN are introduced in such way that they are coherent with the Zadeh's Extension Principle. In line with it, for any pair $(\mathcal{K}, \mathcal{L}) \in \mathbb{F}^2$ represented by their membership functions $\mu_K, \mu_L \in [0,1]^{\mathbb{R}}$, the FN

$$\mathcal{M} = \mathcal{K} \circledast \mathcal{L} \tag{6}$$

is described by its membership function $\mu_M \in [0,1]^{\mathbb{R}}$ determined by means of the identity:

$$\mu_M(z) = \sup\{\min\{\mu_K(x), \mu_L(y)\} : z = x * y, (x,y) \in \mathbb{R}\}. \tag{7}$$

Thanks to the results obtained in [5], we have that any FN can be equivalently defined as follows:

Theorem 1 [71]. *For any FN $\mathcal{L}$ there exists such a non-decreasing sequence $(a,b,c,d) \subset \mathbb{R}$ that $\mathcal{L}(a,b,c,d,L_L,R_L) = \mathcal{L} \in \mathcal{F}(\mathbb{R})$ is determined by its membership function $\mu_L(\cdot|a,b,c,d,L_L,R_L) \in [0,1]^{\mathbb{R}}$ described by the identity*

$$\mu_L(x|a,b,c,d,L_L,R_L) = \begin{cases} 0, & x \notin [a,d], \\ L_L(x), & x \in [a,b], \\ 1, & x \in [b,c], \\ R_L(x), & x \in [c,d], \end{cases} \tag{8}$$

where the left reference function $L_L \in [0,1]^{[a,b]}$ and the right reference function $R_L \in [0,1]^{[c,d]}$ are upper semi-continuous monotonic ones meeting the conditions:

$$L_L(b) = R_L(c) = 1, \tag{9}$$

$$[\mathcal{L}]_{0+} = [a,d]. \tag{10}$$

The FN $\mathcal{L}(a,a,a,a,L_L,R_L) = [\![a]\!]$ represents the real number $a \in \mathbb{R}$. Therefore, we can say $\mathbb{R} \subset \mathbb{F}$. For any $z \in [b,c]$, a FN $\mathcal{L}(a,b,c,d,L_L,R_L)$ is a formal model of linguistic variable "about z". Understanding the phrase "about z" depends on the applied pragmatics of the natural language. Let us note that FN may be replaced by generalized FN [72] which does not meet the condition (4).

In line with the identity (7), the unary minus operator "–" on $\mathbb{R}$ is extended to the minus operator $\ominus$ on $\mathbb{F}$ by the identity

$$\ominus \mathcal{L}(a,b,c,d,L_L,R_L) = \mathcal{L}\left(-d,-c,-b,-a,R_L^{(-)},\, L_L^{(-)}\right), \tag{11}$$

where

$$R_L^{(-)}(x) = R_L(-x), \tag{12}$$

$$L_L^{(-)}(x) = L_L(-x). \tag{13}$$

In further considerations, we will use the following concepts.

Definition 2. For *any upper semi-continuous non-decreasing function* $L \in [0,\, 1]^{[u,\, v]}$, *its cut-function* $L^\star \in [u,v]^{[0;1]}$ *is determined by the identity*

$$L^\star(\alpha) = \min\{x \in [u,v] :\ L(x) \ge \alpha\}. \tag{14}$$

Definition 3. *For any upper semi-continuous non-increasing function* $R \in [0,\, 1]^{[u,\, v]}$ *its cut-function* $R^\star \in [0,\, 1]^{[u,\, v]}$ *is determined by the identity*

$$R^\star(\alpha) = \max\{x \in [u,v] :\ R(x) \ge \alpha\}. \tag{15}$$

Definition 4. For *any bounded continuous and non-decreasing function* $l \in [l(0),\, l(1)]^{[0,1]}$ *its pseudo inverse* $l^\lhd \in [0,1]^{[l(0),\, l(1)]}$ *is determined by the identity*

$$l^\lhd(x) = \max\{\alpha \in [0,1] : l(\alpha) = x\}. \tag{16}$$

Definition 5. *For any bounded continuous and non-increasing function* $r \in [r(0),\, r(1)]^{[0,1]}$ *its pseudo inverse* $r^\lhd \in [0;1]^{[r(1),\, r(0)]}$ *of is determined by the identity*

$$r^\lhd(x) = \min\{\alpha \in [0,1] : r(\alpha) = x\}. \tag{17}$$

In reference [5], it is proved that FNs' sum $\oplus$ is given by the identity

$$\mathcal{L}\left(a+e,b+f,c+g,d+h,L_J,R_J\right) = \mathcal{L}(a,b,c,d,L_K,R_K) \oplus \mathcal{L}(e,f,g,h,L_M,R_M), \tag{18}$$

where

$$\forall_{\alpha\in[0,1]}\ l_J(\alpha) = L_K^\star(\alpha) + L_M^\star(\alpha) \tag{19}$$

$$\forall_{\alpha\in[0,1]}\ r_J(\alpha) = R_K^\star(\alpha) + R_M^\star(\alpha) \tag{20}$$

$$\forall_{x\in[a+e,b+f]}\ L_J(x) = l_J^\lhd(x), \tag{21}$$

$$\forall_{x\in[c+g,d+h]}\ R_J(x) = r_J^\lhd(x). \tag{22}$$

The difference $\ominus$ between FNs is determined in determined in the following way

$$\mathcal{L}(a,b,c,d,L_K,R_K) \ominus \mathcal{L}(e,f,g,h,L_M,R_M) = \mathcal{L}(a,b,c,d,L_K,R_K) \oplus (\ominus\mathcal{L}(e,f,g,h,L_M,R_M)). \quad (23)$$

Then identities (11)–(13) and (18)–(23) imply that

$$\mathcal{L}(a-h,b-g,c-f,d-e,L_W,R_W) = \mathcal{L}(a,b,c,d,L_K,R_K) \ominus \mathcal{L}(e,f,g,h,L_M,R_M), \quad (24)$$

where

$$\forall_{\alpha\in[0,1]}\ l_W(\alpha) = L_K^{\star}(\alpha) - R_M^{\star}(\alpha) \quad (25)$$

$$\forall_{\alpha\in[0,1]}\ r_W(\alpha) = R_K^{\star}(\alpha) - L_M^{\star}(\alpha) \quad (26)$$

$$\forall_{x\in[a-h,b-g]}\ L_W(x) = l_W^{\triangleleft}(x), \quad (27)$$

$$\forall_{x\in[c-f,d-e]}\ R_W(x) = r_W^{\triangleleft}(x). \quad (28)$$

The above arithmetic operators may be generalized to the case of intuitionistic FNs [73]. On the other hand, the dependencies (18)–(28) are not met for discrete FNs [74]. All above identities show a high complexity of arithmetic operations on the space $\mathbb{F}$. Due to that, in many practical applications researchers limit the use of FNs only to their kind distinguished below [75].

Definition 6. *For any non-decreasing sequence* $(a,b,c,d) \subset \mathbb{R}$, *a trapezoidal FN (TrFN) is the FN* $\mathcal{T} = Tr(a,b,c,d) \in \mathbb{F}$ *defined by its membership functions* $\mu_T \in [0,1]^{\mathbb{R}}$ *in the following way*

$$\mu_T(x) = \mu_{Tr}(x|a,b,c,d) = \begin{cases} 0, & x \notin [a,d], \\ \frac{x-a}{b-a}, & x \in [a,b[, \\ 1, & x \in [b,c], \\ \frac{x-d}{c-d}, & x \in]c,d]. \end{cases} \quad (29)$$

The space of all TrFNs is denoted by the symbol $\mathbb{F}_{Tr}$. For any TrFN we have

$$Tr(-d,-c,-b,-a) = \ominus Tr(a,b,c,d) \quad (30)$$

$$Tr(a+e,b+f,c+g,d+h) = Tr(a,b,c,d) \oplus Tr(e,f,g,h), \quad (31)$$

$$Tr(a-h,b-g,c-f,d-e) = Tr(a,b,c,d) \ominus Tr(e,f,g,h). \quad (32)$$

2.2. Ordered Fuzzy Numbers—Some Basic Facts

The notion of OFN is intuitively introduced by Kosiński [1–4], as such model of imprecise number that subtraction of OFNs is the inverse operator to addition of OFNs. Therefore, OFNs can contribute to specific problems concerning the solution of fuzzy linear equations of the form or help with the interpretation of specific improper fuzzy arithmetic results.

An important disadvantage of Kosiński's theory is that there exist such OFNs which are not linked to any membership function [4]. For this reason, the Kosiński's theory is revised in [6] where OFNs are defined as follows:

Definition 7 [6]. *For any monotonic sequence* $(a,b,c,d) \subset \mathbb{R}$, *the ordered fuzzy number OFN* $\overleftrightarrow{\mathcal{L}}(a,b,c,d,S_L,E_L) = \overleftrightarrow{\mathcal{L}}$ *is the pair of orientation* $\overrightarrow{a,d} = (a,d)$ *and fuzzy set* $\mathcal{L} \in \mathcal{F}(\mathbb{R})$ *described by membership function* $\mu_L(\cdot|a,b,c,d,S_L,E_L) \in [0,1]^{\mathbb{R}}$ *given by the identity*

$$\mu_L(x|a,b,c,d,S_L,E_L) = \begin{cases} 0, & x \notin [a,d] \equiv [d,a], \\ S_L(x), & x \in [a,b] \equiv [b,a], \\ 1, & x \in [b,c] \equiv [c,b], \\ E_L(x), & x \in [c,d] \equiv [d,c]. \end{cases} \tag{33}$$

where the starting function $S_L \in [0,1]^{[a,b]}$ *and the ending function* $E_L \in [0,1]^{[c,d]}$ *are upper semi-continuous monotonic ones meeting the conditions (6) and*

$$S_L(b) = E_L(c) = 1 \tag{34}$$

The identity (33) additionally describes such modified notation of numerical intervals which is applied in this work.

Discussion about the terminology: We see above that the notion of "ordered fuzzy number" is defined without applying any ordering relation between FNs. In original Kosińki's works "ordered fuzzy number" is also defined without use of any ordering relation between FN. In each of these cases, "ordered fuzzy number" is defined as FN completed by orientation. Therefore, in my opinion term "ordered fuzzy number" should be replaced by the term "oriented fuzzy number". The following premises support such a proposal for change:

- Any discussion about the ordering of "oriented fuzzy numbers" is clearer than a discussion of ordering of "ordered fuzzy numbers".
- Professor Kosinski's mother language is Polish. In Polish OFNs is called "skierowana liczba rozmyta". This term was proposed by Professor Kosiński. Against, the quoted Polish term is translated into English as "oriented fuzzy number" or "directed fuzzy number". Moreover, the English term "ordered fuzzy number" is translated into Polish as "uporządkowana liczba rozmyta". All this allows us to state that the meanings of the Polish term "skierowana liczba rozmyta" and the English term "ordered fuzzy number" are different.

"Ordered fuzzy numbers" are the most important work of life for Professor Kosiński. Therefore, the proposed change to the term OFN should be discussed with him. Because of Professor Kosiński passed away, this is not possible. Therefore, I agree with other scientists [76,77] that the OFN may be called the "Kosiński's number". Future scientific discussion will allow us to choose a "oriented fuzzy number" or "Kosiński number" or another term. Today we still use the term "ordered fuzzy number". No less in this work, the abbreviation OFN can be read "ordered fuzzy number" or "oriented fuzzy numbers". The use of the second term makes easier to read the section on the ordering relationship between OFNs.

The symbol $\mathbb{K}$ denotes the space of all OFNs. Any OFN describes an imprecise number with additional information about the location of the approximated number. This information is given as orientation of OFN. If $a < d$ then OFN $\overleftrightarrow{\mathcal{L}}(a,b,c,d,S_L,E_L)$ has the positive orientation $\overrightarrow{a,d}$. For any $z \in [b,c]$, the positively oriented OFN $\overleftrightarrow{\mathcal{L}}(a,b,c,d,S_L,E_L)$ is a formal model of linguistic variable "about or slightly above z". The symbol $\mathbb{K}^+$ denotes the space of all positively oriented OFN. If $a > d$, then OFN $\overleftrightarrow{\mathcal{L}}(a,b,c,d,S_L,E_L)$ has the negative orientation $\overrightarrow{a,d}$. For any $z \in [c,b]$, the negatively oriented TrOFN $\overleftrightarrow{\mathcal{L}}(a,b,c,d,S_L,E_L)$ is a formal model of linguistic variable "about or slightly below z". The symbol $\mathbb{K}^-$ denotes the space of all negatively oriented OFN. Understanding the phrases "about or slightly above z" and "about or slightly below z" depend on the applied pragmatics of the natural language. If $a = d$, OFN $\overleftrightarrow{\mathcal{L}}(a,a,a,a,S_L,E_L) = [\![a]\!]$ describes unoriented number $a \in \mathbb{R}$. Summing up, we see that

$$\mathbb{K} = \mathbb{K}^+ \cup \mathbb{R} \cup \mathbb{K}^-. \tag{35}$$

The minus operator "−" on $\mathbb{R}$ is extended by Kosiński [4] to the minus operator ⊟ on $\mathbb{K}$ by means of the identity

$$\boxminus \overleftrightarrow{\mathcal{L}}(a,b,c,d,S_L,E_L) = \overleftrightarrow{\mathcal{L}}\left(-a,-b,-c,-d,S_L^{(-)},\ E_L^{(-)}\right), \tag{36}$$

where

$$S_L^{(-)}(x) = S_L(-x) \tag{37}$$

$$E_L^{(-)}(x) = E_L(-x) \tag{38}$$

Kosiński [1] defines the addition operator $\boxplus_K$ on $\mathbb{K}$ as the extension of operator ⊕ from $\mathbb{F}$ to $\mathbb{K}$. This extension is determined by extension the domain identities (18)–(22) from $\mathbb{F}$ to $\mathbb{K}$. In this way, Kosiński defines addition of OFNs as an extension of results obtained by Goetschel and Voxman [5] for addition of FNs. Moreover, Kosiński [4] have shown that there exist such OFNs that their sum $\boxplus_K$ does not exist. For this reason, Kosiński's operator $\boxplus_K$ is replaced by addition operator ⊞ defined on $\mathbb{K}$ by the identity [6]

$$\overleftrightarrow{\mathcal{L}}(a_K,b_K,c_K,d_K,S_K,E_K) \boxplus \overleftrightarrow{\mathcal{L}}(a_M,b_M,c_M,d_M,S_M,E_M) = \overleftrightarrow{\mathcal{J}} = \overleftrightarrow{\mathcal{L}}\left(a_J,b_J,c_J,d_J,S_J,E_J\right), \tag{39}$$

where we have

$$\check{a}_J = a_K + a_M, \tag{40}$$

$$b_J = b_K + b_M, \tag{41}$$

$$c_J = c_K + c_M, \tag{42}$$

$$\check{d}_J = d_K + d_M, \tag{43}$$

$$a_J = \begin{cases} \min\{\check{a}_J, b_J\}, & (b_J < c_J) \vee (b_J = c_J \wedge \check{a}_J \le \check{d}_J), \\ \max\{\check{a}_J, b_J\}, & (b_J > c_J) \vee (b_J = c_J \wedge \check{a}_J > \check{d}_J), \end{cases} \tag{44}$$

$$d_J = \begin{cases} \max\{\check{d}_J, c_J\}, & (b_J < c_J) \vee (b_J = c_J \wedge \check{a}_J \le \check{d}_J), \\ \min\{\check{d}_J, c_J\}, & (b_J > c_J) \vee (b_J = c_J \wedge \check{a}_J > \check{d}_J), \end{cases} \tag{45}$$

$$\forall_{\alpha\in[0;1]}\ s_J(\alpha) = \begin{cases} S_K^{\star}(\alpha) + S_M^{\star}(\alpha), & a_J \ne b_J, \\ b_J, & a_J = b_J. \end{cases} \tag{46}$$

$$\forall_{\alpha\in[0;1]}\ e_J(\alpha) = \begin{cases} E_K^{\star}(\alpha) + E_M^{\star}(\alpha), & c_J \ne d_J, \\ c_J, & c_J = d_J. \end{cases} \tag{47}$$

$$\forall_{x\in[a_J,b_J]}\ S_J(x) = s_J^{\triangleleft}(x), \tag{48}$$

$$\forall_{x\in[c_J,d_J]}\ E_J(x) = e_J^{\triangleleft}(x). \tag{49}$$

In [6], the definition of addition operator ⊞ is justified in detail. Then, difference ⊟ between OFNs is given as follows

$$\overleftrightarrow{\mathcal{L}}(a,b,c,d,S_K,E_K) \boxminus \overleftrightarrow{\mathcal{L}}(e,f,g,h,S_M,E_M) = \overleftrightarrow{\mathcal{L}}(a,b,c,d,L_K,R_K) \boxplus \left(\boxminus\overleftrightarrow{\mathcal{L}}(e,f,g,h,S_M,E_M)\right). \tag{50}$$

In [1,6], it is shown that for any $\overleftrightarrow{\mathcal{L}} \in \mathbb{K}$ we have

$$\overleftrightarrow{\mathcal{L}} \boxplus_K (\boxminus \overleftrightarrow{\mathcal{L}}) = [\![0]\!] = \overleftrightarrow{\mathcal{L}} \boxminus \overleftrightarrow{\mathcal{L}}. \quad (51)$$

We see that subtraction is inverse operator for both addition operators $\boxplus_K$ and $\boxplus$. We can say that OFNs meet the intuitive postulate put forward by Kosiński.

Due to high complexity of arithmetic operations of OFN, in many practical applications researchers limit the use of OFNs only to their kind distinguished below.

Definition 8 [6]. *For any monotonic sequence $(a,b,c,d) \subset \mathbb{R}$, the trapezoidal OFN (TrOFN) $\overleftrightarrow{Tr}(a,b,c,d) = \overleftrightarrow{\mathcal{T}}$ is the pair of the orientation $\overrightarrow{a,d} = (a,d)$ and fuzzy set $\mathcal{T} \in \mathcal{F}(\mathbb{R})$ determined explicitly by its membership functions $\mu_T \in [0,1]^{\mathbb{R}}$ as follows*

$$\mu_T(x) = \mu_{Tr}(x|a,b,c,d) = \begin{cases} 0, & x \notin [a,d] \equiv [d,a], \\ \frac{x-a}{b-a}, & x \in [a,\, b[\ \equiv\ ]b,a], \\ 1, & x \in [b,c] \equiv [c,b], \\ \frac{x-d}{c-d}, & x \in\ ]c,d] \equiv [d,c[. \end{cases} \quad (52)$$

The symbol $\mathbb{K}_{Tr}$ denotes the space of all TrOFNs. Identity (36) implies that the minus operator $\boxminus$ on $\mathbb{K}_{Tr}$ is given by the identity

$$\overleftrightarrow{Tr}(-a,-b,-c,-d) = \boxminus \overleftrightarrow{Tr}(a,b,c,d). \quad (53)$$

In line with (39), the sum of TrOFNs is determined as follows

$$\begin{aligned} &\overleftrightarrow{Tr}(a,b,\ c,d) \boxplus \overleftrightarrow{Tr}(p-a,q-b,\ r-c,s-d) = \\ &= \begin{cases} \overleftrightarrow{Tr}(\min\{p,q\},q,\ r,\max\{r,s\}), & (q<r) \vee (q=r \wedge p \le s), \\ \overleftrightarrow{Tr}(\max\{p,q\},q,\ r,\min\{r,s\}), & (q>r) \vee (q=r \wedge p > s). \end{cases} \end{aligned} \quad (54)$$

Then the difference $\boxminus$ between TrOFNs is the TrOFN given as follows

$$\begin{aligned} &\overleftrightarrow{Tr}(a,b,\ c,d) \boxminus \overleftrightarrow{Tr}(a-p,b-q,c-r,d-s) = \\ &= \begin{cases} \overleftrightarrow{Tr}(\min\{p,q\},q,\ r,\max\{r,s\}) & (q<r) \vee (q=r \wedge p \le s) \\ \overleftrightarrow{Tr}(\max\{p,q\},q,\ r,\min\{r,s\}) & (q>r) \vee (q=r \wedge p > s). \end{cases} \end{aligned} \quad (55)$$

2.3. *Ordered Fuzzy Numbers vs. Fuzzy Numbers*

For the case $a \ge d$ the membership function of OFN $\overleftrightarrow{\mathcal{L}}(a,b,c,d,S_L,E_L)$ is equal to the membership function of FN $\mathcal{L}(a,b,c,d,S_L,E_L)$. This fact implies the existence of isomorphism $\Psi : (\mathbb{K}^+ \cup \mathbb{R}) \to \mathbb{F}$ given by the identity

$$\mathcal{L}(a,b,c,d,S_L,E_L) = \Psi\Big(\overleftrightarrow{\mathcal{L}}(a,b,c,d,S_L,E_L)\Big). \quad (56)$$

This isomorphism may be extended to the space $\mathbb{K}$ by disorientation map $\overline{\overline{\Psi}} : \mathbb{K} \to \mathbb{F}$ given as follows

$$\overline{\overline{\Psi}}(\overleftrightarrow{\mathcal{L}}) = \begin{cases} \Psi(\overleftrightarrow{\mathcal{L}})\ \overleftrightarrow{\mathcal{L}} \in \mathbb{K}^+ \cup \mathbb{R}, \\ \ominus\Psi(\boxminus\overleftrightarrow{\mathcal{L}})\ \overleftrightarrow{\mathcal{L}} \in \mathbb{K}^-. \end{cases} \quad (57)$$

Let us note, that the disorientation map $\overline{\overline{\Psi}}: \mathbb{K} \to \mathbb{F}$ may be equivalently defined by the identity

$$\overline{\overline{\Psi}}\left(\overleftrightarrow{\mathcal{L}}(a,b,c,d,S_L,E_L)\right) = \begin{cases} \mathcal{L}(a,b,c,d,S_L,E_L) & \overleftrightarrow{\mathcal{L}}(a,b,c,d,S_L,E_L) \in \mathbb{K}^+ \cup \mathbb{R}, \\ \mathcal{L}(d,c,b,a,E_L,S_L) & \overleftrightarrow{\mathcal{L}}(a,b,c,d,S_L,E_L) \in \mathbb{K}^-. \end{cases} \tag{58}$$

Example 1. *Let us consider the OFN* $\overleftrightarrow{\mathcal{X}} = \overleftrightarrow{\mathcal{L}}(12,14,18,20,S_X,E_X)$, *where*

$$\mu_X(x) = \begin{Bmatrix} 0, & x \notin [12,20], \\ S_X(x), & x \in [12,14], \\ 1, & x \in [14,18], \\ E_X(x), & x \in [18,20], \end{Bmatrix} = \begin{cases} 0, & x \notin [12,20], \\ \frac{2 \cdot x - 24}{x - 10}, & x \in [12,14], \\ 1, & x \in [14,18], \\ \frac{7 \cdot x - 140}{2 \cdot x - 50}, & x \in [18,20]. \end{cases} \tag{59}$$

and the OFN $\overleftrightarrow{\mathcal{Y}} = \overleftrightarrow{\mathcal{L}}(13,11,6,5,S_Y,E_Y)$, *where*

$$\mu_Y(x) = \begin{Bmatrix} 0, & x \notin [13,5], \\ S_Y(x), & x \in [13,11], \\ 1, & x \in [11,6], \\ E_Y(x), & x \in [6,5], \end{Bmatrix} = \begin{Bmatrix} 0, & x \notin [13,5], \\ \frac{6 \cdot x - 30}{x}, & x \in [13,11], \\ 1, & x \in [11,6], \\ \frac{2 \cdot x - 26}{x - 15}, & x \in [6,5]. \end{Bmatrix} \tag{60}$$

Since $\overleftrightarrow{\mathcal{X}} \in \mathbb{K}^+$, *using (57) we get*

$$\overline{\overline{\Psi}}(\overleftrightarrow{\mathcal{X}}) = \overline{\overline{\Psi}}\left(\overleftrightarrow{\mathcal{L}}(12,14,18,20,S_X,E_X)\right) = \mathcal{L}(12,14,18,20,S_X,E_X) = \mathcal{L}(12,14,18,20,L_U,R_U) = \mathcal{U}, \tag{61}$$

where FN $\mathcal{U}$ *is explicitly determined by the following membership function*

$$\mu_U(x) = \begin{Bmatrix} 0, & x \notin [12,20], \\ L_U(x), & x \in [12,14], \\ 1, & x \in [14,18], \\ R_U(x), & x \in [18,20], \end{Bmatrix} = \begin{cases} 0, & x \notin [12,20], \\ \frac{2 \cdot x - 24}{x - 10}, & x \in [12,14], \\ 1, & x \in [14,18], \\ \frac{7 \cdot x - 140}{2 \cdot x - 50}, & x \in [18,20]. \end{cases} \tag{62}$$

Because $\overleftrightarrow{\mathcal{Y}} \in \mathbb{K}^-$, *using (57) we get*

$$\overline{\overline{\Psi}}(\overleftrightarrow{\mathcal{Y}}) = \overline{\overline{\Psi}}\left(\overleftrightarrow{\mathcal{L}}(13,11,6,5,S_Y,E_Y)\right) = \mathcal{L}(5,6,11,13,E_Y,S_Y) = \mathcal{L}(5,6,11,13,L_V,R_V) = \mathcal{V}, \tag{63}$$

where FN $\mathcal{V}$ *is described by the membership function*

$$\mu_V(x) = \begin{Bmatrix} 0, & x \notin [5,13], \\ L_V(x), & x \in [5,6], \\ 1, & x \in [6,11], \\ R_V(x), & x \in [11,13], \end{Bmatrix} = \begin{Bmatrix} 0, & x \notin [5,13], \\ \frac{6 \cdot x - 30}{x}, & x \in [5,6], \\ 1, & x \in [6,11], \\ \frac{2 \cdot x - 26}{x - 15}, & x \in [11,13]. \end{Bmatrix} \tag{64}$$

The above example shows that the disorientation map is a simple transformation the space $\mathbb{K}$ on the space $\mathbb{F}$. This simplicity is apparent. It follows from the fact that the arithmetic operations on $\mathbb{F}$ are consistent with the Zadeh's Extension Principle when the arithmetic operations on $\mathbb{K}$ are not consistent with this principle. The main difficulties arise from the difference between the definition (11)–(13) of minus operator $\ominus$ on $\mathbb{F}$ and the definition (36)–(38) of minus operator $\boxminus$ on $\mathbb{K}$.

Let us compare the semigroups $\langle \mathbb{F}, \oplus \rangle$ and $\langle \mathbb{K}, \boxplus \rangle$. The identities (18)–(22) and (39)–(46) imply that the number $[\![0]\!]$ is the identity element in both these semigroups.

In [6], it is shown that addition $\boxplus$ is not associative. It implies that semigroup $\langle \mathbb{K}, \boxplus \rangle$ is not group. Moreover, the identity (51) implies that subtraction $\boxminus$ is the inverse operator to addition $\boxplus$.

The identity (18–22) implies that the addition $\oplus$ is associative. On the other hand, for any TrFN $\mathcal{T} = Tr(a,b,c,d) \in (\mathbb{F}_{Tr} \backslash \mathbb{R}) \subset \mathbb{F}$ we have

$$\mathcal{T} \ominus \mathcal{T} = Tr(a-d, b-c, c-b, d-a) \neq [\![0]\!]. \quad (65)$$

It shows that subtraction $\ominus$ is not inverse operator to addition $\oplus$. It proves that semigroup $\langle \mathbb{F}, \oplus \rangle$ is not group.

All above simple conclusions imply that:

- additive semigroup $\langle \mathbb{F}, \oplus \rangle$ and additive semigroup $\langle \mathbb{K}, \boxplus \rangle$ cannot be considered as homomorphic algebraic structures;
- any theorems on FNs cannot automatically extended to the case of OFNs.

3. Relation "Greater than or Equal to" for Fuzzy Numbers

We consider the pair $(\mathcal{K}, \mathcal{L}) \in \mathbb{F}^2$ of FNs determined by membership functions $\mu_K, \mu_L \in [0,1]^{\mathbb{R}}$. On the space $\mathbb{F}$, we can consider the relation $\mathcal{K}.GE.\mathcal{L}$, which reads:

$$\text{"FN } \mathcal{K} \text{ is greater than or equal to FN } \mathcal{L}\text{."} \quad (66)$$

Orlovsky [61] shows that in agreement with the Zadeh's Extension Principle, this relation is a fuzzy preorder $[GE] \in \mathcal{F}(\mathbb{F}^2)$ described by membership function $\nu_{[GE]} \in [0,1]^{\mathbb{F}^2}$ determined as follows

$$\nu_{[GE]}(\mathcal{K}, \mathcal{L}) = \sup\{\min\{\mu_K(x), \mu_L(y)\} : x \geq y\}. \quad (67)$$

From the multivalued logic point of view, the value $\nu_{[GE]}(\mathcal{K}, \mathcal{L})$ is considered as a truth-value of the sentence (66). It means that we have

$$\nu_{[GE]}(\mathcal{K}, \mathcal{L}) = tv(\text{"}\mathcal{K}.GE.\mathcal{L}\text{"}). \quad (68)$$

We prove that the fuzzy preorder $[GE] \in \mathcal{F}(\mathbb{F}^2)$ fulfils the following well-known properties.

Theorem 2. *For any pair $(\mathcal{K}, \mathcal{L}) \in \mathbb{F}^2$, we have:*

$$\nu_{[GE]}(\mathcal{K}, \mathcal{L}) = \nu_{[GE]}(\ominus\mathcal{L}, \ominus\mathcal{K}), \quad (69)$$

$$\nu_{[GE]}(\mathcal{K}, \mathcal{L}) = \nu_{[GE]}(\mathcal{K} \ominus \mathcal{L}, [\![0]\!]). \quad (70)$$

Proof. Take into account the quadruple $(\mathcal{K}, \mathcal{L}, \mathcal{M}, \mathcal{N}) \in \mathbb{F}^4$ of FNs represented respectively by their membership functions $\mu_K, \mu_L, \mu_M, \mu_N \in [0,1]^{\mathbb{R}}$.

Let us assume that $\mathcal{M} = \ominus\mathcal{K}$ and $\mathcal{N} = \ominus\mathcal{L}$. Using the identities (11), (12), and (13) we obtain:

$$\mu_M(y) = \mu_K(-y) \text{ and } \mu_N(x) = \mu_L(-x).$$

Then the identity (67) implies

$$\begin{aligned} \nu_{[GE]}(\ominus\mathcal{L}, \ominus\mathcal{K}) &= \nu_{[GE]}(\mathcal{N}, \mathcal{M}) = \sup\{\min\{\mu_N(x), \mu_M(y)\} : x \geq y\} = \\ &= \sup\{\min\{\mu_L(-x), \mu_K(-y)\} : -x \leq -y\} = \sup\{\min\{\mu_L(u), \mu_K(v)\} : u \leq v\} = \nu_{[GE]}(\mathcal{K}, \mathcal{L}). \end{aligned}$$

Let us assume now that $\mathcal{M} = \mathcal{K} \ominus \mathcal{L}$. Using the identity (7) we obtain:

$$\mu_M(z) = \sup\{\min\{\mu_K(x), \mu_L(y)\} : z = x - y, (x, y) \in \mathbb{R}\}.$$

Then the identity (67) implies

$$\begin{aligned} \nu_{[GE]}(\mathcal{K} \ominus \mathcal{L}, [\![0]\!]) = \nu_{[GE]}(\mathcal{M}, [\![0]\!]) = \sup\{\mu_M(z) : z \geq 0\} = \\ = \sup\{\sup\{\min\{\mu_K(x), \mu_L(y)\} : z = x - y, (x, y) \in \mathbb{R}\} : z \geq 0\} = \\ = \sup\{\min\{\mu_K(x), \mu_L(y)\} : x - y \geq 0\} = \nu_{[GE]}(\mathcal{K}, \mathcal{L}). \text{ QED} \end{aligned}$$

Theorem 3. *For any FNs $\mathcal{L}(a,b,c,d,L_K,R_K)$, $\mathcal{L}(e,f,g,h,L_M,R_M) \in \mathbb{F}$ we have*

$$\nu_{[GE]}(\mathcal{L}(a,b,c,d,L_K,R_K),\ \mathcal{L}(e,f,g,h,L_M,R_M)) = \begin{cases} 0 & 0 < d - e, \\ R_W(0) & d - e \leq 0 < c - f, \\ 1 & 0 \leq c - f, \end{cases} \tag{71}$$

where the function $R_W : [d - e, c - f] \to [0, 1]$ is given by identity (28).

Proof. For $e > d$, using (67) and (8) we get

$$\begin{aligned} \nu_{[GE]}(\mathcal{L}(a,b,c,d,L_K,R_K),\ \mathcal{L}(e,f,g,h,L_M,R_M)) = \sup\{\min\{\mu_K(x), \mu_M(y)\} : x \geq y\} = \\ = \max\{\sup\{\min\{0, \mu_M(x)\} : x \geq y \geq e\}, \sup\{\min\{\mu_K(x), 0\} : x \geq y \in]e, -\infty[\}\} = 0. \end{aligned} \tag{72}$$

For $c \geq f$ we have

$$\begin{aligned} 1 \geq \nu_{[GE]}(\mathcal{L}(a,b,c,d,L_K,R_K),\ \mathcal{L}(e,f,g,h,L_M,R_M)) = \sup\{\min\{\mu_K(x), \mu_M(y)\} : x \geq y\} \geq \\ \geq \sup\{\min\{\mu_K(x), \mu_M(y)\} : c \geq x \geq y \geq f\} = \sup\{\min\{1, 1\}\} = 1. \end{aligned} \tag{73}$$

For $d \leq e$ and $f < c$ we have $d - e \leq 0 < c - f$. Then from (24), (67) and (70) we obtain

$$\begin{aligned} \nu_{[GE]}(\mathcal{L}(a,b,c,d,L_K,R_K),\ \mathcal{L}(e,f,g,h,L_M,R_M)) = \nu_{[GE]}(\mathcal{L}(a-h, b-g, c-f, d-e, L_W, R_W), [\![0]\!]) = \\ = R_W(0). \text{ QED} \end{aligned} \tag{74}$$

Example 2. *Let us take into account the FNs $\mathcal{U} = \mathcal{L}(12, 14, 18, 20, L_U, R_U)$ and $\mathcal{V} = \mathcal{L}(5, 6, 11, 13, L_V, R_V)$ respectively determined by identities (62) and (64). We compare these FNs with using of fuzzy preorder $[GE] \in \mathcal{F}(\mathbb{F}^2)$. We have here*

$$-1 = 12 - 13 \leq 0 \leq 14 - 11 = 3.$$

Therefore, we should establish the variability of the function $R_W \in [0, 1]^{[-1,3]}$ determined by the identity (28). First, by using identities (14) and (15), we assign functions

$$L_U^{\star}(\alpha) = \min\{x \in [12, 14] :\ L_U(x) \geq \alpha\} = L_U^{-1}(\alpha) = \frac{10\alpha - 24}{\alpha - 2}, \tag{75}$$

$$R_V^{\star}(\alpha) = \max\{x \in [11, 15] :\ R_V(x) \geq \alpha\} = R_V^{-1}(\alpha) = \frac{15\alpha - 26}{\alpha - 2}. \tag{76}$$

In the next step, applying (25) and (27), we obtain

$$r_W(\alpha) = R_V^{\star}(\alpha) - L_U^{\star}(\alpha) = \frac{15\alpha - 26}{\alpha - 2} - \frac{10\alpha - 24}{\alpha - 2} = \frac{5\alpha - 2}{\alpha - 2}, \tag{77}$$

$$R_W(x) = r_W^{\triangleleft}(x) = \min\{\alpha \in [0; 1] : l_W(\alpha) = x\} = l_W^{-1}(x) = \frac{2 \cdot (x - 1)}{x - 5}. \tag{78}$$

Finally, using identity (71), we get

$$\nu_{[GE]}(\mathcal{V},\mathcal{U}) = R_W(0) = \frac{2}{5}. \tag{79}$$

The above example together with Theorem 3 shows that fuzzy preorder $[\mathrm{GE}] \in \mathcal{F}(\mathbb{F}^2)$ depends only on the interaction between the right reference function of the first compared FN and the left reference function of the second compared FN.

Moreover, Theorem 3 immediately implies that for any TrFNs we have:

Theorem 4. *For any TrFNs $Tr(a,b,c,d),\ Tr(e,f,g,h) \in \mathbb{F}_{Tr}$ we have*

$$\nu_{[GE]}(Tr(a,b,c,d),\ Tr(e,f,g,h)) = \begin{cases} 0, & 0 < d-e, \\ \frac{d-e}{d+f-c-e}, & d-e \le 0 < c-f, \\ 1, & 0 \le c-f. \end{cases} \tag{80}$$

4. Relation "Greater than or Equal to" for Ordered Fuzzy Numbers

Let us consider the pair $(\overleftrightarrow{\mathcal{K}},\overleftrightarrow{\mathcal{L}}) \in \mathbb{K}^2$ represented by the pair $(\mu_K,\mu_L) \in \left([0,1]^{\mathbb{R}}\right)^2$ of their membership functions. On the space $\mathbb{K}$, we introduce the relation $\overleftrightarrow{\mathcal{K}}.\widetilde{GE}.\overleftrightarrow{\mathcal{L}}$, which reads:

$$\text{"OFN } \overleftrightarrow{\mathcal{K}} \text{ is greater than or equal to OFN } \overleftrightarrow{\mathcal{L}}.\text{"} \tag{81}$$

This relation is a fuzzy preorder $\widetilde{GE} \in \mathcal{F}(\mathbb{K}^2)$ defined by its membership function $\nu_{GE} \in [0,1]^{\mathbb{K}^2}$. From the point view of the multivalued logic, the value $\nu_{GE}(\mathcal{K},\mathcal{L})$ is considered as a truth-value of the sentence (81). It means that we have

$$\nu_{GE}(\overleftrightarrow{\mathcal{K}},\overleftrightarrow{\mathcal{L}}) = tv(\text{"}\overleftrightarrow{\mathcal{K}}.\widetilde{GE}.\overleftrightarrow{\mathcal{L}}\text{"}). \tag{82}$$

The fuzzy preorder $\widetilde{GE} \in \mathcal{F}(\mathbb{K}^2)$ cannot be determined with use of the Zadeh's Extension Principle because of this principle is not valid for OFNs. Therefore, we additionally assume that any membership function $\nu_{GE} \in [0,1]^{\mathbb{K}^2}$ meets the following well-known conditions:

- for any pair $(\overleftrightarrow{\mathcal{K}},\overleftrightarrow{\mathcal{L}}) \in (\mathbb{K}^+ \cup \mathbb{R})^2$ the extension principle

$$\nu_{GE}(\overleftrightarrow{\mathcal{K}},\overleftrightarrow{\mathcal{L}}) = \nu_{[GE]}(\Psi(\overleftrightarrow{\mathcal{K}}),\Psi(\overleftrightarrow{\mathcal{L}})), \tag{83}$$

- for any pair $(\overleftrightarrow{\mathcal{K}},\overleftrightarrow{\mathcal{L}}) \in (\mathbb{K}^- \cup \mathbb{R})^2$ the sign exchange law

$$\nu_{GE}(\overleftrightarrow{\mathcal{K}},\overleftrightarrow{\mathcal{L}}) = \nu_{GE}(\boxminus \overleftrightarrow{\mathcal{L}}, \boxminus \overleftrightarrow{\mathcal{K}}), \tag{84}$$

- for any pair $(\overleftrightarrow{\mathcal{K}},\overleftrightarrow{\mathcal{L}}) \in (\mathbb{K}^+ \cup \mathbb{R}) \times (\mathbb{K}^- \cup \mathbb{R})$ the law of subtraction of parties

$$\nu_{GE}(\overleftrightarrow{\mathcal{K}},\overleftrightarrow{\mathcal{L}}) = \nu_{GE}(\overleftrightarrow{\mathcal{K}} \boxminus \overleftrightarrow{\mathcal{L}}, [\![0]\!]). \tag{85}$$

Among other things, we prove here:

Lemma 1. *Any pair* $(\overleftrightarrow{\mathcal{K}},\overleftrightarrow{\mathcal{L}}) \in (\mathbb{K}^+ \cup \mathbb{R}) \times (\mathbb{K}^- \cup \mathbb{R})$ *satisfies the condition*

$$\Psi(\overleftrightarrow{\mathcal{K}} \boxminus \overleftrightarrow{\mathcal{L}}) = \Psi(\overleftrightarrow{\mathcal{K}}) \ominus (\ominus \Psi(\boxminus \overleftrightarrow{\mathcal{L}})) \tag{86}$$

Proof. Let $\overleftrightarrow{\mathcal{K}} = \overleftrightarrow{\mathcal{L}}(a,b,c,d,S_K,E_K) \in \mathbb{K}^+ \cup \mathbb{R}$ and $\overleftrightarrow{\mathcal{L}} = \overleftrightarrow{\mathcal{L}}(e,f,g,h,S_L,E_L) \in \mathbb{K}^- \cup \mathbb{R}$. Then, we have $\boxminus \overleftrightarrow{\mathcal{L}}$, $\overleftrightarrow{\mathcal{K}} \boxminus \overleftrightarrow{\mathcal{L}} \in \mathbb{K}^+$ because of the sequences $(-e,-f,-g,-h)$ and $(a-e,b-f,c-g,d-h)$ are nondecreasing. Then, from (39), (50), and (58) we get

$$\Psi(\overleftrightarrow{\mathcal{K}} \boxminus \overleftrightarrow{\mathcal{L}}) = \Psi\left(\overleftrightarrow{\mathcal{L}}(a-e,b-f,c-g,d-h,S_M,E_M)\right) = \mathcal{L}(a-e,b-f,c-g,d-h,S_M,E_M), \tag{87}$$

where

$$\forall_{\alpha\in[0;1]}\ s_M(\alpha) = \begin{cases} S_L^\star(\alpha) + S_L^\star(\alpha), & a-e \neq b-f, \\ b-f, & a-e = b-f. \end{cases} \tag{88}$$

$$\forall_{\alpha\in[0;1]}\ e_M(\alpha) = \begin{cases} E_K^\star(\alpha) + E_L^\star(\alpha), & c-g \neq d-h, \\ c-g, & c-g = d-h. \end{cases} \tag{89}$$

$$\forall_{x\in[a-e,b-f]}\ S_M(x) = s_M^{\triangleleft}(x), \tag{90}$$

$$\forall_{x\in[c-g,d-h]}\ E_M(x) = e_M^{\triangleleft}(x). \tag{91}$$

On the other hand, successively from (36), (57), (11), and (22), we obtain

$$\begin{aligned} \Psi(\overleftrightarrow{\mathcal{K}}) \ominus (\ominus \Psi\left(\boxminus \overleftrightarrow{\mathcal{L}}\right)) &= \Psi\left(\overleftrightarrow{\mathcal{L}}(a,b,c,d,S_K,E_K)\right) \ominus \left(\ominus \Psi\left(\boxminus \overleftrightarrow{\mathcal{L}}(e,f,g,h,S_L,E_L)\right)\right) = \\ &= \mathcal{L}(a,b,c,d,S_K,E_K) \ominus \left(\ominus \Psi\left(\overleftrightarrow{\mathcal{L}}(-e,-f,-g,-h,S_L^{(-)},E_L^{(-)})\right)\right) = \\ &= \mathcal{L}(a,b,c,d,S_K,E_K) \ominus \left(\ominus \mathcal{L}(-e,-f,-g,-h,S_L^{(-)},E_L^{(-)})\right) = \\ = \mathcal{L}(a,b,c,d,S_K,E_K) \ominus \mathcal{L}(h,g,f,e,E_L^{(-)},S_L^{(-)}) &= \mathcal{L}(a-e,b-f,c-g,d-h,S_M,E_M). \text{ QED} \end{aligned} \tag{92}$$

The conjunction of assumptions (83)–(85) is a sufficient condition for the formulation of the following theorem:

Theorem 5. *For any pair* $(\overleftrightarrow{\mathcal{K}},\overleftrightarrow{\mathcal{L}}) \in \mathbb{K}^2$ *we have*

$$\nu_{GE}(\overleftrightarrow{\mathcal{K}},\overleftrightarrow{\mathcal{L}}) = \nu_{[GE]}(\overline{\overline{\Psi}}(\overleftrightarrow{\mathcal{K}}),\overline{\overline{\Psi}}(\overleftrightarrow{\mathcal{L}})). \tag{93}$$

Proof. For any pair $(\overleftrightarrow{\mathcal{K}},\overleftrightarrow{\mathcal{L}}) \in (\mathbb{K}^+ \cup \mathbb{R})^2$ the identity (93) is obvious.

Let us assume that $(\overleftrightarrow{\mathcal{K}},\overleftrightarrow{\mathcal{L}}) \in (\mathbb{K}^- \cup \mathbb{R})^2$. Then, $(\boxminus\overleftrightarrow{\mathcal{K}},\boxminus\overleftrightarrow{\mathcal{L}}) \in (\mathbb{K}^+ \cup \mathbb{R})^2$ and successively from (84), (83), (69) and (56), we get

$$\begin{aligned} \nu_{GE}(\overleftrightarrow{\mathcal{K}},\overleftrightarrow{\mathcal{L}}) &= \nu_{GE}(\boxminus \overleftrightarrow{\mathcal{L}}, \boxminus \overleftrightarrow{\mathcal{K}}) = \nu_{[GE]}(\Psi(\boxminus \overleftrightarrow{\mathcal{L}}),\Psi(\boxminus \overleftrightarrow{\mathcal{K}})) = \nu_{[GE]}(\ominus \Psi(\boxminus \overleftrightarrow{\mathcal{K}}),\ominus \Psi(\boxminus \overleftrightarrow{\mathcal{L}})) = \\ &= \nu_{[GE]}(\overline{\overline{\Psi}}(\overleftrightarrow{\mathcal{K}}),\overline{\overline{\Psi}}(\overleftrightarrow{\mathcal{L}})). \end{aligned} \tag{94}$$

Let us assume now that $(\overleftrightarrow{\mathcal{K}},\overleftrightarrow{\mathcal{L}}) \in (\mathbb{K}^{+} \cup \mathbb{R}) \times (\mathbb{K}^{-} \cup \mathbb{R})$. Then $\overleftrightarrow{\mathcal{K}} \boxminus \overleftrightarrow{\mathcal{L}} \in \mathbb{K}^{+}$ and successively from (85), (83), (86), (70) and (57), we get

$$\begin{aligned} &\nu_{GE}(\overleftrightarrow{\mathcal{K}},\overleftrightarrow{\mathcal{L}}) = \nu_{GE}(\overleftrightarrow{\mathcal{K}} \boxminus \overleftrightarrow{\mathcal{L}}, \llbracket 0 \rrbracket) = \nu_{[GE]}(\Psi(\overleftrightarrow{\mathcal{K}} \boxminus \overleftrightarrow{\mathcal{L}}), \llbracket 0 \rrbracket) = \nu_{[GE]}(\Psi(\overleftrightarrow{\mathcal{K}}) \ominus (\ominus \Psi(\boxminus \overleftrightarrow{\mathcal{L}})), \llbracket 0 \rrbracket) = \\ &= \nu_{[GE]}(\Psi(\overleftrightarrow{\mathcal{K}}), \ominus \Psi(\boxminus \overleftrightarrow{\mathcal{L}})) = \nu_{[GE]}(\overset{=}{\Psi}(\overleftrightarrow{\mathcal{K}}), \overset{=}{\Psi}(\overleftrightarrow{\mathcal{L}})). \end{aligned} \tag{95}$$

Let us assume now that $(\overleftrightarrow{\mathcal{K}},\overleftrightarrow{\mathcal{L}}) \in (\mathbb{K}^{-} \cup \mathbb{R}) \times (\mathbb{K}^{+} \cup \mathbb{R})$. Then $\overleftrightarrow{\mathcal{L}} \boxminus \overleftrightarrow{\mathcal{K}} \in \mathbb{K}^{+}$ and successively from (85), (84), (83), (86), (69), (70), and (57), we get

$$\begin{aligned} &\nu_{GE}(\overleftrightarrow{\mathcal{K}},\overleftrightarrow{\mathcal{L}}) = \nu_{GE}(\overleftrightarrow{\mathcal{K}} \boxminus \overleftrightarrow{\mathcal{L}}, \llbracket 0 \rrbracket) = \nu_{GE}(\llbracket 0 \rrbracket, \overleftrightarrow{\mathcal{L}} \boxminus \overleftrightarrow{\mathcal{K}}) = \nu_{[GE]}(\llbracket 0 \rrbracket, \Psi(\overleftrightarrow{\mathcal{L}} \boxminus \overleftrightarrow{\mathcal{K}})) = \\ &= \nu_{[GE]}(\llbracket 0 \rrbracket, \Psi(\overleftrightarrow{\mathcal{L}}) \ominus (\ominus \Psi(\boxminus \overleftrightarrow{\mathcal{K}}))) = \nu_{[GE]}(\ominus \Psi(\boxminus \overleftrightarrow{\mathcal{K}}) \ominus \Psi(\overleftrightarrow{\mathcal{L}}), \llbracket 0 \rrbracket) = \\ &\nu_{[GE]}(\ominus \Psi(\boxminus \overleftrightarrow{\mathcal{K}}), \Psi(\overleftrightarrow{\mathcal{L}})) = \nu_{[GE]}(\overset{=}{\Psi}(\overleftrightarrow{\mathcal{K}}), \overset{=}{\Psi}(\overleftrightarrow{\mathcal{L}})). \text{ QED} \end{aligned} \tag{96}$$

Example 3. *Let us compare the OFN* $\overleftrightarrow{\mathcal{X}} = \overleftrightarrow{\mathcal{L}}(12, 14, 18, 20, S_X, E_X)$ *determined by (59) and the OFN* $\overleftrightarrow{\mathcal{Y}} = \overleftrightarrow{\mathcal{L}}(13, 11, 6, 5, S_Y, E_Y)$ *determined by (60). Using (93), (62), (64), and (41), we get*

$$\begin{aligned} \nu_{GE}(\overleftrightarrow{\mathcal{Y}},\overleftrightarrow{\mathcal{X}}) &= \nu_{[GE]}(\overset{=}{\Psi}(\overleftrightarrow{\mathcal{Y}}), \overset{=}{\Psi}(\overleftrightarrow{\mathcal{X}})) = \nu_{[GE]}(\mathcal{L}(5, 6, 11, 13, E_Y, S_Y),\ \mathcal{L}(12, 14, 18, 20, S_X, E_X)) = \\ &= \nu_{[GE]}(\mathcal{L}(5, 6, 11, 13, L_V, R_Y), \mathcal{L}(12, 14, 18, 20, L_U, R_U)) = \nu_{[GE]}(\mathcal{V}, \mathcal{U}) = \tfrac{2}{5}. \end{aligned} \tag{97}$$

The simplicity of the calculations in the above example is apparent. In fact, Example 3 together with Theorem 5 shows that:

- if compared OFNs are both positively oriented then the fuzzy preorder $\widetilde{GE} \in \mathcal{F}(\mathbb{K}^2)$ depends only on the interaction between the ending function of the first compared OFN and the starting function of the second compared OFN;
- if the first compared OFN is positively oriented and the second compared OFN is negatively oriented then the fuzzy preorder $\widetilde{GE} \in \mathcal{F}(\mathbb{K}^2)$ depends only on the interaction between the ending functions of compared OFN;
- if the first compared OFN is negatively oriented and the second compared OFN is positively oriented then the fuzzy preorder $\widetilde{GE} \in \mathcal{F}(\mathbb{K}^2)$ depends only on the interaction between the starting functions of compared OFN;
- if compared OFNs are both negatively oriented, then the fuzzy preorder $\widetilde{GE} \in \mathcal{F}(\mathbb{K}^2)$ depends only on the interaction between the starting function of the first compared OFN and the ending function of the second compared OFN.

5. Relations "Greater Than" and "Equal to" for Ordered Fuzzy Numbers

In the last section, we explicitly define the preorder "greater than or equal to" $\widetilde{GE}$ on the space $\mathbb{K}$ of all OFNs. This relation may be applied as start point for determining other basic relations on $\mathbb{K}$.

Let us consider any pair $(\overleftrightarrow{\mathcal{K}},\overleftrightarrow{\mathcal{L}}) \in \mathbb{K}^2$. On the space $\mathbb{K}$ we introduce the relation $\overleftrightarrow{\mathcal{K}}.\widetilde{GT}.\overleftrightarrow{\mathcal{L}}$, which reads:

$$\text{"OFN } \overleftrightarrow{\mathcal{K}} \text{ is greater than OFN } \overleftrightarrow{\mathcal{L}}.\text{"} \tag{98}$$

This relation is a fuzzy strict order $\widetilde{GT} \in \mathcal{F}(\mathbb{K}^2)$ defined by its membership function $\nu_{GT} \in [0,1]^{\mathbb{K}^2}$. From the point view of the multivalued logic, the value $\nu_{GT}(\overleftrightarrow{\mathcal{K}}, \overleftrightarrow{\mathcal{L}})$ is considered as a truth-value of the sentence (98) which is equivalent to the sentence:

$$\text{"OFN } \overleftrightarrow{\mathcal{L}} \text{ is not greater than or equal to OFN } \overleftrightarrow{\mathcal{K}}.\text{"} \tag{99}$$

It means that we have

$$\nu_{GT}(\overleftrightarrow{\mathcal{K}}, \overleftrightarrow{\mathcal{L}}) = tv(\text{"}\overleftrightarrow{\mathcal{K}}.\widetilde{GT}.\overleftrightarrow{\mathcal{L}}\text{"}) = tv(\text{"}\neg\overleftrightarrow{\mathcal{L}}.\widetilde{GE}.\overleftrightarrow{\mathcal{K}}\text{"}) = 1 - tv(\text{"}\overleftrightarrow{\mathcal{L}}.\widetilde{GE}.\overleftrightarrow{\mathcal{K}}\text{"}). \tag{100}$$

Therefore, the membership function $\nu_{GT} \in [0,1]^{\mathbb{K}^2}$ is determined by the identity

$$\nu_{GT}(\overleftrightarrow{\mathcal{K}}, \overleftrightarrow{\mathcal{L}}) = 1 - \nu_{GE}(\overleftrightarrow{\mathcal{L}}, \overleftrightarrow{\mathcal{K}}). \tag{101}$$

Moreover, on the space $\mathbb{K}$ we introduce the relation $\overleftrightarrow{\mathcal{K}}.\widetilde{EQ}.\overleftrightarrow{\mathcal{L}}$, which reads:

$$\text{"OFN } \overleftrightarrow{\mathcal{K}} \text{ is equal to OFN } \overleftrightarrow{\mathcal{L}}.\text{"} \tag{102}$$

The relation $\widetilde{EQ} \in \mathcal{F}(\mathbb{K}^2)$ is fuzzy equivalence determined by membership function $\nu_{EQ} \in [0,1]^{\mathbb{K}^2}$. From the point view of the multivalued logic, the value $\nu_{EQ}(\overleftrightarrow{\mathcal{K}}, \overleftrightarrow{\mathcal{L}})$ is considered as a truth-value of the sentence (102) which is equivalent to the sentence:

$$\text{"OFN } \overleftrightarrow{\mathcal{K}} \text{ is greater than or equal to OFN } \overleftrightarrow{\mathcal{L}} \text{ and OFN } \overleftrightarrow{\mathcal{L}} \text{ is greater than or equal to OFN } \overleftrightarrow{\mathcal{K}}\text{"} \tag{103}$$

It means that we have

$$\begin{aligned}\nu_{EQ}(\overleftrightarrow{\mathcal{K}}, \overleftrightarrow{\mathcal{L}}) &= tv(\text{"}\overleftrightarrow{\mathcal{K}}.\widetilde{EQ}.\overleftrightarrow{\mathcal{L}}\text{"}) = tv(\text{"}\overleftrightarrow{\mathcal{K}}.\widetilde{EQ}.\overleftrightarrow{\mathcal{L}}\text{"} \wedge \text{"}\overleftrightarrow{\mathcal{L}}.\widetilde{GE}.\overleftrightarrow{\mathcal{K}}\text{"}) = \\ &= \min\{tv(\text{"}\overleftrightarrow{\mathcal{K}}.\widetilde{EQ}.\overleftrightarrow{\mathcal{L}}\text{"}), tv(\text{"}\overleftrightarrow{\mathcal{L}}.\widetilde{GE}.\overleftrightarrow{\mathcal{K}}\text{"})\}.\end{aligned} \tag{104}$$

Therefore, the membership function $\nu_{EQ} \in [0,1]^{\mathbb{K}^2}$ is determined by the identity

$$\nu_{EQ}(\overleftrightarrow{\mathcal{K}}, \overleftrightarrow{\mathcal{L}}) = \min\{\nu_{GE}(\overleftrightarrow{\mathcal{K}}, \overleftrightarrow{\mathcal{L}}), \nu_{GE}(\overleftrightarrow{\mathcal{L}}, \overleftrightarrow{\mathcal{K}})\}. \tag{105}$$

For any finite set $A = \{\overleftrightarrow{\mathcal{K}}_1, \overleftrightarrow{\mathcal{K}}_2, \ldots, \overleftrightarrow{\mathcal{K}}_n\} \subset \mathbb{K}_{Tr}$ we can distinguish set of maximal elements $\text{Max}\{A\} \in \mathcal{F}(A)$ which is described by membership function $\mu_{\text{Max}\{A\}} \in [0,1]^A$ determined in the following way [62]

$$\mu_{\text{Max}\{A\}}(\overleftrightarrow{\mathcal{K}}_i) = \min\{\nu_{GE}(\overleftrightarrow{\mathcal{K}}_i, \overleftrightarrow{\mathcal{K}}_j) : \overleftrightarrow{\mathcal{K}}_j \in A\}. \tag{106}$$

This set may be applied as solution of optimization tasks using OFNs. Moreover, let us note, that the set $\text{Max}\{A\}$ of maximal elements may be used as a fuzzy choice function [78].

In [17], the relation $\widetilde{GE} \in \mathcal{F}(\mathbb{K}^2_{Tr})$ is applied for ordering negotiation packages [79]. The considered case study is fully described there. Moreover, let us look on a short case study of applying the relation $\widetilde{GE} \in \mathcal{F}(\mathbb{K}^2)$ for financial effectivity analysis.

6. Financial Effectivity Determined by Imprecise Return—A Numerical Example

Let any financial security $\mathcal{Z} \in \mathbb{Z}$ be represented by the pair (R_z, σ_Z^2), where $R_z \in \mathbb{R}$ is an expected return rate from this security and $\sigma_Z^2 \in \mathbb{R}$ is a variance of its return rate. The symbol $\mathbb{Z}$ denotes the family of all considered securities.

We introduce the relation $\mathcal{P}.NLE.\mathcal{Q}$ which reads

$$\text{"The security } \mathcal{P} \in \mathbb{Z} \text{ is no less effective than the security } \mathcal{Q} \in \mathbb{Z}\text{"}. \tag{107}$$

In financial practice, this relation is defined by the equivalence

$$\mathcal{P}.NLE.\mathcal{Q} \Leftrightarrow (R_P \geq R_Q \vee \sigma_P^2 \leq \sigma_Q^2). \tag{108}$$

In [15], it is justified that return rate may be evaluated OFN. In this case, any financial security $\mathcal{Z}$ be represented by the pair $(\overset{\leftrightarrow}{R}_z, \sigma_Z^2)$, where $\overset{\leftrightarrow}{R}_z \in \mathbb{K}$ is an expected return rate evaluated by OFN. Therefore, the relation $\mathcal{P}.NLE.\mathcal{Q}$ should be replaced by the relation $\mathcal{P}.\widetilde{NLE}.\mathcal{Q}$ defined by the equivalency

$$\mathcal{P}.\widetilde{NLE}.\mathcal{Q} \Leftrightarrow (\overset{\leftrightarrow}{R}_P.\widetilde{GE}.\overset{\leftrightarrow}{R}_Q \vee \sigma_P^2 \leq \sigma_Q^2). \tag{109}$$

The relation $\mathcal{P}.\widetilde{NLE}.\mathcal{Q}$ also reads as the sentence (107). The relation $\widetilde{NLE} \in \mathcal{F}(\mathbb{K}^2)$ is fuzzy one determined by membership function $\nu_{NLE} \in [0,1]^{\mathbb{Z}^2}$. From the point view of the multivalued logic, the value $\nu_{NLE}(\mathcal{P},\mathcal{Q})$ is considered as a truth-value of the sentence (105). It means that we have

$$\begin{aligned} \nu_{NLE}(\mathcal{P},\mathcal{Q}) &= \boldsymbol{tv}(\text{"}\overset{\leftrightarrow}{R}_P.\widetilde{GE}.\overset{\leftrightarrow}{R}_Q \vee \sigma_P^2 \leq \sigma_Q^2\text{"}) = \max\left\{\boldsymbol{tv}(\text{"}\overset{\leftrightarrow}{R}_P.\widetilde{GE}.\overset{\leftrightarrow}{R}_Q\text{"}), \boldsymbol{tv}(\text{"}\sigma_P^2 \leq \sigma_Q^2\text{"})\right\} = \\ &= \left\{\begin{array}{c} \boldsymbol{tv}(\text{"}\overset{\leftrightarrow}{R}_P.\widetilde{GE}.\overset{\leftrightarrow}{R}_Q\text{"})\ \sigma_P^2 > \sigma_Q^2, \\ 1\ \sigma_P^2 \leq \sigma_Q^2. \end{array}\right\} = \left\{\begin{array}{c} \nu_{GE}(\overset{\leftrightarrow}{R}_P, \overset{\leftrightarrow}{R}_Q)\ \sigma_P^2 > \sigma_Q^2, \\ 1\ \sigma_P^2 \leq \sigma_Q^2. \end{array}\right. . \end{aligned} \tag{110}$$

In order to increase the transparency of the considerations, we restrict our future considerations to the case of return rate evaluated by TrOFNs. We consider the securities $\mathcal{P}$, $\mathcal{Q}$ and $\mathcal{R}$ respectively represented by the pairs $(\overset{\leftrightarrow}{R}_P, \sigma_P^2) = (\overset{\leftrightarrow}{Tr}(0.010,\ 0.010, 0.035, 0.040), 0.00023)$, $(\overset{\leftrightarrow}{R}_Q, \sigma_Q^2) = (\overset{\leftrightarrow}{Tr}(0.020,\ 0.025, 0.030, 0.045), 0.00024)$ and $(\overset{\leftrightarrow}{R}_R, \sigma_R^2) = (\overset{\leftrightarrow}{Tr}(0.065,\ 0.055, 0.050, 0.035), 0.00012)$. The return rates $\overset{\leftrightarrow}{R}_P$ and $\overset{\leftrightarrow}{R}_Q$ are positively oriented TrOFNs. Therefore, we can anticipate an increase in the rates of return from the securities $\mathcal{P}$ and $\mathcal{Q}$. Moreover, we can predict a decrease in the rate of return from the security $\mathcal{R}$ because of the return rate $\overset{\leftrightarrow}{R}_R$ is negatively oriented TrOFN. For these reasons, we consider two investment decisions:

(A) We sell the security $\mathcal{R}$ and for the funds obtained we buy the security $\mathcal{P}$,
(B) We sell the security $\mathcal{R}$ and for the funds obtained we buy the security $\mathcal{Q}$.

Let us compare a financial effectivity of considered securities $\mathcal{P}$ and $\mathcal{R}$. In line with (108), (93), (58) and (71), we get

$$\begin{aligned} \nu_{NLE}(\mathcal{P},\mathcal{R}) &= \nu_{GE}\left(\overset{\leftrightarrow}{Tr}(0.010,\ 0.010, 0.035, 0.040), \overset{\leftrightarrow}{Tr}(0.065,\ 0.055, 0.050, 0.035)\right) = \\ &\nu_{[GE]}\left(\overline{\overline{\Psi}}\left(\overset{\leftrightarrow}{Tr}(0.010,\ 0.010, 0.035, 0.040)\right), \overline{\overline{\Psi}}\left(\overset{\leftrightarrow}{Tr}(0.065,\ 0.055, 0.050, 0.035)\right)\right) = \\ &= \nu_{[GE]}(Tr(0.010,\ 0.010, 0.035, 0.040), Tr(0.035,\ 0.050, 0.055, 0.065)) = \\ &= \frac{0.040-0.035}{0.040+0.050-0.035-0.035} = \frac{1}{4}. \end{aligned} \tag{111}$$

In the same way, we can compare a financial effectivity of considered securities $\mathcal{Q}$ and $\mathcal{R}$. We obtain

$$\begin{aligned} \nu_{NLE}(\mathcal{Q},\mathcal{R}) &= \nu_{GE}\left(\overset{\leftrightarrow}{Tr}(0.020,\ 0.025, 0.030, 0.045), \overset{\leftrightarrow}{Tr}(0.065,\ 0.055, 0.050, 0.035)\right) = \\ &\nu_{[GE]}\left(\overline{\overline{\Psi}}\left(\overset{\leftrightarrow}{Tr}(0.020,\ 0.025, 0.030, 0.045)\right), \overline{\overline{\Psi}}\left(\overset{\leftrightarrow}{Tr}(0.065,\ 0.055, 0.050, 0.035)\right)\right) = \\ &= \nu_{[GE]}(Tr(0.020,\ 0.025, 0.030, 0.045), Tr(0.035,\ 0.050, 0.055, 0.065)) = \\ &= \frac{0.045-0.035}{0.045+0.050-0.030-0.035} = \frac{1}{3}. \end{aligned} \tag{112}$$

Therefore, we can say that the investment decisions (A) and (B) are both partially justified. Because of $\nu_{NLE}(\mathcal{Q},\mathcal{R}) > \nu_{NLE}(\mathcal{P},\mathcal{R})$, we ultimately recommend the investment decision (B).

7. Final Remarks

Relation "greater than or equal to" $\widetilde{GE}$ is explicitly defined on the space of all OFNs. In my best knowledge, it will be the first fuzzy order determined for OFNs. Determined relation $\widetilde{GE}$ compares OFNs without losing information about the imprecision and orientation of evaluated OFNs. From the point-view of application needs, this approach is desirable. Nevertheless, I proved that the relation $\widetilde{GE}$ is independent of the orientation of the numbers being compared. This conclusion may be applied for simplification of many OFN applications.

The first application of relation $\widetilde{GE}$ is cited in Section 5. The next application is described in Section 6. Meanwhile, we will employ the proposed relation to model some imprecision decision making problems from some concrete applied fields, such as medical decision making, behavioural economic [11], management [15,16], telecommunication, and financial assessment [7,9–14]. Then these relations may be used for decision making problems with scoring function evaluated by OFNs. In [15,16], such evaluation of scoring function follows from the fact that partial ratings are evaluated by OFNs. Moreover, studying multi criterial group decision making problems, we should take into account some imprecise weights of criteria [80]. Then these weights may be evaluated by OFNs what implies that also the scoring function is evaluated by OFNs. In general, the relation $\widetilde{GE}$ can be applied in any such quantitative model of the real world that a comparison of imprecise numbers is used.

In Section 2.2, I point out some terminology problems connected with the notion of OFN. I believe that this is a very important problem from an ethical point of view. I invite people of science to discuss this topic.

For any OFN we can determine the family of oriented α-cuts defined as a pair of usual α-cut and OFN orientation. An important direction for further development is to propose such fuzzy order of OFNs which is determined by the family of all α-cuts for FNs. At present, the oriented α-cuts theory is unknown.

Funding: This research did not receive external funding.

Acknowledgments: Author is very grateful to the anonymous reviewers for their insightful and constructive comments and suggestions. Using these comments allowed me to improve this article.

Conflicts of Interest: The author declares no conflict of interest.

References

1. Kosiński, W.; Słysz, P. Fuzzy numbers and their quotient space with algebraic operations. *Bull. Pol. Acad. Sci.* **1993**, *41*, 285–295.
2. Kosiński, W.; Prokopowicz, P.; Ślęzak, D. Fuzzy Numbers with Algebraic Operations: Algorithmic Approach. In *Proc.IIS'2002 Sopot, Poland*; Klopotek, M., Wierzchoń, S.T., Michalewicz, M., Eds.; Physica Verlag: Heidelberg, Germany, 2002; pp. 311–320.
3. Kosiński, W.; Prokopowicz, P.; Ślęzak, D. Ordered fuzzy numbers. *Bull. Pol. Acad. Sci.* **2003**, *51*, 327–339.
4. Kosiński, W. On fuzzy number calculus. *Int. J. Appl. Math. Comput. Sci.* **2006**, *16*, 51–57.
5. Goetschel, R.; Voxman, W. Elementary fuzzy calculus. *Fuzzy Set. Syst.* **1986**, *18*, 31–43. [CrossRef]
6. Piasecki, K. Revision of the Kosiński's Theory of Ordered Fuzzy Numbers. *Axioms* **2018**, *7*, 16. [CrossRef]
7. Prokopowicz, P.; Czerniak, J.; Mikołajewski, D.; Apiecionek, Ł.; Slezak, D. *Theory and Applications of Ordered Fuzzy Number. Tribute to Professor Witold Kosiński*; Studies in Fuzziness and Soft Computing, 356; Springer: Berlin, Germany, 2017.
8. Kacprzak, D. A doubly extended TOPSIS method for group decision making based on ordered fuzzy numbers. *Expert Syst. Appl.* **2018**, *116*, 243–254. [CrossRef]
9. Kacprzak, D.; Kosiński, W. Optimizing Firm Inventory Costs as a Fuzzy Problem. *Stud. Log. Gramm. Rhetor.* **2014**, *37*, 17. [CrossRef]

10. Kacprzak, D.; Kosiński, W.; Kosiński, W.K. Financial Stock Data and Ordered Fuzzy Numbers. In *Proceedings of the Artificial Intelligence and Soft Computing: 12th International Conference, 9–13 June 2013, Zakopane, Poland*; IEEE: Piscataway, NJ, USA; pp. 259–270. [CrossRef]
11. Łyczkowska-Hanćkowiak, A. Behavioural present value determined by ordered fuzzy number. *SSRN Electr. J.* **2017**, 6. [CrossRef]
12. Łyczkowska-Hanćkowiak, A. Sharpe's Ratio for Oriented Fuzzy Discount Factor. *Mathematics* **2019**, *7*, 272. [CrossRef]
13. Łyczkowska-Hanćkowiak, A.; Piasecki, K. The expected discount factor determined for present value given as ordered fuzzy number. In *9th International Scientific Conference Analysis of International Relations 2018. Methods and Models of Regional Development. Winter Edition, Katowice, Poland, 12 January 2018*; Szkutnik, W., Sączewska-Piotrowska, A., Hadaś-Dyduch, M., Acedański, J., Eds.; Publishing House of the University of Economics in Katowice: Katowice, Poland, 2018; pp. 69–75.
14. Łyczkowska-Hanćkowiak, A.; Piasecki, K. Present value of portfolio of assets with present values determined by trapezoidal ordered fuzzy numbers. *Oper. Res. Decis.* **2018**, *28*, 41–56. [CrossRef]
15. Piasecki, K. Expected return rate determined as oriented fuzzy number. In *35th International Conference Mathematical Methods in Economics Conference Proceedings*; Pražak, P., Ed.; Gaudeamus; University of Hradec Králové: Hradec Kralove, Czech Republic, 2017; pp. 561–565.
16. Piasecki, K.; Roszkowska, E. On application of ordered fuzzy numbers in ranking linguistically evaluated negotiation offers. *Adv. Fuzzy Syst.* **2018**. [CrossRef]
17. Piasecki, K.; Roszkowska, E.; Łyczkowska-Hanćkowiak, A. Simple Additive Weighting Method Equipped with Fuzzy Ranking of Evaluated Alternatives. *Symmetry* **2019**, *11*, 482. [CrossRef]
18. Roszkowska, E.; Kacprzak, D. The fuzzy SAW and fuzzy TOPSIS procedures based on ordered fuzzy numbers. *Inf. Sci.* **2016**, *369*, 564–584. [CrossRef]
19. Rudnik, K.; Kacprzak, D. Fuzzy TOPSIS method with ordered fuzzy numbers for flow control in a manufacturing system. *Appl. Soft Comput.* **2016**, *21*. [CrossRef]
20. Fortemps, P.; Roubens, M. Ranking and Defuzzification Methods Based on Area Compensation. *Fuzzy Sets Syst.* **1996**, *82*, 319–330. [CrossRef]
21. Zadeh, L.A. Similarity relations and fuzzy orderings. *Inf. Sci.* **1971**, *3*, 177–200.
22. Jain, R. Decision-making in the presence of fuzzy variables. *IEEE Trans. Syst. Man Cybern.* **1976**, *6*, 698–703.
23. Bortolani, G.; Degani, R. A review of some methods for ranking fuzzy subsets. *Fuzzy Sets Syst.* **1985**, *15*, 1–19. [CrossRef]
24. Lee, E.S.; Li, R.J. Comparison of Fuzzy Numbers Based on the Probability Measure of Fuzzy Events. *Comput. Math. Appl.* **1988**, *15*, 887–896. [CrossRef]
25. Campos, L.A.; Munoz, A. A subjective approach for ranking fuzzy numbers. *Fuzzy Sets Syst.* **1989**, *29*, 145–153.
26. Kim, K.; Park, K.S. Ranking Fuzzy Numbers with Index of Optimism. *Fuzzy Sets Syst.* **1990**, *35*, 143–150. [CrossRef]
27. Liou, T.S.; Wang, M.J.J. Ranking Fuzzy Numbers with Integral Value. *Fuzzy Sets Syst.* **1992**, *50*, 247–255. [CrossRef]
28. Facchmetti, G.; Ricci, R.G.; Muzzloh, S. Note on ranking fuzzy triangular numbers. *Int. J. Intell. Syst.* **1998**, *13*, 613–622. [CrossRef]
29. Cheng, C.H. A new approach for ranking fuzzy numbers by distance method. *Fuzzy Sets Syst.* **1998**, *95*, 307–317. [CrossRef]
30. Sarna, M. Fuzzy Relation on Fuzzy and Non-Fuzzy Numbers—Fast computational formulas: II. *Fuzzy Sets Syst.* **1998**, *93*, 63–74. [CrossRef]
31. Yao, J.S.; Wu, K. Ranking fuzzy numbers based on decomposition principle and signed distance. *Fuzzy Sets Syst.* **2000**, *116*, 275–288. [CrossRef]
32. Lim, X. Measuring the satisfaction of constraints in fuzzy linear programming. *Fuzzy Sets Syst.* **2001**, *122*, 263–275.
33. Modarres, M.; Sadi-Nezhad, S. Ranking Fuzzy Numbers by Preference Ratio. *Fuzzy Sets Syst.* **2001**, *118*, 429–436. [CrossRef]
34. Chu, T.C.; Tsao, C.T. Ranking fuzzy numbers with an area between the centroid point and original poin. *Comput. Math. Appl.* **2002**, *43*, 111–117. [CrossRef]

35. Abbasbandy, S.; Asady, B. Ranking of fuzzy numbers by sign distance. *Inf. Sci.* **2006**, *176*, 2405–2412. [CrossRef]
36. Abbasbandy, S.; Hajjari, T. A new approach for ranking of trapezoidal fuzzy numbers. *Comput. Math. Appl.* **2009**, *57*, 413–419. [CrossRef]
37. Wang, Y.J.; Lee, H.S. The Revised Method of Ranking Fuzzy Numbers with an Area Between the Centroid and Original Points. *Comput. Math. Appl.* **2008**, *55*, 2033–2042. [CrossRef]
38. Saeidifar, A. Application of weighting functions to the ranking of fuzzy numbers. *Comput. Math. Appl.* **2011**, *62*, 2246–2258. [CrossRef]
39. Kumar, A.; Singh, P.; Kaur, P.; Kaur, A. A new approach for ranking of L–R type generalized fuzzy numbers. *Expert Syst. Appl.* **2011**, *38*, 10906–10910. [CrossRef]
40. Dat, L.Q.; Yu, V.F.; Chou, S.-Y. An improved ranking method for fuzzy numbers based on the centroid-index. *Fuzzy Sets Syst.* **2012**, *14*, 413–419.
41. Asady, B.; Zendehnam, A. Ranking Fuzzy Numbers by Distance Minimization. *Appl. Math. Model.* **2007**, *31*, 2589–2598. [CrossRef]
42. Tran, L.; Duckstein, L. Comparison of fuzzy numbers using a fuzzy distance measure. *Fuzzy Sets Syst.* **2002**, *130*, 331–341. [CrossRef]
43. Sevastianov, P. Numerical methods for interval and fuzzy number comparison based on the probabilistic approach and Dempster–Shafer theory. *Inf. Sci.* **2007**, *177*, 4645–4661. [CrossRef]
44. Wang, X.; Kerre, E.E. Reasonable properties for the ordering of fuzzy quantities (I). *Fuzzy Sets Syst.* **2001**, *118*, 375–385. [CrossRef]
45. Wang, X.; Kerre, E.E. Reasonable properties for the ordering of fuzzy quantities (II). *Fuzzy Sets Syst.* **2001**, *118*, 387–405. [CrossRef]
46. Deng, Y.; Zhenfu, Z.; Qi, L. Ranking Fuzzy Numbers with an area Method using Radius of Gyration. *Comput. Math. Appl.* **2006**, *51*, 1127–1136. [CrossRef]
47. Nojavan, M.; Ghazanfari, M. A Fuzzy Ranking Method by Desirability Index. *J. Intell. Fuzzy Syst.* **2006**, *17*, 27–34.
48. Chen, C.C.; Tang, H.C. Ranking Non-normal p-Norm Trapezoidal Fuzzy Numbers with Integral Value. *Comput. Math. Appl.* **2008**, *56*, 2340–2346. [CrossRef]
49. Wang, Z.X.; Liu, Y.-J.; Fan, Z.-P.; Feng, B. Ranking L-R Fuzzy Number Based on Deviation Degree. *Inf. Sci.* **2009**, *179*, 2070–2077. [CrossRef]
50. Asady, B. Revision of distance minimization method for ranking of fuzzy numbers. *Appl. Math. Model.* **2011**, *35*, 1306–1313. [CrossRef]
51. Detyniecki, M.R.R.; Yager, R.R. Ranking fuzzy numbers using -weighted valuations. *Int. J. Uncertain. Fuzziness Knowl. Based Syst.* **2000**, *8*, 573–591. [CrossRef]
52. Matarazzo, B.; Munda, G. New approaches for the comparison of L–R fuzzy numbers: A theoretical and operational analysis. *Fuzzy Sets Syst.* **2001**, *118*, 407–418. [CrossRef]
53. Garcia, M.S.; Lamata, M.T. A modification of the index of liou and wang for ranking fuzzy number. *Int. J. Uncertain. Fuzziness Knowl. Based Syst.* **2007**, *15*, 411–424. [CrossRef]
54. Liu, X.-W.; Han, S.-L. Ranking fuzzy numbers with preference weighting function expectations. *Comput. Math. Appl.* **2005**, *49*, 1731–1753. [CrossRef]
55. Huynh, V.N.; Nakamori, Y.; Lawry, J. A probability-based approach to comparison of fuzzy numbers and applications to target-oriented decision making. *IEEE Trans. Fuzzy Syst.* **2008**, *16*, 371–387. [CrossRef]
56. Hajjari, S.; Abbasbandy, S. A note on "The revised method of ranking LR fuzzy number based on deviation degree". *Expert Syst. Appl.* **2011**, *38*, 13491–13492. [CrossRef]
57. Freeling, S. Fuzzy sets and decision analysis. *IEEE Trans. Syst. Man Cybern.* **1980**, *10*, 341–354. [CrossRef]
58. Kosiński, W.; Wilczyńska-Sztyma, D. Defuzzyfication and Implication within Ordered Fuzzy Numbers. In Proceedings of the IEEE World Congress on Computational Intelligence, Barcelona, Spain, 18–23 July 2010; pp. 1073–1079.
59. Ramik, J.; Rimanek, J. Inequality relation between fuzzy numbers and its use in fuzzy optimization. *Fuzzy Sets Syst.* **1985**, *16*, 123–138. [CrossRef]
60. Nejad, A.M.; Mashinchi, M. Ranking fuzzy numbers based on the areas on the left and the right sides of fuzzy number. *Comput. Math. Appl.* **2011**, *61*, 431–442. [CrossRef]

61. Chen, L.H.; Lu, H.W. An Approximate Approach for Ranking Fuzzy Numbers Based on Left and Right Dominance. *Comput. Math. Appl.* **2001**, *41*, 1589–1602. [CrossRef]
62. Orlovsky, S.A. Decision making with a fuzzy preference relation. *Fuzzy Sets Syst.* **1978**, *1*, 155–167. [CrossRef]
63. Zadeh, L.A. The concept of a linguistic variable and its application to approximate reasoning. Part, I. Information linguistic variable. *Expert Syst. Appl.* **1975**, *36*, 3483–3488.
64. Zadeh, L.A. The concept of a linguistic variable and its application to approximate reasoning. Part II. *Inf. Sci.* **1975**, *8*, 301–357. [CrossRef]
65. Zadeh, L.A. The concept of a linguistic variable and its application to approximate reasoning. Part III. *Inf. Sci.* **1975**, *9*, 43–80. [CrossRef]
66. Piasecki, K. The Relations "Less or Equal" and "Less Than" for Ordered Fuzzy Number. In *Analysis of International Relations 2018, Methods and Models of Regional Development, Summer Edition. Proceedings of the 10th International Scientific Conference, Katowice, Poland, 19–20 June 2018*; Publishing House of the University of Economics in Katowice: Katowice, Poland, 2018; pp. 32–39.
67. Zadeh, L.A. Fuzzy sets. *Inf. Control* **1965**, *8*, 338–353. [CrossRef]
68. Łukasiewicz, J. Interpretacja liczbowa teorii zdań, Ruch Filozoficzny **1922/23**, 7, pp. 92–93. Translated as 'A numerical interpretation of the theory of propositions' In *Jan Łukasiewicz-Selected Works*, Borkowski, L. Ed., North-Holland, Amsterdam, Polish Scientific Publishers: Warszawa, Poland, 1970.
69. Dubois, D.; Prade, H. Operations on fuzzy numbers. *Int. J. Syst. Sci.* **1978**, *9*, 613–629. [CrossRef]
70. Dubois, D.; Prade, H. Fuzzy real algebra: Some results. *Fuzzy Sets Syst.* **1979**, *2*, 327–348. [CrossRef]
71. Delgado, M.; Vila, M.A.; Voxman, W. On a canonical representation of fuzzy numbers. *Fuzzy Sets Syst.* **1998**, *93*, 125–135. [CrossRef]
72. Mondal, S.P.; Khan, N.A.; Vishwakarma, D.; Saha, A.K. Existence and Stability of Difference Equation in Imprecise Environment. *Nonlinear Eng.* **2018**, *7*, 263–271. [CrossRef]
73. Mondal, S.P. Interval Valued Intuitionistic Fuzzy Number and its Application in Differential equation. *J. Intell. Fuzzy Syst.* **2018**, *34*, 677–687. [CrossRef]
74. Wang, G.; Wen, C.L. A New Fuzzy Arithmetic for Discrete Fuzzy Numbers. In *Proceedings of the Fourth International Conference on Fuzzy Systems and Knowledge Discovery (FSKD 2007), Haikou, China, 24–27 August 2007*; IEEE: Piscataway, NJ, USA, 2007. [CrossRef]
75. Shyi-Ming, C. Fuzzy system reliability analysis using fuzzy number arithmetic operations. *Fuzzy Sets Syst.* **1994**, *64*, 31–38. [CrossRef]
76. Prokopowicz, P.; Pedrycz, W. The Directed Compatibility Between Ordered Fuzzy Numbers—A Base Tool for a Direction Sensitive Fuzzy Information Processing. *Artif. Intell. Soft Comput.* **2015**, *119*, 249–259.
77. Prokopowicz, P. The Directed Inference for the Kosinski's Fuzzy Number Model. In *Proceedings of the Second International Afro-European Conference for Industrial Advancement, Villejuif, France, 9–11 September 2015*; Abraham, A., Wegrzyn-Wolska, K., Hassanien, A.E., Snasel, V., Alimi, A.M., Eds.; Advances in Inteligent Systems and Computing, Vol. 427; Springer: Cham, Switzerland, 2015; pp. 493–505. [CrossRef]
78. Herrera, F.; Herrera-Viedma, E. Choice functions and mechanisms for linguistic preference relations. *Eur. J. Oper. Res.* **2000**, *120*, 144–161. [CrossRef]
79. Raiffa, H.; Richardson, J.; Metcalfe, D. *Negotiation Analysis*; Harvard University Press: Cambridge, UK, 2002.
80. Zhuosheng, J.; Zhang, H. Interval-Valued Intuitionistic Fuzzy Multiple Attribute Group Decision Making with Uncertain Weights. *Math. Probl. Eng.* **2019**. [CrossRef]

applied system innovation

MDPI

Article

Mathematical Apparatus of Optimal Decision-Making Based on Vector Optimization

Yury K. Mashunin

Far Eastern Federal University, 690068 Vladivostok, Russia; mashunin@mail.ru; Tel.: +79-143277508

Received: 17 April 2019; Accepted: 29 June 2019; Published: 11 October 2019

Abstract: We present a problem of "acceptance of an optimal solution" as a mathematical model in the form of a vector problem of mathematical programming. For the solution of such a class of problems, we show the theory of vector optimization as a mathematical apparatus of acceptance of optimal solutions. Methods of solution of vector problems are directed to problem solving with equivalent criteria and with the given priority of a criterion. Following our research, the analysis and problem definition of decision making under the conditions of certainty and uncertainty are presented. We show the transformation of a mathematical model under the conditions of uncertainty into a model under the conditions of certainty. We present problems of acceptance of an optimal solution under the conditions of uncertainty with data that are represented by up to four parameters, and also show geometrical interpretation of results of the decision. Each numerical example includes input data (requirement specification) for modeling, transformation of a mathematical model under the conditions of uncertainty into a model under the conditions of certainty, making optimal decisions with equivalent criteria (solving a numerical model), and, making an optimal decision with a given priority criterion.

Keywords: modeling; vector optimization; methods of solution of vector problems; optimal decision-making; numerical realization of decision-making

1. Introduction

The problem of making an optimal decision that meets the modern achievements of science and technology is connected, firstly, with the release of high-quality products, and, secondly, with the solution of problems of social and economic human development. The decision making can be undertaken under the conditions of certainty (when the functional dependence of the purpose of the parameters of the studied object and systems is known) [1–4], and under the conditions of uncertainty (when there is not sufficient information on the functional dependence of the purpose of the parameters of the studied object and systems) [5–8]. The conditions of uncertainty are characterized by the fact that input data for decision-making, can be presented as random, fuzzy or incomplete data, [1,2,9,10]. Research on this problem of decision-making began with the work of Keeney and Raiffa [11]. Analyses of modern decision-making approaches (i.e., "simple" methods) are submitted in [6,12]. One of the areas of decision-making automation is associated with the creation of mathematical models and the adoption of an optimal solution based on them [12–14]. Currently, the most common mathematical apparatus for model-based decision making is vector optimization [6,12,14–19]. The purpose of this work is to build a mathematical model for an object or system of decision making in the form of a vector problem of mathematical programming. Vector optimization is considered as a mathematical apparatus of a solution to the problem of acceptance of an optimal solution.

For the realization of the goal of this work, the study considered and solved the following problems.

The construction of a mathematical model of the problem of finding an optimal solution in the form of a vector problem of mathematical programming has been shown previously [4,6,15,20]. In the current paper, the theory and a mathematical apparatus of problem solving using vector optimization are presented. The theory includes an axiomatic principle of optimality of the solution of vector problems. The mathematical apparatus of the solution of vector problems is intended for the solution of vector problems with equivalent criteria [6,13,21] and with the given priority of a criterion [15,20]. The research, analysis and problem definition of decision making under the conditions of certainty and uncertainty are conducted. The realization of a mathematical apparatus of vector optimization is presented for numerical problems of decision making with one, two, three and four parameters. The solution of the problem of decision making includes creation of a numerical model of an object in the form of a vector problem, the solution of a problem of decision making with equivalent criteria, and, the solution of a vector problem of decision making with a criterion priority.

2. Statement of a Problem: Creation of the Mathematical Model "Acceptance of an Optimal Solution"

As an "object for optimal decision-making," we use a "technical system". The problem of the choice of optimum parameters of technical (engineering) systems according to functional characteristics arises during the study, analysis, and design of technical systems, and is connected with quality production.

The problem includes the solution of the following tasks:

- Creation of a mathematical model, which defines the inter-relation of each functional characteristic from parameters of the technical system, i.e., it is formed from the vector problem of mathematical programming;
- Choice of methods of the decision: we suggest using the methods based on normalization of criteria and the principle of the guaranteed result with equivalent criteria and with the set criterion priority;
- Development of the software which realizes these methods;
- The statement of a problem is executed according to [4,6,20,22].

The technical system depends on N, a set of design data: $X = \{x_1, x_2, \ldots, x_N\}$, where N is the number of parameters, each of which lies in the set limits:

$$x_j^{\min} \leq x_j \leq x, j = \overline{1,N}, or\, X^{min} \leq X \leq X^{max}$$

where $x_j^{\min}, x_j^{\max}, \forall j \in N$ are the minimum and maximum limits of change of the vector of parameters of the technical system.

The result of the functioning of the technical system is defined by a set K of technical characteristics of $f_k(X), k = \overline{1,K}$ which functionally depend on design data $X = \{x_j, j = \overline{1,N}\}$, in total these represent a vector function: $F(X) = (f_1(X)\, f_2(X) \ldots f_K\,(X))^{\mathrm{T}}$.

The set of characteristics (criteria) is subdivided into two subsets $\boldsymbol{K_1}$ and $\boldsymbol{K_2}$: $\mathrm{K} = \boldsymbol{K_1} \cup \boldsymbol{K_2}$.

$\boldsymbol{K_1}$ is a subset of technical characteristics, the numerical values of which are desired to be as high as possible: $f_k(X) \rightarrow \mathbf{max}, k = \overline{1,K_1}$.

$\boldsymbol{K_2}$ is a subset of technical characteristics, the numerical values of which are desired to be as low as possible: $f_k(X) \rightarrow \mathbf{min}, k = \overline{K_1+1,K}, \boldsymbol{K_2} \equiv \overline{K_1+1,K}$.

The mathematical model should consider, firstly, the purposes of the technical system which are represented by the characteristics of $F(X)$, and, secondly, the $X^{min} \leq X \leq X^{max}$ restrictions. The mathematical model of the technical system which solves in general a problem of the choice of the optimum design decision (a choice of optimum parameters) is presented in the form of a vector problem of mathematical programming.

$$Opt\,F(X) = \{\mathbf{max}\,F_1(X) = \{\mathbf{max}\,f_k(X), k = \overline{1,K_1}\}, \quad (1)$$

$$\mathbf{min}F_2(X) = \{\mathbf{min}f_k(X), k = \overline{1,K_2}\}\}, \quad (2)$$

$$G(X) \leq 0, \quad (3)$$

$$x_j^{\min} \leq x_j \leq x_j^{\max}, j = \overline{1,N}, \quad (4)$$

where X is the vector of controlled variables (constructive parameters) in Equation (1), $F(X) = \{f_k(X), k = \overline{1,K}\}$ represents the vector criterion for each component of the characteristics of the technical system in Equation (2), which functionally depends on the vector of variables X, $G(X) = (g_1(X)\ g_2(X) \ldots g_M(X))^T$ represents a vector of the restrictions imposed on the functioning of the technical system, and $\boldsymbol{M}$ is a set of restrictions.

Restrictions are defined in terms of technological, physical or similar processes, and can be presented by functional restrictions, for example, $f_k^{\min} \leq f_k(X) \leq f_k^{\max}$, $k = \overline{1,K}$.

It is supposed that the $f_k(X)$, $k = \overline{1,K}$ functions are differentiated and convex, $g_i(X)$, $i = \overline{1,M}$ are continuous, and Equations (3)–(4) represent a non-empty set of admissible points of S restrictions, which can be represented as $S = \{X \in \boldsymbol{R}^N | G(X) \leq 0, X^{min} \leq X \leq X^{max}\} \neq \varnothing$.

Criteria and restrictions in Equations (1)–(4) form the mathematical model of a technical system. It is required to find a vector of the $X^0 \in S$ parameters at which every component of the vector-functions $F_1(X) = \{f_k(X), k = \overline{1,K_1}\}$ accepts the greatest possible value, and vector-functions $F_2(X) = \{f_k(X), k = \overline{1,K_2}\}$ are accepted by the minimum value.

For a substantial class of technical systems which can be represented by the vector problem of Equations (1)–(4), it is possible to refer to their large number of applications in various fields, such as electro-engineering [23], airspace [10,13], metallurgical (choice of optimal structure of material), and chemical [24].

In this article, the technical system is considered to be static. However, technical systems can be considered to be dynamic [23], using differential-difference methods of transformation, conducted for a small discrete period $\Delta t \in T$.

3. Theory. Axioms, the Principle of Optimality and Methods for Solving Vector Problems of Mathematical Programming

The theory of vector optimization includes theoretical foundations (axioms) and methods of the solution of vector problems with equivalent criteria and with the given criterion priority. The theory is a basis of mathematical apparatus of modeling of an "object for optimal decision-making", which allows selection of any point from a set of points that is Pareto optimal, and shows why the selection is optimal.

We have presented, first, axioms and methods of the solution of problems of vector optimization with equivalent criteria (Section 3.1) and, second, the specified priority criteria (Section 3.2).

3.1. Vector Optimization with Equivalent Criteria

3.1.1. Axioms and the Principle of Optimality of Vector Optimization with Equivalent Criteria

Definition 1. (Definition of the relative assessment of criteria).

In the vector problem of Equations (1)–(4), definitions are as follows: $\lambda_k(X) = \frac{f_k(X)-f_k^0}{f_k^*-f_k^0}$, $\forall k \in \boldsymbol{K}$ is the relative estimate of a point $X \in S$ kth criterion, $f_k(X)$ is the kth criterion at the point $X \in S$, f_k^* is the value of the kth criterion at the point of optimum X_k^*, obtained in the vector problem of Equations (1)–(4) of the individual kth criterion, f_k^0 is the worst value of the kth criterion (anti-optimum) at the point X_k^0 (superscript 0) on the admissible set S in Equations (1)–(4), at the task at *max* (3), (5), (6), the value of f_k^0 is the lowest value of the kth criterion, $f_k^0 = \min_{X \in S} f_k(X)$ $\forall k \in \boldsymbol{K}_1$ and the task min f_k^0 is the greatest,

$f_k^0 = \max_{X \in S} f_k(X)\ \forall k \in \mathbf{K}_2$. The relative estimate of $\lambda_k(X)$, $\forall k \in K$ is firstly measured in relative units and, secondly, the relative assessment of $\lambda_k(X)\ \forall k \in K$ on the admissible set is changed from zero at a point of X_k^0: $\forall k \in \mathbf{K}\ \lim_{X \to X_k^o} \lambda_k(X) = 0$, to the unit at the point of an optimum of X_k^*:$\forall k \in \mathbf{K}$, $\lim_{X \to X_k^*} \lambda_k(X) = 1$, i.e.,: $\forall k \in \mathbf{K}, 0 \le \lambda_k(X) \le 1, X \in S$. This allows the comparison of the criteria, measured in relative units, by joint optimization.

Axiom 1. (About equality and equivalence of criteria at an admissible point of vector problems of mathematical programming)

In vector problems of mathematical programming two criteria with the indexes $k \in \mathbf{K}, q \in \mathbf{K}$ shall be considered as equal at point $X \in \mathbf{S}$ if relative estimates of the kth and qth criterion are equal at this point, i.e., $\lambda_k(X) = \lambda_q(X), k, q \in \mathbf{K}$.

We will consider criteria equivalent in vector problems of mathematical programming at a point $X \in S$ if, when comparing the numerical size of relative estimates of $\lambda_k(X)$, $k = \overline{1, K}$, for each criterion of $f_k(X)$, $k = \overline{1, K}$, and, respectively, relative estimates of $\lambda_k(X)$, conditions are not imposed about the priorities of criteria.

Definition 2. (Definition of a minimum level among all relative estimates of criteria).

The relative level λ in a vector problem represents the lower assessment of a point of $X \in S$ among all relative estimates of $\lambda_k(X)$, $k = \overline{1, K}$:

$$\forall X \in S,\ \lambda \le \lambda_k(X), k = \overline{1, K}, \tag{5}$$

The lower level for the performance of the condition of Equation (5) at an admissible point of $X \in \mathbf{S}$ is defined as:

$$\forall X \in S,\ \lambda = \min_{k \in K} \lambda_k(X). \tag{6}$$

Equations (5) and (6) are interconnected. They serve as a transition from Equation (6) of the definition of a minimum to the restrictions of Equation (5), and vice versa.

The level λ allows the union of all criteria in a vector problem with one numerical characteristic of λ, made over certain operations, thereby carrying out these operations over all criteria measured in relative units. The level λ functionally depends on the $X \in \mathbf{S}$ variable, by changing X, we can change the lower level, λ. From here we will formulate the rules of searching for the optimum decision.

Definition 3. (The principle of optimality with equivalent criteria).

The vector problem of mathematical programming with equivalent criteria is solved if the point of $X^0 \in S$ and a maximum level of λ^0 (the top index optimum) among all relative estimates is found such that:

$$\lambda^0 = \max_{X \in S} \min_{k \in K} \lambda_k(X). \tag{7}$$

Using the interrelation of Equations (5) and (6), we will transform the maximine problem of Equation (7) into an extreme problem:

$$\lambda^0 = \max_{X \in S} \lambda, \tag{8}$$

$$\lambda \le \lambda_k(X), k = \overline{1, K}. \tag{9}$$

We can call the resulting problem of Equations (8) and (9) the λ-problem.

The λ-problem of Equations (8) and (9) has $(N + 1)$ dimensions. As a consequence, the solution of the λ-problem represents an optimum vector of $\mathbf{X}^\circ \in \mathbf{R}^{N+1}$, where $(N + 1)$ is a component which has the essence of the value of λ^0, i.e., $\mathbf{X}^0 = \{x_1^o, x_2^o, \ldots, x_N^o, x_{N+1}^o\}$, thus $x_{N+1}^o = \lambda^0$, and $(N + 1)$ is a component of a vector of $\mathbf{X}^0$ selected in view of its specificity.

The obtained pair of $\{\lambda^0, X^0\} = \mathbf{X}^0$ characterizes the optimum solution of the λ-problem and according to the vector problem of mathematical programming in Equations (1)–(4) with equivalent criteria, can be solved on the basis of normalization of criteria and the principle of the guaranteed result. In the optimum solution of $\mathbf{X}^0 = \{X^0, \lambda^0\}$, X^0 is an optimal point and λ^0 is a maximum level.

An important result of the algorithm for solving the vector problems of Equations (1)–(4) with equivalent criteria is the following theorem.

Theorem 1. (The theorem of the two most contradictory criteria in the vector problem of mathematical programming with equivalent criteria).

In convex vector problems of mathematical programming with equivalent criteria that are solved on the basis of normalization of criteria and the principle of the guaranteed result, for an optimum point of $\mathbf{X}^0 = \{\lambda^0, X^0\}$, two criteria are denoted by their indexes $q \in \mathbf{K}, p \in \mathbf{K}$ (which in a sense are the most contradictory of the criteria $k = \overline{1, K}$), for which an equality is carried out:

$$\lambda^0 = \lambda_q(X^0) = \lambda_p(X^0),\ q, p \in \mathbf{K},\ X \in S, \tag{10}$$

and other criteria are defined by inequalities:

$$\lambda^0 \leq \lambda_k(X^0)\ \forall k \in \mathbf{K},\ q \neq p \neq k. \tag{11}$$

3.1.2. Mathematical Algorithm of the Solution of a Vector Problem with Equivalent Criteria

To solve the vector problems of mathematical programming of Equations (1)–(4), the methods based on axioms of the normalization of criteria and the principle of the guaranteed result [12,21] are offered. Methods follow from Axiom 1 and the principle of optimality (Definition 3). We will present this as a number of steps as follows:

The method of the solution of the vector problem of Equations (1)–(4) with equivalent criteria is presented in the form of the sequence of steps [25].

Step 1. The problem of Equations (1)–(4) is solved separately for each criterion, i.e., for $\forall k \in K_1$ is solved at the maximum, and for $\forall k \in K_2$ is solved at a minimum. As a result of the decision, we obtain X_k^*, an optimum point by the corresponding criterion, $k = \overline{1, K}; f_k^* = f_k(X_k^*)$, the kth criterion size at this point, $k = \overline{1, K}$.

Step 2. We define the worst value of each criterion on $\mathbf{S}$: f_k^0, $k = \overline{1, K}$. For the problem of Equations (1)–(4) for each criterion of $k = \overline{1, K}$, a minimum is solved as: $f_k^0 = \min f_k(X)$, $G(X) \leq B$, $X \geq 0$, $k = \overline{1, K}$.

In addition, for Equations (1)–(4) for each criterion, a maximum is solved as: $f_k^0 = \max f_k(X)$, $G(X) \leq B$, $X \geq 0$, $k = \overline{1, K}$.

As a result of the decision, we obtain $X_k^0 = \{x_j, j = \overline{1, N}\}$, an optimum point by the corresponding criterion, $k = \overline{1, K}; f_k^0 = f_k(X_k^0)$, the kth criterion size at the point, X_k^0, $k = \overline{1, K}$.

Step 3. For the system analysis of a set of Pareto optimal points, for this purpose optimum points of $X^* = \{X_k^*, k = \overline{1, K}\}$, are defined as sizes of criterion functions of $F(X^*)$ and relative estimates $\lambda(X^*)$, $\lambda_k(X) = \frac{f_k(X) - f_k^0}{f_k^* - f_k^0}$, $\forall k \in \mathbf{K}$:

$$\begin{aligned} F(X^*) &= \{f_q(X_k^*), q = \overline{1, K}, k = \overline{1, K}\} = \begin{vmatrix} f_1(X_1^*), \ldots, f_k(X_1^*), \\ \ldots \\ f_1(X_k^*), \ldots, f_k(X_k^*) \end{vmatrix}, \\ \lambda(X^*) &= \{\lambda_q(X_k^*), q = \overline{1, K}, k = \overline{1, K}\} = \begin{vmatrix} \lambda_1(X_1^*), \ldots, \lambda_k(X_1^*), \\ \ldots \\ \lambda_1(X_k^*), \ldots, \lambda_k(X_k^*) \end{vmatrix}. \end{aligned} \tag{12}$$

As a whole, for the problem $\forall k \in K$ the relative assessment of $\lambda_k(X)$, $k = \overline{1,K}$ lies within $0 \le \lambda_k(X) \le 1$, $\forall k \in K$.

Step 4. Creation of the λ-problem.

Creation of the λ-problem is carried out in two stages: initially build the maximine problem of optimization with the normalized criteria, which at the second stage will be transformed into the standard problem of mathematical programming called the λ-problem.

For the construction of the maximine problem of optimization we use Definition 2, relative level $\forall X \in S$, $\lambda = \min_{k \in K} \lambda_k(X)$.

The bottom λ level is maximized on $X \in S$, as a result, we obtain a maximine problem of optimization with the normalized criteria:

$$\lambda^0 = \max_x \min_k \lambda_k(X), G(X) \le B, X \ge 0. \quad (13)$$

At the second stage we transform the problem of Equation (13) into a standard problem of mathematical programming:

$$\lambda^0 = \max \lambda, \rightarrow \lambda^0 = \max \lambda, \quad (14)$$

$$\lambda - \lambda_k(X) \le 0, k = \overline{1,K}, \rightarrow \lambda - \frac{f_k(X) - f_k^0}{f_k^* - f_k^0}, k = \overline{1,K}, \quad (15)$$

$$G(X) \le B, X \ge 0, \rightarrow G(X) \le B, X \ge 0, \quad (16)$$

where the vector of unknowns of X has the dimension $N + 1$: $X = \{\lambda, x_1, \ldots, x_N\}$.

Step 5. Solution of the λ-problem.

The λ-problem of Equations (14)–(16) is a standard problem of convex programming for which decision standard methods are used.

As a result of the solution of the λ-problem, the following are obtained:

$\mathbf{X^0} = \{\lambda^0, X^0\}$, an optimum point,

$f_k(X^0)$, $k = \overline{1,K}$, values of the criteria at this point,

$\lambda_k(X^0) = \frac{f_k(X^0) - f_k^0}{f_k^* - f_k^0}$, $k = \overline{1,K}$, sizes of the relative estimates,

λ^0, the maximum relative estimates which represent the maximum bottom level for all relative estimates of $\lambda_k(X^0)$, or the guaranteed result in relative units. λ^0 guarantees that all relative estimates of $\lambda_k(X^\circ)$ are equal λ^0: $\lambda_k(X^0) \ge \lambda^0$, $k = \overline{1,K}$ or $\lambda^0 \le \lambda_k(X^0)$, $k = \overline{1,K}$, $X^0 \in S$, and according to Theorem 1 [12,21], the point of $\mathbf{X}^0 = \{\lambda^0, x_1, \ldots, x_N\}$ is Pareto optimal.

3.1.3. Implementation of the Decision using the Example of a Vector Problem of Linear Programming with Equivalent Criteria

The use of the vector problem of Equations (1)–(4) for decision making is carried out in four stages: statement of the problem, construction of the mathematical model, software development for solving the vector problem, and, solution of the vector problem.

These stages are carried out on the example of a model of an economic system presented by a vector problem of linear programming with equivalent criteria.

Stage 1. Statement of the problem.

As an economic system, a model of the production schedule of an enterprise is considered.

It is given that the company, which produces heterogeneous products of four types, $N = 4$, uses resources of three types, $M = 3$, in production: labor (various specialties), material (different types of materials), power (equipment: welding, turning, etc.).

The technological matrix of production is presented in Table 1. It also indicates the potential of the enterprise for each type of the resource b_i, $i = \overline{1,3}$, as well as income c_j^1 and profit c_j^2 from the sale of a unit of each type of product.

Table 1. The consumption of resources and operational performance.

Type Resources	Costs of Resources of One Product				Possibilities of Firm on Resources
	Type 1	Type 2	Type 3	Type 4	
Labor (people/week)	1	1	1	1	15
Material (in kg)	7	5	3	2	120
Capacity (per hour)	3	5	10	15	100
Income from a unit of production c_j^1	4.0	5.0	9.0	11.0	maximize
Profit $c_j^2, j = 1,...,4$	2	10	6	20	maximize
Output	x_1	x_2	x_3	x_4	To define

It is required to make the production schedule of the enterprise, which includes indicators according to the nomenclature (by types of products) and on a volume basis, i.e., how many products of the corresponding type should be made by the enterprise so that income and profit can be realized as shown above. Construction and solution of the mathematical model follow.

Stage 2. Construction of a mathematical model.

As variables, we take the volume of products that the company produces: $X = \{x_1, \ldots, x_N\}$, $N = 4$. We express target orientation of the production schedule by means of a vector problem of linear programming (VPLP) which will take the form:

opt $F(X) = \{\max f_1(X) = (4.0x_1 + 5.0x_2 + 9.0x_3 + 11.0x_4)$,

$\max f_2(X) = (2x_1 + 10x_2 + 6x_3 + 20x_4)\}$,

with restrictions $x_1 + x_2 + x_3 + x_4 \leq 15$,

$7x_1 + 5x_2 + 3x_3 + 2x_4 \leq 120$,

$3x_1 + 5x_2 + 10x_3 + 15x_4 \leq 100$, $x_1 \geq 0$, $x_2 \geq 0$, $x_3 \geq 0$, $x_4 \geq 0$.

In this VPLP the following is formulated: it is required to find the non-negative solution of $x_1, \ldots, x_4$ in the system of inequalities at which the $f_1(X)$ and $f_2(X)$ functions obtain maximum values.

Stage 3. The software engineering of the solution of the VPLP.

The "Solution of a Vector Problem of Linear Programming" program is presented in the annex to this section.

Stage 4. Solution of a vector problem of linear programming.

We show the solution of a problem of linear programming in the MATLAB system according to an algorithm of the solution of the VPLP on the basis of normalization of criteria and the principle of the guaranteed result. At first input data are prepared (the italicized font indicates the text of the program in the MATLAB system). The vector target function in the form of a matrix is formed:

disp ('Solution of a vector problem of the linear programming')

cvec = [−4.0 −5.0 −9.0 −11.0, % Sales volume

−2. −10. −6. −20.] % profit Volume

a = [1.0 1.0 1.0 1.0,

7.0 5.0 3.0 2.0,

3.0 5.0 10.0 15.0], % matrix of linear restrictions

b= [15. 120. 100] % the vector containing restrictions (b_i)

Aeq = [], beq = [] % restriction like equality

X0 = [0. 0. 0. 0.], % a vector of variables

The algorithm of the solution of VPLP is represented as a sequence of steps.

Step 1. A decision on each criterion.

The decision on the first criterion of the VPLP: [*x1,f1*] = *linprog(cvec(1,:),a,b,Aeq,beq,lb,ub)*

Decision on the first criterion: $X_1^* = x1 = \{x_1 = 7.14, x_2 = x_4 = 0, x_3 = 7.85\}$, $f_1^* = f1 = -99.286$.

Decision on the second criterion: $X_2^* = \{x_1 = 0, x_2 = 12.5, x_3 = 0, x_4 = 2.5\}$, $f_2^* = f_2 = 175$.

Step 2. The worst point of an optimum is determined for each criterion (anti-optimum) by multiplication of criterion by a minus unit. For the decisions on the first and second criterion:

$X_1^0 = x1min = \{x_1 = 0, \ldots, x_4 = 0\}$, $f_1^0 = f1min = 0$.
$X_2^0 = x2min = \{x_1 = 0, \ldots, x_4 = 0\}$, $f_2^0 = f2min = 0$.

Step 3. The system analysis of the criteria in the VPLP is undertaken (i.e., the system of two criteria at optimum points is analyzed). For this purpose, optimum points of X_1^*, X_2^* are defined sizes of criterion functions and relative estimates of: $F(X^*) = \|f_q(X_k^*)\|_{k=\overline{1,K}}^{q=\overline{1,K}}$, $\lambda(X^*) = \|\lambda_q(X_k^*)\|_{k=\overline{1,K}}^{q=\overline{1,K}}$, $\lambda(X^*) = \frac{f_k(X*)-f_k^0}{f_k^*-f_k^0}$, $k = \overline{1,K}$, $F(X^*) = \begin{vmatrix} f_1(X_1^*)\ f_1(X_1^*) \\ f_1(X_2^*)\ f_2(X_2^*) \end{vmatrix} = \begin{vmatrix} 99.29\ 61.43 \\ 90.0\ 175.0 \end{vmatrix}$, $\lambda(X^*) = \begin{vmatrix} \lambda_1(X_1^*)\ \lambda_2(X_1^*) \\ \lambda_1(X_2^*)\ \lambda_2(X_2^*) \end{vmatrix} = \begin{vmatrix} 1.0\ 0.351 \\ 0.907\ 1.0 \end{vmatrix}$.

Step 4. The λ-problem is constructed as: $\lambda^0 = max\ \lambda$,

with restrictions $\lambda - (f_1(X) - f_1^0)/(f_1^* - f_1^0) \le 0$, $\lambda - (f_2(X) - f_2^0)/(f_2^* - f_2^0) \le 0$, $G(X) \le B$, $X \ge 0$.

Substituting numerical data, we obtain:

$\lambda^0 = \max \lambda$,

with restrictions: $\lambda - (4.0x_1 + 5.0x_2 + 9.0x_3 + 11.0x_4 - f_1^0)/(f_1^* - f_1^0) \le 0$,

$\lambda - (2x_1 + 10x_2 + 6x_3 + 20x_4 - f_2^0)/(f_2^* - f_2^0) \le 0$,

$x_1 + x_2 + x_3 + x_4 \le 15$,

$7x_1 + 5x_2 + 3x_3 + 2x_4 \le 120$,

$3x_1 + 5x_2 + 10x_3 + 15x_4 \le 100$,

$x_1 \ge 0, x_2 \ge 0, x_3 \ge 0, x_4 \ge 0$.

Step 5. Solution of the λ-problem.

Results of the solution of the λ-problem:

$X^0 = \{x_1 = 0.9217914, x_2 = 0.0, x_3 = 11.73964, x_4 = 1.520722, x_5 = 1.739639\}$ - optimum values of variables,

$\lambda^0 = 0.9218$, the optimum value of the criterion function

We execute a check, at an optimum point of X^0 we determine sizes of criterion functions of $F(X^0) = \{f_k(X^0), k = \overline{1,K}\}$, relative estimates of $\lambda(X^0) = \{\lambda_k(X^0), k = \overline{1,K}\}$.

As a result of the decision we obtain: $fX^0 = [f_1(X^0) = 91.52, f_2(X^0) = 161.3]$,

$\lambda_1(X^0) = 0.\ 9218$, $\lambda_2(X^0) = 0.9218$, i.e., $\lambda^0 \le \lambda_k(X^0)$, $k = 1,2$.

These results show that at point X^0 both criteria in the relative units reached $\lambda^0 = 0.92$ from the optimum sizes. Any increase in one of the criteria of this level leads to a decrease in the other criterion, i.e., the point X^0 is Pareto optimal.

Here we present the text of the program in the MATLAB system.

The application.

%Vector linear programming problem, 2 criteria

% Author: Машунин Юрий Константинович -Mashunin Yury. K.

%The program is designed for the training and research, for the commercial purposes please contact:

% Mashunin@mail.ru

```
disp(Vector linear programming problem - 2 criteria')
disp(' opt F(X) = {max f1(X) = (4.0x1 + 5.0x2 +9.0x3 + 11.0x4), ')
disp(' max f2(X) = (2x1 + 10x2 + 6x3 + 20x4),')
disp(' x1 + x2 + x3 + x4 <= 15,')
disp('7x1 +5x2 + 3x3 + 2x4 <= 120,')
disp('3x1 +5x2 +10x3 +15x4 <= 100, x1 >= 0,..., x42 >= 0 ')
cvec = [-4.0 -5.0 -9.0 -11.0.; -2. -10. -6. -20.];
disp(''Step 0. Input data of the vector problem')
cvec = [-4.0 -5.0 -9.0 -11.0; -2. -10. -6. -20.];
a = [1. 1. 1. 1.; 7. 5. 3. 2.; 3. 5. 10. 15.];
b = [15. 120. 100.]; Aeq = []; beq = []; x0 = [0. 0. 0. 0.];
disp('Step 1.The solution for each criterion is the best')
[x1,f1] = linprog(cvec(1,:),a,b,Aeq,beq,x0)
```

```
[x2,f2] = linprog(cvec(2,:),a,b,Aeq,beq,x0)
disp('Step 2. Solution by criterion-best-worst')
[x1min,f1min] = linprog(-1*cvec(1,:),a,b,Aeq,beq,x0)
[x2min,f2min] = linprog(-1*cvec(2,:),a,b,Aeq,beq,x0)
disp('Step 3. System analysis of the results of the decision')
d1 = -f1- f1min; d2 = -f2- f2min;
f = [cvec(1,:)*x1 cvec(2,:)*x1; cvec(1,:)*x2 cvec(2,:)*x2]
L = [(-f(1,1) - f1min)/d1 (-f(1,2) - f2min)/d2;
(-f(2,1) – f1min)/d1 (-f(2,2) - f2min)/d2]
disp('Step 4. Solution of L-problem');
cvec0 = [-1. 0. 0. 0. 0.];
a0 = [1. -4.0/d1 -5.0/d1 -9.0/d1 -11.0/d1;
1. -2./d2 -10./d2 -6./d2 -20./d2;
0. 1. 1. 1. 1.;
0. 7. 5. 3. 2.;
0. 3. 5. 10. 15.];
b0 = [- f1min/d1 – f2min)/d2 15. 120. 100.]; x00 = [0. 0. 0. 0. 0.];
[X0,L0] = linprog(cvec0,a0,b0,Aeq,beq,x00)
fX0 = [cvec(1,:)*X0(2:5) cvec(2,:)*X0(2:5)]
LX0 = [(-fX0(1)- f1min)/ d1 (-fX0(2) - f2min)/d2]
```

3.2. *Vector Optimization with a Criterion Priority*

3.2.1. Axioms and the Principle of Optimality of Vector Optimization with a Criterion Priority

For development of methods of the solution of problems of vector optimization with a priority of criterion we use definitions as follows:

- priority of one criterion of vector problems, with a criterion priority over other criteria,
- numerical expression of a priority,
- the set priority of a criterion,
- the lower (minimum) level from all criteria with a priority of one of them,
- a subset of points with priority by criterion (Axiom 2),
- the principle of optimality of the solution of problems of vector optimization with the set priority of one of the criteria, and related theorems. For more details see [7,25].

Definition 4. (About the priority of one criterion over the other).

The criterion of $q \in \mathbf{K}$ in the vector problem of Equations (12) and (13) in a point of $X \in S$ has priority over other criteria of $k = \overline{1,K}$, and the relative estimate of $\lambda_q(X)$ by this criterion is greater than or equal to relative estimates of $\lambda_k(X)$ of other criteria, i.e.:

$$\lambda_q(X) \geq \lambda_k(X),\ k = \overline{1,K},$$

and a strict priority for at least one criterion of $t \in \mathbf{K}$,

$$\lambda_q(X) > \lambda_t(X),\ t \neq q,\ \text{and for other criteria of } \lambda_q(X) \geq \lambda_k(X),\ k = \overline{1,K},\ k \neq t \neq q.$$

Introduction of the definition of a priority of criterion in the vector problem of Equations (1)–(4) executed the redefinition of the early concept of a priority. Earlier the intuitive concept of the importance of this criterion was outlined, now this "importance" is defined as a mathematical concept: the higher the relative estimate of the qth criterion compared to others, the more it is important (i.e., more priority), and the highest priority at a point of an optimum is X_k^*, $\forall q \in \mathbf{K}$.

From the definition of a priority of criterion of $q \in \mathbf{K}$ in the vector problem of Equations (1)–(4), it follows that it is possible to reveal a set of points $\mathbf{S}_q \subset \mathbf{S}$ that is characterized by $\lambda_q(X) \geq \lambda_k(X)$, $\forall k \neq q$, $\forall X \in S_q$. However, the answer to whether a criterion of $q \in \mathbf{K}$ at a point of the set S_q has more priority than others remains open. For clarification of this question, we define a communication coefficient between a couple of relative estimates of q and k that, in total, represent a vector: $P^q(X) = \{p_k^q(X)|k = \overline{1, K}\}$, $q \in \mathbf{K}$ $\forall X \in S_q$.

Definition 5. (About numerical expression of a priority of one criterion over another).

In the vector problem of Equations (12) and (13), with priority of the qth criterion over other criteria of $k = \overline{1, K}$, for $\forall X \in S_q$, and a vector of $P^q(X)$ which shows how many times a relative estimate of $\lambda_q(X)$, $q \in \mathbf{K}$, is more than other relative estimates of $\lambda_k(X)$, $k = \overline{1, K}$, we define a numerical expression of the priority of the qth criterion over other criteria of $k = \overline{1, K}$ as:

$$\begin{aligned} P^q(X) &= \{p_k^q(X) = \lambda_q(X)/\lambda_k(X), k = \overline{1, K}\}, p_k^q(X) \geq 1, \\ \forall X \in S_q \subset S, k &= \overline{1, K}, \forall q \in K. \end{aligned} \tag{17}$$

Definition 6. (About the set numerical expression of a priority of one criterion over another).

In the vector problem of Equations (1)–(4) with a priority of criterion of $q \in \mathbf{K}$ for $\forall X \in \mathbf{S}$, vector $P^q = \{p_k^q, k = \overline{1, K}\}$ is considered to be set by the person making decisions (i.e., decision-maker) if everyone is set a component of this vector. Set by the decision-maker, component p_k^q, from the point of view of the decision-maker, shows how many times a relative estimate of $\lambda_k(X)$, $k = \overline{1, K}$ is greater than other relative estimates of $\lambda_k(X)$, $k = \overline{1, K}$. The vector of p_k^q, $k = \overline{1, K}$ is the numerical expression of the priority of the qth criterion over other criteria of $k = \overline{1, K}$:

$$P^q(X) = \{p_k^q, k = \overline{1, K}\}, p_k^q \geq 1, \forall X \in S_q \subset S, k = \overline{1, K}, \forall q \in K. \tag{18}$$

The vector problem of Equations (1)–(4), in which the priority of any criteria is set, is called a vector problem with the set priority of criterion. The problem of a task of a vector of priorities arises when it is necessary to determine the point $X^0 \in S$ by the set vector of priorities. In the comparison of relative estimates with a priority of criterion of $q \in \mathbf{K}$, as well as in a task with equivalent criteria, we define the additional numerical characteristic of λ which we call the level.

Definition 7. (About the lower level among all relative estimates with a criterion priority).

The λ level is the lowest among all relative estimates with a priority of criterion of $q\in$ such that:

$$\lambda \leq p_k^q \lambda_k(X), k = \overline{1, K}, q \in K, \forall X \in S_q \subset S; \tag{19}$$

The lower level for the performance of the condition in Equation (19) is defined as:

$$\lambda = \min_{k \in K} p_k^q \lambda_k(X), q \in K, \forall X \in S_q \subset S. \tag{20}$$

Equations (19) and (20) are interconnected and serve as a further transition from the operation of the definition of the minimum to restrictions, and vice versa. In Section 3.1, we gave the definition of a Pareto optimal point $X^0 \in S$ with equivalent criteria. Considering this definition as an initial one, we will construct a number of the axioms dividing an admissible set of $\mathbf{S}$ into, first, a subset of Pareto optimal points S^0, and, secondly, a subset of points $S_q \subset S$, $q \in \mathbf{K}$, with priority for the qth criterion.

Axiom 2. (About a subset of points, priority by criterion).

In the vector problem of Equations (12)–(13), the subset of points $S_q \subset S$ is called the area of priority of criterion of $q \in K$ over other criteria, if $\forall X \in S_q\ \forall k{\in}K\ \lambda_q(X) \geq \lambda_k(X),\ q \neq k$.

This definition extends to a set of Pareto optimal points S^0 that is given by the following definition.

Axiom 2a. (About a subset of points, priority by criterion, on Pareto's great number in a vector problem). In a vector problem of mathematical programming the subset of points $S_q^o \subset S^0 \subset S$ is called the area of a priority of criterion of q∈K over other criteria, if $\forall X \in S_q^o\ \forall k \in K\ \lambda_q(X) \geq \lambda_k(X),\ q \neq k$.

In the following we provide explanations.

Axiom 2 and 2a allow the breaking of the vector problem in Equations (1)–(4) into an admissible set of points S, including a subset of Pareto optimal points, $S^0 \subset S$, and subsets:

One subset of points $S' \in S$ where criteria are equivalent, and a subset of points of S' crossed with a subset of points $S°$, allocated to a subset of Pareto optimal points at equivalent criteria $S^{00} = S' \cap S^0$. As will be shown further, this consists of one point of $X^0 \in S$, i.e., $X^0 = S^{00} = S' \cap S^0,\ S' \in S,\ S^0 \in S$.

"K" subsets of points where each criterion of $q = \overline{1,K}$ has a priority over other criteria of $k = \overline{1,K}$, $q \neq k$, and thus breaks, first, sets of all admissible points S, into subsets $S_q \subset S$, $q = \overline{1,K}$ and, second, a set of Pareto optimal points, S^0, into subsets $S_q^o \subset S^0 \subset S$, $q = \overline{1,K}$. This yields: $S'U(\underset{q\in K}{U} S_q^o) = S^0$, $S_q^o \subset S^0 \subset S$, $q = \overline{1,K}$.

We note that the subset of points S_q^o, on the one hand, is included in the area (a subset of points) of priority of criterion of $q{\in}K$ over other criteria: $S_q^o \subset S_q \subset S$, and, on the other, in a subset of Pareto optimal points: $S_q^o \subset S^0 \subset S$.

Axiom 2 and the numerical expression of priority of criterion (Definition 5) allow the identification of each admissible point of $X{\in}S$ (by means of vector $P^q(X) = \{p_k^q(X) = \lambda_q(X)/\lambda_k(X),\ k = \overline{1,K}\}$), to form and choose:

- a subset of points by priority criterion S_q, which is included in a set of points S, $\forall q \in K\ X \in S_q \subset S$, (such a subset of points can be used in problems of clustering, but is beyond this article);
- a subset of points by priority criterion S_q^o, which is included in a set of Pareto optimal points $S°$, $\forall q{\in}K,\ X{\in}S_q^o \subset S^0$.

Thus, full identification of all points in the vector problem of Equations (12) and (13) is executed in sequence as:

Set of admissible points of $X{\in}S$ →	Subset of points, optimum across Pareto, $X{\in}S^0 \subset S$ →	Subset of points, optimum across Pareto $X{\in}S_q^o \subset S^0 \subset S$ →	Separate point of a $\forall X{\in}S$ $X{\in}S_q^o \subset S^0 \subset S$

This is the most important result which allows the output of the principle of optimality and to construct methods of a choice of any point of Pareto's great number.

Definition 8. (Principle of optimality 2. The solution of a vector problem with the set criterion priority).

The vector problem of Equations (12) and (13) with the set priority of the qth criterion of p_k^q, $k = \overline{1,K}$ is considered solved if the point X^0 and maximum level λ^0 among all relative estimates is found such that:

$$\lambda^0 = \max_{X\in S}\min_{k\in K} p_k^q \lambda_k(X), q \in K. \tag{21}$$

Using the interrelation of Equations (19) and (20), we can transform the maximine problem of Equation (33) into an extreme problem of the form:

$$\lambda^0 = \max_{X\in S} \lambda, \tag{22}$$

$$\lambda \leq p_k^q \lambda_k(X), k = \overline{1,K}. \tag{23}$$

We call Equations (22) and (23) the λ-problem with a priority of the qth criterion.

The solution of the λ-problem is the point $\mathbf{X}^0 = \{X^0, \lambda^0\}$ This is also the result of the solution of the vector problem of Equations (1)–(4) with the set priority of the criterion, solved on the basis of normalization of criteria and the principle of the guaranteed result.

In the optimum solution $\mathbf{X}^0 = \{X^0, \lambda^0\}$, X^0, an optimum point, and λ^0, the maximum bottom level, the point of X^0 and the λ^0 level correspond to restrictions of Equation (15), which can be written as: $\lambda^0 \leq p_k^q \lambda_k(X^0)$, $k = \overline{1, K}$.

These restrictions are the basis of an assessment of the correctness of the results of a decision in practical vector problems of optimization.

From Definitions 1 and 2, "Principles of optimality", follows the opportunity to formulate the concept of the operation "opt".

Definition 9. (Mathematical operation "opt").

In the vector problem of Equations (1)–(4), in which "max" and "min" are part of the criteria, the mathematical operation "opt" consists of the definition of a point X^0 and the maximum λ^0 bottom level to which all criteria measured in relative units are lifted:

$$\lambda^0 \leq \lambda_k\left(X^0\right) = \frac{f_k(X) - f_k^o}{f_k^* - f_k^o}, k = \overline{1, K}, \tag{24}$$

i.e., all criteria of $\lambda_k(X^0)$, $k = \overline{1, K}$ are equal to or greater than the maximum level of λ^0 (therefore λ^0 is also called the guaranteed result).

Theorem 2. (The theorem of the most inconsistent criteria in a vector problem with the set priority).

If in the convex vector problem of mathematical programming of Equations (1)–(4) the priority of the qth criterion of p_k^q, $k = \overline{1, K}$, $\forall q \in \mathbf{K}$ over other criteria is set, at a point of an optimum $X^0 \in S$ obtained on the basis of normalization of criteria and the principle of guaranteed result, there will always be two criteria with the indexes $r \in \mathbf{K}$, $t \in \mathbf{K}$, for which the following strict equality holds:

$$\lambda^0 = p_k^r \lambda_r\left(X^0\right) = p_k^t \lambda_t\left(X^0\right), r, t, \in K, \tag{25}$$

and other criteria are defined by inequalities:

$$\lambda^0 \leq p_k^q\left(X^0\right), k = \overline{1, K}, \forall q \in K, q \neq r \neq t. \tag{26}$$

Criteria with the indexes $r \in \mathbf{K}$, $t \in \mathbf{K}$ for which the equality of Equation (38) holds are called the most inconsistent.

Proof. Similar to Theorem 2 [25].

We note that in Equations (25) and (26), the indexes of criteria $r, t \in \mathbf{K}$ can coincide with the $q \in \mathbf{K}$ index.

Consequence of Theorem 1, about equality of an optimum level and relative estimates in a vector problem with two criteria with a priority of one of them.

In a convex vector problem of mathematical programming with two equivalent criteria, solved on the basis of normalization of criteria and the principle of the guaranteed result, at an optimum point X^o equality is always carried out at a priority of the first criterion over the second:

$$\lambda^0 = \lambda_1\left(X^0\right) = p_2^1\left(X^0\right)\lambda_2\left(X^0\right), X^0 \in S, \text{ where } p_2^1\left(X^0\right) = \lambda_1\left(X^0\right)/\lambda_2\left(X^0\right), \tag{27}$$

and at a priority of the second criterion over the first:

$\lambda^0 = p_1^2(X^0)\lambda_1(X^0) = \lambda_2(X^0)$, $X^0 \in \mathbf{S}$, where $p_1^2(X^0) = \lambda_2(X^0)/\lambda_1(X^0)$.

3.2.2. Mathematical Method of the Solution of a Vector Problem with Criterion Priority

(Method of the decision in problems of vector optimization with a criterion priority) [25].

Step 1. We solve a vector problem with equivalent criteria. The algorithm of the decision is presented in Section 3.1.2. As a result of the decision we obtain:

- optimum points by each criterion separately X_k^*, $k = \overline{1,K}$ and sizes of criterion functions in these points of $f_k^* = f_k(X_k^*)$, $k = \overline{1,K}$, which represent the boundary of a set of Pareto optimal points,
- anti-optimum points by each criterion of $X_k^0 = \{x_j, j = \overline{1,N}\}$ and the worst unchangeable part of each criterion of $f_k^0 = f_k(X_k^0)$, $k = \overline{1,K}$,
- $X^0 = \{\lambda^0, X^0\}$, an optimum point, as a result of the solution of VPMP at equivalent criteria, i.e., the result of the solution of a maximine problem and the λ-problem constructed on its basis,
- λ^0, the maximum relative assessment which is the maximum lower level for all relative estimates of $\lambda_k(X^0)$, or the guaranteed result in relative units, λ^0 guarantees that all relative estimates of $\lambda_k(X^0)$ are equal to or greater than λ^0:

$$\lambda^0 \le \lambda_k\left(X^0\right), k = \overline{1,K}, X^0 \in S. \tag{28}$$

The person making the decision carries out the analysis of the results of the solution of the vector problem with equivalent criteria. If the received results satisfy the decision maker, then the process concludes, otherwise subsequent calculations are performed.

In addition, we calculate:

- in each point X_k^*, $k = \overline{1,K}$ we determine sizes of all criteria of $q = \overline{1,K}$: $\{f_q(X_k^*), q = \overline{1,K}\}, k = \overline{1,K}$, and relative estimates $\lambda(X^*) = \{\lambda_q(X_k^*), q = \overline{1,K}, k = \overline{1,K}\}$, $\lambda_k(X) = \frac{f_k(X)-f_k^0}{f_k^*-f_k^0}$, $\forall k \in \mathbf{K}$:

$$F(X^*) = \begin{vmatrix} f_1(X_1^*), \ldots, f_k(X_1^*), \\ \ldots \\ f_1(X_k^*), \ldots, f_k(X_k^*) \end{vmatrix}, \lambda(X^*) = \begin{vmatrix} \lambda_1(X_1^*), \ldots, \lambda_k(X_1^*), \\ \ldots \\ \lambda_1(X_k^*), \ldots, \lambda_k(X_k^*) \end{vmatrix} \tag{29}$$

Matrices of criteria of $F(X^*)$ and relative estimates of $\lambda(X^*)$ show the sizes of each criterion of $k = \overline{1,K}$ upon transition from one optimum point X_k^*, $k \in \mathbf{K}$ to another X_q^*, $q \in \mathbf{K}$, i.e., on the border of a great number of Pareto.

- at an optimum point at equivalent criteria X^0 we calculate sizes of criteria and relative estimates:

$$f_k\left(X^0\right), k = \overline{1,K}; \lambda_k\left(X^0\right), k = \overline{1,K}, \tag{30}$$

which satisfy the inequality of Equation (28). In other points $X \in \mathbf{S}^0$, in relative units the criteria of $\lambda = \min_{k \in K} \lambda_k(X)$ are always less than λ^0, given the λ-problem of Equations (22) and (23).

This information is also a basis for further study of the structure of a great number of Pareto.

Step 2. Choice of priority criterion of $q \in \mathbf{K}$.

From theory (see Theorem 1) it is known that at an optimum point X^0 there are always two most inconsistent criteria, $q \in \mathbf{K}$ and $v \in \mathbf{K}$, for which in relative units an exact equality holds: $\lambda^0 = \lambda_q(X^0) = \lambda_p(X^0)$, $q, v \in \mathbf{K}$, $X \in \mathbf{S}$. Others are subject to inequalities: $\lambda^0 \le \lambda_k(X^0)$ $\forall k \in \mathbf{K}$, $q \ne v \ne k$.

As a rule, the criterion which the decision-maker would like to improve is part of this couple, and such a criterion is called a priority criterion, which we designate $q \in \mathbf{K}$.

Step 3. Numerical limits of the change of the size of a priority of criterion $q \in \mathbf{K}$ are defined.

For priority criterion $q \in \mathbf{K}$ from the matrix of Equation (29) we define the numerical limits of the change of the size of criterion:

- in physical units of $f_q\left(X^0\right) \leq f_q(X) \leq f_q\left(X_q^*\right),\ k \in \mathbf{K},$ (31)

where $f_q(X_q^*)$ derives from the matrix of Equation (29) $F(X^*)$, all criteria showing sizes measured in physical units, $f_q(X^0)$ from Equation (30), and,

- in relative units of $\lambda_q\left(X^0\right) \leq \lambda_q(X) \leq \lambda_q\left(X_q^*\right),\ k \in \mathbf{K},$ (32)

where $\lambda_q(X_q^*)$ derives from the matrix $\lambda(X^*)$, all criteria showing sizes measured in relative units (we note that $\lambda_q(X_q^*) = 1$), $\lambda_q(X^0)$ from Equation (29).

As a rule, Equations (31) and (32) are given for the display of the analysis.

Step 4. Choice of the size of priority criterion (decision-making).

The person making the decision carries out the analysis of the results of calculations of Equation (42) and from the inequality of Equation (31) chooses the numerical size f_q of the criterion of $q \in \mathbf{K}$:

$$f_q\left(X^0\right) \leq f_q \leq f_q\left(X_q^*\right),\ q \in \mathbf{K}. \tag{33}$$

For the chosen size of the criterion of f_q it is necessary to define a vector of unknown X^o. For this purpose, we carry out the subsequent calculations.

Step 5. Calculation of a relative assessment.

For the chosen size of the priority criterion of f_q the relative assessment is calculated as:

$$\lambda_q - \frac{f_q - f_q^o}{f_q^* - f_q^o}, \tag{34}$$

which upon transition from point X^0 to X_q^*, according to Equation (32), lies in the limits: $\lambda_q(X^0) \leq \lambda_q \leq \lambda_q(X_q^*) = 1$.

Step 6. Calculation of the coefficient of linear approximation.

Assuming a linear nature of the change of criterion of $f_q(X)$ in Equation (31) and according to the relative assessment of $\lambda_q(X)$ in Equation (32), using standard methods of linear approximation we calculate the proportionality coefficient between $\lambda_q(X^o)$, λ_q, which we call ρ:

$$\rho = \frac{\lambda_q - \lambda_q(X^o)}{\lambda_q(X_q^*) - \lambda_q(X^o)}, q \in K \tag{35}$$

Step 7. Calculation of coordinates of priority criterion with the size f_q.

In accordance with Equation (33), the coordinates of the X^q priority criterion point lie within the following limits: $X^0 \leq X^q \leq X_q^*$, $q \in \mathbf{K}$. Assuming a linear nature of change of the vector $X^q = \{x_1^q, \ldots, x_N^q\}$ we determine coordinates of a point of priority criterion with the size f_q with the relative assessment of Equation (32):

$$\begin{aligned} X^q &= \{x_1^q = x_1^0 + \rho(x_q^*(1) - x_1^0), \\ &\ldots, \\ x_N^q &= x_N^0 + \rho(x_q^*(N) - x_N^0)\}. \end{aligned} \tag{36}$$

where $X^0 = \{x_1^0, \ldots, x_N^0\}$, $X_q^* = \{x_q^*(1), \ldots, x_q^*(N)\}$.

Step 8. Calculation of the main indicators of a point x_q.

For the obtained point x_q, we calculate:

- all criteria in physical units $f_k(x^q) = \{f_k(x^q), k = \overline{1, K}\}$,
- all relative estimates of criteria $\lambda^q = \{\lambda_k^q, k = \overline{1, K}\}$, $\lambda_k(x^q) = \frac{f_k(x^q) - f_k^o}{f_k^* - f_k^o}$, $k = \overline{1, K}_{\overline{1,K}}$,

- the vector of priorities $P^q = \{p_k^q = \frac{\lambda_q(x^q)}{\lambda_k(x^q)}, k = \overline{1,K}\}$,
- the maximum relative assessment $\lambda^{oq} = \min\,(p_k^q \lambda_k(x^q), k = \overline{1,K})$.

Any point from Pareto's set $X_t^o = \{\lambda_t^o, X_t^o\} \in S^o$ can be similarly calculated.

Analysis of results. The calculated size of criterion $f_q(X_t^o)$, $q \in \mathbf{K}$ is usually not equal to the set f_q. The error of the choice of $\Delta f_q = |f_q(X_t^o) - f_q|$ is defined by the error of linear approximation.

4. Research, Analysis, and Formulation of the Problem of Decision-Making with Uncertain Data

4.1. Investigation of the Model of the "Object for Making an Optimal Decision" under Certainty and Uncertainty

4.1.1. Characteristics of Certainty and Uncertainty

Building a mathematical model of an "object or system for making an optimal decision" (Equations (1)–(4)) is possible, under conditions of certainty and uncertainty, which happen often. Conditions of certainty are characterized by the fact that the functional dependence of each criterion f_k, $k = \overline{1,K}$ (1) and the constraints G(3) on the system parameters x_j, $j = \overline{1,N}$, is known [6,12,22].

To build a mathematical model of a system under certainty, studies of the physical processes occurring in the system are conducted. At the creation of a mathematical model of such processes, fundamental laws of physics are used, for example, models of magnetic and temperature profiles, and laws of conservation of energy and movement. A complete list of all functional characteristics of technical systems and parameters on which these characteristics depend is formed. Their verbal description is given. The technical and information interrelationships of all components of a technical system is established, i.e., the structure is under construction. At this stage, the problem of the choice of the best structure of a technical system (a problem of structural optimization) is solved [4,23].

As a result of the conducted research, the functional interrelationship of a set of characteristics of $F_1(X)$, $F_2(X)$ and restrictions of $G(X)$ from parameters X has to be constructed.

Conditions of uncertainty are characterized by that there is no sufficient information on the functional dependence of each characteristic and the restrictions on the parameters [8,9,12,20].

At the same time, there are two problems associated with decision making.

The first problem is characterized by the fact that only data for some of the indicators are known (such a task is presented in the following section, see Equation (37)). The second problem is that data on some set of parameters, as well as relevant data on some set of characteristics (criteria), of a problem are known (38).

Both problems arise when carrying out pilot studies based on the principle of " input-output". On the basis of the conducted pilot studies, there is a problem with the adoption of the acceptable decision. We present the analysis of these problems and decision making on their basis in the following sections.

Thus, under conditions of certainty the function $f_k(X)$, $k \in \mathbf{K}$ is known, for the infinite set of parameters there is a corresponding infinite set of estimates of the function (criterion). Conversely, under conditions of uncertainty, only a finite set of parameters and the corresponding set of function (criterion) estimates are known, the smaller the set of parameters, the higher the uncertainty.

4.1.2. Investigation of Condition of Uncertainty

Conditions of uncertainty are considered in two aspects. The first relates to a lack of sufficient information. This is the uncertainty associated with the variety of characteristics (criteria) of the object under study.

This aspect is defined by the fact that there is not sufficient information on the functional dependence of the characteristic f and restrictions g from parameters X of the studied object. In this case, input data characterizing the object are presented as:

(a) random data,

(b) fuzzy data, or
(c) incomplete data, which are usually obtained from experimental data.

For options (a) and (b), input data have to be transformed into option (c), and are presented in table form: 1 column—parameter size, 2 columns—characteristic size. The methodology of the transformation of random and fuzzy data into tabular form is presented in the magazine "Fuzzy Decision Making and Soft Computing Applications". Tabular (experimental) data, using regression analysis, will be transformed into the $f(X)$ function, i.e., in terms of certainty.

In the future, only variant c) is considered in this work, and its transformations by regression methods into certainty conditions, i.e., into the function $f(X)$.

The second aspect of decision-making uncertainty is related to the fact that an object is characterized by many characteristics: $f_1(X), \ldots, f_K(X)$. The set of characteristics $\boldsymbol{K}$ is divided into two subsets $\boldsymbol{K}_1$ and $\boldsymbol{K}_2$. The subset of characteristics $\boldsymbol{K}_1$ in a numerical value is desired to be as high as possible (maximum), and the subset of characteristics $\boldsymbol{K}_2$ is desired to be as low as possible (minimum). Below it will be shown that the decision-making problem with a set of characteristics is reduced to a vector problem of mathematical programming, the solution of which is presented in Section 3.

4.2. Conceptual Problem Definition of Decision Making under the Conditions of Uncertainty

Initially, from a general view the conceptual problem definition of decision making was presented in work of R. L. Keeney and H. Raiffa [11], according to which we denote a_i, $i = \overline{1, M}$, for the admissible decision-making alternatives, and $A = (a_1\ a_2 \ldots a_M)$ for the vector of the set of admissible alternatives.

We match each alternative $a \in A$ to K numerical indices (criteria) $f_1(a), \ldots, f_K(a)$ that characterize the system. We can assume that this set of indices maps each alternative onto the point of the K-dimensional space of outcomes (consequences) of decisions made, $F(a) = (f_1(a)\, f_2(a) \ldots f_K(a))^T$. We use the same symbol $f_k(a)$ both for the criterion and for the function that performs the estimation with respect to this criterion. Note that we cannot directly compare the variables $f_v(a)$ and $f_k(a)$, $v \neq k$ at any point $F(a)$ of the K-dimensional space of consequences since it would most likely have no sense because these criteria are generally measured in different units. Using these data, we can state the decision-making problem.

The decision maker is to choose an alternative $a \in A$ to obtain the most suitable result, i.e., $F(a) \to \min$.

This definition means that the required estimating function should reduce the vector $F(a)$ to a scalar preference or "value" criterion. In other words, it is equivalent to setting a scalar function V given the space of consequences and possessing the following property:

$$V(F(a)) \geq V(F(a')) \Leftrightarrow F(a) \gg F(a'),$$

where the symbol $\gg$ means "no less preferable than" [24]. We call the function $V(F(a))$ the value function. The name of this function in other publications may vary from an order value function to a preference function to a value function. Thus, the decision maker is to choose $a \in A$ such that $V(F(a))$ is maximized. The value function allows an indirect comparison of the importance of certain values of various criteria of the system. Thus, the matrix $F(a)$ of admissible outcomes of alternatives takes the form:

$$F = \begin{bmatrix} a_1\, f_1^1 \ldots f_1^K \\ \ldots \\ a_M\, f_M^1 \ldots f_M^K \end{bmatrix}. \tag{37}$$

where $f_i^{\ j} = f_i\ (a_i)$ and all alternatives in it are represented by the vector of indices $F(a)$. For the sake of definiteness and without loss of generality, we assume that the first criterion (any criterion can be the first) is arranged in increasing (decreasing) order, with the alternatives re-numbered $i = \overline{1, M}$.

Problem 1 (Equation (37)) implies that the decision maker is to choose the alternative $a^0 \in A$ such that it will yield the "most suitable (optimal) result" [24].

For an engineering system, we can represent each alternative a_i by the N-dimensional vector $X_i = \{x_{ij}, j = \overline{1,N}\}, i = \overline{1,M}\}$ of its parameters, and its outcomes by the K-dimensional vector criterion $\{f_1(X_i), \ldots, f_K(X_i), i = \overline{1,M}\}$. Taking this into account, the matrix of outcomes (Equation (37)) takes the form:

$$I = \begin{bmatrix} X_1\, f_1(X_1) \ldots f_K(X_1) \\ \ldots \\ X_M\, f_1(X_M) \ldots f_K(X_M) \end{bmatrix}, \tag{38}$$

Problem 2 (Equation (38)) for decision makers consists of the choice of the set of design data $X^0 = \{x_{ij}, j = \overline{1,N}\}, i = \overline{1,M}\}$ that would allow the optimal result [24].

4.3. The Analysis of Modern Methods of Decision Making to the Experimental Data

At present, the problems of Equations (37) and (38) are solved by a number of "simple" methods based on special criteria, such as Wald, Savage, Hurwitz, and Bayes–Laplace criteria, which provide the basis for decision making.

The Wald criterion of maximizing the minimal component helps make the optimal decision that ensures the maximal gain among minimal ones, $\max\limits_{k=\overline{1,K}} \min\limits_{i=\overline{1,M}} f_i^k$.

The Savage minimal risk criterion chooses the optimal strategy so that the value of the risk r_i^k is minimal among maximal values of risks over the columns, $\min\limits_{i=\overline{1,M}} \max\limits_{k=\overline{1,K}} r_i^k$. The value of the risk r_i^k is chosen from the minimal difference between the decision that yields maximal profit $\max\limits_{i=\overline{1,M}} f_i^k, k = \overline{1,K}$, and the current value f_i^k, $r_i^k = (\max\limits_{i=\overline{1,M}} f_i^k) - f_i^k$, with their set being the matrix of risks $R = \|r_i^k\|_{i=\overline{1,M}}^{k=\overline{1,K}}$.

The Hurwitz criterion helps choose the strategy that lies somewhere between absolutely pessimistic and optimistic (i.e., the most considerable risk):

$$\max_{k=\overline{1,K}}\left(\alpha \min_{i=\overline{1,M}} f_i^k + (1-\alpha)\max_{i=\overline{1,M}} f_i^k\right),$$

where α is the pessimistic coefficient chosen in the interval $0 \leq \alpha \leq 1$.

The Bayes–Laplace criterion takes into account each possible consequence of all decision options, given their probabilities $\max\limits_{i=\overline{1,M}} \sum\limits_{k=1}^{K} f_i^k p_i$.

All these and other methods are widely described in publications on decision making [11]. All have certain drawbacks. For instance, if we analyze the Wald maximin criterion, we can see that by the problem's hypothesis all criteria are in different units. Hence, the first step, which is to choose the minimal component $f_k^{\min} = \min\limits_{i=\overline{1,M}} f_i^k$, is quite reasonable. However, all $f_k^{\min}, k = \overline{1,K}$, are measured in different units, therefore the second step, which is to maximize the minimal component $\max\limits_{k=\overline{1,K}} f_k^{\min}$, is pointless. Although it brings us slightly closer to a solution, the criteria measurement scale fails to solve the problem since the chosen criteria scales are judgmental.

We believe that to solve the problem of Equations (37) and (38), we need to form a measure that would allow the evaluation of any decision to be made, including the optimal one. In other words, we need to construct an axiom that shows, based on the set of K criteria, what makes one alternative better than the other. In turn, the axiom can help derive a principle that determines whether the chosen alternative is optimal. The optimality principle should become the basis for the constructive methods of choosing optimal decisions. We propose such an approach for the vector mathematical programming problem that is essentially close to the decision-making problem of Equations (37) and (38).

4.4. Transforming the Decision-Making Problem into Vector Problem

We compare the decision-making problem (DMP) of Equations (37) and (38) with the vector problem mathematical programming (VPMP). Table 2 shows the comparison.

Table 2. Comparing vector problem mathematical programming (VPMP) to decision-making problems (DMP).

VPMP—Equations (1)–(4)	DMP_1—Equation (37)	DMP_2—Equation (38)
2. Common Objective: Making the best decision		
3. Find the vector X from the admissible set, where the vector criterion $F(X)$ is optimal	Objective: Find the alternative a_i, $i = \overline{1,M}$, such that the set of criteria f_i^k, $k = \overline{1,K}$ is optimal	Objective: Find the vector alternative X_i, $i = \overline{1,M}$, such that the set of criteria $F(X)$, is optimal
4. The vector of parameters X and the dependence of criteria $F(X)$ on it are given completely, the set of admissible points is finite: $G(X) \leq 0$, $X^{\min} \leq X \leq X^{\max}$.	Parameters are not given, the criteria are represented as the finite set of values, the set of admissible alternatives is finite	Parameters are given, the criteria are represented as separate values so that the functional dependence between them is not given, the set of admissible points is finite
Opt $F(X) = \{\max F_1(X) = \{\max f_k(X), k = \overline{1,K_1}\}$, $\min F_2(X) = \{\min f_k(X), k = \overline{1,K_2}\}\}$, $G(X) \leq 0$, $x_j^{\min} \leq x_j \leq x_j^{\max}, j = \overline{1,N}$	5. Transforming the Decision Making Problem into VMPP Using regression analysis, we transform each k-th $k = \overline{1,K}$ set of values of criteria, f_i^k, $i = \overline{1,M}$ into the criterion function $f_k(x)$	Using multiple regression, we transform each k-th $k = \overline{1,K}$ set of values of criteria $f_k(X_i)$, $i = \overline{1,M}$ into the criterion function $f_k(X)$
6. ↑ Problems are Equivalent ›	$Opt\ F(x) = (f_1(x) \ldots f_K(x))^T$. $f_k^{\min} \leq f_k(x) \leq f_k^{\max}, k = \overline{1,K}$, $x^{\min} < x < x^{\max}$,	$Opt\ F(X) = (f_1(X) \ldots f_K(X))^T$. $f_k^{\min} \leq f_k(X) \leq f_k^{\max}, k = \overline{1,K}$, $X^{\min} \leq X \leq X^{\max}$

The first, second and third rows of Table 2 show that all three problems have the common objective of "making the best (optimal) decision". Both types of decision-making problems (row 4) have some uncertainty, functional dependences of criteria and restrictions on the problem's parameters are not known. At present, many mathematical methods of regression analysis are implemented in software (such as MATLAB) that allow using some set of initial data (as in Equations (37) and (38)) to construct the functional dependences $f_k(X)$, $k = \overline{1,K}$. For this reason, we use regression methods, including multiple regression, to construct criteria and restrictions in decision-making problems of both types (row 5) [26]. Combining criteria and restrictions, we represent decision-making problems of both types as a vector mathematical programming problem (row 6).

We perform these transformations. We use methods of regression analysis for the problem of Equation (37) and multiple regression for the problem of Equation (38) to transform each kth column of the matrix Ψ into the criterion function $f_k(X)$. We combine them in the vector function $F(X)$: max $F_1(X) = \{f_k(X), k = \overline{1,K}\}$ in Equation (1), and, min $F_2(X) = \{f_k(X), k = \overline{1,K}\}$ in Equation (2). The inequalities:

$$f_k^{\min} \leq f_k(X) \leq f_k^{\max}, k = \overline{1,K}$$

where $f_k^{\min} = \min\limits_{i=\overline{1,M}} f_k(X_i)$, $f_k^{\max} = \max\limits_{i=\overline{1,M}} f_k(X_i)$, are the minimal and maximal values of each function, and the parameters bounded by minimal and maximal values of each of them serve as functional restrictions (Equation (3)). The result is a VPMP (Equations (1)–(4)) and uses the same methods based on normalizing criteria and maximin principles as for the ES model under complete certainty to solve it for tantamount criteria.

5. Statement and Optimal Decision Making with Experimental Data in Problems with One Parameter

We use a particular example to illustrate the decision-making problem of the first type (Equation (37)) and the choice of the optimal decision. We also show the proposed method is independent of the form of the sought extremum of partial criteria.

5.1. Problem Definition of Decision Making of the First Type

The problem definition is carried out by the designer of the system on experimental data.

It is given that the system is defined by one parameter $X = \{x\}$, a vector of (operated) variables. The experimental data for the task of decision making are provided in Table 3.

Table 3. Experimental data (matrix I).

x	f_1	f_2	f_3	f_4
630	4200	1950	1628	245
1580	6000	2100	1577	230
2662	8850	2090	1377	210
3704	11,000	2050	1200	200
4800	12,900	1950	1100	190
5929	14,730	1750	977	170
7284	16,310	1560	1050	151
9353	19,000	1350	1100	150
14,505	23,250	540	1457	100
18,810	29,970	400	2088	55

The system comprises one parameter {x} and four characteristics (criteria):

$$f_1(x) \to \max, f_2(x) \to \max, f_3(x) \to \min, f_4(x) \to \max,$$

used to make a choice. Taken together, they constitute a decision-making problem of the first type.

The requirement is to make the best (optimal) decision given the experimental data available.

5.2. The Solution of the Problem

We find the optimal solution in two stages.

Stage 1. We transform the decision-making problem into a VPMP.

Step 1. We prepare the initial data in Table 2 in the form of the matrix I (Equation (38)). These watch points are shown in Figure 1, using MATLAB operators:

```
xlabel('X'); ylabel('Y'); hold on; plot(I(:,1),I(:,2)/10,'k.');
plot(I(:,1)I(:,3),'go'); plot(I(:,1),I(:,4),'bp'); plot(I(:,1)I(:,5)*10,'r*').
```

For the sake of visualization, the order of the first criterion is decreased by one while the order of the fourth criterion is increased by one.

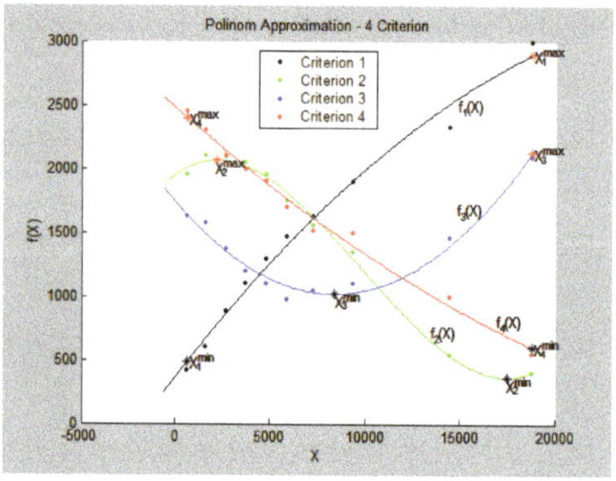

Figure 1. Polynomial approximation of four criteria.

Step 2. Using a method of the smallest square deviations [12], we calculate the coefficients of the approximating polynomial of the second degree:

$$\min_{A} f(A, X) \equiv \sum_{i=1}^{M} (y_j - (a_0 + a_1 x_{1i} + a_2 x_{1i}^2))^2. \quad (39)$$

An approximation is carried out in the MATLAB system using the polyfit function (X, Y, N) where X is a vector of tabular values (nodes), and Y represents preset values of assessment.

Limits of the change of parameter x of the lower and top scale in Figure 1 are set. In the MATLAB system, this is presented as $x = -600.:100.:19000$.

We calculate the first criterion by means of the function: c1=polyfit(I(:,1),I(:,2)/10,2). The result is $c_1(1) = -3.1937 \times 10^{-6}$, $c_1(2) = 0.1947$, $c_1(3) = 365.1$, which corresponds to the polynomial of the second degree:

$$f_1(x) = -3.1937 \times 10^{-6} x^2 + 0.1947x + 365.1. \quad (40)$$

We calculate the values of the polynomial y5 = polyval(c1,x) and show it on a graph using plot(x,y5,'k-'), hold on.

Similarly, the rest of the criteria are:

$$f_2(x) = 9.467 \times 10^{-10} x^3 - 2.7968 \times 10^{-5} x^2 + 0.1090 x + 1949.2, \quad (41)$$

$$f_3(x) = 1.0174 \times 10^{-5} x^2 - 0.1707x + 1737.4, \quad (42)$$

$$f_4(x) = 1.6458 \times 10^{-7} x^2 - 0.01309x + 247.83. \quad (43)$$

All the resulting points and functions are also shown in Figure 1.

Step 3. We form and solve the VPMP. Using the results of the previous stage, we represent the decision making problem as the VPMP of Equations (1)–(4) with the vector criterion $F(x) = (-f_1(x) - f_2(x) f_3(x) - f_4(x))^{\mathrm{T}}$ and restrictions $630 \leq x \leq 18{,}810$:

$$Opt\, F(X) = \{\mathbf{max}\, F_1(X) = \{\mathbf{max}\, f_1(X), \mathbf{max}\, f_2(X), \mathbf{max}\, f_4(X)\}, \quad (44)$$

$$\mathbf{min}\, F_2(X) = \{\mathbf{min}\, f_3 X)\}, \quad (45)$$

$$at\ restrictions\ x_j^{\min} \leq x_j \leq x_j^{\max}, j = \overline{1, N}. \quad (46)$$

Stage 2. We solve the VPMP of Equations (44)–(46) similarly to that shown in Section 2.

Step 1. We solve the problem of Equations (44)–(46) for each criterion separately. Since each is a unimodal function, we use the function [x,f] = fminbnd(c,a,b) to find its minimum or maximum on the segment (Equation (46)). Here, c, a, b are the input parameters, c is the given function, a and b are the beginning and the end of the interval, respectively, and x and f are the output parameters (the optimum point and the value of the objective function at the optimum, respectively). It takes the form:

[x1max,f1max]=fminbnd('−(3.1937 × 10^{-5}×x^2+1.9467×x+3651.1)',I(1,1),I(10,1))

for the first criterion. We thus derive the optimum point with respect to the first criterion $x_1^{\max}$ = 18,810 and the value of the criterion at this point $f_1^* = f_1(x_1^{\max})$ = −28,969. Similarly, we have $x_2^{\max}$ = 2192.8, $f_2^* = f_2(x_2^{\max})$ = −2063.7, $x_4^{\max}$ = 630.0, $f_4^* = f_4(x_4^{\max})$ = −239.65, $x_3^{\min}$ = 8389.0, $f_3^* = f_3(x_3^{\min})$ = 1021.4 for other criteria.

Step 2. We find the worst part for each criterion. We end up with the optimum point $x_1^{\min}$ = 630 with respect to the first criterion and the value of the criterion at the optimum $f_1^o = f_1(x_1^o)$ = 4864.9 (the worst constant part with respect to the first criterion). We use the operator plot(x1min, f1min/10,'k×') to represent this and other points in Figure 2. Similarly, we have for other criteria $x_2^{\min}$ = 17,502, $f_2^o = f_2(x_2^{\min})$ = 365.22, $x_4^{\min}$ = 18,810, $f_4^o = f_4(x_4^{\min})$ = 59.838, $x_3^{\max}$ =18,810, $f_3^o = f_3(x_3^{\max})$ = −2126.3.

Step 3. We analyze the set of Pareto optimal points. In points at an optimum $X^* = \{X_1^*, X_2^*, X_3^*, X_4^*, X_5^*\}$, sizes of criterion functions of $F(X^*) = \|f_q(X_k^*)\|_{q=\overline{1,K}}^{k=\overline{1,K}}$ are determined. We calculate a vector $D = (d_1\ d_2\ d_3\ d_4\ d_5)^{\mathrm{T}}$ of deviations by each criterion on an admissible set S: $d_k = f_k^* - f_k^0$, $k = \overline{1,5}$, and a matrix of relative estimates $\lambda(X^*) = \|\lambda_q(X_k^*)\|_{q=\overline{1,K}}^{k=\overline{1,K}}$, where $\lambda_k(X) = (f_k^* - f_k^0)/d_k$.

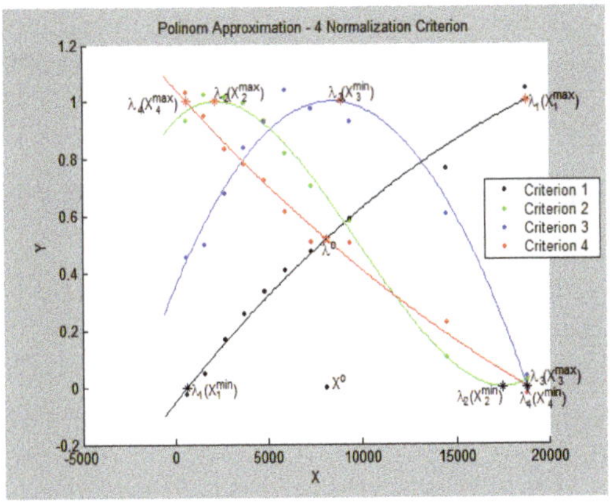

Figure 2. Approximations of four normalized criteria.

In the problem of Equations (44)–(46), criteria in the normalized form $\lambda_k(X^o)$, $k = \overline{1,K}$ can be represented as shown in Table 4. We calculate the coefficients of the approximating polynomial for the normalized criteria to obtain:

$\lambda_1(x) = -1.325 \times 10^{-9}\ x^2 + 8.0762 \times 10^{-5}\ x \times 0.0504,$
$\lambda_2(x) = 5.5735 \times 10^{-13}\ x^3 - 1.6467 \times 10^{-8}\ x^2 + 6.4179 \times 10^{-5}\ x + 0.9325,$
$\lambda_3(x) = -9.2080 \times 10^{-9}\ x^2 + 1.5451 \times 10^{-4}\ x + 0.3519,$
$\lambda_4\ (x) = 9.1530 \times 10^{-10}\ x^2 - 7.2811 \times 10^{-5}\ x + 1.0455.$

Table 4. Normalized criteria.

x	λ_1–max	λ_2–max	λ_3–min	λ_4–max
630	-0.0276	0.933	0.451	1.0298
1580	0.0471	1.0214	0.4971	0.9463
2662	0.1653	1.0155	0.6781	0.8351
3704	0.2545	0.9919	0.8383	0.7795
4800	0.3334	0.933	0.9289	0.7239
5929	0.4093	0.8153	1.0402	0.6127
7284	0.4748	0.7034	0.9741	0.507
9353	0.5864	0.5798	0.9289	0.5014
14,505	0.7627	0.1029	0.6057	0.2234
18,810	1.0415	0.0205	0.0346	−0.0269

Figure 2 shows the optimum points X^0 and normalized criteria.

The optimal point X^0 in Figure 2 can be chosen manually.

We solve the λ-problem to find the exact value of X^0.

Step 4. We construct the λ-problem. Using the obtained function and relative evaluations $\lambda_1(x)$, $\lambda_2(x)$, $\lambda_3(x)$, $\lambda_4(x)$, we construct the λ-problem:

$$\lambda^0 = \max \lambda, \tag{47}$$

$$\begin{aligned} \lambda - (-3.1937 \times 10^{-5}x^2 + 1.9467x + 3651.1 - f_1^o)/d_1 \le 0, \\ \lambda - (9.467 \times 10^{-10}x^3 - 2.7968 \times 10^{-5}x^2 + 0.1090x + 1949.2 - f_2^o)/d_2 \le 0, \\ \lambda - (1.0174 \times 10^{-5}x^2 - 0.1707x + 1737.4 + f_3^o)/d_3 \le 0, \\ \lambda - (1.6458 \times 10^{-7}x^2 - 0.01309x + 247.83 - f_1^o)/d_4 \le 0, \end{aligned} \tag{48}$$

$$630 \le x \le 18810 \tag{49}$$

Step 5. We solve the λ-problem of Equations (47)–(49). We use standard methods, in particular the MATLAB function fmincon (...). From the solution we obtain:

- the optimum point $\mathbf{X}^0 = \{\lambda^0 = 0.5163, x^0 = 8090\}$ (labeled by an asterisk in Figure 2),
- the values of the criteria at this point of $f_k(X^0)$, $k = \overline{1,K}$: $f_1(X^0) = 17{,}310$, $f_2(X^0) = 1501.8$, $f_3(X^0) = 1022.3$, $f_4(X^0) = 152.7$,
- the values of the relative estimates $\lambda_k(X^0)$, $k = \overline{1,K}$: $\lambda_1(X^0) = 0.5163$, $\lambda_2(X^0) = 0.6692$, $\lambda_3(X^0) = 0.9992$, $\lambda_4(X^0) = 0.5165$ (Figure 2).

It follows from this result that the first and fourth criteria are equal: $\lambda_1(X^0) = \lambda_4(X^0) = \lambda^0 = 0.5163$.

According to Theorem 1, the first and fourth criteria are most contradictory. Other criteria are greater than or equal to the maximum relative assessment of λ^o which is the guaranteed result in relative units.

6. Statement and Optimal Decision Making with Experimental Data in Problems with Two Parameters

We use a particular example to illustrate the decision-making problem of the second type (Equation (38)) and the choice of the optimal decision. The solution to the problem of making decisions of the second type and the choice of the optimal solution will be shown on a concrete example. The decision-making problem of the second type with two parameters is solved in three stages: problem statement—the formation of the initial data, transformation of experimental data into the vector problem of mathematical programming (VPMP), and, the VPMP decision—making the best decision.

6.1. Problem Definition of Decision Making of the Second Type with Two Parameters

The problem setting is performed by the system designer based on experimental data.

It is given that we have an engineering system functioning according to a vector of controlled variables $X = (x_1, x_2)$ with two parameters that take values: $x_1, x_2 \in \{0\ 2.5\ 5.\ 7.5\ 10\}$.

Decision-making criteria are represented by five functions $f_1(X), \ldots, f_5(X)$. For the first two, it is desirable to obtain values as high as possible (maximum) while for the other three it is desirable to obtain values as small as possible (minimum). Experimental data are given in Table 5.

Table 5. Experimental data (matrix I).

x_1	x_2	f_1	f_2	f_3	f_4	f_5
0	0	−80	−150	232	278.4	500
0	2.5	−102.5	−121.875	215.125	222.15	556.25
0	5	−130	−97.5	204.5	173.4	625
0	7.5	−162.5	−76.875	200.125	132.15	706.25
0	10	−200	−60	202	98.4	800
2.5	0	−102.5	−166.875	185.125	258.15	406.25
2.5	2.5	−125	−138.75	168.25	201.9	462.5
2.5	5	−152.5	−114.375	157.625	153.15	531.25
2.5	7.5	−185	−93.75	153.25	111.9	612.5
2.5	10	−222.5	−76.875	155.125	78.15	706.25
5	0	−130	−187.5	144.5	245.4	325
5	2.5	−152.5	−159.375	127.625	189.15	381.25
5	5	−180	−135	117	140.4	450
5	7.5	−212.5	−114.375	112.625	99.15	531.25
5	10	−250	−97.5	114.5	65.4	625
7.5	0	−162.5	−211.875	110.125	240.15	256.25
7.5	2.5	−185	−183.75	93.25	183.9	312
7.5	5	−212.5	−159.375	82.625	135.15	381.25
7.5	7.5	−245	−138.75	78.25	93.9	462.5
7.5	10	−282.5	−121.875	80.125	60.15	556.25
10	0	−200	−240	82	242.4	200
10	2.5	−222.5	−211.875	65.125	186.15	256.25
10	5	−250	−187.5	54.5	137.4	325
10	7.5	−282.5	−166.875	50.125	96.15	406.25
10	10	−320	−150	52	62.4	500

The requirement is to make the best (optimal) decision given the experimental data available.

6.2. Transformation of Experimental Data into a Vector Problem of Mathematical Programming

We construct a regression model and the regression problem based on it—the decision-making model of Equation (38)—and solve it. This is done in two stages, each of which consists of a number of steps.

Stage 1. We represent the decision-making problem as a VPMP.

Step 0. We prepare the initial data in MATLAB, forming the matrix I (Table 5).

Step 1. We approximate the function initial data (the third column) by cubic splines using: zz=interp2(X,Y,Z,xx,yy,zz,'linear').

We use the function surf(xx,yy,zz), hold on to represent the piecewise polynomial function $f_1(X)$ in Figure 3. Similarly, we perform the same approximation of the four other functions that we also represent (in natural units) in Figure 3 together with their minimal and maximal values. Although it appears to be difficult to choose the optimal solution visually using Figure 3, it is easier than using the initial data of the matrix I to do so.

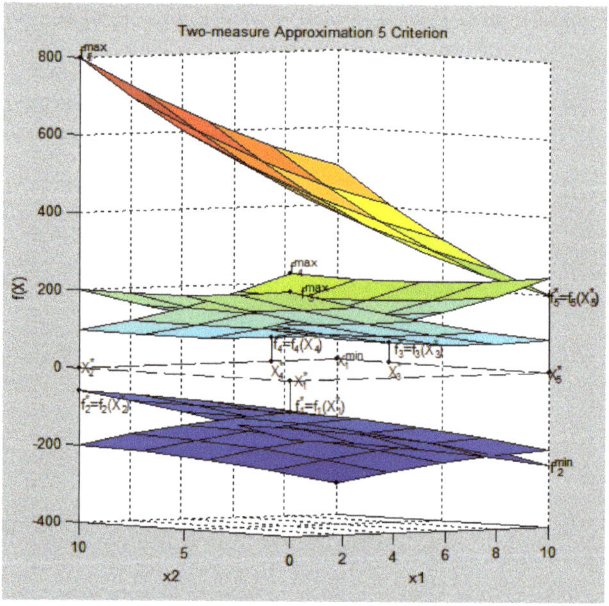

Figure 3. Approximation of five criteria of $f_1(X), \ldots, f_5(X)$.

Step 2. Using a method of the smallest square deviations [12], we calculate the coefficients of the approximating polynomial of the second degree:

$$\min_A f(A,X) \equiv \sum_{i=1}^{M} \left(y_j - (a_0 + a_1x_{1i} + a_2x_{1i}^2 + a_3x_{2i} + a_4x_{2i}^2 + a_5x_{1i} * x_{2i})\right)^2. \tag{50}$$

Based on data from columns 3–7 of the matrix *I*, we calculate the coefficients of the best approximating polynomial in the sense of minimum quadratic deviation at the nodes. This yields the polynomials of the second degree with two variables (four factors):

$$\begin{aligned} f_1(x) &= -0.4{x_1}^2 - 8x_1 - 0.4{x_2}^2 - 8x_2 - 80, \\ f_2(x) &= -0.3{x_1}^2 - 6x_1 - 0.3{x_2}^2 + 12x_2 - 150, \\ f_3(x) &= 0.5{x_1}^2 - 20x_1 + 0.5{x_2}^2 - 8x_2 + 232, \\ f_4(x) &= 0.6{x_1}^2 - 9.6\,x_1 + 0.6{x_2}^2 - 24x_2 + 278.4, \\ f_5(x) &= {x_1}^2 - 40x_1 + {x_2}^2 + 20x_2 + 500, \end{aligned} \tag{51}$$

$$\text{the restrictions } 0 \le x_1 \le 10,\ 0 \le x_2 \le 10. \tag{52}$$

Stage 2. We form and solve the VPMP.

Step 0. Using the results of the previous stage, we represent the decision-making problem as the VPMP of Equations (1)–(4) with the vector criterion $F(X) = (-f_1(X)\ -f_2(X)\ f_3(X)\ f_4(X)\ f_5(X))^{\mathrm{T}}$ for the stated restrictions:

$$Opt\ F(X) = \{\mathbf{max}\ F_1(X) = \{\mathbf{max}\ f_1(X), \mathbf{max}\ f_2(X), \tag{53}$$

$$\mathbf{min}\ F_2(X) = \{\mathbf{min}\ f_3X), \mathbf{min}\ f_4(X)\}, \mathbf{min}\ f_5(X)\}\}, \tag{54}$$

$$\text{at restrictions } 0 \le x_1 \le 10,\ 0 \le x_2 \le 10. \tag{55}$$

6.3. The Solution of a Vector Problem of Mathematical Programming–Decision-Making

For the solution of a vector problem of mathematical programming using the algorithm based on normalization of criteria, the above is used.

Step 1. We solve the problem of Equations (53)–(55) with respect to each criterion separately using the function fmincon(...), resulting in the optimum points:

$$X_1^* = \{x_1 = 0, x_2 = 0\}, X_2^* = \{x_1 = 0, x_2 = 10\}, X_3^* = \{x_1 = 10, x_2 = 8\},$$
$$X_4^* = \{x_1 = 8, x_2 = 10\}, X_5^* = \{x_1 = 10, x_2 = 0\}. \quad (56)$$

and values of the criteria at these points:

$$(1) f_1^* = -80,\ (2) f_2^* = -60,\ (3) f_3^* = 50,\ (4) f_4^* = 60,\ (5) f_5^* = 200. \quad (57)$$

Step 2. We find the worst constant part by solving the problem of Equations (53)–(55) with respect to each criterion separately, i.e., we minimize the first two criteria and maximize the other criteria. The result is: (1) $X_1^o = (10, 10), f_1^o = -320$, (2), $X_2^o = (10, 0), f_2^o = -240$, (3) $X_3^o = (0, 0), f_3^o = 232$, (4) $X_4^o = (0, 0)$, $f_4^o = 278.4$, (5) $X_5^o = (0, 10), f_5^o = 800$.

We represent the domain of admissible points **S** given by the restrictions of Equation (55) and the optimum points $X_1^*, \ldots, X_5^*$ in Figure 4.

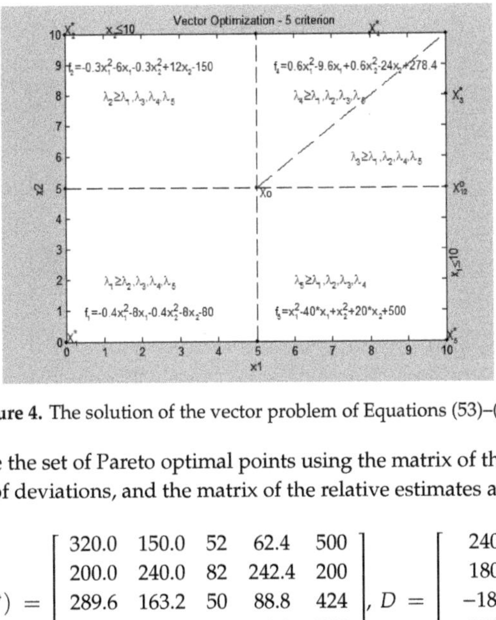

Figure 4. The solution of the vector problem of Equations (53)–(55).

Step 3. We analyze the set of Pareto optimal points using the matrix of the values of all objective functions, the column of deviations, and the matrix of the relative estimates at the optimal points:

$$F(X^*) = \begin{bmatrix} 320.0 & 150.0 & 52 & 62.4 & 500 \\ 200.0 & 240.0 & 82 & 242.4 & 200 \\ 289.6 & 163.2 & 50 & 88.8 & 424 \\ 289.6 & 127.2 & 74 & 60.0 & 544 \\ 200.0 & 240.0 & 82 & 242.4 & 200 \end{bmatrix}, D = \begin{bmatrix} 240 \\ 180 \\ -182 \\ -218.4 \\ -600 \end{bmatrix},$$

$$\Lambda(X^*) = \begin{bmatrix} 1.0000 & 0.5000 & 0.9890 & 0.9890 & 0.5000 \\ 0.5000 & 1.0000 & 0.8242 & 0.1648 & 1.0000 \\ 0.8733 & 0.5733 & 1.0000 & 0.8681 & 0.6267 \\ 0.8733 & 0.3733 & 0.8681 & 1.0000 & 0.4267 \\ 0.5000 & 1.0000 & 0.8242 & 0.1648 & 1.0000 \end{bmatrix}.$$

We also show the relative estimates in Figure 4.

Step 4. We construct the λ-problem for the VPMP of Equations (53)–(55)

$$\lambda^0 = \max \lambda, \tag{58}$$

$$\begin{aligned}
&\lambda - (-0.4{x_1}^2 - 8x_1 - 0.4{x_2}^2 - 8x_2 - 80 - f_1^o)/d_1 \le 0,\\
&\lambda - (-0.3x_1^2 - 6x_1 - 0.3x_2^2 + 12x_2 - 150 - f_2^o)/d_2 \le 0,\\
&\lambda - (0.5x_1^2 - 20x_1 + 0.5x_2^2 - 8x_2 + 232 - f_3^o)/d_3 \le 0,\\
&\lambda - (0.6x_1^2 - 9.6x_1 + 0.6x_2^2 - 2x_2 + 278.4 - f_4^o)/d_4 \le 0,\\
&\lambda - (x_1^2 - 40x_1 + x_2^2 + 20x_2 + 500 - f_5^o)/d_5 \le 0,
\end{aligned} \tag{59}$$

$$0 \le \lambda \le 1, 0 \le x_1 \le 10, 0 \le x_2 \le 10. \tag{60}$$

Step 5. We solve the λ-problem of Equations (58)–(60) using the same MATLAB function fmincon(...). This results in:

- the optimum point $X^0 = \{x_1 = 5.0, x_2 = 5.0, \lambda^0 = 0.5833\}$,
- the values of criteria: $f_1(X^0) = -180, f_2(X^0) = -135, f_3(X^0) = 117, f_4(X^0) = 140.4, f_4(X^0) = 450$,
- the values of the relative estimates $\lambda_k(X^0), k = \overline{1,K}$: $\lambda_1(X^0) = 0.5833, \lambda_2(X^0) = 0.5833, \lambda_3(X^0) = 0.6319$, $\lambda_4(X^0) = 0.6319, \lambda_5(X^0) = 0.5833$),
- the maximal relative estimate $\lambda^0 = 0.5833$ that is the maximal lower level for all relative estimates $\lambda^0 = \min\{\lambda_1(X^0), \lambda_2(X^0), \lambda_3(X^0), \lambda_4(X^0), \lambda_5(X^0)\} = 0.5833$.

Figure 5 shows all found points and relative estimates.

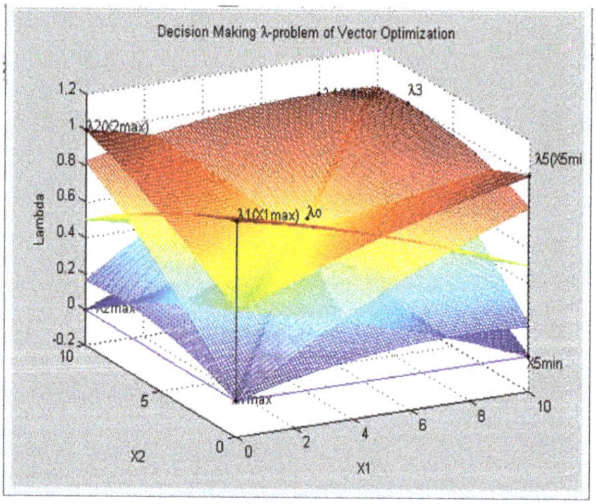

Figure 5. Solution of the λ-problem of Equations (58)–(60).

From the result of the solution of the λ-problem of Equations (58)–(60), the optimal point $\mathbf{X}^0$ and the maximum relative assessment λ^0 represent the results of the decision with equivalent criteria. For the solution of the VPMP with a priority of criterion coordinate $\mathbf{X}^{00} = \{x_1, x_2\}$, in Figure 4 the area where the corresponding assessment is higher than other relative estimates is chosen.

For example, if the second criterion has priority, then point $\mathbf{X}^{00} = \{x_1, x_2\}$ is chosen where $\lambda_2 \ge \lambda_1, \lambda_3, \lambda_4, \lambda_5$. More difficult technology illustrating the choice of a priority of criterion is presented in problems of decision-making with three and four criteria in the following sections.

7. Statement and Optimal Decision Making with Experimental Data in Problems with Three Parameters

The conditional object, namely, the technical system for which data on some set of the functional characteristics (certainty conditions), the discrete values of characteristics (certainty conditions), and the restrictions imposed on the functioning of the system [10] are known is considered. The numerical problem of model operation of the system is considered with equivalent criteria and with the given priority of criterion and proceeds as:

Statement of the problem of decision making in a system with three parameters,

Construction of a numerical model of a system with three parameters in the form of a vector problem,

The solution of the vector problem and decision making with equivalent criteria,

Decision making in a system with three parameters with a criterion priority,

Analysis of the results of the final decision.

7.1. Statement of the Problem of Decision Making in a System with Three Parameters

It is given that the technical system is defined by three parameters. (Practical problems of the simulation of technical systems using this algorithm can be solved with the dimensionality of parameters X greater than two, $N > 2$. The structure of the software becomes complicated and geometric interpretation of $N = 3,4\ldots$ is not possible.) $X = \{x_1, x_2, x_3\}$ represents a vector of operating variables. The basic data for the solution of the problem are the characteristics (criterion) of $F(X) = \{f_1(X), f_2(X), f_3(X), f_4(X)\}$, whose size of assessment depends on the vector X. For characteristics $f_3(X), f_4(X)$, functional dependence on parameters X (a definiteness condition) is known:

$$f_3(X) = 55.7188 - 0.1187 \times x_1 + 0.1844 \times x_2 - 0.0438 \times x_3 - 0.0002 \times x_1 \times x_2 - 0.0023 \times x_1 \times x_3 - 0.0011 \times x_2 \times x_3 + 0.0032 \times {x_1}^2 + 0.0634 \times x - 0 \times {x_3}^2 \quad (61)$$

$$f_4(X) = 25.6484 - 0.2967 \times x_1 - 0.3384 \times x_2 + 0.1433 \times x_3 - 0.0048 \times x_1 \times x_2 + 0.0169 \times x_1 \times x_3 + 0.0009 \times x_2 \times x_3 + 0.012 \times {x_1}^2 + 0.0014 \times {x_2}^2 - 0.0018 \times {x_3}^2 \quad (62)$$

$$\text{Parametrical restrictions: } 25 \le x_1 \le 100,\ 25 \le x_2 \le 100,\ 25 \le x_3 \le 100. \quad (63)$$

For the first and second characteristic results of experimental data, sizes of parameters and corresponding characteristics are known (uncertainty condition).

The numerical values of parameters X and characteristics of $y_1(X), y_2(X)$ are presented in Table 6.

Table 6. Numerical values of parameters and characteristics of the system.

x_1	x_2	x_3	$y_1(X) \to$ **max**	$y_2(X) \to$ **min**
25	25	25	412.5	1197.2
25	25	50	437.5	1232.8
25	25	75	462.5	1393.3
25	25	100	87.5	1303.8
25	50	25	312.5	2232.3
25	50	50	37.5	2267.7
25	50	75	62.5	2303.2
25	50	100	87.5	2338.8
25	75	25	212.5	3077.2
25	75	50	237.5	2862.8
25	75	75	262.5	3148.3
25	75	100	287.5	3183.7
25	100	25	12.5	3732.3
25	100	50	37.5	3767.7
25	100	75	62.5	3803.2
25	100	100	87.5	3838.8

Table 6. *Cont.*

x_1	x_2	x_3	$y_1(X) \to$ max	$y_2(X) \to$ min
50	25	25	512.5	1245.3
50	25	50	537.5	1303.8
50	25	75	562.5	1374.7
50	25	100	587.5	1445.8
50	50	25	512.5	2267.7
50	50	50	537.5	2338.8
50	50	75	562.5	2409.7
50	50	100	587.5	2480.8
50	75	25	412.5	3112.8
50	75	50	437.5	3183.7
50	75	75	462.5	3379.8
50	75	100	487.5	3325.8
50	100	25	212.5	3767.7
50	100	50	237.5	3838.8
50	100	75	262.5	3909.7
50	100	100	287.5	3980.8
75	25	25	612.5	1268.3
75	25	50	637.5	1374.7
75	25	75	662.5	1481.3
75	25	100	687.5	1587.8
75	50	25	612.5	2303.2
75	50	50	637.5	2409.7
75	50	75	662.5	2516.2
75	50	100	687.5	2622.7
75	75	25	512.5	3148.3
75	75	50	537.5	3254.8
75	75	75	562.5	3361.3
75	75	100	587.5	3467.8
75	100	25	312.5	3803.2
75	100	50	337.5	3909.7
75	100	75	362.5	4016.3
75	100	100	387.5	4122.7
100	25	25	612.5	1303.8
100	25	50	637.5	1445.8
100	25	75	662.5	1587.8
100	25	100	687.5	1729.7
100	50	25	612.5	2338.8
100	50	50	637.5	2480.8
100	50	75	662.5	2622.7
100	50	100	687.5	2764.7
100	75	25	512.5	3183.7
100	75	50	537.5	3325.8
100	75	75	562.5	3467.8
100	75	100	587.5	3609.8
100	100	25	312.5	3838.8
100	100	50	337.5	3980.8
100	100	75	362.5	4122.7
100	100	100	387.5	4264.8

In the decision, in the assessment size of the first and the third characteristic (criterion), it is possible to obtain: $f_1(X) \to$ **max** $y_3(X) \to$ **max**, for the second and fourth characteristic: $y_2(X) \to$ **min** $y_4(X) \to$ **min**. Parameters $X = \{x_1, x_2, x_3\}$ change according to the following limits: $x_1, x_2, x_3 \in$ (25. 50. 75. 100.).

The following is required: to construct a model of the technical system in the form of a vector problem, to solve the vector problem with equivalent criteria, to choose a priority criterion, to establish a numerical value of the priority criterion, to make the best decision (optimum).

7.2. Construction of a Numerical Model of a System with Three Parameters in the Form of a Vector Problem

The construction of a numerical model of the system in the form of a vector problem includes three stages:

- Building a model under the conditions of certainty;
- Building a model under the conditions of uncertainty;
- Construction of a mathematical model of a technical system (i.e., the general part for the conditions of certainty and uncertainty).

7.2.1. Building a Model under the Conditions of Certainty

Construction under the conditions of definiteness is defined by functional dependence of each characteristic and restrictions on the parameters of the technical system. In our example, two characteristics (Equations (61) and (62)) and the restrictions of Equation (63) are known. Uniting these, we obtain a vector task with two criteria:

$$opt\ F(X) = \{max\ F_1(X)\} = \{max\ f_3(X)\} = 55.7188 - 0.1187 \times x_1 + 0.1844 \times x_2 - 0.0438 \times x_3 - 0.0002 \times x_1 \times x_2 - 0.0023 \times x_1 \times x_3 - 0.0011 \times x_2 \times x_3 + 0.0032 \times {x_1}^2 + 0.0634 \times x - 0 \times {x_3}^2 \quad (64)$$

$$min\ F_2(X) = \{min\ f_4(X)\} = 25.6484 - 0.2967 \times x_1 - 0.3384 \times x_2 + 0.1433 \times x_3 - 0.0048 \times x_1 \times x_2 + 0.0169 \times x_1 \times x_3 + 0.0009 \times x_2 \times x_3 + 0.012 \times {x_1}^2 + 0.0014 \times {x_2}^2 - 0.0018 \times {x_3}^2 \quad (65)$$

$$\text{Parametrical restrictions: } 25 \le x_1 \le 100,\ 25 \le x_2 \le 100,\ 25 \le x_3 \le 100. \quad (66)$$

These data are used further to create a mathematical model of the technical system.

7.2.2. Building a Model under the Conditions of Uncertainty

Construction under the conditions of uncertainty entails the use of the qualitative and quantitative descriptions of the technical system obtained by the "input-output" principle in Table 5. Transformation of information (basic data $y_3(X)$, $y_4(X)$) into functional types $f_3(X), f_4(X)$ is carried out by the use of mathematical methods (i.e., regression analysis).

The basic data of Table 1 are created in the MATLAB system in the form of a matrix:

$$I = |X, Y| = \{y_{i1}\ y_{i2},\ i = \overline{1, M}\}. \quad (67)$$

For each experimental set function y_k, $k = \overline{1,2}$, regression using the method of least squares $\min \sum_{i=1}^{M} (y_i - \overline{y_i})^2$ in MATLAB is performed. A_k, a polynomial defining the interrelationship of factors $X_i = \{y_{1i}, y_{2i}\}$ (67) and functions $\overline{y}_{ki} = f(X_i, A_k)$, $k = \overline{1,2}$ is constructed. As a result, we obtain a system of coefficients $A_k = \{A_{0k}, A_{1k}, \ldots, A_{9k}\}$ which define the coefficients of a polynomial (function):

$$f_k(X, A) = A_{0k} + A_{1k}x_1 + A_{2k}x_1^2 + A_{3k}x_2 + A_{4k}x_2^2 + A_{5k}x_3 + A_{6k}x_3^2 + A_{7k}x_1 * x_2 + A_{8k}x_1 * x_3 + A_{9k}x_2 * x_3,\ k = \overline{1,2}. \quad (68)$$

As a result of the calculation of the coefficients A_k, $k = 1$, we obtain the $f_1(X)$ function:

$$f_1(X) = 50.0 + 11.55 \times x_1 + 3.55 \times x_2 + 1.0 \times x_3 + 0.0144 \times x_1 \times x_2 - 0 \times x_1 \times x_3 + 0 \times x_2 \times x_3 - 0.07 \times x_1^2 - 0.07 \times x_2^2 - 0 \times x_3^2. \quad (69)$$

As a result of the calculations of the coefficients A_k, $k = 2$, we obtain the $f_2(X)$ function:

$$f_2(X) = -53.875 + 0.7359 \times x_1 + 51.3703 \times x_2 + 0.3516 \times x_3 + 0.0072 \times x_1 \times x_2 + 0.0519 \times x_1 \times x_3 + 0.0005 \times x_2 \times x_3 - 0.0066 \times x_1^2 - 0.1454 \times x_2^2 + 0.0003 \times x_3^2 \quad (70)$$

Parametric restrictions are similar to those of Equation (8).

7.2.3. Creation of a Mathematical Model of a Technical System under the Conditions of Definiteness and Uncertainty

For the creation of a mathematical model of the technical system we used: the functions obtained from conditions of definiteness (Equations (64) and (65)) and uncertainty (Equations (69) and (70)), and parametric restrictions (Equation (66)).

Block 4. We consider the functions of Equations (64), (65), (69) and (70) as the criteria defining the functioning of the technical system. A set of criteria $K = 4$ includes three criteria of $f_1(X), f_3(X) \rightarrow$ max and two of $f_2(X), f_4(X) \rightarrow$ min. As a result, the model of the functioning of the technical system is presented as a vector problem of mathematical programming:

$$\begin{aligned} & opt\ F(X) = \{max\ F_1(X) = \{max\ f_1(X) \equiv 50.0 + 11.55 \times x_1 \\ & + 3.55 \times x_2 + 1.0 \times x_3 + 0.0144 \times x_1 \times x_2 - 0.0 \times x_1 \times x_3 \\ & + 0.0 \times x_2 \times x_3 - 0.07 \times x_1^2 - 0.07 \times x_2^2 - 0.0 \times x_3^2, \\ & max\ f_3(X) = 55.7188 - 0.1187 \times x_1 + 0.1844 \times x_2 - \\ & 0.0438 \times x_3 - 0.0002 \times x_1 \times x_2 - 0.0023 \times x_1 \times x_3 - \\ & 0.0011 \times x_2 \times x_3 + 0.0032 \times x_1^2 + 0.0634 \times x - 0 \times x_3^2\}, \end{aligned} \quad (71)$$

$$\begin{aligned} & min\ F_2(X) = \{min\ f_2(X) \equiv -53.875 + 0.7359 \times x_1 \\ & + 51.3703 \times x_2 + 0.3516 \times x_3 + 0.0072 \times x_1 \times x_2 \\ & + 0.0519 \times x_1 \times x_3 + 0.0005 \times x_2 \times x_3 - 0.0066 \times x_1^2 \\ & - 0.1454 \times x_2^2 + 0.0003 \times x_3^2, \\ & min\ f_4(X) = 25.6484 - 0.2967 \times x_1 - 0.3384 \times x_2 \\ & + 0.1433 \times x_3 - 0.0048 \times x_1 \times x_2 + 0.0169 \times x_1 \times x_3 \\ & + 0.0009 \times x_2 \times x_3 + 0.012 \times x_1^2 + 0.0014 \times x_2^2 - 0.0018 \times x_3^2\}\} \end{aligned} \quad (72)$$

$$\text{restrictions: } 25 \le x_1 \le 100,\ 25 \le x_2 \le 100,\ 25 \le x_3 \le 100. \quad (73)$$

The vector problem of mathematical programming in Equations (71)–(73) represents the model of optimal decision making under conditions of certainty and uncertainty in the aggregate.

7.3. The Solution of the Vector Problem and Decision Making with Equivalent Criteria

(Algorithm1 of decision making in problems of vector optimization with equivalent criteria).

The solution of the vector problem of Equations (71)–(73) is undertaken as a sequence of steps.

Step 1. Equations (71)–(73) are solved for each criterion separately, using the function *fmincon* (...) of the MATLAB system, the use of the function *fmincon* (...) is considered in [12].

As a result, we obtain optimum points: X_k^* and $f_k^* = f_k(X_k^*)$, $k = \overline{1, K}$, the sizes of the criteria at this point, i.e., the best decision for each criterion:

$X_1^* = \{x_1 = 86.02,\ x_2 = 34.2,\ x_3 = 100\}$, $f_1^* = f_1(X_1^*) = -707.47$, $X_2^* = \{x_1 = 25,\ x_2 = 25,\ x_3 = 25\}$, $f_2^* = f_2(X_2^*) = 1200.0$,

$X_3^* = \{x_1 = 100,\ x_2 = 100,\ x_3 = 25\}$, $f_3^* = f_3(X_3^*) = -724.69$, $X_4^* = \{x_1 = 25,\ x_2 = 100,\ x_3 = 25\}$, $f_4^* = f_4(X_4^*) = 9.16$.

The restrictions in Equation (73) and points of an optimum of coordinates $\{x_1, x_2\}$ are presented in Figure 6.

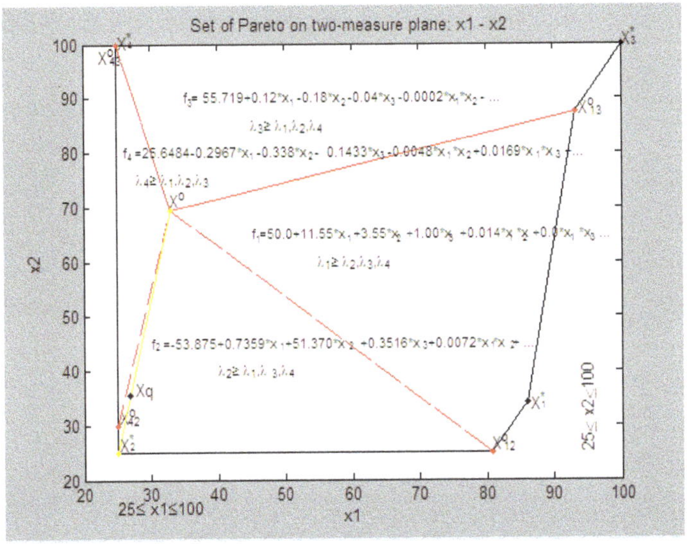

Figure 6. Pareto's great number, $S^0 \subset S$ in a two-dimensional system of coordinates.

Step 2. We define the worst unchangeable part of each criterion (anti-optimum):

$X_1^0 = \{x_1 = 25, x_2 = 100, x_3 = 25\}$, $f_1^0 = f_1(X_1^0) = 11.0$, $X_2^0 = \{x_1 = 100, x_2 = 100, x_3 = 100\}$, $f_2^0 = f_2(X_2^0) = -4270.9$,

$X_3^0 = \{x_1 = 43.5, x_2 = 20, x_3 = 80\}$, $f_3^0 = f_3(X_3^0) = 85.0$, $X_4^0 = \{x_1 = 100, x_2 = 25, x_3 = 100\}$, $f_4^0 = f_2(X_4^0) = -263.97$.

(Top index zero).

Step 3. The system analysis of a set of Pareto optimal points is conducted, (i.e., analysis by each criterion). At the optimal points $X^* = \{X_1^*, X_2^*, X_3^*, X_4^*\}$, the sizes of the criterion functions of $F(X^*) = \|f_q(X_k^*)\|_{q=\overline{1,K}}^{k=\overline{1,K}}$ are determined. We calculated a vector of $D = (d_1\ d_2\ d_3\ d_4)^T$, deviations of each criterion on an admissible set S: $d_k = f_k^* - f_k^0$, $k = \overline{1,4}$, and a matrix of relative estimates of $\lambda(X^*) = \|\lambda_q(X_k^*)\|_{q=\overline{1,K}}^{k=\overline{1,K}}$, where $\lambda_k(X) = (f_k^* - f_k^0)/d_k$:

$$F(X^*) = \begin{vmatrix} 707.5 & 2055.1 & 127.1 & 209.6 \\ 374.0 & 1200.0 & 96.1 & 28.7 \\ 329.0 & 3848.7 & 724.7 & 95.1 \\ 11.0 & 3704.1 & 701.9 & 9.2 \end{vmatrix}, \quad D = \begin{vmatrix} 696.5 \\ -3070.9 \\ 639.7 \\ -254.8 \end{vmatrix},$$

$$\lambda(X^*) = \begin{vmatrix} 1.0000 & 0.7216 & 0.0658 & 0.2132 \\ 0.5212 & 1.0000 & 0.0174 & 0.9232 \\ 0.4566 & 0.1375 & 1.0000 & 0.6628 \\ 0 & 0.1846 & 0.9644 & 1.0000 \end{vmatrix}$$

Discussion. The analysis of sizes of criteria in relative estimates shows that at optimal points $X^* = \{X_1^*, X_2^*, X_3^*, X_4^*\}$ the relative assessment is equal to unity. Other criteria there are much less than unity. It is required to find such points (parameters) at which relative estimates are closest to unity. The following steps 4 and 5 are directed to the solution of this problem.

Step 4. Creation of the λ-problem is carried out in two stages: first, the maximum problem of optimization with normalized criteria is constructed:

$$\lambda^0 = \max_x \min_k \lambda_k(X), G(X) \leq 0, X \geq 0,$$

Second, this is transformed into a standard problem of mathematical programming (the λ-problem):

$$\lambda^0 = \max \lambda, \tag{74}$$

$$\begin{aligned} \text{restrictions}:\ &\lambda - \frac{50.0+11.55{*}x_1 \ldots +0.014{*}x_1{*}x_2 \ldots -0.07{*}x_1^2 \ldots -f_1^o}{f_1^*-f_1^o} \le 0 \\ &\lambda - \frac{55.71 - 0.118{*}x_1 \ldots - 0.002{*}x_1{*}x_2 \ldots -0.0032{*}x_1^2 \ldots -f_3^o}{f_3^*-f_3^o} \le 0 \\ &\lambda - \frac{53.87+0.7359{*}x_1 + \ldots -0.0519{*}x_1{*}x_2 \ldots +0.0066{*}x_1^2 \ldots -f_2^o}{f_2^*-f_2^o} \le 0 \\ &\lambda - \frac{25.6484-0.2967\times x_1 \ldots -0.0048\times x_1\times x_2 \ldots + 0.012\times x_1^2 \ldots -f_4^o}{f_4^*-f_4^o} \le 0 \end{aligned} \tag{75}$$

$$25 \le x_1 \le 100, 25 \le x_2 \le 100, 25 \le x_3 \le 100. \tag{76}$$

where the vector of unknowns has the dimension $N + 1$: $\mathbf{X} = \{x_1, \ldots, x_N, \lambda\}$.

Step 5. The λ-problem solution.

Using the function fmincon(…), [12,15]:

[Xo,Lo] = fmincon('Z_TehnSist_4Krit_L',X0,Ao,bo,Aeq,beq,lbo,ubo,'Z_TehnSist_LConst',options).

As a result, the solutions of the vector problem of mathematical programming of Equations (71)–(73) with equivalent criteria and the λ-problem corresponding to Equations (74)–(75) are obtained:

$\mathbf{X}^0 = \{X^0, \lambda^0\} = \{X^0 = \{x_1 = 33.027, x_2 = 69.54, x_3 = 25.0, \lambda^0 = 0.4459\}\}$ is an optimum point of the design data of the technical system. Point $\mathbf{X}^0$ is presented in Figure 6. $f_k(X^0)$, $k = \overline{1,K}$ represents sizes of criteria (characteristics of technical system):

$$\{f_1(X^0) = 321.5, f_2(X^0) = 2901.7, f_3(X^0) = 370.2, f_4(X^0) = 19.1\}, \tag{77}$$

and $\lambda_k(X^0)$, $k = \overline{1,K}$ represents sizes of relative estimates:

$$\{\lambda_1(X^0) = 0.4459, \lambda_2(X^0) = 0.4459, \lambda_3(X^0) = 0.4459, \lambda_4(X^0) = 0.9609\}, \tag{78}$$

$\lambda^0 = 0.4459$ is the maximum lower level among all relative estimates measured in relative units: $\lambda^0 = min\ (\lambda_1(X^0), \lambda_2(X^0), \lambda_3(X^0), \lambda_4(X^0)) = 0.4459$. A relative assessment, λ^0, is called the guaranteed result in relative units, i.e., $\lambda_k(X^0)$. According to the characteristics of the technical system $f_k(X^0)$, it is impossible to improve, without worsening other characteristics.

Discussion. We note that according to Theorem 1, at point $\mathbf{X}^0$ criteria 1, 2, 3 are contradictory. This contradiction is defined by the equality of $\lambda_1(X^0) = \lambda_2(X^0) = \lambda_3(X^0) = \lambda^0 = 0.4459$, and for other criteria an inequality of $\{\lambda_4(X^0) = 0.9609\} > \lambda^0$.

Thus, Theorem 1 forms a basis for the determination of correctness of the solution of a vector problem. In a vector problem of mathematical programming, as a rule, for two criteria the equality holds: $\lambda^0 = \lambda_q(X^0) = \lambda_p(X^0)$, $q, p \in \mathbf{K}$, $X \in S$, (in our example, three criteria), and for other criteria is defined as an inequality: $\lambda^0 \le \lambda_k(X^0)$ "$k \in \mathbf{K}, q \neq p \neq k$.

In an admissible set of points **S** formed by the restrictions of Equation (76), the optimum points $X_1{}^*$, $X_2{}^*$, $X_3{}^*$, $X_4{}^*$ are united in a contour and presented as a set of Pareto optimal points, $\mathbf{S}^0 \subset \mathbf{S}$. For specification of the border of a great number of Pareto additional points are calculated: X^o_{12}, X^o_{13}, X^o_{42}, X^o_{34} which lie between the corresponding criteria. For definition of a point X^o_{12}, the vector problem was solved with two criteria of Equations (71), (72): $\lambda_1 X$, $\lambda_2 X$, (Equation (76)).

Results of the decision are:

$X^o_{12} = \{80.78\ 25.0\ 55.89\}$, $\lambda^0(X^o_{12}) = 0.9264$, $F_{12} = \{656.2\ 1426.0\ 101.7\ 142.7\}$,

$L_{12} = \{\mathbf{0.9264}\ \mathbf{0.9264}\ 0.0261\ 0.4761\}$.

Other points X^o_{13}, X^o_{42}, X^o_{43} were similarly defined:

$X^o_{13} = \{93.29\ 87.49\ 100.0\}$, $\lambda^o(X^o_{13}) = 0.\ 7173$, $F_{13} = \{510.6\ 3924.4\ 543.8\ 206.2\}$,

$L_{13} = (\mathbf{0.7173}\ 0.1128\ \mathbf{0.7173}\ 0.2267\}$,

$X^o_{42} = \{25.0\ 29.92\ 25.0\}$, $\lambda^0(X^o_{42}) = 0.\ 9301$, $F_{42} = \{374.3\ 1414.5\ 114.0\ 27.0\}$,

L_{42} = {0.5217 **0.9301** 0.0454 **0.9301**},
X^o_{43} = {25.0 100.0 56.02}, $\lambda^0(X^o_{43})$ = 0. 8366, F_{43} = {42.0 3757.6 695.4 25.0},
L_{43} = {0.0445 0.1672 **0.9541 0.9541**},

Points: X^o_{12}, X^o_{13}, X^o_{42}, X^o_{43} are presented in Figure 6. Coordinates of these points and the characteristics of the technical system in relative units of $\lambda_1(X)$, $\lambda_2(X)$, $\lambda_3(X)$, $\lambda_4(X)$, $\lambda_5(X)$ are shown in Figure 7 in three-dimensional measured space $\{x_1, x_2, \lambda\}$, where the third axis λ is a relative assessment.

The solution of the λ-problem of Equations (74)–(76) is the optimal point $\mathbf{X^o}$ and the maximum relative assessment of λ^0 represents the result of the decision with equivalent criteria.

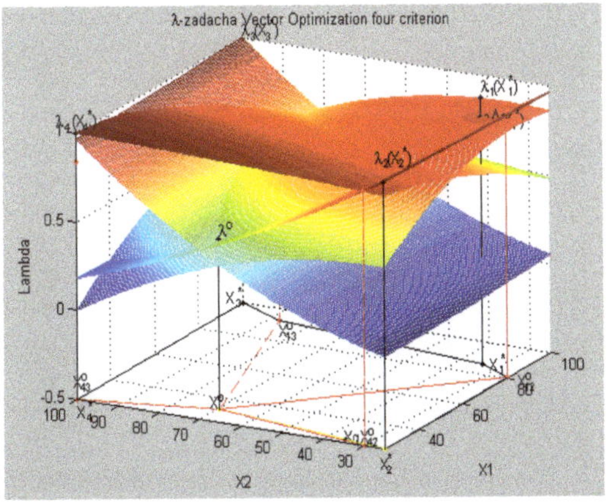

Figure 7. The solution of the λ-problem in a three-dimensional system of coordinates of x1, x2 and λ.

7.4. Decision Making in a System with Three Parameters with a Criterion Priority

(Method of decision making in problems of vector optimization with a criterion priority)

Step1. We solve a vector problem with equivalent criteria. The numerical results of the solution of the vector problem are given above. Pareto's great number $S^o \subset S$ lies between optimum points:

$$S^0 = \{X_1^* X_3^0 X_3^* X_{43}^0 \; X_4^* X_{42}^0 X_2^* X_{12}^0 X_1^*\}.$$

We carry out the analysis of the great number of Pareto $S^0 \subset S$. For this purpose, we will connect auxiliary points: X^o_{12}, X^o_{13}, X^o_{43}, X^o_{42}, with a point X^o which conditionally represents the center of a great number of Pareto. As a result, we obtain four subsets of points $X \in S^o_q \subset S^0 \subset \mathbf{S}$, $q = \overline{1,4}$. The subset of $S^o_1 \subset S^0 \subset \mathbf{S}$ is characterized by the fact that in the relative assessment, $\lambda_1 \geq \lambda_2, \lambda_3, \lambda_{45}$, i.e., in the field of $\mathbf{S}$ the first criterion has priority over the others. This applies similarly for the S^o_2, S^o_3, S^o_4, subsets of points where the second, third or fourth criterion has a priority over the others, respectively. We designate the set of Pareto optimal points $\mathbf{S^0} = S^o_1 \cup S^o_2 \cup S^o_3 \cup S^o_4$. Coordinates of all obtained points and relative estimates are presented in two-dimensional space in Figure 6. These coordinates are shown in three-dimensional space $\{x_1, x_2, \lambda\}$ from a point X^*_4 in Figure 7, where the third axis λ is a relative assessment. Restrictions of the set of Pareto optimal points in Figure 7 is lowered to −0.5 (so that restrictions are visible). This information is also a basis for further research on the structure of a great number of Pareto. The person making decisions, as a rule, is the designer of the technical system. If results of the solution of the vector problem with equivalent criteria do not satisfy the person making the decision, then the choice of the optimal solution is taken from any subset of points S^o_1, S^o_2, S^o_3, S^o_4.

Step 2. Choice of priority criterion of $q \in \mathbf{K}$. From theory (see Theorem 2) it is known that at an optimum point $\mathbf{X^0}$ there are always two most inconsistent criteria, $q \in \mathbf{K}$ and $v \in \mathbf{K}$, for which in relative units an equality holds: $\lambda^0 = \lambda_q(X^0) = \lambda_p(X^0)$, $q, v \in \mathbf{K}$, $X \in S$. Others are subject to inequalities: $\lambda^0 \leq \lambda_k(\mathbf{X^0})$ "$k \in \mathbf{K}, q \neq v \neq k$.

In the model of the technical system of Equations (71)–(73) and the corresponding λ-problem of Equations (74)–(76), such criteria are the first, second and third:

$$\lambda^0 = \lambda_1(X^0) = \lambda_2(X^0) = \lambda_3(X^0) = 0.4459. \quad (79)$$

These are shown in Figure 8.

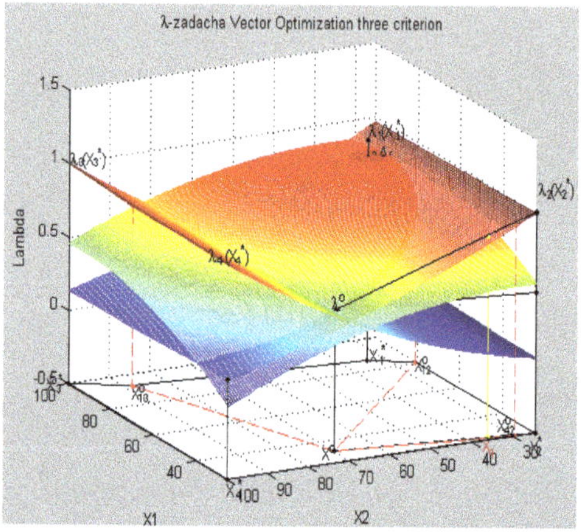

Figure 8. The solution of the λ-problem (1, 2, 3 criterion) in a three-dimensional system of coordinates of x1, x2 and λ.

As a rule, the criterion which the decision-maker would like to improve is chosen from a couple of contradictory criteria. Such a criterion is called the "priority criterion", which we designate $q = 2 \in$ $\boldsymbol{K}$. This criterion is investigated in interaction with the first criterion of $k = 1 \in \boldsymbol{K}$.

On the display the message is given:

q = input ('Enter priority criterion (number) of $q =$ '), Have entered: $q = 2$.

Step 3. Numerical limits of the change of the size of a priority of criterion of $q = 2 \in \boldsymbol{K}$ are defined.

For priority criterion $q = 2$, the numerical limits in physical units upon transition from an optimal point X^0 to the point X_q^* obtained in the first step are defined. Information about the criteria for $q = 2$ is given on the screen:

$$f_q\left(X^0\right) \;=\; 2901.68 \le f_q(X) \le 1200.0 \;=\; f_q(X_q^*), q \in K. \quad (80)$$

In relative units the criterion of $q = 2$ changes according to the following limits:

$\lambda_q(X^0) = 0.4459 \le \lambda_q(X) \le 1 = \lambda_q(X_q^*)$, $q = 2 \in \boldsymbol{K}$.

These data are analyzed.

Step 4. Choice of the size of priority criterion $q \in \boldsymbol{K}$ (decision making). The message is displayed: "Enter the size of priority criterion $f_q =$", we enter, for example, $f_q = 1500$.

Step 5. Calculation of relative assessment.

For the chosen size of priority criterion $f_q = 1600$ the relative assessment is calculated:

$$\lambda_q \;=\; \frac{f_q - f_q^o}{f_q^* - f_q^o} \;=\; \frac{1600 - 4279.9}{1200.0 - 4279.9} \;=\; 0.8697, \quad (81)$$

which upon transition from point X^0 to X_q^* according to Equation (78) lies in the limits:

$$0.4459 \;=\; \lambda_2\left(X^0\right) \le \lambda_2 \;=\; 0.8697 \le \lambda_2(X_2^*) \;=\; 1, q \in K.$$

Step 6. Calculation of the coefficient of linear approximation.

Assuming a linear nature of the change of the criterion $f_q(X)$ in Equation (80) and according to a relative assessment of $\lambda_q(X)$, using standard methods of linear approximation we calculate the proportionality coefficient between $\lambda_q(X^0)$, λ_q, which we call ρ:

$$\rho = \frac{\lambda_q - \lambda_q(X^o)}{\lambda_q(X_q^*) - \lambda_q(X^o)} = \frac{0.8697 - 0.4459}{1 - 0.4459} = 0.7649, q = 2. \quad (82)$$

Step 7. Calculation of coordinates of the priority criterion with the size f_q.

Assuming a linear nature of the change of a vector $X^q = \{x_1\ x_2\}$, $q = 2$ we determine coordinates of a point of priority criterion with the size f_q =1600 with a relative assessment (Equation (81)):

$X^q = \{x_1 = X^0(1) + \rho(X_q^*(1) - X^0(1))\ x_2 = X^0\ (2) + \rho(X_q^*(2) - X^0(2))\}$,

where $X^0 = \{x_1 = 33.02, x_2 = 69.54\}$, $X_2^* = \{x_1 = 25, x_2 = 25\}$.

As a result of these calculations we obtain the point coordinates:

$$X^q = \{x_1 = 26.88, x_2 = 69.54\}. \quad (83)$$

Step 8. Calculation of the main indicators of a point of X^q.

For the obtained X^q point, we calculate:

- all criteria in physical units $f_k(X^q) = \{f_k(X^q), k = \overline{1,K}\}$: $f(X^q) = \{f_1(X^q) = 386.5, f_2(X^q) = 1651.5, f_3(X^q) = 137.9, f_4(X^q) = 26.1\}$,
- all relative estimates of criteria $\lambda^q = \{\lambda_k^q,\ k = \overline{1,K}\}$, $\lambda_k(X^q) = \frac{f_k(X^q) - f_k^o}{f_k^* - f_k^o}$, $k = \overline{1,K}_{\overline{1,K}}$: $\lambda_k(x^q) = \{\lambda_1(x^q) = 0.5392, \lambda_2(x^q) = 0.8530, \lambda_3(x^q) = 0.0827, \lambda_4(x^q) = 0.9334\}$,
- vector of priorities $P^q = \{p_k^q = \frac{\lambda_q(X^q)}{\lambda_k(X^q)}, k = \overline{1,K}\}$: $P^q = [p_1^2 = 1.5820, p_2^2 = 1.0, p_3^2 = 10.3123, p_4^2 = 0.9139]$,
- the minimum relative assessment: min LXq = min (LXq): min LXq = min $(\lambda_k(X^q)) = 0.0827$,
- the relative assessment taking into account a criterion priority: $\lambda^{00} = \min\ (p_1^2\lambda_1(X^q) = 0.7564, p_2^2\lambda_2(X^q) = 0.7564, p_3^2\lambda_3(X^q) = 0.7564, p_4^2\lambda_4(X^q)) = 0.7564$.

Any point from Pareto's set $X_t^o = \{\lambda_t^o, X_t^o\} \in S^o$ can be similarly calculated.

7.5. Analysis of the Results of the Final Decision

The calculated size of criterion $f_q(X_t^o)$, $q \in \mathbf{K}$ is usually not equal to the set f_q. The error of the choice of $\Delta f_q = |f_q(X_t^o) - f_q| = |1651.5 - 1600| = 51.5$ is defined by an error of linear approximation, $\Delta f_{q\%} = 3.2\%$.

In the course of modeling, parametrical restrictions of Equation (73) can be changed, i.e., some set of optimum decisions is obtained. We can choose a final version, which in our example includes this set of optimum decisions:

- parameters of the technical system $X^0 = \{x_1 = 33.03, x_2 = 69.54, x_3 = 25.0\}$,
- the parameters of the technical system at a given priority criterion $q = 2$: $X^q = \{x_1 = 26.88, x_2 = 35.47, x_3 = 25.0\}$.

We represent these parameters in two-dimensional (x_1, x_2) and three-dimensional $(x_1, x_2$and $\lambda)$ coordinate systems in Figures 6–8, and also in physical units for each function $f_1(X), \ldots, f_4(X)$ in Figures 9–12, respectively.

The first characteristic $f_1(X)$ in physical units is shown in Figure 9.

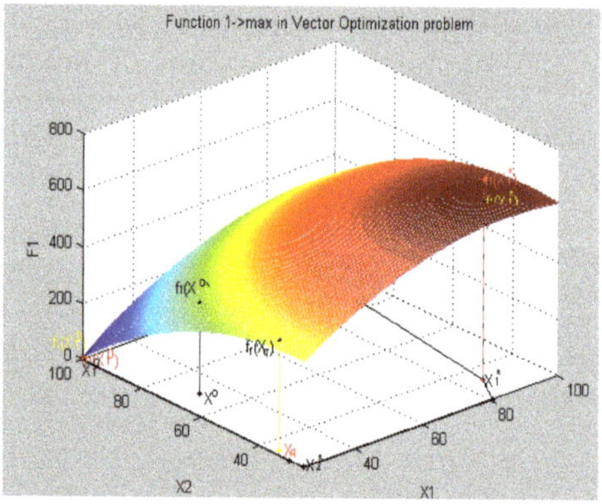

Figure 9. The first characteristic $f_1(X)$ of the technical system in a natural indicator.

At point $X^{\mathbf{0}}$, X^q of the second characteristic $f_2(X)$ will appear as presented in Figure 10.

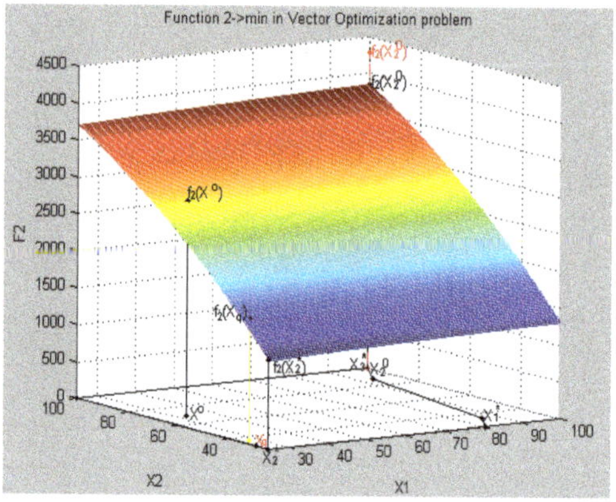

Figure 10. The second characteristic $f_2(X)$ of the technical system in a natural indicator.

At point $X^{\mathbf{0}}$, X^q of the third characteristic $f_3(X)$ will appear as presented in Figure 11.

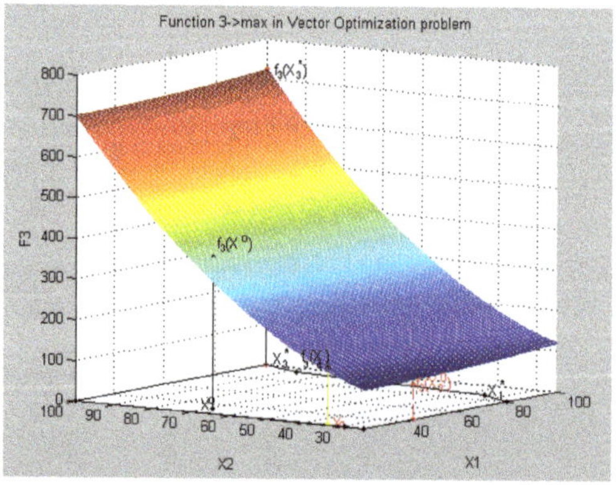

Figure 11. The third characteristic $f_3(X)$ of the technical system in a natural indicator.

At point X^0, X^q of the fourth characteristic $f_4(X)$ will appear as presented in Figure 12.

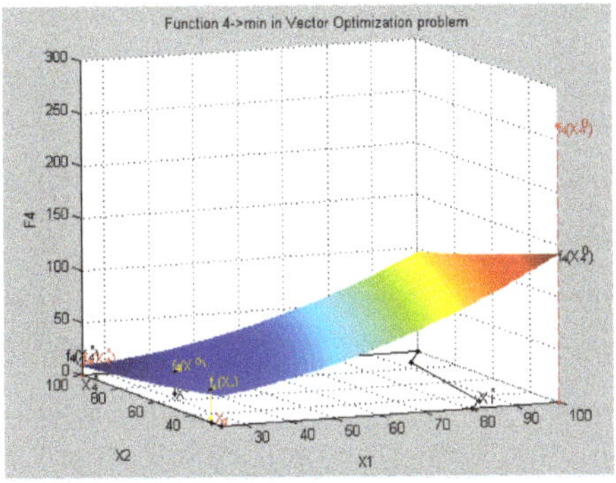

Figure 12. The fourth characteristic $f_4(X)$ of the technical system in a natural indicator.

Collectively, for the submitted version, at the point X^0 there exist characteristics of $f_1(X^0)$, $f_2(X^0)$, $f_3(X^0)$, $f_4(X^0)$, relative estimates of $\lambda_1(X^0)$, $\lambda_2(X^0)$, $\lambda_3(X^0)$, $\lambda_4(X^0)$, and maximum λ^o relative level $\lambda^0 \leq \lambda_k(X^0)$ "$k \in \mathbf{K}$ such that there is an optimal solution with equivalent criteria (characteristics), and the procedure for obtaining acceptance of the optimal solution with equivalent criteria (characteristics).

At point X^q there exist: characteristics of $f_1(X^q)$, $f_2(X^q)$, $f_3(X^q)$, $f_4(X^q)$, relative estimates of $\lambda_1(X^q)$, $\lambda_2(X^q)$, $\lambda_3(X^q)$, $\lambda_4(X^q)$, maximum λ^0 relative level $\lambda^0 \leq \lambda_k(X^q)$ "$k \in \mathbf{K}$ such that there is an optimal solution at the set priority of the second criterion (characteristic) in relation to other criteria. The procedure of obtaining a point X^q is the adoption of the optimal solution at the set priority of the second criterion.

Based on the theory of vector optimization, methods of solution of vector problems with equivalent criteria and a given priority of criterion allow the choice of any point from the set of Pareto optimal points and demonstration of the optimality of this point.

Conclusions. The problem of adoption of the optimum decision in a difficult technical system based on some set of functional characteristics is one of the most important problems of system analysis and design.

8. The Methodology of Making Optimal Decisions with the Functional and Experimental Data (For Example, a Problem with Four Parameters)

In the studied object, a system is known. Data on the functional characteristics, discrete values of separate characteristics, and data on restrictions that are imposed on the functioning of a system. The process of model operation of such a system is presented in the methodology form: "The methodology of making the optimal decision based on the functional and experimental data".

The methodology includes a number of stages.

- Formation of the requirement specification (source data) for numerical modeling and choice of optimum parameters of a system. The initial data is determined by the designer who operates the system.
- Creation of a mathematical and numerical model of the system under the conditions of definiteness and indeterminacy.
- The solution of the vector problem of mathematical programming (VPMP), i.e., a model of the system with equivalent criteria.
- Geometric interpretation of the results of the decision in a three-dimensional coordinate system in relative units.
- The solution of a vector problem of mathematical programming, i.e., a model of the system at the given priority of the criterion.
- Geometric interpretation of the results of the decision in a three-dimensional coordinate system in physical units.

8.1. Formation of Technical Specifications (Source Data) for the Numerical Simulation of the System

We will consider a problem "Numerical modeling of the system" in which data on some set of functional characteristics (definiteness conditions), discrete values of characteristics (an uncertainty condition) and the restrictions imposed on the functioning of the technical system are known [6–10,13,15,20,22].

It is given that the system function is defined by four parameters $X = \{x_1, x_2, x_3, x_4\}$, a vector of (operated) variables. Basic data for the solution of the problem are the four characteristics (criterion) of: $F(X) = \{f_1(X), f_2(X), f_3(X), f_4(X)\}$, whose size of assessment depends on a vector of X.

The definiteness condition. For the first and third characteristics of $f_1(X)$ and $f_3(X)$ functional dependence on parameters X is known (indexing of formulas within the individual section (methods)):

$$\begin{aligned} f_1(X) \equiv\ & 269.867 - 1.8746 \times x_1 - 1.7469 \times x_2 + 0.8939 \times x_3 \\ & +1.0937 \times x_4 + 0.0484 \times x_1 \times x_2 - 0.0052 \times x_1 \times x_3 - \\ & 0.0141 \times x_1 \times x_4 + 0.0037 \times x_2 \times x_3 - 0.0052 \times x_2 \times x_4 - \\ & 0.0002 \times x_3 \times x_4 + 0.0119 \times x_1^2 + 0.0035 \times x_2^2 - 0.002 \times x_3^2 \\ & - 0.0042 \times x_4^2, \end{aligned} \quad (84)$$

$$\begin{aligned} f_4(X) =\ & 19.253 - 0.0081 \times x_1 - 0.7005 \times x_2 - 0.3605 \times x_3 \\ & + 0.9769 \times x_4 + 0.0126 \times x_1 \times x_2 + 0.0644 \times x_1 \times x_3 - 0 \times x_1 \times x_4 \\ & + 0.0396 \times x_2 \times x_3 + 0.0002 \times x_2 \times x_4 + 0.0004 \times x_3 \times x_4 - \\ & 0.0016 \times x_1^2 + 0.0027 \times x_2^2 + 0.0045 \times x_3^2 - 0.0235 \times x_4^2, \end{aligned} \quad (85)$$

$$\text{restrictions: } 22 \le x_1 \le 88,\ 0 \le x_2 \le 66,\ 2.2 \le x_3 \le 8.8,\ 2.2 \le x_4 \le 8.8 \quad (86)$$

The uncertainty condition. For the second and fourth characteristic results of the experimental data, the sizes of parameters and corresponding characteristics are known. Numerical values of parameters X and characteristics of $y_2(X)$ and $y_4(X)$ are presented in Table 7.

Table 7. Numerical values of parameters and characteristics of the system.

x_1	x_2	x_3	x_4	$y_2(X) \rightarrow min$	$y_3(X) \rightarrow max$
22	0	2.2	2.2	1053.8	47.7
22	0	2.2	5.5	1067	47.3
22	0	2.2	8.8	1078	47.2
22	0	5.5	2.2	1111	50.7
22	0	5.5	5.5	1155	46.8
22	0	5.5	8.8	1152.8	46.3
22	0	8.8	2.2	1151.7	44.2
22	0	8.8	5.5	1148.4	43
22	0	8.8	8.8	1147.3	42.5
22	33	2.2	2.2	1964.6	58.3
22	33	2.2	5.5	1974.5	57.5
22	33	2.2	8.8	1983.3	57.1
22	33	5.5	2.2	1995.4	56.5
22	33	5.5	5.5	2003.1	55.1
22	33	5.5	8.8	2015.2	54.9
22	33	8.8	2.2	2027.3	54.8
22	33	8.8	5.5	2046	52.8
22	33	8.8	8.8	2058.1	53
22	66	2.2	2.2	2708.2	75.9
22	66	2.2	5.5	2585	71.5
22	66	2.2	8.8	2541	68.2
22	66	5.5	2.2	2519	66.4
22	66	5.5	5.5	2596	68.2
22	66	5.5	8.8	2662	70.4

Table 7. *Cont.*

x_1	x_2	x_3	x_4	$y_2(X) \rightarrow min$	$y_3(X) \rightarrow max$
22	66	8.8	2.2	2770.9	72.4
22	66	8.8	5.5	2783	71.5
22	66	8.8	8.8	2801.7	70.6
55	0	2.2	2.2	3284.6	100.5
55	0	2.2	5.5	3301.1	100.1
55	0	2.2	8.8	3307.7	99
55	0	5.5	2.2	3315.4	98.8
55	0	5.5	5.5	3320.9	97.9
55	0	5.5	8.8	3334.1	97.6
55	0	8.8	2.2	3347.3	97
55	0	8.8	5.5	3366	95.7
55	0	8.8	8.8	3378.1	95.3
55	33	2.2	2.2	1095.6	54.6
55	33	2.2	5.5	1111	50.6
55	33	2.2	8.8	1133	48.4
55	33	5.5	2.2	1147.3	47.7
55	33	5.5	5.5	1166	46.2
55	33	5.5	8.8	1188	45.1
55	33	8.8	2.2	1208.9	44.2
55	33	8.8	5.5	1232	42.2
55	33	8.8	8.8	1272.7	40.7
55	66	2.2	2.2	1995.4	61.8
55	66	2.2	5.5	2013	60.5
55	66	2.2	8.8	2035	59.4
55	66	5.5	2.2	2058.1	58.3
55	66	5.5	5.5	2095.5	57.2
55	66	5.5	8.8	2103.2	56.1
55	66	8.8	2.2	2120.8	54.8
55	66	8.8	5.5	2145	47.3
55	66	8.8	8.8	2183.5	51.3
88	0	2.2	2.2	2739	79.4
88	0	2.2	5.5	2761	78.1
88	0	2.2	8.8	2783	77
88	0	5.5	2.2	2801.7	75.9
88	0	5.5	5.5	2849	76.1
88	0	5.5	8.8	2893	76.6
88	0	8.8	2.2	2974.4	76.8
88	0	8.8	5.5	2959	715
88	0	8.8	8.8	2927.1	682
88	33	2.2	2.2	3315.4	1041
88	33	2.2	5.5	3336.3	1023
88	33	2.2	8.8	3355	1012
88	33	5.5	2.2	3378.1	1005
88	33	5.5	5.5	3399	990
88	33	5.5	8.8	3421	979
88	33	8.8	2.2	3440.8	970
88	33	8.8	5.5	3366	957
88	33	8.8	8.8	3503.5	935
88	66	2.2	2.2	1116.5	583
88	66	2.2	5.5	1144	561
88	66	2.2	8.8	1166	550
88	66	5.5	2.2	1208.9	530
88	66	5.5	5.5	1232	506
88	66	5.5	8.8	1276	484
88	66	8.8	2.2	1303.5	477
88	66	8.8	5.5	1342	440
88	66	8.8	8.8	1397	425

In the decision, from the assessment size of the first and third characteristic (criterion), it is possible to obtain: $f_1(X)\rightarrow\max f_3(X)\rightarrow\max$, and for the second and fourth characteristic: $y_2(X)\rightarrow\min$ $y_4(X)\rightarrow\min$. Parameters $X = \{x_1, x_2, x_3, x_4\}$ change according to the following limits:

$$x_1 \in [22.\ 55.\ 88.], x_2 \in [0.\ 33.\ 66.], x_3 \in [2.2\ 5.5\ 8.8], x_4 \in [2.2\ 5.5\ 8.8].$$

The requirements are as follows: to construct a model of the system in the form of a vector problem, to solve a vector problem with equivalent criteria, to choose a priority criterion, to establish the numerical value of the priority criterion, and to make the best decision (optimum) with a specified priority criterion.

Note that using the MATLAB system, the author developed the software for the decision making of a vector problem of mathematical programming. The vector problem includes four variables (parameters of the technical system): $X = \{x_1, x_2, x_3, x_4\}$ and four criteria (characteristics) of $F(X) = \{f_1(X), f_2(X), f_3(X), f_4(X)\}$. However, for each new set of data (new system) the program is configured individually. In the software criteria $F(X) = \{f_1(X), f_2(X), \ldots f_6(X)\}$ with uncertainty conditions (provided as a part of $y_2(X)$, $y_4(X)$ in Table 6) can vary between zero (i.e., all criteria are constructed under the conditions of determinacy) and six (i.e., all criteria are constructed under the conditions of uncertainty).

8.2. *Creation of a Mathematical and Numerical Model of the System under the Conditions of Definiteness and Indeterminacy*

Creating a numerical model of the system includes the following sections:

- Choosing a mathematical model of the system,
- Building a model under certainty conditions,
- Construction under the conditions of uncertainty,
- Construction of a numerical model of the system under certainty and uncertainty.

8.2.1. Mathematical Model of the System

We will present the model of the system under the conditions of definiteness and uncertainty in total:

$$Opt\ F(X) = \{\mathbf{max} F_1(X) = \{\mathbf{max} f_k(X), k = \overline{1, K_1^{def}}\}, \quad (87)$$

$$\mathbf{max} I_1(X) \equiv \{\mathbf{max}\{f_k(X_i, i = \overline{1, M})\}^T, k = \overline{1, K_1^{unc}}\}, \quad (88)$$

$$\mathbf{min} F_2(X) = \{\mathbf{min} f_k(X), k = \overline{1, K_2^{def}}\}, \quad (89)$$

$$\mathbf{min} I_2(X) o \{\mathbf{min}\{f_k(X_i, i = \overline{1, M})\}^T, k = \overline{1, K_2^{unc}}\}\}, \quad (90)$$

$$\text{at restrictions } f_k^{\min} \le f_k(X) \le f_k^{\max}, k = \overline{1, K}, x_j^{\min} \le x_j \le x_j^{\max}, j = \overline{1, N} \quad (91)$$

where $X = \{x_j, j = \overline{1, N}\}$ is a vector of operated variable (design data), $F(X) = \{F_1(X)\ F_2(X)\ I_1(X), I_2(X)\}$ represents the vector criterion of Equations (87)–(91) in which each component represents a vector of criteria (characteristics) of the system that functionally depend on the discrete values of a vector of variables X, $F_1(X) = \{f_k(X), k = \overline{1, K_1^{def}}\}$, $F_2(X) = \{f_k(X), k = \overline{1, K_2^{def}}\}$ is a set of the max and min functions, respectively, $I_1(X) = \{\{f_k(X_i, i = \overline{1, M})\}^T, k = \overline{1, K_1^{unc}}\}$, $I_2(X) = \{\{f_k(X_i, i = \overline{1, M})\}^T, k = \overline{1, K_2^{unc}}\}$ is a set of matrices of max and min, respectively, K_1^{def}, K_2^{def} (*definiteness*), K_1^{unc}, K_2^{unc} (*uncertainty*) are the sets of criteria of *max* and *min* created under the conditions of definiteness and uncertainty. In Equation (91), $f_k^{\min} \le f_k(X) \le f_k^{\max}$, $k = \overline{1, K}$ is a vector function of the restrictions imposed on the functioning of the technical system, and $x_j^{\min} \le x_j \le x_j^{\max}$, $j = \overline{1, N}$ represent the parametrical restrictions.

It is assumed that the functions $f_k(X)$, $k = \overline{1,K}$ are differentiable and convex, $g_i(X)$, $i = \overline{1,M}$ are continuous, and the set of admissible points S given by constraints of Equation (5) is non–empty and is a compact: $S = \{X \in R^N | G(X) \leq 0, X^{\min} \leq X \leq X^{\max}\} \neq \emptyset$.

8.2.2. Building a Model under the Conditions of Certainty

Construction under the conditions of definiteness is defined by the functional dependence of each characteristic and the restrictions on the parameters of the technical system. In our example, three characteristic (92) and (93) and restrictions (94) are known:

$$\begin{aligned} f_1(X) \equiv\ & 269.867 - 1.8746 \times x_1 - 1.7469 \times x_2 + 0.8939 \times x_3 + \\ & 1.0937 \times x_4 + 0.0484 \times x_1 \times x_2 - 0.0052 \times x_1 \times x_3 - 0.0141 \times x_1 \times x_4 \\ & + 0.0037 \times x_2 \times x_3 - 0.0052 \times x_2 \times x_4 - 0.0002 \times x_3 \times x_4 + 0.0119 \times x_1^2 \\ & + 0.0035 \times x_2^2 - 0.002 \times x_3^2 - 0.0042 \times x_4^2, \end{aligned} \tag{92}$$

$$\begin{aligned} f_4(X) =\ & 19.253 - 0.0081 \times x_1 - 0.7005 \times x_2 - 0.3605 \times x_3 + \\ & 0.9769 \times x_4 + 0.0126 \times x_1 \times x_2 + 0.0644 \times x_1 \times x_3 \\ & - 0 \times x_1 \times x_4 + 0.0396 \times x_2 \times x_3 + 0.0002 \times x_2 \times x_4 + \\ & 0.0004 \times x_3 \times x_4 - 0.0016 \times x_1^2 + 0.0027 \times x_2^2 + 0.0045 \times x_3^2 \\ & - 0.0235 \times x_4^2, \end{aligned} \tag{93}$$

$$\text{restrictions: } 22 \leq x_1 \leq 88,\ 0 \leq x_2 \leq 66,\ 2.2 \leq x_3 \leq 8.8,\ 2.2 \leq x_4 \leq 8.8 \tag{94}$$

These data are further used in the creation of the mathematical model of the technical system.

8.2.3. Construction under the Conditions of Uncertainty

Construction under the conditions of uncertainty involves the use of the qualitative and quantitative descriptions of the technical system obtained by the "input–output" principle shown in Table 6. Transformation of information (basic data of $y_2(X)$, $y_3(X)$) to a functional type of $f_2(X)$, $f_3(X)$ is carried out by the use of mathematical methods (i.e., regression analysis).

The basic data of Table 1 are created in the MATLAB system in the form of a matrix:

$$I = |X, Y| = \{x_{i1} x_{i2} y_{i3} y_{i4}, i = \overline{1, M}\}. \tag{95}$$

For each experimental set function y_k, $k = 2, 3$ regression was performed using the method of least squares $\min \sum_{i=1}^{M} (y_i - \overline{y_i})^2$ in MATLAB. A_k, a polynomial defining the interrelationship of the parameters $X_i = \{x_{1i}, x_{2i}, x_{3i}, x_{4i}\}$ and functions $\overline{y}_{ki} = f(X_i, A_k)$, $k = 2, 3$, is formed for this purpose.

As a result of the calculations, we obtain the system of coefficients $A_k = \{A_{0k}, A_{1k}, \ldots, A_{14k}\}$ which define the coefficients of the quadratic polynomial (function):

$$\begin{aligned} f_k(X, A) =\ & A_{0k} + A_{1k}x_1 + A_{2k}x_2 + A_{3k}x_3 + A_{4k}x_4 + A_{5k}x_1 * x_2 + A_{6k}x_1 * x_3 + A_{7k}x_1 * x_4 + A_{8k}x_2 * x_3 \\ & + A_{9k}x_2 * x_4 + A_{10k}x_3 * x_4 + A_{11k}x_1^2 + A_{12k}x_2^2 + A_{13k}x_3^2 + A_{14k}x_4^2,\ k = 2, 3. \end{aligned} \tag{96}$$

As a result of the calculations of the coefficients A_k, $k = 2$, we obtain the $f_2(X)$ function:

$$\begin{aligned} f_2(X) =\ & 875.3 + 23.893 \times x_1 - 30.866 \times x_2 - 25.858 \times x_3 - 45 \times x_4 \\ & - 0.6984 \times x_1 \times x_2 + 0.4276 \times x_1 \times x_3 + 0.6793 \times x_1 \times x_4 \\ & - 0.1167 \times x_2 \times x_3 + 0.2969 \times x_2 \times x_4 - 0.0093 \times x_3 \times x_4 \\ & + 0.0362 \times x_1^2 + 0.0331 \times x_2^2 + 2.9158 \times x_3^2 + 2.4052 \times x_4^2 \end{aligned} \tag{97}$$

As a result of the calculations of coefficients A_k, k =3, we obtain the $f_3(X)$ function:

$$\begin{aligned} f_3(X) &= 43.734 + 0.6598 \times x_1 + 0.4493 \times x_2 - 0.3094 \times x_3 - \\ &1.8334 \times x_4 - 0.01 \times x_1 \times x_2 - 0.0062 \times x_1 \times x_3 + 0.0146 \times x_1 \times x_4 \\ &-0.013 \times x_2 \times x_3 + 0.0121 \times x_2 \times x_4 - 0.0004 \times x_3 \times x_4 - 0.0003 \times x_1^2 \\ &-0.0002 \times x_2^2 + 00.0254 \times x_3^2 + 0.0939 \times x_4^2 \end{aligned} \quad (98)$$

$$\text{restrictions } 22 \le x_1 \le 88, 0 \le x_2 \le 66, 2.2 \le x_3 \le 8.8, 2.2 \le x_4 \le 8.8 \quad (99)$$

The minimum and maximum values of experimental data $y_1(X)$, $y_2(X)$, $y_4(X)$ are presented in the lower part of Table 1. The minimum and maximum values of the functions $f_1(X)$, $f_2(X)$, $f_4(X)$ slightly differ from experimental data. For comparison, the settlement of these $f_4(X)$ functions at the specified points of X presented in the right part of the eighth column of Table 7 are given. The index of correlation and coefficients of determination are presented in the lower lines of Table 7.

Results of the regression analysis of Equations (97)–(99) are further used in the creation of the mathematical model of the technical system.

8.2.4. Construction of a Numerical Model of the System under Certainty and Uncertainty

For the creation of a numerical model of the system, we use the functions obtained under conditions of definiteness (Equations (92) and (93)) and uncertainty (Equations (97) and (98)), and parametric restrictions (Equations (94) and (99)).

We consider the functions of Equations (92), (93), (97) and (98) as the criteria defining the functioning of the system. A set of criteria $\boldsymbol{K}$ = 4 includes two criteria $f_1(X), f_3(X) \rightarrow$ max and two $f_2(X), f_4(X) \rightarrow$ min. As a result, the model of the functioning of the system is presented as a vector problem of mathematical programming:

$$\begin{aligned} opt\ F(X) &= \{max\ F_1(X) = \{max f_1(X) \equiv 269.867 - 1.8746 \times x_1 \\ &-1.7469 \times x_2 + 0.8939 \times x_3 + 1.0937 \times x_4 + 0.0484 \times x_1 \times x_2 - \\ &0.0052 \times x_1 \times x_3 - 0.0141 \times x_1 \times x_4 + 0.0037 \times x_2 \times x_3 - 0.0052 \times x_2 \times x_4 \\ &- 0.0002 \times x_3 \times x_4 + 0.0119 \times x_1^2 + 0.0035 \times x_2^2 - 0.002 \times x_3^2 - 0.0042 \times x_4^2, \end{aligned} \quad (100)$$

$$\begin{aligned} &max\ f_3(X) \equiv 43.734 + 0.659 \times x_1 + 0.4493 \times x_2 - 0.3094 \times x_3 \\ &-1.8334 \times x_4 - 0.01 \times x_1 \times x_2 - 0.0062 \times x_1 \times x_3 + 0.0146 \times x_1 \times x_4 \\ &-0.013 \times x_2 \times x_3 + 0.0121 \times x_2 \times x_4 - 0.0004 \times x_3 \times x_4 - 0.0003 \times x_1^2 \\ &-0.0002 \times x_2^2 + 0.0254 \times x_3^2 + 0.0939 \times x_4^2 \end{aligned} \quad (101)$$

$$\begin{aligned} min\ F_2(X) &= \{min\ f_2(X) \equiv 875.3 + 23.893 \times x_1 - 30.866 \times x_2 - \\ &25.858 \times x_3 - 45 \times x_4 - 0.6984 \times x_1 \times x_2 + 0.4276 \times x_1 \times x_3 + \\ &0.6793 \times x_1 \times x_4 - 0.1167 \times x_2 \times x_3 + 0.2969 \times x_2 \times x_4 - \\ &0.0093 \times x_3 \times x_4 + 0.0362 \times x_1^2 + 0.0331 \times x_2^2 + 2.9158 \times x_3^2 + 2.4052 \times x_4^2. \end{aligned} \quad (102)$$

$$\begin{aligned} &min\ f_4(X) \equiv 19.25 - 0.008 \times x_1 - 0.7005 \times x_2 - 0.3605 \times x_3 + \\ &0.977 \times x_4 + 0.0126 \times x_1 \times x_2 + 0.0644 \times x_1 \times x_3 - 0 \times x_1 \times x_4 + \\ &0.0396 \times x_2 \times x_3 + 0.0002 \times x_2 \times x_4 + 0.0004 \times x_3 \times x_4 - 0.0016 \times x_1^2 \\ &+0.0027 \times x_2^2 + 0.0045 \times x_3^2 - 0.0235 \times x_4^2\}\}, \end{aligned} \quad (103)$$

$$\text{restrictions: } 22 \le x_1 \le 88, 0 \le x_2 \le 66, 2.2 \le x_3 \le 8.8, 2.2 \le x_4 \le 8.8 \quad (104)$$

The vector problem of mathematical programming of Equations (100)–(104) represents the model of decision making under certainty and uncertainty in the aggregate.

8.3. The Solution of the Vector Problem of Mathematical Programming (VPMP)—Model of the System with Equivalent Criteria

To solve the vector problem of mathematical programming of Equations (14)–(18), methods based on the axioms of the normalization of criteria and the principle of guaranteed results are presented, which follow from Axiom 1 and the principle of optimality 1.

The solution of the vector problem of Equations (14)–(18) follows a sequence of steps.

Step 1. Equations (100)–(104) are solved for each criterion separately, using the function *fmincon* (...) of the MATLAB system, the use of the function *fmincon* (...) is considered in [7–10,20,22].

As a result of the calculation for each criterion we obtain optimum points: X_k^* and $f_k^* = f_k(X_k^*)$, $k = \overline{1,K}$, the sizes of criteria at this point, i.e., the best decision for each criterion:

$$X_1^* = \{x_1 = 88.0, x_2 = 66.0, x_3 = 8.8, x_4 = 2.2\}, f_1^* = f_1(X_1^*) = -535.06,$$
$$X_2^* = \{x_1 = 22.0, x_2 = 0.0, x_3 = 2.83, x_4 = 6.25\}, f_2^* = f_2(X_2^*) = 1301.2,$$
$$X_3^* = \{x_1 = 88.0, x_2 = 0.0, x_3 = 2.2, x_4 = 8.8\}, f_3^* = f_3(X_3^*) = -100.15,$$
$$X_4^* = \{x_1 = 22.0, x_2 = 62.17, x_3 = 2.2, x_4 = 2.2\}, f_4^* = f_4(X_4^*) = 12.247.$$

The restrictions of Equation (104) and the points of an optimum X_1^*, X_2^*, X_3^*, X_4^* in coordinates $\{x_1, x_2\}$ are presented in Figure 13.

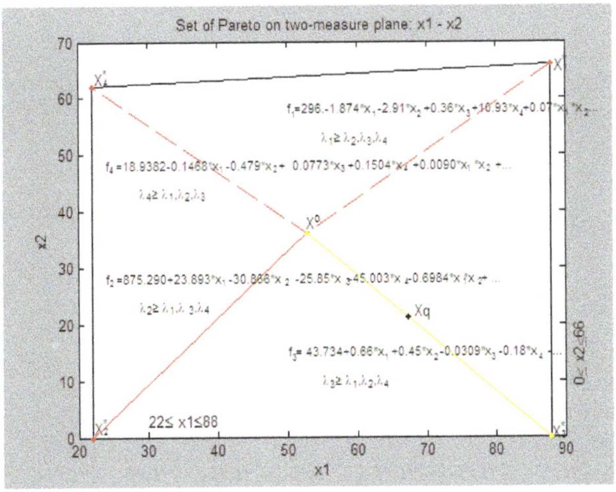

Figure 13. Pareto's great number, $S^o \subset S$ in a two−dimensional system of coordinates $\{x_1, x_2\}$.

Step 2. We define the worst unchangeable part of each criterion (anti-optimum):

$$
\begin{aligned}
X_1^0 &= \{x_1 = 22.0, x_2 = 66.0, x_3 = 2.2, x_4 = 2.2\}, f_1^0 = f_1(X_2^0) = 243.25,\\
X_2^0 &= \{x_1 = 88.0, x_2 = 0.0, x_3 = 8.8, x_4 = 8.8\}, f_2^0 = f_2(X_2^0) = -3903.1,\\
X_3^0 &= \{x_1 = 22.0, x_2 = 0.0, x_3 = 8.8, x_4 = 8.07\}, f_3^0 = f_3(X_3^0) = 50.03,\\
X_4^0 &= \{x_1 = 88.0, x_2 = 66.0, x_3 = 8.8, x_4 = 8.8\}, f_4^0 = f_2(X_4^0) = -121.83.
\end{aligned}
$$

Step 3. We analyze the set of Pareto optimal points. At optimal points $X^* = \{X_1^*, X_2^*, X_3^*, X_4^*\}$, the sizes of the criterion functions of $F(X^*) = \|f_q(X_k^*)\|_{q=\overline{1,K}}^{k=\overline{1,K}}$ are determined. We calculate a vector

$D = (d_1\ d_2\ d_3\ d_4)^T$, deviations by each criterion on an admissible set S: $d_k = f_k^* - f_k^0$, $k = \overline{1,4}$, and a matrix of relative estimates $\lambda(X^*) = \|\lambda_q(X_k^*)\|_{q=\overline{1,K}}^{k=\overline{1,K}}$, where $\lambda_k(X) = (f_k^* - f_k^0)/d_k$.

$$F(X^*) = \begin{vmatrix} 535.1 & 1731.9 & 58.1 & 117.0 \\ 317.6 & 1301.2 & 51.3 & 26.5 \\ 192.5 & 3614.3 & 100.2 & 24.6 \\ 244.0 & 2458.2 & 67.7 & 12.2 \end{vmatrix},\ D = \begin{vmatrix} 291.8 \\ -2602.0 \\ 50.12 \\ -109.58 \end{vmatrix},$$
$$\lambda(X^*) = \begin{vmatrix} 1.0000 & 0.8345 & 0.1603 & 0.0443 \\ 0.2548 & 1.0000 & 0.0244 & 0.8697 \\ -0.1740 & 0.1110 & 1.0000 & 0.8870 \\ 0.0027 & 0.5553 & 0.3532 & 1.0000 \end{vmatrix} \quad (105)$$

The analysis of the sizes of the criteria in relative estimates shows that at optimal points $X^* = \{X_1^*, X_2^*, X_3^*, X_4^*\}$ the relative assessment is equal to unity. Other criteria are much less than unity. It is required to find such points (parameters) at which the relative estimates are closest to unity. Step 4 is directed to the solution of this problem.

Step 4. Creation of the λ-problem is carried out in two stages: first, the maximine problem of optimization with the normalized criteria is constructed:

$$\lambda^0 = \max_x \min_k \lambda_k(X), G(X) \leq 0, X \geq 0, \quad (106)$$

Second, this is transformed into a standard problem of mathematical programming (the λ-problem):

$$\lambda^0 = \max \lambda, \quad (107)$$

$$\text{at restrictions } \lambda - (f_1(X) - f_1^o)/(f_1^* - f_1^o) \leq 0, \quad (108)$$

$$\lambda - (f_2(X) - f_2^o)/(f_2^* - f_2^o) \leq 0 \quad (109)$$

$$\lambda - (f_3(X) - f_3^o)/(f_3^* - f_3^o) \leq 0 \quad (110)$$

$$\lambda - (f_4(X) - f_4^o)/(f_4^* - f_4^o) \leq 0 \quad (111)$$

$$0 \leq \lambda \leq 1, 22 \leq x_1 \leq 88, 0 \leq x_2 \leq 66, 2.2 \leq x_3 \leq 8.8, 2.2 \leq x_4 \leq 8.8, \quad (112)$$

where the vector of unknowns had the dimension $N + 1$: $\mathbf{X} = \{x_1, \ldots, x_N, \lambda\}$, the functions $f_1(X), f_2(X), f_3(X), f_4(X)$ correspond to Equations (100)–(104), respectively. Substituting the numerical values of the functions $f_1(X), f_2(X), f_3(X), f_4(X)$, we obtain the λ-problem in the following form:

$$\lambda^0 = \max \lambda, \quad (113)$$

$$\text{at restrictions } \lambda - \frac{296.85 - 1.875 \times x_1 \ldots + 0.0734 \times x_1 \times x_2 \ldots - 0.0108 \times x_1^2 \ldots - f_1^o}{f_1^* - f_1^o} \leq 0, \quad (114)$$

$$\lambda - \frac{43.73 + 0.659 \times x_1 \ldots - 0.01 \times x_1 \times x_2 \ldots - 0.0003 \times x_1^2 \ldots - f_3^o}{f_3^* - f_3^o} \leq 0, \quad (115)$$

$$\lambda - \frac{875.3 + 23.893 \times x_1 + \ldots . -0.6984 \times x_1 \times x_2 \ldots + 0.036 \times x_1^2 \ldots - f_2^o}{f_2^* - f_2^o} \leq 0, \quad (116)$$

$$\lambda - \frac{19.253 - 0.0081 \times x_1 \ldots + 0.0126 \times x_1 \times x_2 \ldots + (-0.0016 \times x_1^2) \ldots - f_4^o}{f_4^* - f_4^o} \leq 0, \quad (117)$$

$$0 \leq \lambda \leq 1, 22 \leq x_1 \leq 88, 0 \leq x_2 \leq 66, 2.2 \leq x_3 \leq 8.8, 2.2 \leq x_4 \leq 8.8, \quad (118)$$

Using function fmincon(...):

[Xo,Lo] = fmincon('Z_TehnSist_4Krit_L',X0,Ao,bo,Aeq,beq,lbo,ubo,'Z_TehnSist_LConst',options).

As a result of the solution of the vector problem of mathematical programming in Equations (14)–(18) with equivalent criteria and the λ-problem corresponding to Equations (113)–(118), we obtain:

$$\mathbf{X}^0 = \{X^0, \lambda^0\} = \{X^0 = \{x_1 = 52.9, x_2 = 36.097, x_3 = 8.8, x_4 = 2.2, \lambda^0 = 0.3179\} \quad (119)$$

i.e., the optimum point of the design data of the system, point $\mathbf{X}^0$, which is presented in Figure 8,

$f_k(X^0)$, $k = \overline{1,K}$, the sizes of criteria (characteristics of technical system):

$$\{f_1(X^0) = 336.0, f_2(X^0) = 2239.5, f_3(X^0) = 65.962, f_4(X^0) = 58.435\}, \quad (120)$$

And $\lambda_k(X^o)$, $k = \overline{1,K}$, the sizes of the relative estimates:

$$\{\lambda_1(X^0) = 0.3179, \lambda_2(X^0) = 0.6394, \lambda_3(X^0) = 0.3179, \lambda_4(X^0) = 0.5785\}, \quad (121)$$

$\lambda^0 = 0.3179$ is the maximum lower level among all relative estimates measured in relative units: $\lambda^0 = min\ (\lambda_1(X^0), \lambda_2(X^0), \lambda_3(X^0), \lambda_4(X^0), \lambda_5(X^0)) = 0.3179$. A relative assessment, λ^0, is called the guaranteed result in relative units, i.e., $\lambda_k(X^o)$, and according to the characteristics of the technical $f_k(X^0)$ system is impossible to improve upon, without worsening thus characteristics.

We note that according to Theorem 1, at point $\mathbf{X^o}$ criteria 1 and 3 are contradictory. This contradiction is defined by the equality of $\lambda_1(X^0) = \lambda_3(X^0) = \lambda^0 = 0.3179$, and other criteria are subject to an inequality of $\{\lambda_2(X^0) = 0.7954, \lambda_4(X^0) = 0.5557\} > \lambda^0$.

Thus, Theorem 1 forms a basis for the determination of the correctness of the solution of a vector problem. In a vector problem of mathematical programming, as a rule, for two criteria an equality holds: $\lambda^0 = \lambda_q(X^0) = \lambda_p(X^0)$, $q, p \in \mathbf{K}$, $X \in S$, and other criteria are subject to an inequality: $\lambda^0 \leq \lambda_k(X^0)$ $\forall k \in \mathbf{K}, q \neq p \neq k$.

8.4. Geometric Interpretation of Results of the Decision in a Three-Dimensional Coordinate System in Relative Units

In an admissible set of points, S formed by restrictions of Equation (32), the optimum points X_1^*, X_2^*, X_3^*, X_4^* are united in a contour and presented as a set of Pareto optimal points $S^0 \subset S$ in Figure 13. Coordinates of these points, and the characteristics of the technical system in relative units of $\lambda_1(X)$, $\lambda_2(X)$, $\lambda_3(X)$, $\lambda_4(X)$ are shown in Figure 14 in three-dimensional space, where the third axis λ is a relative assessment.

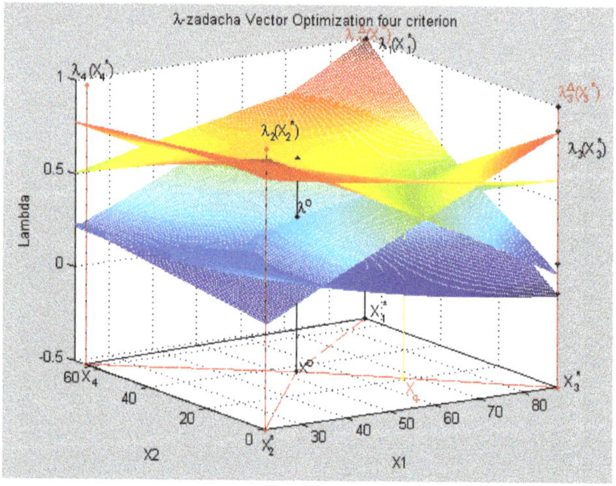

Figure 14. The solution of the λ-problem in a three-dimensional system of coordinates of x_1, x_2 and λ.

Discussion. Looking at Figure 9, we can provide changes of all functions of $\lambda_1(X)$, $\lambda_2(X)$, $\lambda_3(X)$, $\lambda_4(X)$ in four-dimensional space. We consider, for example, an optimum point X_3^*. The $\lambda_3(X)$ function is created from the functions $f_3(X)$ with variable coordinates $\{x_1, x_2\}$ and with constant coordinates $\{x_3 = 8.8, x_4 = 2.2\}$, taken from an optimum point $\mathbf{X}^0$ (33). At point X_3^* the relative assessment of $\lambda_3(X_3^*) = 0.83$ is shown in Figure 9 by a black point. However, we know that the relative assessment of $\lambda_3(X_3^*)$ obtained from the function $f_3(X_3^*)$ in the third step is equal to unity, which we designate as $\lambda_3^\Delta(X_{31}^{**}) = 1$, and is shown in Figure 9 by a red point. The difference between $\lambda_3^\Delta(X_{31}^{**}) = 1$ and $\lambda_3(X_3^*) = 0.83$ is an error $\Delta = 0.17$ due to transitioning from four-dimensional (and generally N-dimensional) to two-dimensional space.

The point X_1^* and appropriate relative estimates of $\lambda_1(X_1^*)$ and $\lambda_1^\Delta(X_1^*)$ are similarly shown.

Thus, for the first time in domestic and foreign practice, the transition and its geometric illustration from an N-dimensional to a two-dimensional measurement of function is shown in vector problems of mathematical programming with the appropriate errors.

8.5. The Solution of a Vector Problem of Mathematical Programming—Model of the System at the Given Priority of the Criterion

The decision maker is usually the system designer.

Step 1. We solve a vector problem with equivalent criteria. The algorithm of the decision is presented in Section 8.3. The numerical results of the solution of the vector problem are given above.

Pareto's great number $S^0 \subset S$ lies between optimum points $X_1^*\ X^0\ X_3^*\ X^0\ X_4^*\ X^0\ X_2^*\ X^0\ X_1^*$. We carry out the analysis of a great number of Pareto $S^o \subset S$. For this purpose, we will connect auxiliary points: $X_1^* X_3^* X_4^* X_2^* X_1^*$ with a point X^0 which conditionally represents the center of a great number of Pareto. As a result, we obtain four subsets of points $X \in \mathbf{S}_q^o \subset \mathbf{S}^0 \subset \mathbf{S}$, $q = \overline{1,4}$. The subset $\mathbf{S}_1^o \subset \mathbf{S}^0 \subset \mathbf{S}$ is characterized by the fact that the relative assessment $\lambda_1 \geq \lambda_2, \lambda_3, \lambda_4$, i.e., in the field $\mathbf{S}_1^o$, the first criterion has priority over the others. Similarly, $\mathbf{S}_2^o$, $\mathbf{S}_3^o$, $\mathbf{S}_4^o$ are the subsets of points where the second, third and fourth criterion has a priority over the others, respectively. We designate the set of Pareto optimal points $\mathbf{S}^0 = \mathbf{S}_1^o \cup \mathbf{S}_2^o \cup \mathbf{S}_3^o \cup \mathbf{S}_4^o$. Coordinates of all obtained points and relative estimates are presented in two-dimensional space $\{x_1, x_2\}$ in Figure 13. These coordinates are shown in three-dimensional space $\{x_1, x_2, \lambda\}$ in Figure 14 where the third axis λ is a relative assessment. The restrictions of the set of Pareto optimal points in Figure 14 are lowered to −0.5 (so that restrictions are visible). This information is also a basis for further research on the structure of a great number of Pareto. The person making decisions, as a rule, is the designer of the system. If results of the solution of a vector problem with equivalent criteria do not satisfy the person making the decision, then the choice of the optimal solution is taken from any subset of points $\mathbf{S}_1^o$, $\mathbf{S}_2^o$, $\mathbf{S}_3^o$, $\mathbf{S}_4^o$. These subsets of Pareto points are shown in Figure 8 in the form of functions $f_1(X), f_2(X), f_3(X), f_4(X)$.

Step 2. Choice of priority criterion of $q \in \mathbf{K}$. From the theory (see Theorem 1) it is known that at an optimum point X^0 there are always two most inconsistent criteria, $q \in \mathbf{K}$ and $v \in \mathbf{K}$ for which in relative units an equality holds: $\lambda^0 = \lambda_q(X^0) = \lambda_p(X^0)$, $q, v \in \mathbf{K}$, $X \in \mathbf{S}$. Others are subject to inequalities: $\lambda^0 \leq \lambda_k(X^0)$ $\forall k \in \mathbf{K}$, $q \neq v \neq k$.

In a model of the system in Equations (100)–(104) and the corresponding λ-problem in Equations (113)–(117), such criteria are the first and third:

$$\lambda^0 = \lambda_1(X^0) = \lambda_3(X^0) = 0.3179. \tag{122}$$

We show the $\lambda_1(X)$ and$\lambda_3(X)$ functions separately in Figure 15 for an optimum point $\boldsymbol{X^o} = \{X^o, \lambda^o\}$.

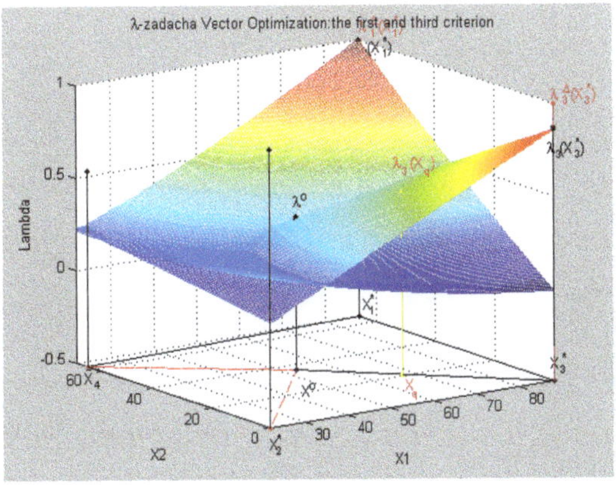

Figure 15. The solution of the λ-problem (first and third criteria) in a three-dimensional system of coordinates of x_1, x_2 and λ.

All points and data are shown in Figure 14.

As a rule, the criterion which the decision-maker would like to improve is taken from a couple of contradictory criteria. Such a criterion is called the "priority criterion", which we designate q=3∈$\mathbf{K}$. This criterion is investigated in interaction with the first criterion of $k = 1 \in \mathbf{K}$. We allocate these two criteria from the set of all criteria $\mathbf{K} = 4$ shown in Figure 15.

On the display the message is given:

q=input ('Enter priority criterion (number) of q ='), have entered: $q = 3$.

Step 3. Numerical limits of the change of the size of a priority of criterion of $q = 3 \in \mathbf{K}$ are defined.

For priority criterion $q = 3$, the numerical limits in physical units upon transition from an optimal point X^o (119) to the point X_q^* obtained in the first step are defined.

Information about the criteria for $q = 3$ are given on the screen:

$$f_q\left(X^0\right) \quad = \quad 65.96 \leq f_q(X) \leq 100.15 \quad = \quad f_q(X_q^*), q \in K. \tag{123}$$

In relative units the criterion of $q = 2$ changes according to the following limits:

$\lambda_q(X^0) = 0.3179 \leq \lambda_q(X) \leq 1 = \lambda_q(X_q^*)$, $q = 3 \in \mathbf{K}$.

These data are analyzed.

Step 4. Choice of the size of priority criterion $q \in \mathbf{K}$ (decision making).

The message is displayed: "Enter the size of priority criterion f_q = ", we enter, for example, $f_q = 80$.

Step 5. Calculation of a relative assessment.

For the chosen size of the priority criterion of f_q =80 the relative assessment is calculated:

$$\lambda_q = \frac{f_q - f_q^o}{f_q^* - f_q^o} = \frac{80 - 50.03}{100.15 - 50.03} = 0.5979, \quad (124)$$

which upon transition from point X^0 to X_q^* according to Equation (38) lies in the limits:

$$0.3179 = \lambda_3(X^0) \le \lambda_3 = 0.5979 \le \lambda_3(X) = 1, q \in K. \quad (125)$$

Step 6. Calculation of the coefficient of linear approximation.

Assuming a linear nature of the change of criterion of $f_q(X)$ in Equation (123) and according to a relative assessment of $\lambda_q(X)$, using standard methods of linear approximation we calculate the proportionality coefficient between $\lambda_q(X^0)$, λ_q, which we call ρ:

$$\rho = \frac{\lambda_q - \lambda_q(X^o)}{\lambda_q(X_q^*) - \lambda_q(X^o)} = \frac{0.5979 - 0.3179}{1 - 0.3179} = 0.4106, q = 3 \in K. \quad (126)$$

Step 7. Calculation of the coordinates of priority criterion with the size f_q.

Assuming a linear nature of the change of a vector $X^q = \{x_1\ x_2\}$, $q = 3$ we determine coordinates of a point of priority criterion with the size f_q with a relative assessment (95):

$$X^q = \{x_1 = X^0(1) + \rho(X_q^*(1) - X^0(1))x_2 = X^0(2) + \rho(X_q^*(2) - X^0(2))\}. \quad (127)$$

where $X^0 = \{X^0(1) = 80.0, X^0(2) = 69.11\}$, $X_3^* = \{X^{3^*}(1) = 80.0, X^{3^*}(2) = 0.0\}$.

As a result of the calculations, we obtain point coordinates: $X^q = \{x_1 = 67.31, x_2 = 21.27\}$.

Step 8. Calculation of the main indicators of a point X_q.

For the obtained point X_q, we calculate:

- all criteria in physical units $f_k(X^q) = \{f_k(X^q), k = \overline{1,K}\}$: $f(X^q) = \{f_1(X^q) = 313.45, f_2(x^q) = 2575.7, f_3(x^q) = 74.2, f_4(x^q) = 60.6\}$,
- all relative estimates of criteria $\lambda^q = \{\lambda_k^q, k = \overline{1,K}\}$, $\lambda_k(X^q) = \frac{f_k(X^q) - f_k^o}{f_k^* - f_k^o}$, $k = \overline{1,K}_{\overline{1,K}}$: $\lambda_k(X^q) = \{\lambda_1(X^q) = 0.2405, \lambda_2(x^q) = 0.5102, \lambda_3(x^q) = 0.4825, \lambda_4(x^q) = 0.5586\}_{\overline{1,K}}$,
- the minimum relative assessment: minLXq = min(LXq): minLXq = $\min(\lambda_k(X^q)) = 0.2405$,
- the vector of priorities $P^q = \{p_k^q = \frac{\lambda_q(X^q)}{\lambda_k(X^q)}, k = \overline{1,K}\}$: $P^q = (p_1^3 = 2.0061, p_2^3 = 0.9458, p_3^3 = 1.0, p_4^3 = 0.8637)$,
- the relative assessment taking into account a criterion priority: $LX_qP_q = \lambda_k(X^q).{*}P^q = \{p_1^3\lambda_1(X^q), p_2^3\lambda_2(X^q), p_3^3\lambda_3(X^q), p_4^3\lambda_4(X^q)\}$ $\lambda_k(X^q).{*}P^q = \{0.7968\ 0.\ 7968\ 0.\ 7968\ 0.\ 7968\}$,
- the the minimum relative assessment taking into account a criterion priority: $\lambda^{00} = \min(p_1^2\lambda_1(X^q) = 0.4825, p_2^2\lambda_2(X^q) = 0.4825, p_3^2\lambda_3(X^q) = 0.4825, p_4^2\lambda_4(X^q)) = 0.4825$

Any point from Pareto's set $X_t^o = \{\lambda_t^o, X_t^o\} \in S^o$ can be similarly calculated.

Step 9. Analysis of results. The calculated size of criterion $f_q(X_t^o)$, $q \in \boldsymbol{K}$ is usually not equal to the set f_q. The error of the choice of $\Delta f_q = |f_q(X_t^o) - f_q| = |74.2 - 80| = 5.8$ is defined by an error of linear approximation, $\Delta f_{q\%} = 7.25\%$.

In the course of modeling and simulation, as well as in the previous example, Section 7.5, the parametric restrictions of Equation (118) can be changed, i.e., some set of optimum decisions is obtained. We can choose a final version from, in our example, this set of optimum decisions: parameters of technical system $X^0 = \{x_1 = 52.9, x_2 = 36.097, x_3 = 8.8, x_4 = 2.2, \lambda^0 = 0.3179$, the parameters of the technical system at a given priority criterion $q = 2$: $X^q = \{x_1 = 67.31, x_2 = 21.27\}$.

If the error $\Delta f_q = |f_q(X^{00}) - f_q| = |79.6–80| = 0.4$, measured in physical units or as a percentage $\Delta f_{q\%} = \Delta f_q / f_q \times 100 = 0.5\%$, is more than set Δf, $\Delta f_q > \Delta f$, f, we pass to Step 2, else if $\Delta f_q \le \Delta f$, calculations come to an end.

8.6. Geometric Interpretation of Results of the Decision in a Three-Dimensional Coordinate System in Physical Units

In the course of modeling, the parametric restrictions of Equation (32) and functions can be changed, i.e., some set of optimum decisions is obtained. We can choose a final version from, in our example, this set of optimum decisions:

- parameters of the system $X^0 = \{x_1 = 80.0, x_2 = 69.11, x_3 = 32.58, x_4 = 20.0\}$,
- the parameters of the system at a given priority criterion $q = 3$: $X^q = \{x_1 = 67.313, x_2 = 21.276\}$.

We represent these parameters in a two-dimensional (x_1, x_2) system in Figure 13, in a three-dimensional coordinate (x_1, $x_2$and λ) system in Figure 14, and, in physical units for each function $f_1(X), \ldots, f_4(X)$ in Figures 16–19, respectively. The first characteristic $f_1(X)$ in physical units is shown in Figure 16.

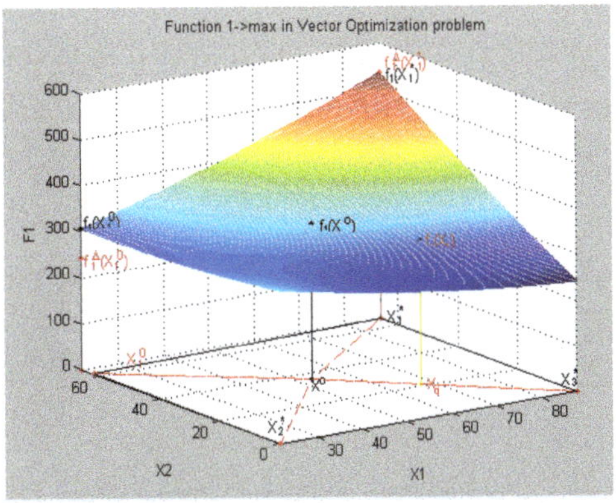

Figure 16. The first characteristic $f_1(X)$ of the system in a natural indicator.

Indicators of the first $f_1^{\Delta}(X_1^*), f_1^{\Delta}(X_1^0)$ characteristic of the system (highlighted in red) define the transition errors from four-dimensional $X^0 = \{x_1, x_2, x_3, x_4\}$ to two-dimensional $X^0 = \{x_1, x_2\}$ systems of coordinates. The second characteristic $f_2(X)$ in physical units is shown in Figure 17.

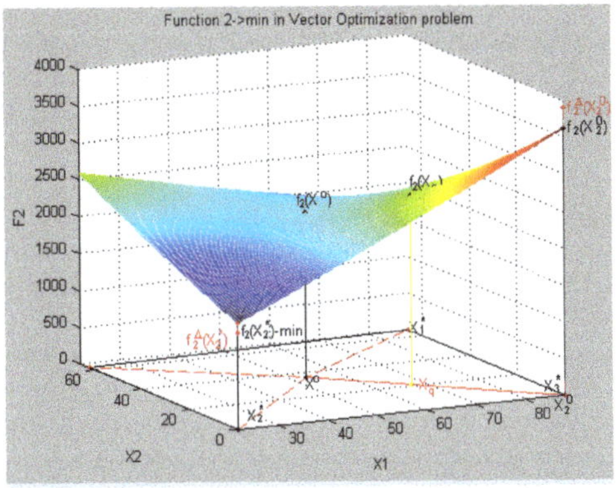

Figure 17. The second characteristic $f_2(X)$ of the system in a natural indicator.

Indicators of the second $f_2^{\Delta}(X_2^*), f_2^{\Delta}(X_2^0)$ characteristic of the system (highlighted in red) define the transition errors from four-dimensional $X^0 = \{x_1, x_2, x_3, x_4\}$ to two-dimensional $X^0 = \{x_1, x_2\}$ systems of coordinates. The third characteristic $f_3(X)$ in physical units is shown in Figure 18.

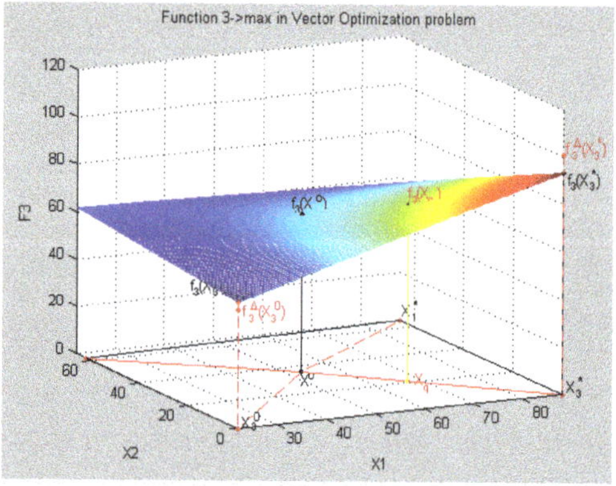

Figure 18. The third characteristic f3(X) of the system in a natural indicator.

Indicators of the third $f_3^{\Delta}(X_3^*)$, $f_3^{\Delta}(X_3^0)$ characteristic of the system (highlighted in red) define transition errors from four-dimensional $X^0 = \{x_1, x_2, x_3, x_4\}$ to two-dimensional $X^0 = \{x_1, x_2\}$ systems of coordinates. The fourth characteristic $f_4(X)$ in physical units is shown in Figure 19.

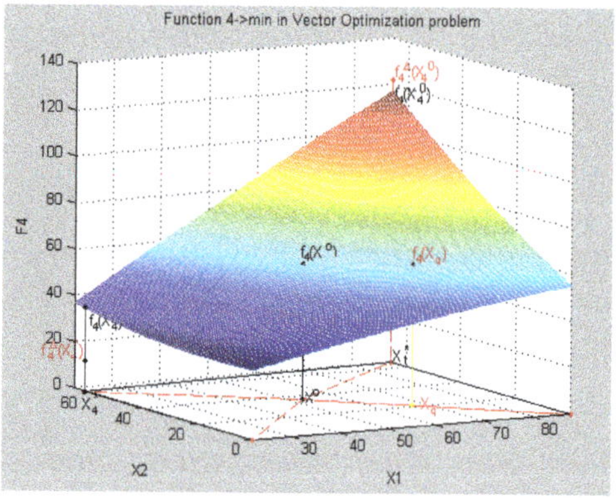

Figure 19. The fourth characteristic f3(X) of the system in a natural indicator.

Indicators of the fourth $f_4^{\Delta}(X_4^*)$, $f_4^{\Delta}(X_4^0)$ characteristic of the system (highlighted in red) define the transition errors from four-dimensional $X^o = \{x_1, x_2, x_3, x_4\}$ to two-dimensional $X^o = \{x_1, x_2\}$ systems of coordinates.

Collectively, for the submitted version with:

- point X^o, characteristics of $f_1(X^0), f_2(X^0), f_3(X^0), f_4(X^0)$,
- relative estimates of $\lambda_1(X^0)$, $\lambda_2(X^0)$, $\lambda_3(X^0)$, $\lambda_4(X^0)$, and
- maximum relative level λ^0 such that $\lambda^0 \leq \lambda_k(X^0)\ \forall k \in \mathbf{K}$

there is an optimum decision with equivalent criteria (characteristics) and, for the procedure of obtaining the optimum decision with equivalent criteria (characteristics):

- point X^q, characteristics of $f_1(X^q), f_2(X^q), f_3(X^q), f_4(X^q)$,
- relative estimates of $\lambda_1(X^q)$, $\lambda_2(X^q)$, $\lambda_3(X^q)$, $\lambda_4(X^q)$,
- maximum relative level λ^o such that $\lambda^o \leq p_k^q \lambda_k(X^q)$, $k = \overline{1, K}$

there is an optimal solution at the set priority of the qth criterion (characteristic) in relation to other criteria. The procedure of obtaining a point X^q is the adoption of the optimal solution at the set priority of the second criterion.

Based on the theory of vector optimization, methods of solution of vector problems with equivalent criteria, and a given priority of criterion, we can choose any point from the set of Pareto optimal points, and show the optimality of this point.

Conclusions. The problem of the development of mathematical methods of vector optimization and the adoption of the optimal solution in a difficult technical system based on some set of experimental data and functional characteristics are some of the most important tasks of system analysis and design.

In this work, the methodology of the creation of a mathematical model of a technical system under the conditions of definiteness and indeterminacy in the form of a vector problem of mathematical programming is developed. New methods of vector optimization based on normalization of criteria and the principle of the guaranteed result are developed for the solution of a vector problem. Methods of vector optimization allow making a decision, first, with equivalent criteria, and second, with a given priority of criterion. In the creation of the characteristics under conditions of indeterminacy, regression methods of transformation of information are used. The practice of "making optimal decisions" on the basis of a mathematical model is shown using a number of numerical examples of solutions of vector problems of optimization. The solution to the problem of "acceptance of an optimal solution" is realized with examples of 1, 2, 3 and 4 variables, respectively.

These methods of processing experimental data and vector optimization can be used in the design of technical systems of various industries: electro-technical, aerospace, metallurgical, etc. This methodology has system characteristics and can be used when modeling technical, economic and other systems. The author is ready to contribute to the solutions of vector problems of linear and nonlinear programming.

Funding: This research received no external funding.

Conflicts of Interest: The authors declare no conflict of interest.

References

1. Germeier, Y.B. *Non_Antagonistic Games*; Nauka: Moscow, Russia, 1976; Springer, 1986.
2. Podinovskii, V.V. Analysis of Multicriteria Choice Problems by Methods of the Theory of Criteria Importance, Based on Computer Systems of Decision_Making Support. *Comput. Syst. Sci. Int.* **2008**, *47*, 221. [CrossRef]
3. Malyshev, V.V.; Piyavskii, B.S.; Piyavskii, S.A. A decision making method under conditions of diversity of means of reducing uncertainty. *Comput. Syst. Sci. Int.* **2010**, *49*, 44–58. [CrossRef]
4. Mashunin, Y.K. Solving composition and decomposition problems of synthesis of complex engineering systems by vector_optimization methods. *Comput. Syst. Sci. Int.* **1999**, *38*, 421.
5. Mashunin, Y.K. *Market Theory and Simulation Based on Vector Optimization*; Universitetskaya kniga: Moscow, Russia, 2010. (In Russian)
6. Mashunin, K.Y.; Mashunin, Y.K. Simulation Engineering Systems under Uncertainty and Optimal Descision Making. *J. Comput. Syst. Sci. Int.* **2013**, *52*, 519–534. [CrossRef]
7. Mashunin, Y.K.; Mashunin, K.Y. Modeling of technical systems on the basis of vector optimization (2. with a Criterion Priority). *Int. J. Eng. Sci. Res. Technol.* **2014**, *3*, 224–240.
8. Mashunin, Y.K.; Mashunin, K.Y. Simulation and Optimal Decision Making the Design of Technical Systems (2. The Decision with a Criterion Priority). *Am. J. Model. Optim.* **2016**, *4*, 51–66.
9. Mashunin, Y.K. Vector optimization in the system optimal Decision Making the Design in economic and technical systems. *Int. J. Emerg. Trends Technol. Comput. Sci.* **2017**, *7*, 42–57.
10. Mashunin, Y.K. Concept of Technical Systems Optimum Designing (Mathematical and Organizational Statement). In Proceedings of the International Conference on Industrial Engineering, Applications and Manufacturing, ICIEAM 2017, St. Petersburg, Russia, 16–19 May 2017; Proceedings 8076394; WOS: 000414282400287. ISBN 978-1-5090-5648-4.

11. Keeney, R.L.; Raiffa, H. *Decisions with Multiple Objectives–Preferences and Value Tradeoffs*; Wiley: New York, NY, USA, 1976; Radio i svyaz': Moscow, Russia, 1981.
12. Mashunin, Y.K. Control Theory. The Mathematical Apparatus of Management of the Economy. Logos. Moscow. 2013. Available online: http://www.sciepub.com/reference/162155 (accessed on 11 October 2019). (In Russian).
13. Mashunin, Y.K. Optimum Designing of the Technical Systems Concept (Numerical Realization). In Proceedings of the International Conference on Industrial Engineering, Applications and Manufacturing, ICIEAM 2017, St. Petersburg, Russia, 16–19 May 2017; ISBN 978-1-5090-5648-4.
14. Cooke, T.; Lingard, H.; Blismas, N. The development and evaluation of a decision support tool for health and safety in construction design. *Eng. Constr. Archit. Manag.* **2008**, *15*, 336–351. [CrossRef]
15. Mashunin, K.Y.; Mashunin, Y.K. Vector Optimization with Equivalent and Priority Criteria. *J. Comput. Syst. Sci. Int.* **2017**, *56*, 975–996. [CrossRef]
16. Johannes, J. *Vector Optimization: Theory, Applications, and Extensions*; Springer: Berlin/Heidelberg, Germany; New York, NY, USA, 2010; p. 510.
17. Ansari, Q.; Jen Chih, Y. *Recent Developments in Vector Optimization*; Springer: Heidelberg, Germany; Dordrecht, The Netherlands; London, UK; New York, NY, USA, 2010.
18. Hirotaka, N.; Yeboon, Y.; Min, Y. *Sequential Approximate Multiobjective Optimization Using Computational Intelligence*; Springer: Berlin/Heidelberg, Germany, 2009; p. 197.
19. Shankar, R. *Decision Making in the Manufacturing Environment: Using Graft Theory and Fuzzy Multiple Attribute Decision Making Methods*; Springer: Berlin/Heidelberg, Germany, 2007.
20. Mashunin, Y.K.; Mashunin, K.Y. Modeling of technical systems on the basis of vector optimization (1. At equivalent criteria). *Int. J. Eng. Sci. Res. Technol.* **2014**, *3*, 84–96.
21. Mashunin, Y.K. *Methods and Models of Vector Optimization*; Nauka: Moscow, Russia, 1986.
22. Mashunin, Y.K.; Mashunin, K.Y. Simulation and Optimal Decision Making the Design of Technical Systems. *Am. J. Model. Optim.* **2015**, *3*, 56–67.
23. Mashunin, Y.K.; Levitskii, V.L. *Methods of Vector Optimization in Analysis and Synthesis of Engineering Systems. Monograph*; DVGAEU: Vladivostok, Russia, 1996. (In Russian)
24. Torgashov, A.Y.; Krivosheev, V.P.; Mashunin, Y.K.; Holland, C.D. Calculation and multiobjective optimization of static modes of mass_exchange processes by the example of absorption in gas separation. *Izv. Vyssh. Uchebn. Zaved. Neft. Gaz.* **2001**, *3*, 82–86.
25. Mashunin, Y.K. Engineering system modeling on the base of vector problem of nonlinear optimization. In *Control Applications of Optimization, Preprint of the Eleventh IFAC International Workshop*; CAO: St._Petersburg, Russia, 2000.
26. Ketkov, Y.L.; Ketkov, A.Y.; Shul'ts, M.M. *MATLAB 6.x.: Numerical Programming*; BKhV_Peterburg: St. Petersburg, Russia, 2004; p. 672. (In Russian)

applied system innovation

MDPI

Article

Using Dual Double Fuzzy Semi-Metric to Study the Convergence

Hsien-Chung Wu

Department of Mathematics, National Kaohsiung Normal University, Kaohsiung 802, Taiwan; hcwu@nknucc.nknu.edu.tw

Received: 6 March 2019; Accepted: 4 April 2019; Published: 11 April 2019

Abstract: Convergence using dual double fuzzy semi-metric is studied in this paper. Two types of dual double fuzzy semi-metric are proposed in this paper, which are called the infimum type of dual double fuzzy semi-metric and the supremum type of dual double fuzzy semi-metric. Under these settings, we also propose different types of triangle inequalities that are used to investigate the convergence using dual double fuzzy semi-metric.

Keywords: dual double fuzzy semi-metric; double fuzzy semi-metric; fuzzy semi-metric space; triangle inequality; triangular norm

1. Introduction

The concept of fuzzy metric space proposed by Kramosil and Michalek [1] was inspired by the Menger space that is a special kind of probabilistic metric space by referring to Schweizer and Sklar [2–4], Hadžić and Pap [5], and Chang et al. [6]. Kaleva and Seikkala [7] proposed another concept of fuzzy metric space by considering the membership degree of the distance between any two different points. George and Veeramani [8,9] studied some properties of fuzzy metric spaces in the sense of Kramosil and Michalek [1]. Gregori and Romaguera [10–12] also extended the study of the properties of fuzzy metric spaces and fuzzy quasi-metric spaces in which the symmetric condition was not assumed.

The Hausdorff topology induced by the fuzzy metric space was studied in Wu [13]. In this paper, we shall propose the concept of double fuzzy-semi metric in fuzzy semi-metric space and study its convergent properties. The potential application for using the convergence of dual double fuzzy semi-metric is to study the new type of fixed point theorems in fuzzy semi-metric space by considering the Cauchy sequences, which will be the future research and may refer to the previous work of Wu [14] for studying the common coincidence points and common fixed points in fuzzy semi-metric spaces. Wu [15] studied the so-called fuzzy semi-metric space without assuming the symmetric condition. In the fuzzy semi-metric space (X, M), the symmetric condition $M(x,y,t) = M(y,x,t)$ for all $x, y \in X$ and $t > 0$ is not assumed to be true. Therefore, four kinds of triangle inequalities should be considered.

In order to obtain the new type of fixed point theorems in fuzzy semi-metric space, we need to study the convergence using dual double fuzzy semi-metric. Based on the concept of t-norm $*$, we shall firstly define the double fuzzy semi-metric by considering the mapping $\zeta : X^4 \times [0, +\infty) \to [0,1]$ that is defined by:

$$\zeta(x,y;u,v,t) = M(x,y,t) * M(u,v,t),$$

where ζ is called a double fuzzy semi-metric.

The convergence using fuzzy semi-metric has been studied in Wu [16], where the infimum type of dual fuzzy semi-metric is the function $\Gamma^{\downarrow}(\lambda,\cdot,\cdot) : X \times X \to [0,+\infty)$ defined by:

$$\Gamma^{\downarrow}(\lambda,x,y) = \inf\{t > 0 : M(x,y,t) \geq 1-\lambda\},$$

and the supremum type of dual fuzzy semi-metric is the function $\Gamma^{\uparrow}(\lambda,\cdot,\cdot) : X \times X \to [0,+\infty)$ defined by:

$$\Gamma^{\uparrow}(\lambda,x,y) = \sup\{t > 0 : M(x,y,t) \leq 1-\lambda\}.$$

In this paper, we shall consider the double fuzzy semi-metric ζ to define the infimum and supremum types of dual double fuzzy semi-metric. The infimum type of dual double fuzzy semi-metric is the function $\Psi^{\downarrow}(\lambda,\cdot,\cdot;\cdot,\cdot) : X^4 \to [0,+\infty)$ defined by:

$$\Psi^{\downarrow}(\lambda,x,y;u,v) = \inf\{t > 0 : \zeta(x,y;u,v,t) \geq 1-\lambda\},$$

and the supremum type of dual double fuzzy semi-metric is the function $\Psi^{\uparrow} : X^4 \to [0,+\infty)$ defined by:

$$\Psi^{\uparrow}(\lambda,x,y;u,v) = \sup\{t > 0 : \zeta(x,y;u,v,t) \leq 1-\lambda\}.$$

Using the infimum and supremum types of dual fuzzy semi-metric $\Gamma^{\downarrow}(\lambda,x,y)$ and $\Gamma^{\uparrow}(\lambda,x,y)$, the convergence of sequences in (X,M) and the concept of Cauchy sequence in (X,M) have been studied in Wu [16]. In this paper, we study the extended convergence of sequences in (X,M) and the concept of joint Cauchy sequence in (X,M) using the infimum and supremum types of dual double fuzzy semi-metric $\Psi^{\downarrow}(\lambda,x,y;u,v)$ and $\Psi^{\uparrow}(\lambda,x,y;u,v)$. As we mentioned above, these convergences will be used in the near future to establish the new types of fixed point theorems in fuzzy semi-metric space (X,M).

In Section 2, we review some basic properties of fuzzy semi-metric space that will be used for further discussion. In Section 3, we introduce the concept of double fuzzy semi-metric and derive the related triangle inequalities. In Sections 4 and 5, the concepts of infimum and supremum types of dual double fuzzy semi-metric are proposed, and their convergent properties and triangle inequalities are studied.

2. Fuzzy Semi-Metric Space

Let X be a nonempty universal set, and let M be a mapping defined on $X \times X \times [0,\infty)$ into $[0,1]$. Then (X,M) is called a fuzzy semi-metric space if and only if the following conditions are satisfied:

- For any $x,y \in X$, $M(x,y,t) = 1$ for all $t > 0$ if and only if $x = y$;
- $M(x,y,0) = 0$ for all $x,y \in X$ with $x \neq y$.

We say that M satisfies the symmetric condition if and only if $M(x,y,t) = M(y,x,t)$ for all $x,y \in X$ and $t > 0$. We say that M satisfies the strongly symmetric condition if and only if $M(x,y,t) = M(y,x,t)$ for all $x,y \in X$ and $t \geq 0$. Since the symmetric condition is not assumed to be true in fuzzy semi-metric space, four kinds of triangle inequalities called $\circ$-triangle inequality for $\circ \in \{\bowtie,\triangleright,\triangleleft,\diamond\}$ were proposed by Wu [15].

Example 1. *Let X be a universal set, and let $d : X \times X \to \mathbb{R}_+$ satisfy the following conditions:*

- *$d(x,y) \geq 0$ for any $x,y \in X$;*
- *$d(x,y) = 0$ if and only if $x = y$ for any $x,y \in X$;*
- *$d(x,y) + d(y,z) \geq d(x,z)$ for any $x,y,z \in X$.*

Note that we do not assume $d(x,y) = d(y,x)$. For example, let $X = [0,1]$. We define:

$$d(x,y) = \begin{cases} y - x & \text{if } y \geq x \\ 1 & \text{otherwise.} \end{cases}$$

Then $d(x,y) \neq d(y,x)$ and the above three conditions are satisfied. Now we take t-norm $*$ as $a * b = ab$ and define:

$$M(x,y,t) = \begin{cases} \dfrac{t}{t+d(x,y)} & \text{if } t > 0 \\ 1 & \text{if } t = 0 \text{ and } d(x,y) = 0 \\ 0 & \text{if } t = 0 \text{ and } d(x,y) > 0 \end{cases} = \begin{cases} \dfrac{t}{t+d(x,y)} & \text{if } t > 0 \\ 1 & \text{if } t = 0 \text{ and } x = y \\ 0 & \text{if } t = 0 \text{ and } x \neq y. \end{cases}$$

It is clear to see that $M(x,y,t) \neq M(y,x,t)$ for $t > 0$, since $d(x,y) \neq d(y,x)$. It is not hard to check that $(X, M, *)$ is a fuzzy semi-metric space satisfying the $\bowtie$-triangle inequality.

The following interesting observations will be used in further study.

Remark 1. *Let (X, M) be a fuzzy semi-metric space.*

- *Suppose that M satisfies the $\bowtie$-triangle inequality. Then:*

$$M(a,b,t_1) * M(b,c,t_2) * M(c,d,t_3) \leq M(a,c,t_1+t_2) * M(c,d,t_3) \leq M(a,d,t_1+t_2+t_3).$$

In general, we have:

$$M(x_1, x_2, t_1) * M(x_2, x_3, t_2) * \cdots * M(x_p, x_{p+1}, t_p) \leq M(x_1, x_{p+1}, t_1 + t_2 + \cdots + t_p). \quad (1)$$

- *Suppose that M satisfies the $\triangleright$-triangle inequality. Since:*

$$M(a,b,t_1) * M(c,b,t_2) \leq \min\{M(a,c,t_1+t_2), M(c,a,t_1+t_2)\},$$

which implies:

$$M(a,b,t_1) * M(c,b,t_2) * M(d,c,t_3) \leq \min\{M(a,d,t_1+t_2+t_3), M(d,a,t_1+t_2+t_3)\}. \quad (2)$$

In general, we have:

$$\begin{aligned} &M(x_1, x_2, t_1) * M(x_3, x_2, t_2) * M(x_4, x_3, t_3) * \cdots * M(x_{p+1}, x_p, t_p) \\ &\quad \leq \min\{M(x_1, x_{p+1}, t_1 + t_2 + \cdots + t_p), M(x_{p+1}, x_1, t_1 + t_2 + \cdots + t_p)\}. \end{aligned}$$

- *Suppose that M satisfies the $\triangleleft$-triangle inequality. Since:*

$$M(b,a,t_1) * M(b,c,t_2) \leq \min\{M(a,c,t_1+t_2), M(c,a,t_1+t_2)\},$$

which implies:

$$M(b,a,t_1) * M(b,c,t_2) * M(c,d,t_3) \leq \min\{M(a,d,t_1+t_2+t_3), M(d,a,t_1+t_2+t_3)\}. \quad (3)$$

In general, we have:

$$\begin{aligned} &M(x_2, x_1, t_1) * M(x_2, x_3, t_2) * M(x_3, x_4, t_3) * \cdots * M(x_p, x_{p+1}) \\ &\quad \leq \min\{M(x_1, x_{p+1}, t_1 + t_2 + \cdots + t_p), M(x_{p+1}, x_1, t_1 + t_2 + \cdots + t_p)\}. \end{aligned}$$

- *Suppose that M satisfies the $\diamond$-triangle inequality. Then:*

$$\begin{aligned} M(a,b,t_1) * M(b,c,t_2) * M(d,c,t_3) &= M(b,c,t_1) * M(a,b,t_2) * M(d,c,t_3) \\ &\leq M(c,a,t_1+t_2) * M(d,c,t_3) \leq M(a,d,t_1+t_2+t_3) \end{aligned} \tag{4}$$

and:

$$\begin{aligned} M(b,a,t_1) * M(c,b,t_2) * M(c,d,t_3) &\leq M(a,c,t_1+t_2) * M(c,d,t_3) \\ &= M(c,d,t_3) * M(a,c,t_1+t_2) \leq M(d,a,t_1+t_2+t_3). \end{aligned} \tag{5}$$

From Equation (4), we also have:

$$\begin{aligned} M(a,b,t_1) * M(c,b,t_2) * M(d,c,t_3) &= M(d,c,t_3) * M(c,b,t_2) * M(a,b,t_1) \\ &\leq M(d,a,t_1+t_2+t_3), \end{aligned} \tag{6}$$

which implies:

$$M(b,a,t_1) * M(b,c,t_2) * M(c,d,t_3) \geq M(a,d,t_1+t_2+t_3) \tag{7}$$

by referring to Equation (5). In general, we have the following cases:

(a) *If p is even, then:*

$$\begin{aligned} &M(x_1,x_2,t_1) * M(x_2,x_3,t_2) * M(x_4,x_3,t_3) * M(x_4,x_5,t_4) * M(x_6,x_5,t_5) \\ &\quad * M(x_6,x_7,t_6) * \cdots * M(x_p,x_{p+1},t_p) \leq M(x_{p+1},x_1,t_1+t_2+\cdots+t_p) \end{aligned}$$

and:

$$\begin{aligned} &M(x_2,x_1,t_1) * M(x_3,x_2,t_2) * M(x_3,x_4,t_3) * M(x_5,x_4,t_4) * M(x_5,x_6,t_5) \\ &\quad * M(x_7,x_6,t_6) * \cdots * M(x_p,x_{p+1},t_p) \leq M(x_1,x_{p+1},t_1+t_2+\cdots+t_p). \end{aligned}$$

(b) *If p is odd, then:*

$$\begin{aligned} &M(x_1,x_2,t_1) * M(x_2,x_3,t_2) * M(x_4,x_3,t_3) * M(x_4,x_5,t_4) * M(x_6,x_5,t_5) \\ &\quad * M(x_6,x_7,t_6) * \cdots * M(x_p,x_{p+1},t_p) \leq M(x_1,x_{p+1},t_1+t_2+\cdots+t_p) \end{aligned}$$

and:

$$\begin{aligned} &M(x_2,x_1,t_1) * M(x_3,x_2,t_2) * M(x_3,x_4,t_3) * M(x_5,x_4,t_4) * M(x_5,x_6,t_5) \\ &\quad * M(x_7,x_6,t_6) * \cdots * M(x_{p+1},x_p\, t_p) \leq M(x_{p+1},x_1,t_1+t_2+\cdots+t_p). \end{aligned}$$

Let (X,M) be a fuzzy semi-metric space.

- We say that M is nondecreasing if and only if, given any fixed $x,y \in X$, $M(x,y,t_1) \geq M(x,y,t_2)$ for $t_1 > t_2 > 0$.
- We say that M is increasing if and only if, given any fixed $x,y \in X$, $M(x,y,t_1) > M(x,y,t_2)$ for $t_1 > t_2 > 0$.
- We say that M is symmetrically nondecreasing if and only if, given any fixed $x,y \in X$, $M(x,y,t_1) \geq M(y,x,t_2)$ for $t_1 > t_2 > 0$.
- We say that M is symmetrically increasing if and only if, given any fixed $x,y \in X$, $M(x,y,t_1) > M(y,x,t_2)$ for $t_1 > t_2 > 0$.

The following interesting results were modified from Wu [15] using the similar argument, which will be used in further discussion.

Proposition 1. *(Wu [15]) Let (X, M) be a fuzzy semi-metric space. Then we have the following properties:*

(i) *If M satisfies the $\bowtie$-triangle inequality, then M is nondecreasing. If M satisfies the strict $\bowtie$-triangle inequality, then M is increasing.*
(ii) *If M satisfies the $\triangleright$-triangle inequality or the $\triangleleft$-triangle inequality, then M is both nondecreasing and symmetrically nondecreasing. If M satisfies the strict $\triangleright$-triangle inequality or the strict $\triangleleft$-triangle inequality, then M is both increasing and symmetrically increasing.*
(iii) *If M satisfies the $\diamond$-triangle inequality, then M is symmetrically nondecreasing. If M satisfies the strict $\diamond$-triangle inequality, then M is symmetrically increasing.*

3. Double Fuzzy Semi-Metric

Let (X, M) be a fuzzy semi-metric space along with a t-norm $*$. Given any four elements $x, y, u, v \in X$, recall that the value $M(x, y, t)$ means the membership degree of the distance that is less than t between x and y, and the value $M(u, v, t)$ means the membership degree of the distance that is less than t between u and v. In this case, we can define a value:

$$\zeta(x, y; u, v, t) = \min\{M(x, y, t), M(u, v, t)\},$$

which means the membership degree of the distance that is simultaneously less than t between x and y and between u and v. In general, instead of considering the min function, we shall use the t-norm. The formal definition is given below.

Definition 1. *Let (X, M) be a fuzzy semi-metric space along with a t-norm $*$. We define the mapping $\zeta : X^4 \times [0, +\infty) \to [0, 1]$ by:*

$$\zeta(x, y; u, v, t) = M(x, y, t) * M(u, v, t).$$

Then ζ is called a double fuzzy semi-metric.

Example 2. *Continued from Example 1, we consider:*

$$M(x, y, t) = \begin{cases} \dfrac{t}{t + d(x, y)} & \text{if } t > 0 \\ 1 & \text{if } t = 0 \text{ and } x = y \\ 0 & \text{if } t = 0 \text{ and } x \neq y. \end{cases}$$

If we take t-norm as $a * b = a \cdot b$, then the double fuzzy semi-metric can be obtained as:

$$\begin{aligned} \zeta(x, y; u, v, t) &= M(x, y, t) \cdot M(u, v, t) \\ &= \begin{cases} \dfrac{t}{t + d(x, y)} \cdot \dfrac{t}{t + d(u, v)} & \text{if } t > 0 \\ 1 & \text{if } t = 0 \text{ and } x = y \text{ and } u = v \\ 0 & \text{if } t = 0 \text{ and } x \neq y \text{ or } u \neq v. \end{cases} \end{aligned}$$

The potential application for considering the double fuzzy semi-metric is to study the new type of fixed point theorems in fuzzy semi-metric space.

Proposition 2. *(Triangle Inequalities for Dual Fuzzy Semi-Metric) Let (X, M) be a fuzzy semi-metric space along with a t-norm $*$. Given any $x, y, z, u, v, w \in X$, we have the following properties:*

(i) *Suppose that M satisfies the* $\bowtie$*-triangle inequality. Then we have the inequality:*

$$\zeta(x,z;u,w,t+s) \geq \zeta(x,y;u,v,t) * \zeta(y,z;v,w,s)$$

for $s,t > 0$.

(ii) *Suppose that M satisfies the* $\triangleright$*-triangle inequality. Then we have the inequality:*

$$\zeta(x,z;u,w,t+s) \geq \zeta(x,y;u,v,t) * \zeta(z,y;w,v,s)$$

for $s,t > 0$.

(iii) *Suppose that M satisfies the* $\triangleleft$*-triangle inequality. Then we have the inequality:*

$$\zeta(x,z;u,w,t+s) \geq \zeta(y,x;v,u,t) * \zeta(y,z;v,w,s)$$

for $s,t > 0$.

(iv) *Suppose that M satisfies the* $\diamond$*-triangle inequality. Then we have the inequality:*

$$\zeta(x,z;u,w,t+s) \geq \zeta(y,x;v,u,t) * \zeta(z,y;w,v,s)$$

for $s,t > 0$.

Proof. It suffices to prove part (i); we have:

$$\begin{aligned}
\zeta(x,z;u,w,t+s) &= M(x,z,t+s) * M(u,w,t+s) \\
&\geq (M(x,y,t) * M(y,z,s)) * M(u,w,t+s) \\
&\quad \text{(using the } \bowtie \text{ triangle inequality and the increasing property of t-norm)} \\
&\geq (M(x,y,t) * M(y,z,s)) * (M(u,v,t) * M(v,w,s)) \\
&\quad \text{(using the } \bowtie\text{-triangle inequality and the increasing property of t-norm)} \\
&= (M(x,y,t) * M(u,v,t)) * (M(y,z,s) * M(v,w,s)) \\
&\quad \text{(using the associative and commutative properties of t-norm)} \\
&= \zeta(x,y;u,v,t) * \zeta(y,z;v,w,s).
\end{aligned}$$

This completes the proof. □

Definition 2. *Let* (X,M) *be a fuzzy semi-metric space along with a t-norm* $*$*, and let* ζ *be a double fuzzy semi-metric given by:*

$$\zeta(x,y;u,v,t) = M(x,y,t) * M(u,v,t).$$

Given any fixed $x,y,u,v \in X$*, we define the following concepts of monotonicity:*

- *The mapping* $\zeta(x,y;u,v,\cdot)$ *is said to be nondecreasing if and only if* $\zeta(x,y;u,v,t_1) \geq \zeta(x,y;u,v,t_2)$ *for* $t_1 > t_2$. *The mapping* $\zeta(x,y;u,v,\cdot)$ *is said to be increasing if and only if* $\zeta(x,y;u,v,t_1) > \zeta(x,y;u,v,t_2)$ *for* $t_1 < t_2$.
- *The mapping* $\zeta(x,y;u,v,\cdot)$ *is is said to be symmetrically nondecreasing if and only if* $\zeta(x,y;u,v,t_1) \geq \zeta(y,x;v,u,t_2)$ *for* $t_1 > t_2$. *The mapping* $\zeta(x,y;u,v,\cdot)$ *is said to be symmetrically increasing if and only if* $\zeta(x,y;u,v,t_1) > \zeta(y,x;v,u,t_2)$ *for* $t_1 < t_2$.
- *The mapping* $\zeta(x,y;u,v,\cdot)$ *is said to be* $\triangleleft$*-semisymmetrically nondecreasing if and only if* $\zeta(x,y;u,v,t_1) \geq \zeta(y,x;u,v,t_2)$ *for* $t_1 > t_2$. *The mapping* $\zeta(x,y;u,v,\cdot)$ *is said to be* $\triangleleft$*-semisymmetrically increasing if and only if* $\zeta(x,y;u,v,t_1) > \zeta(y,x;u,v,t_2)$ *for* $t_1 < t_2$.
- *The mapping* $\zeta(x,y;u,v,\cdot)$ *is said to be* $\triangleright$*-semisymmetrically nondecreasing if and only if* $\zeta(x,y;u,v,t_1) \geq \zeta(x,y;v,u,t_2)$ *for* $t_1 > t_2$. *The mapping* $\zeta(x,y;u,v,\cdot)$ *is said to be* $\triangleright$*-semisymmetrically increasing if and only if* $\zeta(x,y;u,v,t_1) > \zeta(x,y;v,u,t_2)$ *for* $t_1 < t_2$.

Proposition 3. *Let (X, M) be a fuzzy semi-metric space along with a t-norm $*$. Given any fixed $x, y, u, v \in X$, the double fuzzy semi-metric ζ satisfies the following properties:*

(i) *Suppose that M satisfies the $\bowtie$-triangle inequality. Then the mapping $\zeta(x, y; u, v, \cdot)$ from $[0, \infty)$ into $[0, 1]$ is nondecreasing.*
(ii) *Suppose that M satisfies the $\triangleright$-triangle inequality or the $\triangleleft$-triangle inequality. Then the mapping $\zeta(x, y; u, v, \cdot)$ from $[0, \infty)$ into $[0, 1]$ is simultaneously nondecreasing, symmetrically nondecreasing, $\triangleleft$-semisymmetrically nondecreasing, and $\triangleright$-semisymmetrically nondecreasing.*
(iii) *Suppose that M satisfies the $\diamond$-triangle inequality. Then the mapping $\zeta(x, y; u, v, \cdot)$ from $[0, \infty)$ into $[0, 1]$ is symmetrically nondecreasing.*

Proof. Part (i) of Proposition 1 says that the mappings $M(x, y, \cdot)$ and $M(u, v, \cdot)$ from $[0, \infty)$ into $[0, 1]$ are nondecreasing. According to the increasing property of t-norm, we conclude that the mapping $\zeta(x, y; u, v, \cdot)$ from $[0, \infty)$ into $[0, 1]$ is nondecreasing, which proves part (i).

Part (ii) can be obtained from part (ii) of Proposition 1, and part (iii) can be obtained from part (iii) of Proposition 1. This completes the proof. □

By using the strictly increasing property of t-norm, the proof of Proposition 3 is still valid for obtaining the following results.

Proposition 4. *Let (X, M) be a fuzzy semi-metric space along with a t-norm $*$. Suppose that the t-norm satisfies the strictly increasing property. Given any fixed $x, y, u, v \in X$, the double fuzzy semi-metric ζ satisfies the following properties:*

(i) *Suppose that M satisfies the strict $\bowtie$-triangle inequality. Then the mapping $\zeta(x, y; u, v, \cdot)$ from $[0, \infty)$ into $[0, 1]$ is increasing.*
(ii) *Suppose that M satisfies the strict $\triangleright$-triangle inequality or the strict $\triangleleft$-triangle inequality. Then the mapping $\zeta(x, y; u, v, \cdot)$ from $[0, \infty)$ into $[0, 1]$ is simultaneously increasing, symmetrically increasing, $\triangleleft$-semisymmetrically increasing, and $\triangleright$-semisymmetrically increasing.*
(iii) *Suppose that M satisfies the strict $\diamond$-triangle inequality. Then the mapping $\zeta(x, y; u, v, \cdot)$ from $[0, \infty)$ into $[0, 1]$ is symmetrically increasing.*

Let (X, M) be a fuzzy semi-metric space. The motivation for considering the following two concepts can refer to Wu [16].

- M is said to satisfy the canonical condition if and only if:

$$\lim_{t \to +\infty} M(x, y, t) = 1 \text{ for any fixed } x, y \in X.$$

- M is said to satisfy the rational condition if and only if:

$$\lim_{t \to 0+} M(x, y, t) = 0 \text{ for any fixed } x, y \in X.$$

Proposition 5. *Let (X, M) be a fuzzy semi-metric space along with a t-norm $*$.*

(i) *Suppose that M satisfies the canonical condition. If the t-norm $*$ is left-continuous at 1 with respect to the first or second argument, then we have:*

$$\lim_{t \to +\infty} \zeta(x, y; u, v, t) = 1. \tag{8}$$

(ii) *Suppose that M satisfies the rational condition. If the t-norm $*$ is right-continuous at 0 with respect to the first or second argument, then we have:*

$$\lim_{t \to 0+} \zeta(x, y; u, v, t) = 0. \tag{9}$$

Proof. To prove part (i), the canonical condition says that:

$$\lim_{t \to +\infty} M(x,y,t) = 1 = \lim_{t \to +\infty} M(u,v,t).$$

The left-continuity of t-norm $*$ at 1 also says that:

$$\lim_{t \to +\infty} \zeta(x,y;u,v,t) = \left(\lim_{t \to +\infty} M(x,y,t) \right) * \left(\lim_{t \to +\infty} M(u,v,t) \right) = 1 * 1 = 1.$$

To prove part (ii), the rational condition says that:

$$\lim_{t \to 0+} M(x,y,t) = 0 = \lim_{t \to 0+} M(u,v,t).$$

The right-continuity of t-norm $*$ at 0 also says that:

$$\lim_{t \to 0+} \zeta(x,y;u,v,t) = \left(\lim_{t \to 0+} M(x,y,t) \right) * \left(\lim_{t \to 0+} M(u,v,t) \right) = 0 * 0 = 0.$$

This completes the proof. □

Example 3. *Continued from Example 1, it is not hard to check that M satisfies both the canonical and rational conditions. Suppose that we take t-norm as $a * b = a \cdot b$. Then Proposition 5 says that:*

$$\lim_{t \to +\infty} \zeta(x,y;u,v,t) = 1 \text{ and } \lim_{t \to 0+} \zeta(x,y;u,v,t) = 0.$$

4. Convergence Based on the Infimum

From Definition 1, we see that the double fuzzy semi-metric ζ is a mapping from $X^4 \times [0,\infty)$ into $[0,1]$. Here, we shall consider its dual sense by considering the mapping from $(0,1] \times X^4$ into $[0,\infty)$. The formal definition is given below.

Definition 3. *Let (X, M) be a fuzzy semi-metric space along with a t-norm $*$. We also assume that M satisfies the canonical condition, and that the t-norm $*$ is left-continuous at 1 with respect to the first or second argument. Given any fixed $x,y,u,v \in X$ and any fixed $\lambda \in (0,1]$, we consider the following set:*

$$\Pi^{\downarrow}(\lambda,x,y;u,v) = \{t > 0 : \zeta(x,y;u,v,t) \geq 1 - \lambda\},$$

which is used to define a mapping $\Psi^{\downarrow}(\lambda,\cdot,\cdot;\cdot,\cdot) : X^4 \to [0,+\infty)$ by:

$$\Psi^{\downarrow}(\lambda,x,y;u,v) = \inf \Pi^{\downarrow}(\lambda,x,y;u,v) = \inf\{t > 0 : \zeta(x,y;u,v,t) \geq 1 - \lambda\}.$$

In this case, the mapping $\Psi^{\downarrow}$ from $(0,1] \times X^4$ into $[0,\infty)$ is called the infimum type of dual double fuzzy semi-metric.

Example 4. *Continued from Example 2, we have:*

$$\Pi^{\downarrow}(\lambda,x,y;u,v) = \left\{ t > 0 : \frac{t}{t + d(x,y)} \cdot \frac{t}{t + d(u,v)} \geq 1 - \lambda \right\} = \left\{ t > 0 : t \geq \frac{C + \sqrt{C^2 + D}}{2} \right\},$$

where:

$$C = \frac{(d(x,y) + d(u,v))(1-\lambda)}{\lambda} \text{ and } D = \frac{d(x,y) \cdot d(u,v) \cdot (1-\lambda)}{\lambda}.$$

We also have:

$$\Psi^{\downarrow}(\lambda,x,y;u,v)=\inf\Pi^{\downarrow}(\lambda,x,y;u,v)=\left\{t>0:t\geq\frac{C+\sqrt{C^2+D}}{2}\right\}=\frac{C+\sqrt{C^2+D}}{2}.$$

The potential application of dual double fuzzy semi-metric will be used to study the fixed point theorems in fuzzy semi-metric space. However, we first need to claim that the set $\Pi^{\downarrow}(\lambda,x,y;u,v)$ is nonempty. Suppose that $\Pi^{\downarrow}(\lambda,x,y;u,v)=\emptyset$. The definition says that $\zeta(x,y;u,v,t)<1-\lambda$ for all $t>0$; that is:

$$\lim_{t\to+\infty}\zeta(x,y;u,v,t)\leq 1-\lambda<1,$$

which contradicts Equation (8). Therefore, Definition 3 is well-defined and $\Pi^{\downarrow}(\lambda,x,y;u,v)\neq\emptyset$.

Remark 2. *The following observations will be useful for further discussion.*

- *For any $\lambda\in(0,1]$, we have:*

$$\Psi^{\downarrow}(1,x,y;u,v)=\inf\{t>0:\zeta(x,y;u,v,t)\geq 0\}=\inf\{t>0\}=0,$$

and:

$$\begin{aligned}\Psi^{\downarrow}(\lambda,x,x;u,u)&=\inf\{t>0:\zeta(x,x;u,u,t)\geq 1-\lambda\}\\&=\inf\{t>0:1\geq 1-\lambda\}=\inf\{t>0\}=0.\end{aligned}\tag{10}$$

- *Given any fixed $x,y,u,v\in X$, if $\lambda_1>\lambda_2$, then:*

$$\Pi^{\downarrow}(\lambda_2,x,y;u,v)\subseteq\Pi^{\downarrow}(\lambda_1,x,y;u,v)\text{ and }\Psi^{\downarrow}(\lambda_1,x,y;u,v)\leq\Psi^{\downarrow}(\lambda_2,x,y;u,v).\tag{11}$$

Proposition 6. *Let (X,M) be a fuzzy semi-metric space along with a t-norm $*$. We also assume that M satisfies the canonical condition, and that the t-norm $*$ is left-continuous at 1 with respect to the first or second argument. Given any fixed $x,y,u,v\in X$, suppose that the following conditions are satisfied:*

$$\Pi^{\downarrow}(0+,x,y;u,v)\equiv\bigcap_{0<\lambda\leq 1}\Pi^{\downarrow}(\lambda,x,y;u,v)\neq\emptyset,$$

and:

$$\{t>0:\zeta(x,y;u,v,t)=1\}\neq\emptyset.$$

Then we have:

$$\Pi^{\downarrow}(0+,x,y;u,v)=\{t>0:\zeta(x,y;u,v,t)=1\}.\tag{12}$$

Moreover, the following limit exists:

$$\lim_{\lambda\to 0+}\Psi^{\downarrow}(\lambda,x,y;u,v)=\sup_{0<\lambda\leq 1}\Psi^{\downarrow}(\lambda,x,y;u,v).\tag{13}$$

Proof. The assumption $\Pi^{\downarrow}(0+,x,y;u,v)\neq\emptyset$ says that we can consider $t\in\Pi^{\downarrow}(0+,x,y;u,v)$, i.e., $\zeta(x,y;u,v,t)\geq 1-\lambda$ for all $\lambda\in(0,1]$. Then we obtain $\zeta(x,y;u,v,t)\geq 1$ by taking $\lambda\to 0+$, which shows that $\zeta(x,y;u,v,t)=1$, i.e.,

$$\Pi^{\downarrow}(0+,x,y;u,v)=\bigcap_{0<\lambda\leq 1}\Pi^{\downarrow}(\lambda,x,y;u,v)\subseteq\{t>0:\zeta(x,y;u,v,t)=1\}.$$

On the other hand, suppose that $\zeta(x,y;u,v,t)=1$. Then $\zeta(x,y;u,v,t)=1\geq 1-\lambda$ for all $\lambda\in(0,1]$. Therefore, we obtain $t\in\Pi^{\downarrow}(0+,x,y;u,v)$, which implies the desired equality

(Equation (12)). Further, the inequality (Equation (11)) says that the limit (Equation (13)) exists. This completes the proof. □

Proposition 7. *Suppose that (X, M) is a fuzzy semi-metric space along with a t-norm $*$. We also assume that M satisfies the canonical and rational conditions, and that the t-norm $*$ is left-continuous at 1 and right-continuous at 0 with respect to the first or second argument. If M satisfies the $\circ$-triangle inequality for $\circ \in \{\bowtie, \triangleright, \triangleleft, \diamond\}$, then, for any fixed $x, y, u, v \in X$ with $x \neq y$ or $u \neq v$, we have $\Psi^{\downarrow}(\lambda, x, y; u, v) > 0$ for $\lambda \in (0, 1)$.*

Proof. We first consider the case of M, satisfying the $\circ$-triangle inequality for $\circ \in \{\bowtie, \triangleright, \triangleleft\}$. From Equation (10), we need to consider $x \neq y$ or $u \neq v$. We want to assume $\Psi^{\downarrow}(\lambda, x, y; u, v) = 0$ for $\lambda \in (0, 1)$ to obtain a contradiction. Using the concept of infimum from $\Psi^{\downarrow}(\lambda, x, y; u, v)$, given any $\epsilon > 0$, there exists $t_\epsilon > 0$ such that $\zeta(x, y; u, v, t_\epsilon) \geq 1 - \lambda$ and:

$$t_\epsilon < \Psi^{\downarrow}(\lambda, x, y; u, v) + \epsilon = \epsilon.$$

Parts (i) and (ii) of Proposition 3 say that the mapping $\zeta(x, y; u, v, \cdot)$ from $[0, \infty)$ into $[0, 1]$ is nondecreasing. Therefore, we obtain:

$$\zeta(x, y; u, v, \epsilon) \geq \zeta(x, y; u, v, t_\epsilon) \geq 1 - \lambda.$$

Since ϵ can be any positive real number, using Equation (9), we must have:

$$0 = \lim_{\epsilon \to 0+} \zeta(x, y; u, v, \epsilon) \geq \zeta(x, y; u, v, t_\epsilon) \geq 1 - \lambda,$$

which contradicts $0 < \lambda < 1$.

Now we assume that M satisfies the $\diamond$-triangle inequality. Suppose that $\Psi^{\downarrow}(\lambda, y, x; v, u) = 0$ for $\lambda \in (0, 1)$. Part (iii) of Proposition 3 says that the mapping $\zeta(x, y; u, v, \cdot)$ is symmetrically nondecreasing. Therefore, we can similarly obtain:

$$\zeta(x, y; u, v, \epsilon) \geq \zeta(y, x; v, u, t_\epsilon) \geq 1 - \lambda.$$

This completes the proof. □

Proposition 8. *Let (X, M) be a fuzzy semi-metric space along with a t-norm $*$. We also assume that M satisfies the canonical condition, and that the t-norm $*$ is left-continuous at 1 with respect to the first or second argument. Given any fixed $x, y, u, v \in X$ and $\lambda \in (0, 1)$, we have the following properties:*

(i) *If $\epsilon > 0$ is sufficiently small satisfying $\Psi^{\downarrow}(\lambda, x, y; u, v) > \epsilon$, then we have:*

$$\zeta\left(x, y; u, v, \Psi^{\downarrow}(\lambda, x, y; u, v) - \epsilon\right) < 1 - \lambda. \tag{14}$$

(ii) *Suppose that M satisfies the $\circ$-triangle inequality for $\circ \in \{\bowtie, \triangleright, \triangleleft\}$. For any $\epsilon > 0$, we have:*

$$\zeta\left(x, y; u, v, \Psi^{\downarrow}(\lambda, x, y; u, v) + \epsilon\right) \geq 1 - \lambda \tag{15}$$

(iii) *Suppose that M satisfies the $\circ$-triangle inequality for $\circ \in \{\triangleright, \triangleleft\}$. For any $\epsilon > 0$, we have:*

$$\zeta\left(x, y; u, v, \Psi^{\downarrow}(\lambda, y, x; u, v) + \epsilon\right) \geq 1 - \lambda \tag{16}$$

and:

$$\zeta\left(x, y; u, v, \Psi^{\downarrow}(\lambda, x, y; v, u) + \epsilon\right) \geq 1 - \lambda \tag{17}$$

(iv) *Suppose that M satisfies the* $\circ$*-triangle inequality for* $\circ \in \{\triangleright, \triangleleft, \diamond\}$. *For any* $\epsilon > 0$, *we have:*

$$\zeta\left(x, y; u, v, \Psi^{\downarrow}(\lambda, y, x; v, u) + \epsilon\right) \geq 1 - \lambda \tag{18}$$

Proof. To prove part (i), we assume that:

$$\zeta(x, y; u, v, \Psi^{\downarrow}(\lambda, x, y; u, v) - \epsilon) \geq 1 - \lambda.$$

The definition of $\Psi^{\downarrow}$ says that $\Psi^{\downarrow}(\lambda, x, y; u, v) \leq \Psi^{\downarrow}(\lambda, x, y; u, v) - \epsilon$. This contradiction implies $\zeta(x, y; u, v, \Psi^{\downarrow}(\lambda, x, y; u, v) - \epsilon) \leq 1 - \lambda$.

To prove part (ii), using the concept of infimum from $\Psi^{\downarrow}(\lambda, x, y; u, v)$, given any $\epsilon > 0$, there exists $t_\epsilon > 0$ such that $\zeta(x, y; u, v, t_\epsilon) \geq 1 - \lambda$ and $t_\epsilon < \Psi^{\downarrow}(\lambda, x, y; u, v) + \epsilon$. Parts (i) and (ii) of Proposition 3 says that the mapping $\zeta(x, y; u, v, \cdot)$ is nondecreasing. Therefore, we obtain:

$$\zeta\left(x, y; u, v, \Psi^{\downarrow}(\lambda, x, y; u, v) + \epsilon\right) \geq \zeta(x, y; u, v, t_\epsilon) \geq 1 - \lambda.$$

To prove part (iii), using the concept of infimum from $\Psi^{\downarrow}(\lambda, y, x; u, v)$, given any $\epsilon > 0$, there exists $t_\epsilon > 0$ such that $\zeta(y, x; u, v, t_\epsilon) \geq 1 - \lambda$ and $t_\epsilon < \Psi^{\downarrow}(\lambda, y, x; u, v) + \epsilon$. Part (ii) of Proposition 3 says that the mapping $\zeta(y, x; u, v, \cdot)$ is $\triangleleft$-semisymmetrically nondecreasing. Therefore, we obtain:

$$\zeta\left(x, y; u, v, \Psi^{\downarrow}(\lambda, y, x; u, v) + \epsilon\right) \geq \zeta(y, x; u, v, t_\epsilon) \geq 1 - \lambda.$$

Since the mapping $\zeta(x, y; u, v, \cdot)$ is also $\triangleright$-semisymmetrically nondecreasing, we can similarly obtain another inequality.

Since the mapping $\zeta(x, y; u, v, \cdot)$ is semisymmetrically nondecreasing, using parts (ii) and (iii) of Proposition 3, we can similarly obtain part (iv). This completes the proof. □

Proposition 9. *Let* (X, M) *be a fuzzy semi-metric space along with a t-norm* $*$. *We also assume that M satisfies the canonical condition, and that the t-norm* $*$ *is left-continuous at* 1 *with respect to the first or second argument. Given any fixed* $x, y, u, v \in X$ *and* $\lambda \in (0,1)$, *the following statements hold true:*

(i) *Suppose that M satisfies the* $\circ$*-triangle inequality for* $\circ \in \{\bowtie, \triangleright, \triangleleft, \diamond\}$. *If* $t > \Psi^{\downarrow}(\lambda, x, y; u, v)$, *then we have* $\zeta(x, y; u, v, t) \geq 1 - \lambda$.

(ii) *If* $0 < t < \Psi^{\downarrow}(\lambda, x, y; u, v)$, *then we have the following properties:*

- *Suppose that M satisfies the* $\circ$*-triangle inequality for* $\circ \in \{\bowtie, \triangleright, \triangleleft\}$. *Then we have* $\zeta(x, y; u, v, t) < 1 - \lambda$.
- *Suppose that M satisfies the* $\circ$*-triangle inequality for* $\circ \in \{\triangleright, \triangleleft\}$. *Then we have* $\zeta(y, x; u, v, t) < 1 - \lambda$ *and* $\zeta(x, y; v, u, t) < 1 - \lambda$.
- *Suppose that M satisfies the* $\circ$*-triangle inequality for* $\circ \in \{\triangleright, \triangleleft, \diamond\}$. *Then we have* $\zeta(y, x; v, u, t) < 1 - \lambda$.

Proof. To prove part (i), the inequality $t > \Psi^{\downarrow}(\lambda, x, y; u, v)$ says that there exists $\epsilon > 0$, satisfying $t \geq \Psi^{\downarrow}(\lambda, x, y; u, v) + \epsilon$. Therefore, we consider the following cases:

- Suppose that M satisfies the $\circ$-triangle inequality for $\circ \in \{\bowtie, \triangleright, \triangleleft\}$. Parts (i) and (ii) of Proposition 3 say that the mapping $\zeta(x, y; u, v, \cdot)$ is nondecreasing. Therefore, using Equation (15), we obtain:

$$\zeta(x, y; u, v, t) \geq \zeta\left(x, y; u, v, \Psi^{\downarrow}(\lambda, x, y; u, v) + \epsilon\right) \geq 1 - \lambda.$$

- Suppose that M satisfies the $\diamond$-triangle inequality. Part (iii) of Proposition 3 says that the mapping $\zeta(x,y;u,v,\cdot)$ is symmetrically nondecreasing. Therefore, using Equation (18), we obtain:

$$\zeta(x,y;u,v,t) \geq \zeta\left(y,x;v,u,\Psi^{\downarrow}(\lambda,x,y;u,v)+\epsilon\right) \geq 1-\lambda.$$

To prove part (ii), the inequality $0 < t < \Psi^{\downarrow}(\lambda,x,y;u,v)$ says that there exists $\epsilon > 0$, satisfying $t \leq \Psi^{\downarrow}(\lambda,x,y;u,v) - \epsilon$. Therefore, we consider the following cases:

- Suppose that M satisfies the $\circ$-triangle inequality for $\circ \in \{\bowtie,\triangleright,\triangleleft\}$. Parts (i) and (ii) of Proposition 3 say that the mapping $\zeta(x,y;u,v,\cdot)$ is nondecreasing. Therefore, using Equation (14), we obtain:
$$\zeta(x,y;u,v,t) \leq \zeta\left(x,y;u,v,\Psi^{\downarrow}(\lambda,x,y;u,v)-\epsilon\right) < 1-\lambda.$$
- Suppose that M satisfies the $\circ$-triangle inequality for $\circ \in \{\triangleright,\triangleleft\}$. Part (ii) of Proposition 3 says that the mapping $\zeta(x,y;u,v,\cdot)$ is $\triangleleft$-semisymmetrically nondecreasing. Therefore, using Equation (14), we obtain:
$$\zeta(y,x;u,v,t) \leq \zeta\left(x,y;u,v,\Psi^{\downarrow}(\lambda,x,y;u,v)-\epsilon\right) < 1-\lambda.$$

  We can similarly obtain another inequality using the fact that the mapping $\zeta(x,y;u,v,\cdot)$ is also $\triangleright$-semisymmetrically nondecreasing.
- Suppose that M satisfies the $\circ$-triangle inequality for $\circ \in \{\triangleright,\triangleleft,\diamond\}$. Parts (ii) and (iii) of Proposition 3 say that the mapping $\zeta(x,y;u,v,\cdot)$ is symmetrically nondecreasing. Therefore, using Equation (14), we obtain:

$$\zeta(y,x;v,u,t) \leq \zeta\left(x,y;u,v,\Psi^{\downarrow}(\lambda,x,y;u,v)-\epsilon\right) < 1-\lambda.$$

This completes the proof. □

Proposition 10. *Let (X,M) be a fuzzy semi-metric space along with a t-norm $*$. We also assume that M satisfies the canonical condition, and that the t-norm $*$ is left-continuous at 1 with respect to the first or second argument. Given any fixed $x,y,u,v \in X$ and $\lambda \in (0,1)$, the following statements hold true:*

(i) *If $\zeta(x,y;u,v,t) < 1-\lambda$, then we have the following properties:*
   - *Suppose that M satisfies the $\circ$-triangle inequality for $\circ \in \{\bowtie,\triangleright,\triangleleft,\diamond\}$. Then we have $t \leq \Psi^{\downarrow}(\lambda,x,y;u,v)$.*
   - *Suppose that M satisfies the $\circ$-triangle inequality for $\circ \in \{\triangleright,\triangleleft\}$. Then we have $t \leq \Psi^{\downarrow}(\lambda,y,x;u,v)$ and $t \leq \Psi^{\downarrow}(\lambda,x,y;v,u)$.*
   - *Suppose that M satisfies the $\circ$-triangle inequality for $\circ \in \{\triangleright,\triangleleft,\diamond\}$. Then we have $t \leq \Psi^{\downarrow}(\lambda,y,x;v,u)$.*

(ii) *Suppose that the t-norm $*$ satisfies the strictly increasing property. If $\zeta(x,y;u,v,t) = 1-\lambda$ for $t > 0$, then we have the following properties.*
   - *Suppose that M satisfies the strict $\circ$-triangle inequality for $\circ \in \{\bowtie,\triangleright,\triangleleft\}$. If $\Psi^{\downarrow}(\lambda,x,y;u,v) > 0$, then we have $t = \Psi^{\downarrow}(\lambda,x,y;u,v)$.*
   - *Suppose that M satisfies the strict $\circ$-triangle inequality for $\circ \in \{\triangleright,\triangleleft\}$. If $\Psi^{\downarrow}(\lambda,y,x;u,v) > 0$, then we have $t = \Psi^{\downarrow}(\lambda,y,x;u,v)$, and if $\Psi^{\downarrow}(\lambda,x,y;v,u) > 0$, then we have $t = \Psi^{\downarrow}(\lambda,x,y;v,u)$.*
   - *Suppose that M satisfies the strict $\circ$-triangle inequality for $\circ \in \{\triangleright,\triangleleft,\diamond\}$. If $\Psi^{\downarrow}(\lambda,y,x;v,u) > 0$, then we have $t = \Psi^{\downarrow}(\lambda,y,x;v,u)$.*

(iii) *If $\zeta(x,y;u,v,t) \geq 1-\lambda$, then we have the following properties:*

- *Suppose that M satisfies the ◦-triangle inequality for* $\circ \in \{\bowtie, \triangleright, \triangleleft\}$*. Then we have* $t \geq \Psi^{\downarrow}(\lambda, x, y; u, v)$*.*
- *Suppose that M satisfies the ◦-triangle inequality for* $\circ \in \{\triangleright, \triangleleft\}$*. Then we have* $t \geq \Psi^{\downarrow}(\lambda, y, x; u, v)$ *and* $t \geq \Psi^{\downarrow}(\lambda, x, y; v, u)$*.*
- *Suppose that M satisfies the ◦-triangle inequality for* $\circ \in \{\triangleright, \triangleleft, \diamond\}$*. Then we have* $t \geq \Psi^{\downarrow}(\lambda, y, x; v, u)$*.*

Proof. To prove part (i), three cases are separately considered below:

- Suppose that M satisfies the ◦-triangle inequality for $\circ \in \{\bowtie, \triangleright, \triangleleft, \diamond\}$. Using the contrapositive statement of part (i) of Proposition 9, we can obtain the desired result.
- Suppose that M satisfies the ◦-triangle inequality for $\circ \in \{\triangleright, \triangleleft\}$. According to the concept of infimum, given any $\epsilon > 0$, there exists $t_\epsilon > 0$, satisfying $\zeta(y, x; u, v, t_\epsilon) \geq 1 - \lambda$ and $t_\epsilon < \Psi^{\downarrow}(\lambda, y, x; u, v) + \epsilon$. Part (ii) of Proposition 3 says that the mapping $\zeta(x, y; u, v, \cdot)$ is $\triangleleft$-semisymmetrically nondecreasing. Therefore, if $t > t_\epsilon$ then $\zeta(x, y; u, v, t) \geq \zeta(y, x; u, v, t_\epsilon)$, which contradicts $\zeta(x, y; u, v, t) < 1 - \lambda$. It says that:

  $$t \leq t_\epsilon < \Psi^{\downarrow}(\lambda, y, x; u, v) + \epsilon.$$

  Since ϵ can be any positive real number, we must have $t \leq \Psi^{\downarrow}(\lambda, y, x; u, v)$. We can similarly obtain another inequality using the fact of the mapping $\zeta(x, y; u, v, \cdot)$ being $\triangleright$-semisymmetrically nondecreasing.
- Suppose that M satisfies the ◦-triangle inequality for $\circ \in \{\triangleright, \triangleleft, \diamond\}$. According to the concept of infimum, given any $\epsilon > 0$, there exists $t_\epsilon > 0$, satisfying $\zeta(y, x; v, u, t_\epsilon) \geq 1 - \lambda$ and $t_\epsilon < \Psi^{\downarrow}(\lambda, y, x; v, u) + \epsilon$. Parts (ii) and (iii) of Proposition 3 say that if $t > t_\epsilon$ then $\zeta(x, y; u, v, t) \geq \zeta(y, x; v, u, t_\epsilon)$, which contradicts $\zeta(x, y; u, v, t) < 1 - \lambda$. It says that:

  $$t \leq t_\epsilon < \Psi^{\downarrow}(\lambda, y, x; v, u) + \epsilon.$$

  Since ϵ can be any positive real number, we must have $t \leq \Psi^{\downarrow}(\lambda, y, x; v, u)$.

To prove part (ii), three cases are separately considered below:

- Suppose that M satisfies the ◦-triangle inequality for $\circ \in \{\bowtie, \triangleright, \triangleleft\}$. According to the concept of infimum, given any $\epsilon > 0$, there exists $t_\epsilon > 0$, satisfying $\zeta(x, y; u, v, t_\epsilon) \geq 1 - \lambda$ and $t_\epsilon < \Psi^{\downarrow}(\lambda, x, y; u, v) + \epsilon$. Regarding the strict property, parts (i) and (ii) of Proposition 4 say that if $t > t_\epsilon$ then $\zeta(x, y; u, v, t) > \zeta(x, y; u, v, t_\epsilon)$, which contradicts $\zeta(x, y; u, v, t) = 1 - \lambda$. It says that:

  $$t \leq t_\epsilon < \Psi^{\downarrow}(\lambda, x, y; u, v) + \epsilon.$$

  Since ϵ can be any positive real number, we must have $t \leq \Psi^{\downarrow}(\lambda, x, y; u, v)$. Now we assume that $t < \Psi^{\downarrow}(\lambda, x, y; u, v)$. The first case of part (ii) of Proposition 9 says that $\zeta(x, y; u, v, t) < 1 - \lambda$, which also contradicts $\zeta(x, y; u, v, t) = 1 - \lambda$. Therefore, we must have $t = \Psi^{\downarrow}(\lambda, x, y; u, v)$.
- Suppose that M satisfies the ◦-triangle inequality for $\circ \in \{\triangleright, \triangleleft\}$. We can similarly obtain $t \leq \Psi^{\downarrow}(\lambda, y, x; u, v)$. Now we assume that $t < \Psi^{\downarrow}(\lambda, y, x; u, v)$. The second case of part (ii) of Proposition 9 says that $\zeta(x, y; u, v, t) < 1 - \lambda$, which also contradicts $\zeta(x, y; u, v, t) = 1 - \lambda$. Therefore, we must have $t = \Psi^{\downarrow}(\lambda, y, x; u, v)$. Another result can be similarly obtained.
- Suppose that M satisfies the ◦-triangle inequality for $\circ \in \{\triangleright, \triangleleft, \diamond\}$. We can similarly obtain $t \leq \Psi^{\downarrow}(\lambda, y, x; v, u)$. Now we assume that $t < \Psi^{\downarrow}(\lambda, y, x; v, u)$. The third case of part (ii) of Proposition 9 says that $\zeta(x, y; u, v, t) < 1 - \lambda$, which also contradicts $\zeta(x, y; u, v, t) = 1 - \lambda$. Therefore, we must have $t = \Psi^{\downarrow}(\lambda, y, x; v, u)$.

Part (iii) can be obtained from the contrapositive statement of part (ii) of Proposition 9. This completes the proof. □

Proposition 11. *Let* (X, M) *be a fuzzy semi-metric space along with a t-norm* $*$*. We also assume that* M *satisfies the canonical condition, and that the t-norm* $*$ *is left-continuous at* 1 *with respect to the first or second argument. Given any fixed* $x, y, u, v \in X$ *and* $\lambda \in (0,1)$*, the following statements hold true:*

(i) *Suppose that the mapping* $\zeta(x, y; u, v, \cdot) : (0, \infty) \to [0,1]$ *is left-continuous on* $(0, \infty)$*. If* $\Psi^{\downarrow}(\lambda, x, y; u, v) > 0$*, then we have:*

$$\zeta\left(x, y; u, v, \Psi^{\downarrow}(\lambda, x, y; u, v)\right) \leq 1 - \lambda. \tag{19}$$

(ii) *Suppose that the mapping* $\zeta(x, y; u, v, \cdot) : (0, \infty) \to [0,1]$ *is right-continuous on* $(0, \infty)$*. Then the following statements hold true:*

- *Suppose that* M *satisfies the* $\circ$*-triangle inequality for* $\circ \in \{\bowtie, \triangleright, \triangleleft\}$*. If* $\Psi^{\downarrow}(\lambda, x, y; u, v) > 0$*, then we have:*

$$\zeta\left(x, y; u, v, \Psi^{\downarrow}(\lambda, x, y; u, v)\right) \geq 1 - \lambda. \tag{20}$$

- *Suppose that* M *satisfies the* $\circ$*-triangle inequality for* $\circ \in \{\triangleright, \triangleleft\}$*. If* $\Psi^{\downarrow}(\lambda, y, x; u, v) > 0$*, then we have:*

$$\zeta\left(x, y; u, v, \Psi^{\downarrow}(\lambda, y, x; u, v)\right) \geq 1 - \lambda, \tag{21}$$

and if $\Psi^{\downarrow}(\lambda, x, y; v, u) > 0$*, then we have:*

$$\zeta\left(x, y; u, v, \Psi^{\downarrow}(\lambda, x, y; v, u)\right) \geq 1 - \lambda. \tag{22}$$

- *Suppose that* M *satisfies the* $\circ$*-triangle inequality for* $\circ \in \{\triangleright, \triangleleft, \diamond\}$*. If* $\Psi^{\downarrow}(\lambda, y, x; v, u) > 0$*, then we have:*

$$\zeta\left(x, y; u, v, \Psi^{\downarrow}(\lambda, y, x; v, u)\right) \geq 1 - \lambda. \tag{23}$$

Proof. By applying $\epsilon \to 0+$ to the inequality Equation (14), we obtain Equation (19), which proves part (i). By applying $\epsilon \to 0+$ to the inequality (Equation (15)), we obtain Equation (20), which proves part (ii). The other inequalities can be similarly obtained by parts (iii) and (iv) of Proposition 8. This completes the proof. □

In order to establish the triangle inequalities for the infimum type of dual double fuzzy semi-metric, we provide a useful lemma.

Lemma 1. (Wu [16]) *Suppose that the t-norm* $*$ *is left-continuous at* 1 *with respect to the first or second argument. For any* $a \in (0,1)$ *and any* $p \in \mathbb{N}$*, there exists* $r \in (0,1)$ *such that:*

$$\overbrace{r * r * \cdots * r}^{p \text{ times}} > a.$$

Theorem 1. (Triangle Inequalities for Dual Double Fuzzy Semi-Metric) *Let* (X, M) *be a fuzzy semi-metric space along with a t-norm* $*$*. We also assume that* M *satisfies the canonical condition, and that the t-norm* $*$ *is left-continuous at* 1 *with respect to the first or second argument. Given any fixed* $\mu \in (0,1]$ *and any fixed and distinct* $x_1, x_2, \cdots, x_p, y_1, y_2, \cdots, y_p \in X$*, we have the following inequalities:*

(i) *Suppose that M satisfies the* $\bowtie$*-triangle inequality. Then, there exists* $\lambda \in (0,1)$*, satisfying:*

$$\Psi^{\downarrow}(\mu, x_1, x_p; y_1, y_p) \leq \Psi^{\downarrow}(\lambda, x_1, x_2; y_1, y_2) + \Psi^{\downarrow}(\lambda, x_2, x_3; y_2, y_3) + \cdots + \Psi^{\downarrow}(\lambda, x_{p-2}, x_{p-1}; y_{p-2}, y_{p-1}) + \Psi^{\downarrow}(\lambda, x_{p-1}, x_p; y_{p-1}, y_p) \tag{24}$$

$$\Psi^{\downarrow}(\mu, x_1, x_p; y_p, y_1) \leq \Psi^{\downarrow}(\lambda, x_1, x_2; y_2, y_1) + \Psi^{\downarrow}(\lambda, x_2, x_3; y_3, y_2) + \cdots + \Psi^{\downarrow}(\lambda, x_{p-2}, x_{p-1}; y_{p-1}, y_{p-2}) + \Psi^{\downarrow}(\lambda, x_{p-1}, x_p; y_p, y_{p-1}) \tag{25}$$

$$\Psi^{\downarrow}(\mu, x_p, x_1; y_1, y_p) \leq \Psi^{\downarrow}(\lambda, x_p, x_{p-1}; y_{p-1}, y_p) + \Psi^{\downarrow}(\lambda, x_{p-1}, x_{p-2}; y_{p-2}, y_{p-1}) + \cdots + \Psi^{\downarrow}(\lambda, x_3, x_2; y_2, y_3) + \Psi^{\downarrow}(\lambda, x_2, x_1; y_1, y_2)$$

$$\Psi^{\downarrow}(\mu, x_p, x_1; y_p, y_1) \leq \Psi^{\downarrow}(\lambda, x_p, x_{p-1}; y_p, y_{p-1}) + \Psi^{\downarrow}(\lambda, x_{p-1}, x_{p-2}; y_{p-1}, y_{p-2}) + \cdots + \Psi^{\downarrow}(\lambda, x_3, x_2; y_3, y_2) + \Psi^{\downarrow}(\lambda, x_2, x_1; y_2, y_1).$$

(ii) *Suppose that M satisfies the* $\triangleright$*-triangle inequality. Then, there exists* $\lambda \in (0,1)$*, satisfying:*

$$\max\left\{\Psi^{\downarrow}(\mu, x_1, x_p; y_1, y_p), \Psi^{\downarrow}(\mu, x_1, x_p; y_p, y_1), \Psi^{\downarrow}(\mu, x_p, x_1; y_1, y_p), \Psi^{\downarrow}(\mu, x_p, x_1; y_p, y_1)\right\} \leq \Psi^{\downarrow}(\lambda, x_1, x_2; y_1, y_2) + \Psi^{\downarrow}(\lambda, x_3, x_2; y_3, y_2) + \Psi^{\downarrow}(\lambda, x_4, x_3; y_4, y_3) + \cdots + \Psi^{\downarrow}(\lambda, x_p, x_{p-1}; y_p, y_{p-1}).$$

(iii) *Suppose that M satisfies the* $\triangleleft$*-triangle inequality. Then, there exists* $\lambda \in (0,1)$*, satisfying:*

$$\max\left\{\Psi^{\downarrow}(\mu, x_1, x_p; y_1, y_p), \Psi^{\downarrow}(\mu, x_1, x_p; y_p, y_1), \Psi^{\downarrow}(\mu, x_p, x_1; y_1, y_p), \Psi^{\downarrow}(\mu, x_p, x_1; y_p, y_1)\right\} \leq \Psi^{\downarrow}(\lambda, x_2, x_1; y_2, y_1) + \Psi^{\downarrow}(\lambda, x_2, x_3; y_2, y_3) + \Psi^{\downarrow}(\lambda, x_3, x_4; y_3, y_4) + \cdots + \Psi^{\downarrow}(\lambda, x_{p-1}, x_p; y_{p-1}, y_p).$$

(iv) *Suppose that M satisfies the* $\diamond$*-triangle inequality. Then, there exists* $\lambda \in (0,1)$ *such that the following inequalities are satisfied:*

- *If p is even, then:*

$$\Psi^{\downarrow}(\mu, x_1, x_p; y_1, y_p) \leq \Psi^{\downarrow}(\lambda, x_2, x_1; y_2, y_1) + \Psi^{\downarrow}(\lambda, x_3, x_2; y_3, y_2) + \Psi^{\downarrow}(\lambda, x_3, x_4; y_3, y_4) + \Psi^{\downarrow}(\lambda, x_5, x_4; y_5, y_4) + \Psi^{\downarrow}(\lambda, x_5, x_6; y_5, y_6) + \Psi^{\downarrow}(\lambda, x_7, x_6; y_7, y_6) + \cdots + \Psi^{\downarrow}(\lambda, x_{p-1}, x_p; y_{p-1}, y_p), \tag{26}$$

$$\Psi^{\downarrow}(\mu, x_1, x_p; y_p, y_1) \leq \Psi^{\downarrow}(\lambda, x_2, x_1; y_1, y_2) + \Psi^{\downarrow}(\lambda, x_3, x_2; y_2, y_3) + \Psi^{\downarrow}(\lambda, x_3, x_4; y_4, y_3) + \Psi^{\downarrow}(\lambda, x_5, x_4; y_4, y_5) + \Psi^{\downarrow}(\lambda, x_5, x_6; y_6, y_5) + \Psi^{\downarrow}(\lambda, x_7, x_6; y_6, y_7) + \cdots + \Psi^{\downarrow}(\lambda, x_{p-1}, x_p; y_p, y_{p-1}), \tag{27}$$

$$\Psi^{\downarrow}(\mu, x_p, x_1; y_1, y_p) \leq \Psi^{\downarrow}(\lambda, x_1, x_2; y_2, y_1) + \Psi^{\downarrow}(\lambda, x_2, x_3; y_3, y_2) + \Psi^{\downarrow}(\lambda, x_4, x_3; y_3, y_4) + \Psi^{\downarrow}(\lambda, x_4, x_5; y_5, y_4) + \Psi^{\downarrow}(\lambda, x_6, x_5; y_5, y_6) + \Psi^{\downarrow}(\lambda, x_6, x_7; y_7, y_6) + \cdots + \Psi^{\downarrow}(\lambda, x_p, x_{p-1}; y_{p-1}, y_p), \tag{28}$$

$$\Psi^{\downarrow}(\mu, x_p, x_1; y_p, y_1) \leq \Psi^{\downarrow}(\lambda, x_1, x_2; y_1, y_2) + \Psi^{\downarrow}(\lambda, x_2, x_3; y_2, y_3) + \Psi^{\downarrow}(\lambda, x_4, x_3; y_4, y_3) + \Psi^{\downarrow}(\lambda, x_4, x_5; y_4, y_5) + \Psi^{\downarrow}(\lambda, x_6, x_5; y_6, y_5) + \Psi^{\downarrow}(\lambda, x_6, x_7; y_6, y_7) + \cdots + \Psi^{\downarrow}(\lambda, x_p, x_{p-1}; y_p, y_{p-1}). \tag{29}$$

- *If p is odd, then:*

$$\begin{aligned}\Psi^{\downarrow}(\mu, x_1, x_p; y_1, y_p) \le\ & \Psi^{\downarrow}(\lambda, x_1, x_2; y_1, y_2) + \Psi^{\downarrow}(\lambda, x_2, x_3; y_2, y_3) + \Psi^{\downarrow}(\lambda, x_4, x_3; y_4, y_3) \\ & + \Psi^{\downarrow}(\lambda, x_4, x_5; y_4, y_5) + \Psi^{\downarrow}(\lambda, x_6, x_5; y_6, y_5) + \Psi^{\downarrow}(\lambda, x_6, x_7; y_6, y_7) \\ & + \cdots + \Psi^{\downarrow}(\lambda, x_p, x_{p-1}; y_p, y_{p-1}),\end{aligned} \tag{30}$$

$$\begin{aligned}\Psi^{\downarrow}(\mu, x_1, x_p; y_p, y_1) \le\ & \Psi^{\downarrow}(\lambda, x_1, x_2; y_2, y_1) + \Psi^{\downarrow}(\lambda, x_2, x_3; y_3, y_2) + \Psi^{\downarrow}(\lambda, x_4, x_3; y_3, y_4) \\ & + \Psi^{\downarrow}(\lambda, x_4, x_5; y_5, y_4) + \Psi^{\downarrow}(\lambda, x_6, x_5; y_5, y_6) + \Psi^{\downarrow}(\lambda, x_6, x_7; y_7, y_6) \\ & + \cdots + \Psi^{\downarrow}(\lambda, x_p, x_{p-1}; y_{p-1}, y_p),\end{aligned} \tag{31}$$

$$\begin{aligned}\Psi^{\downarrow}(\mu, x_p, x_1; y_1, y_p) \le\ & \Psi^{\downarrow}(\lambda, x_2, x_1; y_1, y_2) + \Psi^{\downarrow}(\lambda, x_3, x_2; y_2, y_3) + \Psi^{\downarrow}(\lambda, x_3, x_4; y_4, y_3) \\ & + \Psi^{\downarrow}(\lambda, x_5, x_4; y_4, y_5) + \Psi^{\downarrow}(\lambda, x_5, x_6; y_6, y_5) + \Psi^{\downarrow}(\lambda, x_7, x_6; y_6, y_7) \\ & + \cdots + \Psi^{\downarrow}(\lambda, x_{p-1}, x_p; y_p, y_{p-1}),\end{aligned} \tag{32}$$

$$\begin{aligned}\Psi^{\downarrow}(\mu, x_p, x_1; y_p, y_1) \le\ & \Psi^{\downarrow}(\lambda, x_2, x_1; y_2, y_1) + \Psi^{\downarrow}(\lambda, x_3, x_2; y_3, y_2) + \Psi^{\downarrow}(\lambda, x_3, x_4; y_3, y_4) \\ & + \Psi^{\downarrow}(\lambda, x_5, x_4; y_5, y_4) + \Psi^{\downarrow}(\lambda, x_5, x_6; y_5, y_6) + \Psi^{\downarrow}(\lambda, x_7, x_6; y_7, y_6) \\ & + \cdots + \Psi^{\downarrow}(\lambda, x_{p-1}, x_p; y_{p-1}, y_p).\end{aligned} \tag{33}$$

Proof. To prove part (i), if $\mu = 1$, then $\Psi(1, x_1, x_p; y_1, y_p) = 0$. Therefore, the result is obvious. Now we assume $\mu \in (0,1)$. Using Lemma 1, there exists $\lambda \in (0,1)$, satisfying:

$$(1-\lambda) * \cdots * (1-\lambda) > 1 - \mu. \tag{34}$$

Given any $\epsilon > 0$, the first observation of Remark 1 says that:

$$\begin{aligned}& M\left(x_1, x_p, \Psi^{\downarrow}(\lambda, x_1, x_2; y_1, y_2) + \Psi^{\downarrow}(\lambda, x_2, x_3; y_2, y_3) + \cdots + \Psi^{\downarrow}(\lambda, x_{p-1}, x_p; y_{p-1}, y_p) + (p-1)\epsilon\right) \\ & \ge M\left(x_1, x_2, \Psi^{\downarrow}(\lambda, x_1, x_2; y_1, y_2) + \epsilon\right) * \cdots * M\left(x_{p-1}, x_p, \Psi^{\downarrow}(\lambda, x_{p-1}, x_p; y_{p-1}, y_p) + \epsilon\right),\end{aligned} \tag{35}$$

and:

$$\begin{aligned}& M\left(y_1, y_p, \Psi^{\downarrow}(\lambda, x_1, x_2; y_1, y_2) + \Psi^{\downarrow}(\lambda, x_2, x_3; y_2, y_3) + \cdots + \Psi^{\downarrow}(\lambda, x_{p-1}, x_p; y_{p-1}, y_p) + (p-1)\epsilon\right) \\ & \ge M\left(y_1, y_2, \Psi^{\downarrow}(\lambda, x_1, x_2; y_1, y_2) + \epsilon\right) * \cdots * M\left(y_{p-1}, y_p, \Psi^{\downarrow}(\lambda, x_{p-1}, x_p; y_{p-1}, y_p) + \epsilon\right).\end{aligned} \tag{36}$$

Now applying the increasing property and commutativity of t-norm to Equations (35) and (36), we obtain:

$$\begin{aligned}& \zeta\left(x_1, x_p; y_1, y_p, \Psi^{\downarrow}(\lambda, x_1, x_2; y_1, y_2) + \Psi^{\downarrow}(\lambda, x_2, x_3; y_2, y_3) + \cdots + \Psi^{\downarrow}(\lambda, x_{p-1}, x_p; y_{p-1}, y_p) + (p-1)\epsilon\right) \\ & \ge \zeta\left(x_1, x_2; y_1, y_2, \Psi^{\downarrow}(\lambda, x_1, x_2; y_1, y_2) + \epsilon\right) * \cdots * \zeta\left(x_{p-1}, x_p; y_{p-1}, y_p, \Psi^{\downarrow}(\lambda, x_{p-1}, x_p; y_{p-1}, y_p) + \epsilon\right) \\ & \ge (1-\lambda) * \cdots * (1-\lambda) \text{ (by Equation (15) and the increasing property of t-norm)} \\ & > 1 - \mu \text{ (by Equation (34)).}\end{aligned}$$

The definition of $\Psi^{\downarrow}$ says that:

$$\begin{aligned}& \Psi^{\downarrow}(\lambda, x_1, x_2; y_1, y_2) + \Psi^{\downarrow}(\lambda, x_2, x_3; y_2, y_3) + \cdots + \Psi^{\downarrow}(\lambda, x_{p-1}, x_p; y_{p-1}, y_p) + (p-1)\epsilon \\ & \ge \Psi^{\downarrow}(\mu, x_1, x_p; y_1, y_p).\end{aligned}$$

By taking $\epsilon \to 0+$, we obtain the desired inequality (Equation (24)).

On the other hand, we also have:

$$\begin{aligned}&M\left(x_1,x_p,\Psi^{\downarrow}(\lambda,x_1,x_2;y_2,y_1)+\Psi^{\downarrow}(\lambda,x_2,x_3;y_3,y_2)+\cdots+\Psi^{\downarrow}(\lambda,x_{p-1},x_p;y_p,y_{p-1})+(p-1)\epsilon\right)\\&\geq M\left(x_1,x_2,\Psi^{\downarrow}(\lambda,x_1,x_2;y_2,y_1)+\epsilon\right)*\cdots*M\left(x_{p-1},x_p,\Psi^{\downarrow}(\lambda,x_{p-1},x_p;y_p,y_{p-1})+\epsilon\right),\end{aligned}\tag{37}$$

and:

$$\begin{aligned}&M\left(y_p,y_1,\Psi^{\downarrow}(\lambda,x_1,x_2;y_2,y_1)+\Psi^{\downarrow}(\lambda,x_2,x_3;y_3,y_2)+\cdots+\Psi^{\downarrow}(\lambda,x_{p-1},x_p;y_p,y_{p-1})+(p-1)\epsilon\right)\\&\geq M\left(y_2,y_1,\Psi^{\downarrow}(\lambda,x_1,x_2;y_2,y_1)+\epsilon\right)*\cdots*M\left(y_p,y_{p-1},\Psi^{\downarrow}(\lambda,x_{p-1},x_p;y_p,y_{p-1})+\epsilon\right).\end{aligned}\tag{38}$$

Now applying the increasing property and commutativity of t-norm to Equations (37) and (38), we also obtain:

$$\begin{aligned}&\zeta\left(x_1,x_p;y_p,y_1,\Psi^{\downarrow}(\lambda,x_1,x_2;y_2,y_1)+\Psi^{\downarrow}(\lambda,x_2,x_3;y_3,y_2)+\cdots+\Psi^{\downarrow}(\lambda,x_{p-1},x_p;y_p,y_{p-1})+(p-1)\epsilon\right)\\&\geq\zeta\left(x_1,x_2;y_2,y_1,\Psi^{\downarrow}(\lambda,x_1,x_2;y_2,y_1)+\epsilon\right)*\cdots*\zeta\left(x_{p-1},x_p;y_p,y_{p-1},\Psi^{\downarrow}(\lambda,x_{p-1},x_p;y_p,y_{p-1})+\epsilon\right)\\&\geq(1-\lambda)*\cdots*(1-\lambda)\text{ (by Equation (15) and the increasing property of t-norm)}\\&>1-\mu\text{ (by Equation (34)).}\end{aligned}$$

The definition of $\Psi^{\downarrow}$ says that:

$$\begin{aligned}&\Psi^{\downarrow}(\lambda,x_1,x_2;y_2,y_1)+\Psi^{\downarrow}(\lambda,x_2,x_3;y_3,y_2)+\cdots+\Psi^{\downarrow}(\lambda,x_{p-1},x_p;y_p,y_{p-1})+(p-1)\epsilon\\&\geq\Psi^{\downarrow}(\mu,x_1,x_p;y_p,y_1).\end{aligned}$$

By taking $\epsilon\to 0+$, we obtain the desired inequality (Equation (25)). Since the other inequalities can be similarly obtained, we omit the details.

The above argument is still valid to obtain part (ii) by referring the second observation of Remark 1. Further, we can use the third observation of Remark 1 to obtain part (iii). Finally, part (iv) can be obtained by referring to the fourth observation of Remark 1. This completes the proof. □

Let (X,M) be a fuzzy semi-metric space, and let $\{x_n\}_{n=1}^{\infty}$ be a sequence in X. We write $x_n\stackrel{M^{\triangleright}}{\longrightarrow}x$ as $n\to\infty$ if and only if:

$$\lim_{n\to\infty}M(x_n,x,t)=1\text{ for all }t>0.$$

We also write $x_n\stackrel{M^{\triangleleft}}{\longrightarrow}x$ as $n\to\infty$ if and only if:

$$\lim_{n\to\infty}M(x,x_n,t)=1\text{ for all }t>0.$$

The main convergence theorem is presented below. We first provide a useful lemma.

Lemma 2. *Let $*$ be a t-norm. If $a*b>k$ then $a>k$ and $b>k$.*

Proof. Since $b\leq 1$, the increasing property and boundary condition show that $b*k\leq 1*k=k$. Suppose that $a\leq k$. Then we have $a*b\leq k*b$ and:

$$k<a*b\leq k*b\leq k.$$

A contradiction occurs. Therefore, we must have $a>k$. We can similarly show that $b>k$. This completes the proof. □

Theorem 2. *Let (X, M) be a fuzzy semi-metric space along with a t-norm $*$. We also assume that M satisfies the canonical condition, and that the t-norm $*$ is left-continuous at 1 with respect to the first or second argument. Let $\{x_n\}_{n=1}^{\infty}$ and $\{y_n\}_{n=1}^{\infty}$ be two sequences in X. Then we have the following properties:*

- $x_n \xrightarrow{M^{\triangleright}} x$ *and* $y_n \xrightarrow{M^{\triangleright}} y$ *as* $n \to \infty$ *if and only if* $\Psi^{\downarrow}(\lambda, x_n, x; y_n, y) \to 0$ *as* $n \to \infty$ *for all* $\lambda \in (0,1)$.
- $x_n \xrightarrow{M^{\triangleright}} x$ *and* $y_n \xrightarrow{M^{\triangleleft}} y$ *as* $n \to \infty$ *if and only if* $\Psi^{\downarrow}(\lambda, x_n, x; y, y_n) \to 0$ *as* $n \to \infty$ *for all* $\lambda \in (0,1)$.
- $x_n \xrightarrow{M^{\triangleleft}} x$ *and* $y_n \xrightarrow{M^{\triangleright}} y$ *as* $n \to \infty$ *if and only if* $\Psi^{\downarrow}(\lambda, x, x_n; y_n, y) \to 0$ *as* $n \to \infty$ *for all* $\lambda \in (0,1)$.
- $x_n \xrightarrow{M^{\triangleleft}} x$ *and* $y_n \xrightarrow{M^{\triangleleft}} y$ *as* $n \to \infty$ *if and only if* $\Psi^{\downarrow}(\lambda, x, x_n; y, y_n) \to 0$ *as* $n \to \infty$ *for all* $\lambda \in (0,1)$.

Proof. For any fixed $\lambda \in (0,1)$, using Lemma 1, it follows that there exists $\lambda_0 \in (0,1)$, satisfying:

$$(1-\lambda_0) * (1-\lambda_0) > 1-\lambda. \tag{39}$$

We just prove the first case, since the other cases can be similarly obtained. Suppose that $M(x_n, x, t) \to 1$ and $M(y_n, y, t) \to 1$ as $n \to \infty$ for all $t > 0$. Then, given any $t > 0$ and $\delta > 0$, there exists $n_{t,\delta}^{(1)}, n_{t,\delta}^{(2)} \in \mathbb{N}$, satisfying $|M(x_n, x, t) - 1| < \delta$ for $n \geq n_{t,\delta}^{(1)}$ and $|M(y_n, y, t) - 1| < \delta$ for $n \geq n_{t,\delta}^{(2)}$. Therefore, given any $\epsilon \in (0,1)$, there exists $n_\epsilon \in \mathbb{N}$, satisfying:

$$\left|M\left(x_n, x, \frac{\epsilon}{2}\right) - 1\right| < \lambda_0 \text{ and } \left|M\left(y_n, y, \frac{\epsilon}{2}\right) - 1\right| < \lambda_0,$$

for $n \geq n_\epsilon$. We also have:

$$M\left(x_n, x, \frac{\epsilon}{2}\right) > 1-\lambda_0 \text{ and } M\left(y_n, y, \frac{\epsilon}{2}\right) > 1-\lambda_0,$$

for $n \geq n_\epsilon$. The increasing property of t-norm says that:

$$\zeta\left(x_n, x; y_n, y, \frac{\epsilon}{2}\right) = M\left(x_n, x, \frac{\epsilon}{2}\right) * M\left(y_n, y, \frac{\epsilon}{2}\right) \geq (1-\lambda_0) * (1-\lambda_0) > 1-\lambda.$$

The definition of $\Psi^{\downarrow}$ says that:

$$\Psi^{\downarrow}(\lambda, x_n, x; y_n, y) \leq \frac{\epsilon}{2} < \epsilon,$$

for $n \geq n_\epsilon$. This shows that $\Psi^{\downarrow}(\lambda, x_n, x; y_n, y) \to 0$ as $n \to \infty$.

Conversely, assume that $\Psi^{\downarrow}(\lambda, x_n, x; y_n, y) \to 0$ as $n \to \infty$ for all $\lambda \in (0,1)$. Now, given any $\delta > 0$ and $\lambda \in (0,1]$, there exists $n_{\delta,\lambda} \in \mathbb{N}$, satisfying $|\Psi^{\downarrow}(\lambda, x_n, x; y_n, y)| < \delta$ for all $n \geq n_{\delta,\lambda}$. Therefore, for any fixed $t > 0$ and given any $\epsilon \in (0,1)$, there exists $n_\epsilon \in \mathbb{N}$, satisfying:

$$\Psi\left(\frac{\epsilon}{2}, x_n, x; y_n, y\right) = \left|\Psi\left(\frac{\epsilon}{2}, x_n, x; y_n, y\right)\right| < t,$$

for $n \geq n_\epsilon$, which implies:

$$\zeta(x_n, x; y_n, y, t) \geq 1 - \frac{\epsilon}{2} > 1-\epsilon,$$

for $n \geq n_\epsilon$ by part (i) of Proposition 9, i.e.,

$$M(x_n, x, t) * M(y_n, y, t) > 1-\epsilon,$$

for $n \geq n_\epsilon$. Lemma 2 says that:

$$M(x_n, x, t) > 1-\epsilon \text{ and } M(y_n, y, t) > 1-\epsilon,$$

for $n \geq n_\epsilon$. This shows that $x_n \xrightarrow{M^{\triangleright}} x$ and $y_n \xrightarrow{M^{\triangleright}} y$ as $n \to \infty$, and the proof is complete. □

Example 5. *From Example 1, we see that:*

$$x_n \stackrel{M^{\triangleright}}{\longrightarrow} x \text{ if and only if } \lim_{n\to\infty} d(x_n, x) = 0,$$

and:

$$x_n \stackrel{M^{\triangleleft}}{\longrightarrow} x \text{ if and only if } \lim_{n\to\infty} d(x, x_n) = 0.$$

From Example 4, we have:

$$\Psi^{\downarrow}(\lambda, x, y; u, v) = \frac{C + \sqrt{C^2 + D}}{2},$$

where:

$$C = \frac{(d(x,y) + d(u,v))(1-\lambda)}{\lambda} \text{ and } D = \frac{d(x,y)\cdot d(u,v)\cdot(1-\lambda)}{\lambda}.$$

It is clear to see that $x_n \stackrel{M^{\triangleright}}{\longrightarrow} x$ *and* $y_n \stackrel{M^{\triangleright}}{\longrightarrow} y$ *as* $n \to \infty$ *if and only if* $\Psi^{\downarrow}(\lambda, x_n, x; y_n, y) \to 0$ *as* $n \to \infty$ *for all* $\lambda \in (0,1)$*. The other convergence presented in Theorem 2 can be similarly verified.*

Definition 4. *Let* (X, M) *be a fuzzy semi-metric space, and let* $\{x_n\}_{n=1}^{\infty}$ *be a sequence in* X*.*

- *The sequence* $\{x_n\}_{n=1}^{\infty}$ *is said to be a* $>$*-Cauchy sequence in a metric sense if and only if, given any pair* (r,t) *with* $t > 0$ *and* $0 < r < 1$*, there exists* $n_{r,t} \in \mathbb{N}$*, satisfying* $M(x_m, x_n, t) > 1 - r$ *for all pairs* (m,n) *of integers* m *and* n *with* $m > n \geq n_{r,t}$*.*
- *The sequence* $\{x_n\}_{n=1}^{\infty}$ *is said to be a* $<$*-Cauchy sequence in a metric sense if and only if, given any pair* (r,t) *with* $t > 0$ *and* $0 < r < 1$*, there exists* $n_{r,t} \in \mathbb{N}$*, satisfying* $M(x_n, x_m, t) > 1 - r$ *for all pairs* (m,n) *of integers* m *and* n *with* $m > n \geq n_{r,t}$*.*
- *The sequence* $\{x_n\}_{n=1}^{\infty}$ *is said to be a Cauchy sequence in a metric sense if and only if, given any pair* (r,t) *with* $t > 0$ *and* $0 < r < 1$*, there exists* $n_{r,t} \in \mathbb{N}$ *satisfying* $M(x_m, x_n, t) > 1 - r$ *and* $M(x_n, x_m, t) > 1 - r$ *for all pairs* (m,n) *of integers* m *and* n *with* $m, n \geq n_{r,t}$ *and* $m \neq n$*.*

Definition 5. *Let* (X, M) *be a fuzzy semi-metric space such that* M *satisfies the canonical condition, and let* $\{x_n\}_{n=1}^{\infty}$ *and* $\{y_n\}_{n=1}^{\infty}$ *be two sequences in* X*.*

- *Given any fixed* $\lambda \in (0,1)$*, the sequences* $\{x_n\}_{n=1}^{\infty}$ *and* $\{y_n\}_{n=1}^{\infty}$ *are said to be the joint* $(\lambda, >, >)$*-Cauchy sequences with respect to* $\Psi^{\downarrow}$ *if and only if, given any* $\epsilon > 0$*, there exists* $n_{\epsilon,\lambda} \in \mathbb{N}$ *such that* $m > n \geq n_{\epsilon,\lambda}$ *implies* $\Psi^{\downarrow}(\lambda, x_m, x_n; y_m, y_n) < \epsilon$*.*
- *Given any fixed* $\lambda \in (0,1)$*, the sequences* $\{x_n\}_{n=1}^{\infty}$ *and* $\{y_n\}_{n=1}^{\infty}$ *are said to be the joint* $(\lambda, >, <)$*-Cauchy sequences with respect to* $\Psi^{\downarrow}$ *if and only if, given any* $\epsilon > 0$*, there exists* $n_{\epsilon,\lambda} \in \mathbb{N}$ *such that* $m > n \geq n_{\epsilon,\lambda}$ *implies* $\Psi^{\downarrow}(\lambda, x_m, x_n; y_n, y_m) < \epsilon$*.*
- *Given any fixed* $\lambda \in (0,1)$*, the sequences* $\{x_n\}_{n=1}^{\infty}$ *and* $\{y_n\}_{n=1}^{\infty}$ *are said to be the joint* $(\lambda, <, >)$*-Cauchy sequences with respect to* $\Psi^{\downarrow}$ *if and only if, given any* $\epsilon > 0$*, there exists* $n_{\epsilon,\lambda} \in \mathbb{N}$ *such that* $m > n \geq n_{\epsilon,\lambda}$ *implies* $\Psi^{\downarrow}(\lambda, x_n, x_m; y_m, y_n) < \epsilon$*.*
- *Given any fixed* $\lambda \in (0,1)$*, the sequences* $\{x_n\}_{n=1}^{\infty}$ *and* $\{y_n\}_{n=1}^{\infty}$ *are said to be the joint* $(\lambda, <, <)$*-Cauchy sequences with respect to* $\Psi^{\downarrow}$ *if and only if, given any* $\epsilon > 0$*, there exists* $n_{\epsilon,\lambda} \in \mathbb{N}$ *such that* $m > n \geq n_{\epsilon,\lambda}$ *implies* $\Psi^{\downarrow}(\lambda, x_n, x_m; y_n, y_m) < \epsilon$*.*

Theorem 3. *Let* (X, M) *be a fuzzy semi-metric space along with a t-norm* $*$*. We also assume that* M *satisfies the canonical condition, and that the t-norm* $*$ *is left-continuous at 1 with respect to the first or second argument. Let* $\{x_n\}_{n=1}^{\infty}$ *and* $\{y_n\}_{n=1}^{\infty}$ *be two sequences in* X*. Then, we have the following properties:*

(i) $\{x_n\}_{n=1}^{\infty}$ *and* $\{y_n\}_{n=1}^{\infty}$ *are two* $>$*-Cauchy sequences in a metric sense if and only if* $\{x_n\}_{n=1}^{\infty}$ *and* $\{y_n\}_{n=1}^{\infty}$ *are the joint* $(\lambda, >, >)$*-Cauchy sequences with respect to* $\Psi^{\downarrow}$ *for any* $\lambda \in (0,1)$*.*

(ii) $\{x_n\}_{n=1}^{\infty}$ *is a* $>$*-Cauchy sequences and* $\{y_n\}_{n=1}^{\infty}$ *is a* $<$*-Cauchy sequences in a metric sense if and only if* $\{x_n\}_{n=1}^{\infty}$ *and* $\{y_n\}_{n=1}^{\infty}$ *are the joint* $(\lambda, >, <)$*-Cauchy sequences with respect to* $\Psi^{\downarrow}$ *for any* $\lambda \in (0,1)$*.*

(iii) $\{x_n\}_{n=1}^{\infty}$ *is a* $<$*-Cauchy sequences and* $\{y_n\}_{n=1}^{\infty}$ *is a* $>$*-Cauchy sequences in a metric sense if and only if* $\{x_n\}_{n=1}^{\infty}$ *and* $\{y_n\}_{n=1}^{\infty}$ *are the joint* $(\lambda,<,>)$*-Cauchy sequences with respect to* $\Psi^{\downarrow}$ *for any* $\lambda \in (0,1)$.

(iv) $\{x_n\}_{n=1}^{\infty}$ *and* $\{y_n\}_{n=1}^{\infty}$ *are two* $<$*-Cauchy sequences if and only if* $\{x_n\}_{n=1}^{\infty}$ *and* $\{y_n\}_{n=1}^{\infty}$ *are the joint* $(\lambda,<,<)$*-Cauchy sequences in a metric sense with respect to* $\Psi^{\downarrow}$ *for any* $\lambda \in (0,1)$.

Proof. It suffices to just prove part (i), since the other cases can be similarly obtained. Suppose that $\{x_n\}_{n=1}^{\infty}$ and $\{y_n\}_{n=1}^{\infty}$ are $>$-Cauchy sequences. Then, given any $t > 0$ and $\delta > 0$, there exists $n_{t,\delta} \in \mathbb{N}$ such that $m > n \geq n_{t,\delta}$ implies $M(x_m, x_n, t) > 1 - \delta$ and $M(y_m, y_n, t) > 1 - \delta$. Now, given any $\epsilon \in (0,1)$, there exists $n_\epsilon \in \mathbb{N}$ such that $m > n \geq n_\epsilon$ implies:

$$M\left(x_m, x_n, \frac{\epsilon}{2}\right) > 1 - \lambda_0 \text{ and } M\left(y_m, y_n, \frac{\epsilon}{2}\right) > 1 - \lambda_0.$$

The increasing property of t-norm says that:

$$\begin{aligned}\zeta\left(x_m, x_n; y_m, y_n, \frac{\epsilon}{2}\right) &= M\left(x_m, x_n, \frac{\epsilon}{2}\right) * M\left(y_m, y_n, \frac{\epsilon}{2}\right)\\ &\geq (1-\lambda_0) * (1-\lambda_0) > 1 - \lambda \text{ (using Equation (39))}.\end{aligned}$$

Further, by referring to the definition of $\Psi^{\downarrow}$, we obtain:

$$\Psi^{\downarrow}(\lambda, x_m, x_n; y_m, y_n) \leq \frac{\epsilon}{2} < \epsilon,$$

for $m > n \geq n_\epsilon$.

Conversely, using the assumption, for any fixed $t > 0$ and given any $\epsilon \in (0,1)$, there exists $n_\epsilon \in \mathbb{N}$ such that $m > n \geq n_\epsilon$ implies $\Psi(\epsilon/2, x_m, x_n; y_m, y_n) < t$. Using Proposition 9, we obtain:

$$\zeta(x_m, x_n; y_m, y_n, t) \geq 1 - \frac{\epsilon}{2} > 1 - \epsilon,$$

for $m > n \geq n_\epsilon$, i.e.,

$$M(x_m, x_n, t) * M(y_m, y_n, t) > 1 - \epsilon,$$

for $m > n \geq n_\epsilon$. Lemma 2 says that:

$$M(x_m, x_n, t) > 1 - \epsilon \text{ and } M(y_m, y_n, t) > 1 - \epsilon,$$

for $m > n \geq n_\epsilon$, which shows that $\{x_n\}_{n=1}^{\infty}$ and $\{y_n\}_{n=1}^{\infty}$ are $>$-Cauchy sequences. This completes the proof. □

5. Convergence Based on the Supremum

Using the infimum and assuming the canonical condition, the infimum type of dual double fuzzy semi-metric was proposed in the previous section. In this section, we shall consider the supremum to propose the so-called supremum type of dual double fuzzy semi-metric.

Recall that the purpose for considering the canonical condition is to guarantee the infimum type of dual fuzzy semi-metric space to be well-defined. Now, we shall consider the rational condition to guarantee the supremum type of dual fuzzy semi-metric space to be well-defined. The formal definition is given below.

Definition 6. *Let* (X, M) *be a fuzzy semi-metric space along with a t-norm* $*$ *such that* M *satisfies the rational condition, and that the t-norm* $*$ *is right-continuous at* 0 *with respect to the first or second argument. Given any fixed* $x, y, u, v \in X$ *with* $x \neq y$ *or* $u \neq v$ *and any fixed* $\lambda \in [0,1)$*, we consider the following set:*

$$\Pi^{\uparrow}(\lambda, x, y; u, v) = \{t > 0 : \zeta(x, y; u, v, t) \leq 1 - \lambda\},$$

which will be used to define a function $\Psi^{\uparrow} : X^4 \to [0,+\infty)$ *by:*

$$\Psi^{\uparrow}(\lambda,x,y;u,v) = \sup \Pi^{\uparrow}(\lambda,x,y;u,v) = \sup\left\{t>0 : \zeta(x,y;u,v,t) \leq 1-\lambda\right\}.$$

The mapping $\Pi^{\uparrow}$ *from* $(0,1] \times X^4$ *into* $[0,\infty)$ *is called the supremum type of dual double fuzzy semi-metric.*

Example 6. *Continued from Example 1, we have:*

$$\Pi^{\uparrow}(\lambda,x,y;u,v) = \left\{t>0 : \frac{t}{t+d(x,y)} \cdot \frac{t}{t+d(u,v)} \leq 1-\lambda\right\} = \left\{t>0 : t \leq \frac{C+\sqrt{C^2+D}}{2}\right\},$$

where:

$$C = \frac{(d(x,y)+d(u,v))(1-\lambda)}{\lambda} \text{ and } D = \frac{d(x,y)\cdot d(u,v)\cdot(1-\lambda)}{\lambda}.$$

We also have:

$$\Psi^{\uparrow}(\lambda,x,y;u,v) = \sup \Pi^{\uparrow}(\lambda,x,y;u,v) = \left\{t>0 : t \leq \frac{C+\sqrt{C^2+D}}{2}\right\} = \frac{C+\sqrt{C^2+D}}{2}.$$

For any $x \neq y$ or $u \neq v$, we need to claim that the set $\Pi^{\uparrow}(\lambda,x,y;u,v)$ is nonempty. Suppose that $\Pi^{\uparrow}(\lambda,x,y;u,v) = \emptyset$. The definition says that $\zeta(x,y;u,v,t) > 1-\lambda$ for all $t>0$. Therefore, we obtain:

$$\lim_{t\to 0+} \zeta(x,y;u,v,t) \geq 1-\lambda,$$

which contradicts Equation (9). This says that Definition 6 is well-defined, which also says that $\Psi^{\uparrow}(\lambda,x,y;u,v) > 0$. We also have:

$$\Psi^{\uparrow}(0,x,y;u,v) = \sup\left\{t>0 : \zeta(x,y;u,v,t) \leq 0\right\} = \sup\{t>0\} = +\infty.$$

Moreover, if $\lambda_1 > \lambda_2$, then:

$$\Pi^{\uparrow}(\lambda_1,x,y;u,v) \subseteq \Pi^{\uparrow}(\lambda_2,x,y;u,v) \text{ and } \Psi^{\uparrow}(\lambda_1,x,y;u,v) \leq \Psi^{\uparrow}(\lambda_2,x,y;u,v). \tag{40}$$

Proposition 12. *Let* (X,M) *be a fuzzy semi-metric space along with a t-norm* $*$. *We also assume that* M *satisfies the rational condition, and that the t-norm* $*$ *is right-continuous at* 0 *with respect to the first or second argument. Given any fixed* $x,y,u,v \in X$ *with* $x \neq y$ *or* $u \neq v$, *suppose that* $\Psi^{\uparrow}(\lambda,x,y;u,v) = +\infty$. *Then, the following statements hold true:*

(i) *Suppose that* M *satisfies the* $\circ$*-triangle inequality for* $\circ \in \{\bowtie,\triangleright,\triangleleft\}$. *Then we have* $\zeta(x,y;u,v,t) \leq 1-\lambda$ *for all* $t>0$.
(ii) *Suppose that* M *satisfies the* $\circ$*-triangle inequality for* $\circ \in \{\triangleright,\triangleleft\}$. *Then we have* $\zeta(y,x;u,v,t) \leq 1-\lambda$ *and* $\zeta(x,y;v,u,t) \leq 1-\lambda$ *for all* $t>0$.
(iii) *Suppose that* M *satisfies the* $\circ$*-triangle inequality for* $\circ \in \{\triangleright,\triangleleft,\diamond\}$. *Then we have* $\zeta(y,x;v,u,t) \leq 1-\lambda$ *for all* $t>0$.

Proof. The fact $\Psi^{\uparrow}(\lambda,x,y;u,v) = +\infty$ says that $\zeta(x,y;u,v,t) \leq 1-\lambda$ for sufficiently large $t>0$ in the sense of $t\to\infty$. To prove part (i), we assume that there exists $t_0>0$, satisfying $\zeta(x,y;u,v,t_0) > 1-\lambda$. Parts (i) and (ii) of Proposition 3 say that the mapping $\zeta(x,y;u,v,\cdot)$ is nondecreasing. Therefore, if $t_1 > t_0$, then:

$$\zeta(x,y;u,v,t_1) \geq \zeta(x,y;u,v,t_0) > 1-\lambda,$$

which contradicts $\zeta(x,y;u,v,t) \leq 1-\lambda$ for sufficiently large $t>0$.

To prove part (ii), we assume that there exists $t_0 > 0$, satisfying $\zeta(y,x;u,v,t_0) > 1-\lambda$. Part (ii) of Proposition 3 says that the mapping $\zeta(x,y;u,v,\cdot)$ is $\triangleleft$-semisymmetrically nondecreasing. Therefore, if $t_1 > t_0$, then:

$$\zeta(x,y;u,v,t_1) \geq \zeta(y,x;u,v,t_0) > 1-\lambda,$$

which contradicts $\zeta(x,y;u,v,t) \leq 1-\lambda$ for sufficiently large $t > 0$. We can similarly obtain another inequality using the fact of the mapping $\zeta(x,y;u,v,\cdot)$ to be $\triangleright$-semisymmetrically nondecreasing.

To prove part (iii), we assume that there exists $t_0 > 0$, satisfying $\zeta(y,x;v,u,t_0) > 1-\lambda$. Parts (ii) and (iii) of Proposition 3 say that the mapping $\zeta(x,y;u,v,\cdot)$ is symmetrically nondecreasing. Therefore, if $t_1 > t_0$, then:

$$\zeta(x,y;u,v,t_1) \geq \zeta(y,x;v,u,t_0) > 1-\lambda,$$

which contradicts $\zeta(x,y;u,v,t) \leq 1-\lambda$ for sufficiently large $t > 0$. This completes the proof. □

Proposition 13. *Let (X,M) be a fuzzy semi-metric space along with a t-norm $*$. We also assume that M satisfies the rational and canonical conditions, and that the t-norm $*$ is right-continuous at 0 and left-continuous at 1 with respect to the first or second argument. Then, given any fixed $x,y,u,v \in X$ with $x \neq y$ or $u \neq v$, we have $\Psi^{\uparrow}(\lambda,x,y;u,v) < +\infty$ for $\lambda \in (0,1)$.*

Proof. We assume that $\Psi^{\uparrow}(\lambda,x,y;u,v) = +\infty$, which means that $\zeta(x,y;u,v,t) \leq 1-\lambda$ for sufficiently large t in the sense of $t \to \infty$. Using Equation (8), we obtain

$$1 = \lim_{t\to\infty} \zeta(x,y;u,v,t) \leq 1-\lambda,$$

which leads to a contradiction for $0 < \lambda < 1$. This completes the proof. □

Proposition 14. *Let (X,M) be a fuzzy semi-metric space along with a t-norm $*$. Assume that M satisfies the canonical and rational conditions. We also assume that the t-norm $*$ is left-continuous at 1 and right-continuous at 0 with respect to the first or second argument, and that the t-norm $*$ also satisfies the strictly increasing property. For any fixed $x,y,u,v \in X$ with $x \neq y$ or $u \neq v$, the following statements hold true:*

(i) *Suppose that M satisfies the strict $\circ$-triangle inequality for $\circ \in \{\bowtie,\triangleright,\triangleleft\}$. Then we have:*

$$\Psi^{\uparrow}(\lambda,x,y;u,v) \leq \Psi^{\downarrow}(\lambda,x,y;u,v)$$

for each $\lambda \in (0,1)$.

(ii) *Suppose that M satisfies the strict $\circ$-triangle inequality for $\circ \in \{\triangleright,\triangleleft\}$. Then we have:*

$$\Psi^{\uparrow}(\lambda,x,y;u,v) \leq \Psi^{\downarrow}(\lambda,y,x;u,v) \text{ and } \Psi^{\uparrow}(\lambda,x,y;u,v) \leq \Psi^{\downarrow}(\lambda,x,y;v,u)$$

for each $\lambda \in (0,1)$.

(iii) *Suppose that M satisfies the strict $\circ$-triangle inequality for $\circ \in \{\triangleright,\triangleleft,\diamond\}$. Then we have:*

$$\Psi^{\uparrow}(\lambda,x,y;u,v) \leq \Psi^{\downarrow}(\lambda,y,x;v,u)$$

for each $\lambda \in (0,1)$.

Proof. Proposition 13 says that $\Psi^{\uparrow}(\lambda,x,y;u,v) < +\infty$ for all $\lambda \in (0,1)$. According to the concept of supremum, given any $\epsilon > 0$, there exists $t_\epsilon > 0$, satisfying $\zeta(x,y;u,v,t_\epsilon) \leq 1-\lambda$ and $\Psi^{\uparrow}(\lambda,x,y;u,v) - \epsilon < t_\epsilon$. To prove part (i), parts (i) and (ii) of Proposition 10 say that $t_\epsilon \leq \Psi^{\downarrow}(\lambda,x,y;u,v)$, which implies $\Psi^{\uparrow}(\lambda,x,y;u,v) - \epsilon < \Psi^{\downarrow}(\lambda,x,y;u,v)$. Since ϵ can be any positive real number, we obtain the desired inequality.

To prove part (ii), parts (i) and (ii) of Proposition 10 say that $t_\epsilon \leq \Psi^{\downarrow}(\lambda, y, x; u, v)$, which implies $\Psi^{\uparrow}(\lambda, x, y; u, v) - \epsilon < \Psi^{\downarrow}(\lambda, y, x; u, v)$. Since ϵ can be any positive real number, we obtain $\Psi^{\uparrow}(\lambda, x, y; u, v) \leq \Psi^{\downarrow}(\lambda, y, x; u, v)$. Another inequality can be similarly obtained.

To prove part (iii), parts (ii) and (iii) of Proposition 10 say that $t_\epsilon \leq \Psi^{\downarrow}(\lambda, y, x; v, u)$, which implies $\Psi^{\uparrow}(\lambda, x, y; u, v) - \epsilon < \Psi^{\downarrow}(\lambda, y, x; v, u)$. Since ϵ can be any positive real number, we obtain the desired inequality. This completes the proof. □

Proposition 15. *Let (X, M) be a fuzzy semi-metric space along with a t-norm $*$. We also assume that M satisfies the rational condition, and that the t-norm $*$ is right-continuous at 0 with respect to the first or second argument. For any fixed $x, y, u, v \in X$ with $x \neq y$ or $u \neq v$, and any fixed $\lambda \in (0, 1)$, we assume $\Psi^{\uparrow}(\lambda, x, y; u, v) < +\infty$.*

(i) *For any $\epsilon > 0$, we have the following inequality:*

$$\zeta\left(x, y; u, v, \Psi^{\uparrow}(\lambda, x, y; u, v) + \epsilon\right) > 1 - \lambda \tag{41}$$

(ii) *If $\epsilon > 0$ is sufficiently small satisfying $\Psi^{\uparrow}(\lambda, x, y; u, v) > \epsilon$, then the following statements hold true:*

- *Suppose that M satisfies the $\circ$-triangle inequality for $\circ \in \{\bowtie, \triangleright, \triangleleft\}$. Then we have:*

$$\zeta\left(x, y; u, v, \Psi^{\uparrow}(\lambda, x, y; u, v) - \epsilon\right) \leq 1 - \lambda. \tag{42}$$

- *Suppose that M satisfies the $\circ$-triangle inequality for $\circ \in \{\triangleright, \triangleleft\}$. Then we have:*

$$\zeta\left(y, x; u, v, \Psi^{\uparrow}(\lambda, x, y; u, v) - \epsilon\right) \leq 1 - \lambda \text{ and } \zeta\left(x, y; v, u, \Psi^{\uparrow}(\lambda, x, y; u, v) - \epsilon\right) \leq 1 - \lambda. \tag{43}$$

- *Suppose that M satisfies the $\circ$-triangle inequality for $\circ \in \{\triangleright, \triangleleft, \diamond\}$. Then we have:*

$$\zeta\left(y, x; v, u, \Psi^{\uparrow}(\lambda, x, y; u, v) - \epsilon\right) \leq 1 - \lambda. \tag{44}$$

Proof. To prove part (i), given any $\epsilon > 0$, we assume that $\zeta(x, y, \Psi^{\uparrow}(\lambda, x, y; u, v) + \epsilon) \leq 1 - \lambda$. The definition of $\Psi^{\uparrow}$ says that $\Psi^{\uparrow}(\lambda, x, y; u, v) \geq \Psi^{\uparrow}(\lambda, x, y; u, v) + \epsilon$. This contradiction shows that $\zeta(x, y; u, v, \Psi^{\uparrow}(\lambda, x, y; u, v) + \epsilon) > 1 - \lambda$.

To prove part (ii), according to the concept of supremum for $\Psi^{\uparrow}(\lambda, x, y; u, v)$, given any $\epsilon > 0$ with $\Psi^{\uparrow}(\lambda, x, y; u, v) > \epsilon$, there exists $t_\epsilon > 0$, satisfying $\zeta(x, y; u, v, t_\epsilon) \leq 1 - \lambda$ and $t_\epsilon > \Psi^{\uparrow}(\lambda, x, y; u, v) - \epsilon$. Therefore, we consider three cases below:

- Suppose that M satisfies the $\circ$-triangle inequality for $\circ \in \{\bowtie, \triangleright, \triangleleft\}$. Parts (i) and (ii) of Proposition 3 say that the mapping $\zeta(x, y; u, v, \cdot)$ is nondecreasing. Therefore, we have:

$$\zeta\left(x, y; u, v, \Psi^{\uparrow}(\lambda, x, y; u, v) - \epsilon\right) \leq \zeta(x, y; u, v, t_\epsilon) \leq 1 - \lambda.$$

- Suppose that M satisfies the $\circ$-triangle inequality for $\circ \in \{\triangleright, \triangleleft\}$. Part (ii) of Proposition 3 says that the mapping $\zeta(x, y; u, v, \cdot)$ is $\triangleleft$-semisymmetrically nondecreasing. Therefore, we have:

$$\zeta\left(y, x; u, v, \Psi^{\uparrow}(\lambda, x, y; u, v) - \epsilon\right) \leq \zeta(x, y; u, v, t_\epsilon) \leq 1 - \lambda.$$

  We can similarly obtain another inequality using the fact of the mapping $\zeta(x, y; u, v, \cdot)$ to be $\triangleright$-semisymmetrically nondecreasing.

- Suppose that M satisfies the $\circ$-triangle inequality for $\circ \in \{\triangleright, \triangleleft, \diamond\}$. Parts (ii) and (iii) of Proposition 3 say that the mapping $\zeta(x,y;u,v,\cdot)$ is symmetrically nondecreasing. Therefore, we have:

$$\zeta\left(y,x;v,u,\Psi^{\uparrow}(\lambda,x,y;u,v)-\epsilon\right) \leq \zeta(x,y;u,v,t_{\epsilon}) \leq 1-\lambda.$$

This completes the proof. □

Proposition 16. *Let (X,M) be a fuzzy semi-metric space along with a t-norm $*$. We also assume that M satisfies the rational condition, and that the t-norm $*$ is right-continuous at 0 with respect to the first or second argument. Given any fixed $x,y,u,v \in X$ with $x \neq y$ or $u \neq v$, and any fixed $\lambda \in (0,1)$, the following statements hold true:*

(i) *Suppose that $t > \Psi^{\uparrow}(\lambda,x,y;u,v)$. Then, we have the following properties:*

- *If M satisfies the $\circ$-triangle inequality for $\circ \in \{\bowtie, \triangleright, \triangleleft\}$, then $\zeta(x,y;u,v,t) > 1-\lambda$.*
- *If M satisfies the $\circ$-triangle inequality for $\circ \in \{\triangleright, \triangleleft\}$, then $\zeta(y,x;u,v,t) > 1-\lambda$ and $\zeta(x,y;v,u,t) > 1-\lambda$.*
- *If M satisfies the $\circ$-triangle inequality for $\circ \in \{\triangleright, \triangleleft, \diamond\}$, then $\zeta(y,x;v,u,t) > 1-\lambda$.*

(ii) *We have the following properties:*

- *Suppose that M satisfies the $\circ$-triangle inequality for $\circ \in \{\bowtie, \triangleright, \triangleleft\}$. If $0 < t < \Psi^{\uparrow}(\lambda,x,y;u,v)$, then $\zeta(x,y;u,v,t) \leq 1-\lambda$.*
- *Suppose that M satisfies the $\circ$-triangle inequality for $\circ \in \{\triangleright, \triangleleft\}$. If $\Psi^{\uparrow}(\lambda,y,x;u,v) = +\infty$ or $\Psi^{\uparrow}(\lambda,x,y;v,u) = +\infty$ or $0 < t < \Psi^{\uparrow}(\lambda,x,y;u,v) < +\infty$, then $\zeta(x,y;u,v,t) \leq 1-\lambda$.*
- *Suppose that M satisfies the $\circ$-triangle inequality for $\circ \in \{\triangleright, \triangleleft, \diamond\}$. If $\Psi^{\uparrow}(\lambda,y,x;v,u) = +\infty$ or $0 < t < \Psi^{\uparrow}(\lambda,x,y;u,v) < +\infty$, then $\zeta(x,y;u,v,t) \leq 1-\lambda$.*

Proof. To prove part (i), the fact $t > \Psi^{\uparrow}(\lambda,x,y;u,v)$ says that there exists $\epsilon > 0$, satisfying $t \geq \Psi^{\uparrow}(\lambda,x,y;u,v)+\epsilon$. We consider three cases below:

- Suppose that M satisfies the $\circ$-triangle inequality for $\circ \in \{\bowtie, \triangleright, \triangleleft\}$. Parts (i) and (ii) of Proposition 3 say that the mapping $\zeta(x,y;u,v,\cdot)$ is nondecreasing. Therefore, using Equation (41), we obtain:

$$\zeta(x,y;u,v,t) \geq \zeta\left(x,y;u,v,\Psi^{\uparrow}(\lambda,x,y;u,v)+\epsilon\right) > 1-\lambda.$$

- Suppose that M satisfies the $\circ$-triangle inequality for $\circ \in \{\triangleright, \triangleleft\}$. Part (ii) of Proposition 3 says that the mapping $\zeta(x,y;u,v,\cdot)$ is both $\triangleleft$-semisymmetrically nondecreasing and $\triangleright$-semisymmetrically nondecreasing.Therefore, using Equation (41), we obtain:

$$\zeta(y,x;u,v,t) \geq \zeta\left(x,y;u,v,\Psi^{\uparrow}(\lambda,x,y;u,v)+\epsilon\right) > 1-\lambda,$$

and:

$$\zeta(x,y;v,u,t) \geq \zeta\left(x,y;u,v,\Psi^{\uparrow}(\lambda,x,y;u,v)+\epsilon\right) > 1-\lambda.$$

- Suppose that M satisfies the $\circ$-triangle inequality for $\circ \in \{\triangleright, \triangleleft, \diamond\}$. Parts (ii) and (iii) of Proposition 3 say that the mapping $\zeta(x,y;u,v,\cdot)$ is symmetrically nondecreasing. Therefore, using Equation (41), we obtain:

$$\zeta(y,x;v,u,t) \geq \zeta\left(x,y;u,v,\Psi^{\uparrow}(\lambda,x,y;u,v)+\epsilon\right) > 1-\lambda.$$

To prove part (ii), we consider three cases below:

- Suppose that M satisfies the $\circ$-triangle inequality for $\circ \in \{\bowtie, \triangleright, \triangleleft\}$. Using part (i) of Proposition 12, if $\Psi^{\uparrow}(\lambda, x, y; u, v) = +\infty$, then it is done. Now, for $\Psi^{\uparrow}(\lambda, x, y; u, v) < +\infty$, the fact $t < \Psi^{\uparrow}(\lambda, x, y; u, v)$ says that there exists $\epsilon > 0$, satisfying $0 < t \leq \Psi^{\uparrow}(\lambda, x, y; u, v) - \epsilon$. Using Equation (42), we obtain:

$$\zeta(x, y; u, v, t) \leq \zeta\left(x, y; u, v, \Psi^{\uparrow}(\lambda, x, y; u, v) - \epsilon\right) \leq 1 - \lambda.$$

- Suppose that M satisfies the $\circ$-triangle inequality for $\circ \in \{\triangleright, \triangleleft\}$. Using part (ii) of Proposition 12, if $\Psi^{\uparrow}(\lambda, y, x; u, v) = +\infty$ or $\Psi^{\uparrow}(\lambda, x, y; v, u) = +\infty$, then it is done. Now, for $\Psi^{\uparrow}(\lambda, x, y; u, v) < +\infty$, using Equation (43), we obtain:

$$\zeta(x, y; u, v, t) \leq \zeta\left(y, x; u, v, \Psi^{\uparrow}(\lambda, x, y; u, v) - \epsilon\right) \leq 1 - \lambda,$$

and:

$$\zeta(x, y; u, v, t) \leq \zeta\left(x, y; v, u, \Psi^{\uparrow}(\lambda, x, y; u, v) - \epsilon\right) \leq 1 - \lambda.$$

- Suppose that M satisfies the $\circ$-triangle inequality for $\circ \in \{\triangleright, \triangleleft, \diamond\}$. Using part (iii) of Proposition 12, if $\Psi^{\uparrow}(\lambda, y, x; v, u) = +\infty$, then it is done. Now, for $\Psi^{\uparrow}(\lambda, x, y; u, v) < +\infty$, using Equation (44), we obtain:

$$\zeta(x, y; u, v, t) \leq \zeta\left(y, x; v, u, \Psi^{\uparrow}(\lambda, x, y; u, v) - \epsilon\right) \leq 1 - \lambda.$$

This completes the proof. □

Proposition 17. *Let (X, M) be a fuzzy semi-metric space along with a t-norm $*$. We also assume that M satisfies the rational condition, and that the t-norm $*$ is right-continuous at 0 with respect to the first or second argument. Given any fixed $x, y, u, v \in X$ with $x \neq y$ or $u \neq v$, and any fixed $\lambda \in (0, 1)$, the following statements hold true:*

(i) *Suppose that $\zeta(x, y; u, v, t) \leq 1 - \lambda$ for $t > 0$. Then, we have the following properties:*

- *If M satisfies the $\circ$-triangle inequality for $\circ \in \{\bowtie, \triangleright, \triangleleft\}$, then $t \leq \Psi^{\uparrow}(\lambda, x, y; u, v)$.*
- *If M satisfies the $\circ$-triangle inequality for $\circ \in \{\triangleright, \triangleleft\}$, then $t \leq \Psi^{\uparrow}(\lambda, y, x; u, v)$ and $t \leq \Psi^{\uparrow}(\lambda, x, y; v, u)$.*
- *If M satisfies the $\circ$-triangle inequality for $\circ \in \{\triangleright, \triangleleft, \diamond\}$, then $t \leq \Psi^{\uparrow}(\lambda, y, x; v, u)$.*

(ii) *We have the following properties:*

- *Suppose that M satisfies the $\circ$-triangle inequality for $\circ \in \{\bowtie, \triangleright, \triangleleft\}$. If $\zeta(x, y; u, v, t) > 1 - \lambda$, then $\Psi^{\uparrow}(\lambda, x, y; u, v) < +\infty$ and $t \geq \Psi^{\uparrow}(\lambda, x, y; u, v)$.*
- *Suppose that M satisfies the $\circ$-triangle inequality for $\circ \in \{\triangleright, \triangleleft\}$. If $\zeta(x, y; u, v, t) > 1 - \lambda$, then $\Psi^{\uparrow}(\lambda, y, x; u, v) < +\infty$, $\Psi^{\uparrow}(\lambda, x, y; v, u) < +\infty$ and $\Psi^{\uparrow}(\lambda, y, x; v, u) < +\infty$.*
- *Suppose that M satisfies the $\diamond$-triangle inequality:*
  - *If $\zeta(x, y; u, v, t) > 1 - \lambda$, then $\Psi^{\uparrow}(\lambda, y, x; v, u) < +\infty$.*
  - *If $\zeta(x, y; u, v, t) > 1 - \lambda$ and $\Psi^{\uparrow}(\lambda, x, y; u, v) < +\infty$, then $t \geq \Psi^{\uparrow}(\lambda, x, y; u, v)$.*

Proof. To prove part (i), we consider three cases below:

- Suppose that M satisfies the $\circ$-triangle inequality for $\circ \in \{\bowtie, \triangleright, \triangleleft\}$. It is clear to see that the fact $\Psi^{\uparrow}(\lambda, x, y; u, v) = +\infty$ implies $t \leq \Psi^{\uparrow}(\lambda, x, y; u, v)$. Now, for $\Psi^{\uparrow}(\lambda, x, y; u, v) < +\infty$, using the contraposition of first property of part (i) of Proposition 16, we see that if $\zeta(x, y; u, v, t) \leq 1 - \lambda$, then $t \leq \Psi^{\uparrow}(\lambda, x, y; u, v)$.

- Suppose that M satisfies the $\circ$-triangle inequality for $\circ \in \{\triangleright, \triangleleft\}$. It is clear to see that the fact $\Psi^{\uparrow}(\lambda, y, x; u, v) = +\infty$ implies $t \leq \Psi^{\uparrow}(\lambda, y, x; u, v)$. Now, for $\Psi^{\uparrow}(\lambda, y, x; u, v) < +\infty$, using the contraposition of second property of part (i) of Proposition 16, we see that if $\zeta(x, y; u, v, t) \leq 1 - \lambda$, then $t \leq \Psi^{\uparrow}(\lambda, y, x; u, v)$. We can similarly show that if $\zeta(x, y; u, v, t) \leq 1 - \lambda$, then $t \leq \Psi^{\uparrow}(\lambda, x, y; v, u)$
- Suppose that M satisfies the $\circ$-triangle inequality for $\circ \in \{\triangleright, \triangleleft, \diamond\}$. It is clear to see that the fact $\Psi^{\uparrow}(\lambda, y, x; v, u) = +\infty$ implies $t \leq \Psi^{\uparrow}(\lambda, y, x; v, u)$. Now, for $\Psi^{\uparrow}(\lambda, y, x; v, u) < +\infty$, using the contraposition of third property of part (i) of Proposition 16, we see that if $\zeta(y, x; v, u, t) \leq 1 - \lambda$, then $t \leq \Psi^{\uparrow}(\lambda, y, x; v, u)$.

To prove part (ii), we consider three cases below:

- Suppose that M satisfies the $\circ$-triangle inequality for $\circ \in \{\bowtie, \triangleright, \triangleleft\}$. Using the contraposition of part (i) of Proposition 12 and the contraposition of first property of part (ii) of Proposition 16, we can obtain the desired result.
- Suppose that M satisfies the $\circ$-triangle inequality for $\circ \in \{\triangleright, \triangleleft\}$. Using part (ii) of Proposition 12, if $\zeta(x, y; u, v, t) > 1 - \lambda$, then $\Psi^{\uparrow}(\lambda, y, x; u, v) < +\infty$ and $\Psi^{\uparrow}(\lambda, x, y; v, u) < +\infty$. Using part (iii) of Proposition 12, if $\zeta(x, y; u, v, t) > 1 - \lambda$, then $\Psi^{\uparrow}(\lambda, y, x; v, u) < +\infty$.
- Suppose that M satisfies the $\circ$-triangle inequality for $\circ \in \{\triangleright, \triangleleft, \diamond\}$. Using part (iii) of Proposition 12, if $\zeta(x, y; u, v, t) > 1 - \lambda$, then $\Psi^{\uparrow}(\lambda, y, x; v, u) < +\infty$. Using the contraposition of third property of part (ii) of Proposition 16, if $\zeta(x, y; u, v, t) > 1 - \lambda$ and $\Psi^{\uparrow}(\lambda, x, y; u, v) < +\infty$ then $t \geq \Psi^{\uparrow}(\lambda, x, y; u, v)$.

This completes the proof. □

Proposition 18. *Let (X, M) be a fuzzy semi-metric space along with a t-norm $*$. We also assume that M satisfies the rational condition, and that the t-norm $*$ is right-continuous at 0 with respect to the first or second argument. Given any fixed $x, y, u, v \in X$ with $x \neq y$ or $u \neq v$, and any fixed $\lambda \in (0, 1)$, the following statements hold true:*

(i) *Suppose that $\Psi^{\uparrow}(\lambda, x, y; u, v) < +\infty$, and that the mapping $\zeta(x, y; u, v, \cdot) : (0, \infty) \to [0, 1]$ is right-continuous on $(0, \infty)$. Then we have:*

$$\zeta\left(x, y; u, v, \Psi^{\uparrow}(\lambda, x, y; u, v)\right) \geq 1 - \lambda. \tag{45}$$

(ii) *Suppose that the mapping $\zeta(x, y; u, v, \cdot) : (0, \infty) \to [0, 1]$ is left-continuous on $(0, \infty)$. Then, the following statements hold true:*

- *Suppose that M satisfies the $\circ$-triangle inequality for $\circ \in \{\bowtie, \triangleright, \triangleleft\}$.*

$$\text{If } \Psi^{\uparrow}(\lambda, x, y; u, v) < +\infty, \text{ then } \zeta\left(x, y; u, v, \Psi^{\uparrow}(\lambda, x, y; u, v)\right) \leq 1 - \lambda.$$

- *Suppose that M satisfies the $\circ$-triangle inequality for $\circ \in \{\triangleright, \triangleleft\}$.*

$$\text{If } \Psi^{\uparrow}(\lambda, y, x; u, v) < +\infty, \text{ then } \zeta\left(x, y; u, v, \Psi^{\uparrow}(\lambda, y, x; u, v)\right) \leq 1 - \lambda,$$

and:

$$\text{if } \Psi^{\uparrow}(\lambda, x, y; v, u) < +\infty, \text{ then } \zeta\left(x, y; u, v, \Psi^{\uparrow}(\lambda, x, y; v, u)\right) \leq 1 - \lambda.$$

- *Suppose that M satisfies the $\circ$-triangle inequality for $\circ \in \{\triangleright, \triangleleft, \diamond\}$.*

$$\text{If } \Psi^{\uparrow}(\lambda, y, x; v, u) < +\infty, \text{ then } \zeta\left(x, y; u, v, \Psi^{\uparrow}(\lambda, y, x; v, u)\right) \leq 1 - \lambda.$$

(iii) *Suppose that M satisfies the* $\circ$*-triangle inequality for* $\circ \in \{\bowtie, \triangleright, \triangleleft\}$, *and that the mapping* $\zeta(x,y;u,v,\cdot) : (0,\infty) \to [0,1]$ *is continuous on* $(0,\infty)$.

$$\text{If } \Psi^{\uparrow}(\lambda,x,y;u,v) < +\infty, \text{ then } \zeta\left(x,y;u,v,\Psi^{\uparrow}(\lambda,x,y;u,v)\right) = 1-\lambda.$$

Proof. To prove part (i), by taking the limit $\epsilon \to 0+$ to the inequality (Equation (41)), we obtain Equation (45). To prove part (ii), by taking the limit $\epsilon \to 0+$ to the inequalities (Equations (42)–(44)), we also obtain the desired results. Part (iii) follows from parts (i) and (ii) immediately. This completes the proof. □

Theorem 4. (Triangle Inequalities for Dual Double Fuzzy Semi-Metric). *Let* (X, M) *be a fuzzy semi-metric space along with a t-norm* $*$. *We also assume that M satisfies the rational condition, and that the t-norm* $*$ *is right-continuous at 0 and left-continuous at 1 with respect to the first or second argument. Given any distinct fixed* $x_1, x_2, \cdots, x_p, y_1, y_2, \cdots, y_p \in X$ *and any fixed* $\mu \in (0,1]$, *we have the following properties:*

(i) *Suppose that M satisfies the* $\bowtie$*-triangle inequality. There exists* $\lambda \in (0,1)$, *satisfying:*

$$\Psi^{\uparrow}(\mu,x_1,x_p;y_1,y_p) \leq \Psi^{\uparrow}(\lambda,x_1,x_2;y_1,y_2) + \Psi^{\uparrow}(\lambda,x_2,x_3;y_2,y_3) + \cdots + \Psi^{\uparrow}(\lambda,x_{p-2},x_{p-1};y_{p-2},y_{p-1}) + \Psi^{\uparrow}(\lambda,x_{p-1},x_p;y_{p-1},y_p), \tag{46}$$

$$\Psi^{\uparrow}(\mu,x_1,x_p;y_p,y_1) \leq \Psi^{\uparrow}(\lambda,x_1,x_2;y_2,y_1) + \Psi^{\uparrow}(\lambda,x_2,x_3;y_3,y_2) + \cdots + \Psi^{\uparrow}(\lambda,x_{p-2},x_{p-1};y_{p-1},y_{p-2}) + \Psi^{\uparrow}(\lambda,x_{p-1},x_p;y_p,y_{p-1}), \tag{47}$$

$$\Psi^{\uparrow}(\mu,x_p,x_1;y_1,y_p) \leq \Psi^{\uparrow}(\lambda,x_p,x_{p-1};y_{p-1},y_p) + \Psi^{\uparrow}(\lambda,x_{p-1},x_{p-2};y_{p-2},y_{p-1}) + \cdots + \Psi^{\uparrow}(\lambda,x_3,x_2;y_2,y_3) + \Psi^{\uparrow}(\lambda,x_2,x_1;y_1,y_2),$$

$$\Psi^{\uparrow}(\mu,x_p,x_1;y_p,y_1) \leq \Psi^{\uparrow}(\lambda,x_p,x_{p-1};y_p,y_{p-1}) + \Psi^{\uparrow}(\lambda,x_{p-1},x_{p-2};y_{p-1},y_{p-2}) + \cdots + \Psi^{\uparrow}(\lambda,x_3,x_2;y_3,y_2) + \Psi^{\uparrow}(\lambda,x_2,x_1;y_2,y_1).$$

(ii) *Suppose that M satisfies the* $\triangleright$*-triangle inequality. There exists* $\lambda \in (0,1)$, *satisfying:*

$$\max\left\{\Psi^{\uparrow}(\mu,x_1,x_p;y_1,y_p), \Psi^{\uparrow}(\mu,x_1,x_p;y_p,y_1), \Psi^{\uparrow}(\mu,x_p,x_1;y_1,y_p), \Psi^{\uparrow}(\mu,x_p,x_1;y_p,y_1)\right\} \leq \Psi^{\uparrow}(\lambda,x_1,x_2;y_1,y_2) + \Psi^{\uparrow}(\lambda,x_3,x_2;y_3,y_2) + \Psi^{\uparrow}(\lambda,x_4,x_3;y_4,y_3) + \cdots + \Psi^{\uparrow}(\lambda,x_p,x_{p-1};y_p,y_{p-1}).$$

(iii) *Suppose that M satisfies the* $\triangleleft$*-triangle inequality. There exists* $\lambda \in (0,1)$, *satisfying:*

$$\max\left\{\Psi^{\uparrow}(\mu,x_1,x_p;y_1,y_p), \Psi^{\uparrow}(\mu,x_1,x_p;y_p,y_1), \Psi^{\uparrow}(\mu,x_p,x_1;y_1,y_p), \Psi^{\uparrow}(\mu,x_p,x_1;y_p,y_1)\right\} \leq \Psi^{\uparrow}(\lambda,x_2,x_1;y_2,y_1) + \Psi^{\uparrow}(\lambda,x_2,x_3;y_2,y_3) + \Psi^{\uparrow}(\lambda,x_3,x_4;y_3,y_4) + \cdots + \Psi^{\uparrow}(\lambda,x_{p-1},x_p;y_{p-1},y_p).$$

(iv) *Suppose that M satisfies the* $\diamond$*-triangle inequality. There exists* $\lambda \in (0,1)$ *such that the following inequalities are satisfied:*

- *If p is even and* $\Psi^{\uparrow}(\mu,x_1,x_p;y_1,y_p) < +\infty$, *then:*

$$\Psi^{\uparrow}(\mu,x_1,x_p;y_1,y_p) \leq \Psi^{\uparrow}(\lambda,x_2,x_1;y_2,y_1) + \Psi^{\uparrow}(\lambda,x_3,x_2;y_3,y_2) + \Psi^{\uparrow}(\lambda,x_3,x_4;y_3,y_4) + \Psi^{\uparrow}(\lambda,x_5,x_4;y_5,y_4) + \Psi^{\uparrow}(\lambda,x_5,x_6;y_5,y_6) + \Psi^{\uparrow}(\lambda,x_7,x_6;y_7,y_6) + \cdots + \Psi^{\uparrow}(\lambda,x_{p-1},x_p;y_{p-1},y_p). \tag{48}$$

- *If p is even and* $\Psi^{\uparrow}(\mu, x_1, x_p; y_p, y_1) < +\infty$, *then:*

$$\begin{aligned}\Psi^{\downarrow}(\mu, x_1, x_p; y_p, y_1) \leq\ & \Psi^{\downarrow}(\lambda, x_2, x_1; y_1, y_2) + \Psi^{\downarrow}(\lambda, x_3, x_2; y_2, y_3) + \Psi^{\downarrow}(\lambda, x_3, x_4; y_4, y_3)\\ & + \Psi^{\downarrow}(\lambda, x_5, x_4; y_4, y_5) + \Psi^{\downarrow}(\lambda, x_5, x_6; y_6, y_5) + \Psi^{\downarrow}(\lambda, x_7, x_6; y_6, y_7)\\ & + \cdots + \Psi^{\downarrow}(\lambda, x_{p-1}, x_p; y_p, y_{p-1}).\end{aligned} \tag{49}$$

- *If p is even and* $\Psi^{\uparrow}(\mu, x_p, x_1; y_1, y_p) < +\infty$, *then:*

$$\begin{aligned}\Psi^{\downarrow}(\mu, x_p, x_1; y_1, y_p) \leq\ & \Psi^{\downarrow}(\lambda, x_1, x_2; y_2, y_1) + \Psi^{\downarrow}(\lambda, x_2, x_3; y_3, y_2) + \Psi^{\downarrow}(\lambda, x_4, x_3; y_3, y_4)\\ & + \Psi^{\downarrow}(\lambda, x_4, x_5; y_5, y_4) + \Psi^{\downarrow}(\lambda, x_6, x_5; y_5, y_6) + \Psi^{\downarrow}(\lambda, x_6, x_7; y_7, y_6)\\ & + \cdots + \Psi^{\downarrow}(\lambda, x_p, x_{p-1}; y_{p-1}, y_p).\end{aligned} \tag{50}$$

- *If p is even and* $\Psi^{\uparrow}(\mu, x_p, x_1; y_p, y_1) < +\infty$, *then:*

$$\begin{aligned}\Psi^{\uparrow}(\mu, x_p, x_1; y_p, y_1) \leq\ & \Psi^{\uparrow}(\lambda, x_1, x_2; y_1, y_2) + \Psi^{\uparrow}(\lambda, x_2, x_3; y_2, y_3) + \Psi^{\uparrow}(\lambda, x_4, x_3; y_4, y_3)\\ & + \Psi^{\uparrow}(\lambda, x_4, x_5; y_4, y_5) + \Psi^{\uparrow}(\lambda, x_6, x_5; y_6, y_5) + \Psi^{\uparrow}(\lambda, x_6, x_7; y_6, y_7)\\ & + \cdots + \Psi^{\uparrow}(\lambda, x_p, x_{p-1}; y_p, y_{p-1}).\end{aligned} \tag{51}$$

- *If p is odd and* $\Psi^{\uparrow}(\mu, x_1, x_p; y_1, y_p) < +\infty$, *then:*

$$\begin{aligned}\Psi^{\uparrow}(\mu, x_1, x_p; y_1, y_p) \leq\ & \Psi^{\uparrow}(\lambda, x_1, x_2; y_1, y_2) + \Psi^{\uparrow}(\lambda, x_2, x_3; y_2, y_3) + \Psi^{\uparrow}(\lambda, x_4, x_3; y_4, y_3)\\ & + \Psi^{\uparrow}(\lambda, x_4, x_5; y_4, y_5) + \Psi^{\uparrow}(\lambda, x_6, x_5; y_6, y_5) + \Psi^{\uparrow}(\lambda, x_6, x_7; y_6, y_7)\\ & + \cdots + \Psi^{\uparrow}(\lambda, x_p, x_{p-1}; y_p, y_{p-1}).\end{aligned} \tag{52}$$

- *If p is odd and* $\Psi^{\uparrow}(\mu, x_1, x_p; y_p, y_1) < +\infty$, *then:*

$$\begin{aligned}\Psi^{\downarrow}(\mu, x_1, x_p; y_p, y_1) \leq\ & \Psi^{\downarrow}(\lambda, x_1, x_2; y_2, y_1) + \Psi^{\downarrow}(\lambda, x_2, x_3; y_3, y_2) + \Psi^{\downarrow}(\lambda, x_4, x_3; y_3, y_4)\\ & + \Psi^{\downarrow}(\lambda, x_4, x_5; y_5, y_4) + \Psi^{\downarrow}(\lambda, x_6, x_5; y_5, y_6) + \Psi^{\downarrow}(\lambda, x_6, x_7; y_7, y_6)\\ & + \cdots + \Psi^{\downarrow}(\lambda, x_p, x_{p-1}; y_{p-1}, y_p).\end{aligned} \tag{53}$$

- *If p is odd and* $\Psi^{\uparrow}(\mu, x_p, x_1; y_1, y_p) < +\infty$, *then:*

$$\begin{aligned}\Psi^{\downarrow}(\mu, x_p, x_1; y_1, y_p) \leq\ & \Psi^{\downarrow}(\lambda, x_2, x_1; y_1, y_2) + \Psi^{\downarrow}(\lambda, x_3, x_2; y_2, y_3) + \Psi^{\downarrow}(\lambda, x_3, x_4; y_4, y_3)\\ & + \Psi^{\downarrow}(\lambda, x_5, x_4; y_4, y_5) + \Psi^{\downarrow}(\lambda, x_5, x_6; y_6, y_5) + \Psi^{\downarrow}(\lambda, x_7, x_6; y_6, y_7)\\ & + \cdots + \Psi^{\downarrow}(\lambda, x_{p-1}, x_p; y_p, y_{p-1}).\end{aligned} \tag{54}$$

- *If p is odd and* $\Psi^{\uparrow}(\mu, x_p, x_1; y_p, y_1) < +\infty$, *then:*

$$\begin{aligned}\Psi^{\uparrow}(\mu, x_p, x_1; y_p, y_1) \leq\ & \Psi^{\uparrow}(\lambda, x_2, x_1; y_2, y_1) + \Psi^{\uparrow}(\lambda, x_3, x_2; y_3, y_2) + \Psi^{\uparrow}(\lambda, x_3, x_4; y_3, y_4)\\ & + \Psi^{\uparrow}(\lambda, x_5, x_4; y_5, y_4) + \Psi^{\uparrow}(\lambda, x_5, x_6; y_5, y_6) + \Psi^{\uparrow}(\lambda, x_7, x_6; y_7, y_6)\\ & + \cdots + \Psi^{\uparrow}(\lambda, x_{p-1}, x_p; y_{p-1}, y_p).\end{aligned} \tag{55}$$

Proof. Lemma 1 says that there exists $\lambda \in (0, 1)$, satisfying:

$$(1-\lambda) * \cdots * (1-\lambda) > 1-\mu. \tag{56}$$

To prove part (i), we assume that $\Psi^{\uparrow}(\lambda, x_i, x_{i+1}; y_i, y_{i+1}) < +\infty$ for all $i = 1, \cdots, p-1$. Given any $\epsilon > 0$, the first observation of Remark 1 says that:

$$\begin{aligned} & M\left(x_1, x_p, \Psi^{\uparrow}(\lambda, x_1, x_2; y_1, y_2) + \Psi^{\uparrow}(\lambda, x_2, x_3; y_2, y_3) + \cdots + \Psi^{\uparrow}(\lambda, x_{p-1}, x_p; y_{p-1}, y_p) + (p-1)\epsilon\right) \\ & \quad \geq M\left(x_1, x_2, \Psi^{\uparrow}(\lambda, x_1, x_2; y_1, y_2) + \epsilon\right) * \cdots * M\left(x_{p-1}, x_p, \Psi^{\uparrow}(\lambda, x_{p-1}, x_p; y_{p-1}, y_p) + \epsilon\right), \end{aligned} \tag{57}$$

and:

$$\begin{aligned} & M\left(y_1, y_p, \Psi^{\uparrow}(\lambda, x_1, x_2; y_1, y_2) + \Psi^{\uparrow}(\lambda, x_2, x_3; y_2, y_3) + \cdots + \Psi^{\uparrow}(\lambda, x_{p-1}, x_p; y_{p-1}, y_p) + (p-1)\epsilon\right) \\ & \quad \geq M\left(y_1, y_2, \Psi^{\uparrow}(\lambda, x_1, x_2; y_1, y_2) + \epsilon\right) * \cdots * M\left(y_{p-1}, y_p, \Psi^{\uparrow}(\lambda, x_{p-1}, x_p; y_{p-1}, y_p) + \epsilon\right). \end{aligned} \tag{58}$$

Now, applying the increasing property and commutativity of t-norm to Equations (57) and (58), we obtain:

$$\begin{aligned} & \zeta\left(x_1, x_p; y_1, y_p, \Psi^{\uparrow}(\lambda, x_1, x_2; y_1, y_2) + \Psi^{\uparrow}(\lambda, x_2, x_3; y_2, y_3) + \cdots + \Psi^{\uparrow}(\lambda, x_{p-1}, x_p; y_{p-1}, y_p) + (p-1)\epsilon\right) \\ & \geq \zeta\left(x_1, x_2; y_1, y_2, \Psi^{\uparrow}(\lambda, x_1, x_2; y_1, y_2) + \epsilon\right) * \cdots * \zeta\left(x_{p-1}, x_p; y_{p-1}, y_p, \Psi^{\uparrow}(\lambda, x_{p-1}, x_p; y_{p-1}, y_p) + \epsilon\right) \\ & \geq (1-\lambda) * \cdots * (1-\lambda) \text{ (by Equation (41) and the increasing property of t-norm)} \\ & > 1 - \mu \text{ (by Equation (56))}. \end{aligned} \tag{59}$$

Therefore, we consider the following cases:

- Suppose that $\Psi^{\uparrow}(\mu, x_1, x_p; y_1, y_p) = +\infty$. We want to show that there exists i_0, satisfying $\Psi^{\uparrow}(\lambda, x_{i_0}, x_{i_0+1}; y_{i_0}, y_{i_0+1}) = +\infty$. Assume that $\Psi^{\uparrow}(\lambda, x_i, x_{i+1}; y_i, y_{i+1}) < +\infty$ for all $i = 1, \cdots, p-1$. Using Equation (59) and part (ii) of Proposition 17, it follows that $\Psi^{\uparrow}(\mu, x_1, x_p; y_1, y_p) < +\infty$. This contradiction says that there exists i_0, satisfying $\Psi^{\uparrow}(\lambda, x_{i_0}, x_{i_0+1}; y_{i_0}, y_{i_0+1}) = +\infty$. In this case, the inequality (Equation (46)) holds true.
- Suppose that $\Psi^{\uparrow}(\mu, x_1, x_p; y_1, y_p) < +\infty$. We also consider the following cases:
  - If there exists i_0, satisfying $\Psi^{\uparrow}(\lambda, x_{i_0}, x_{i_0+1}; y_{i_0}, y_{i_0+1}) = +\infty$, then the inequality (Equation (46)) also holds true.
  - We assume that $\Psi^{\uparrow}(\lambda, x_i, x_{i+1}; y_i, y_{i+1}) < +\infty$ for all $i = 1, \cdots, p-1$. Using Equation (59) and part (ii) of Proposition 17 again, it follows that:

$$\begin{aligned} & \Psi^{\uparrow}(\lambda, x_1, x_2; y_1, y_2) + \Psi^{\uparrow}(\lambda, x_2, x_3; y_2, y_3) + \cdots + \Psi^{\uparrow}(\lambda, x_{p-1}, x_p; y_{p-1}, y_p) + (p-1)\epsilon \\ & \quad \geq \Psi^{\uparrow}(\mu, x_1, x_p; y_1, y_p). \end{aligned}$$

    By taking the limit $\epsilon \to 0+$, we obtain the desired inequality (Equation (46)).

On the other hand, we also have:

$$\begin{aligned} & M\left(x_1, x_p, \Psi^{\uparrow}(\lambda, x_1, x_2; y_2, y_1) + \Psi^{\uparrow}(\lambda, x_2, x_3; y_3, y_2) + \cdots + \Psi^{\uparrow}(\lambda, x_{p-1}, x_p; y_p, y_{p-1}) + (p-1)\epsilon\right) \\ & \quad \geq M\left(x_1, x_2, \Psi^{\uparrow}(\lambda, x_1, x_2; y_2, y_1) + \epsilon\right) * \cdots * M\left(x_{p-1}, x_p, \Psi^{\uparrow}(\lambda, x_{p-1}, x_p; y_p, y_{p-1}) + \epsilon\right), \end{aligned} \tag{60}$$

and:

$$\begin{aligned} & M\left(y_p, y_1, \Psi^{\uparrow}(\lambda, x_1, x_2; y_2, y_1) + \Psi^{\uparrow}(\lambda, x_2, x_3; y_3, y_2) + \cdots + \Psi^{\uparrow}(\lambda, x_{p-1}, x_p; y_p, y_{p-1}) + (p-1)\epsilon\right) \\ & \quad \geq M\left(y_2, y_1, \Psi^{\uparrow}(\lambda, x_1, x_2; y_2, y_1) + \epsilon\right) * \cdots * M\left(y_p, y_{p-1}, \Psi^{\uparrow}(\lambda, x_{p-1}, x_p; y_p, y_{p-1}) + \epsilon\right). \end{aligned} \tag{61}$$

Now, applying the increasing property and commutativity of t-norm to Equations (60) and (61), we obtain:

$$\begin{aligned}
&\zeta\left(x_1, x_p; y_p, y_1, \Psi^{\uparrow}(\lambda, x_1, x_2; y_2, y_1) + \Psi^{\uparrow}(\lambda, x_2, x_3; y_3, y_2) + \cdots + \Psi^{\uparrow}(\lambda, x_{p-1}, x_p; y_p, y_{p-1}) + (p-1)\epsilon\right) \\
&\geq \zeta\left(x_1, x_2; y_2, y_1, \Psi^{\uparrow}(\lambda, x_1, x_2; y_2, y_1) + \epsilon\right) * \cdots * \zeta\left(x_{p-1}, x_p; y_p, y_{p-1}, \Psi^{\uparrow}(\lambda, x_{p-1}, x_p; y_p, y_{p-1}) + \epsilon\right) \\
&\geq (1-\lambda) * \cdots * (1-\lambda) \text{ (by Equation (41) and the increasing property of t-norm)} \\
&> 1 - \mu \text{ (by Equation (56))}.
\end{aligned}$$

The inequality (Equation (47)) can be similarly obtained using the above argument. Further, the other inequalities can be similarly obtained.

The above argument is still valid by applying the second observation of Remark 1 to obtain part (ii). We can also apply the third observation of Remark 1 to obtain part (iii). Finally, part (iv) can be obtained using the fourth observation of Remark 1. This completes the proof. □

Theorem 5. *Let (X, M) be a fuzzy semi-metric space along with a t-norm $*$. We also assume that M satisfies the rational condition, and that the t-norm $*$ is right-continuous at 0 with respect to the first or second argument. Let $\{x_n\}_{n=1}^{\infty}$ and $\{y_n\}_{n=1}^{\infty}$ be two sequences in X. Then we have the following properties:*

(i) *Suppose that M satisfies the $\circ$-triangle inequality for $\circ \in \{\bowtie, \triangleright, \triangleleft\}$. Then the following statements hold true:*

- $x_n \xrightarrow{M^{\triangleright}} x$ *and* $y_n \xrightarrow{M^{\triangleright}} y$ *as* $n \to \infty$ *if and only if* $\Psi^{\uparrow}(\lambda, x_n, x; y_n, y) \to 0$ *as* $n \to \infty$ *for all* $\lambda \in (0,1)$.
- $x_n \xrightarrow{M^{\triangleright}} x$ *and* $y_n \xrightarrow{M^{\triangleleft}} y$ *as* $n \to \infty$ *if and only if* $\Psi^{\uparrow}(\lambda, x_n, x; y, y_n) \to 0$ *as* $n \to \infty$ *for all* $\lambda \in (0,1)$.
- $x_n \xrightarrow{M^{\triangleleft}} x$ *and* $y_n \xrightarrow{M^{\triangleright}} y$ *as* $n \to \infty$ *if and only if* $\Psi^{\uparrow}(\lambda, x, x_n; y_n, y) \to 0$ *as* $n \to \infty$ *for all* $\lambda \in (0,1)$.
- $x_n \xrightarrow{M^{\triangleleft}} x$ *and* $y_n \xrightarrow{M^{\triangleleft}} y$ *as* $n \to \infty$ *if and only if* $\Psi^{\uparrow}(\lambda, x, x_n; y, y_n) \to 0$ *as* $n \to \infty$ *for all* $\lambda \in (0,1)$.

(ii) *Suppose that M satisfies the $\diamond$-triangle inequality. Then the following statements hold true:*

- *If* $x_n \xrightarrow{M^{\triangleright}} x$ *and* $y_n \xrightarrow{M^{\triangleright}} y$ *as* $n \to \infty$, *given any fixed* $\lambda \in (0,1)$, *we have* $\Psi^{\uparrow}(\lambda, x_n, x; y_n, y) < +\infty$ *for all* $n \in \mathbb{N}$ *imply* $\Psi^{\uparrow}(\lambda, x_n, x; y_n, y) \to 0$ *as* $n \to \infty$.
- *If* $x_n \xrightarrow{M^{\triangleright}} x$ *and* $y_n \xrightarrow{M^{\triangleleft}} y$ *as* $n \to \infty$, *given any fixed* $\lambda \in (0,1)$, *we have* $\Psi^{\uparrow}(\lambda, x_n, x; y, y_n) < +\infty$ *for all* $n \in \mathbb{N}$ *imply* $\Psi^{\uparrow}(\lambda, x_n, x; y, y_n) \to 0$ *as* $n \to \infty$.
- *If* $x_n \xrightarrow{M^{\triangleleft}} x$ *and* $y_n \xrightarrow{M^{\triangleright}} y$ *as* $n \to \infty$, *given any fixed* $\lambda \in (0,1)$, *we have* $\Psi^{\uparrow}(\lambda, x, x_n; y_n, y) < +\infty$ *for all* $n \in \mathbb{N}$ *imply* $\Psi^{\uparrow}(\lambda, x, x_n; y_n, y) \to 0$ *as* $n \to \infty$.
- *If* $x_n \xrightarrow{M^{\triangleleft}} x$ *and* $y_n \xrightarrow{M^{\triangleleft}} y$ *as* $n \to \infty$, *given any fixed* $\lambda \in (0,1)$, *we have* $\Psi^{\uparrow}(\lambda, x, x_n; y, y_n) < +\infty$ *for all* $n \in \mathbb{N}$ *imply* $\Psi^{\uparrow}(\lambda, x, x_n; y, y_n) \to 0$ *as* $n \to \infty$.
- *If* $\Psi^{\uparrow}(\lambda, x_n, x; y_n, y) \to 0$ *as* $n \to \infty$ *for all* $\lambda \in (0,1)$, *then* $x_n \xrightarrow{M^{\triangleleft}} x$ *and* $y_n \xrightarrow{M^{\triangleleft}} y$ *as* $n \to \infty$.
- *If* $\Psi^{\uparrow}(\lambda, x, x_n; y_n, y) \to 0$ *as* $n \to \infty$ *for all* $\lambda \in (0,1)$, *then* $x_n \xrightarrow{M^{\triangleright}} x$ *and* $y_n \xrightarrow{M^{\triangleleft}} y$ *as* $n \to \infty$.
- *If* $\Psi^{\uparrow}(\lambda, x_n, x; y, y_n) \to 0$ *as* $n \to \infty$ *for all* $\lambda \in (0,1)$, *then* $x_n \xrightarrow{M^{\triangleleft}} x$ *and* $y_n \xrightarrow{M^{\triangleright}} y$ *as* $n \to \infty$.
- *If* $\Psi^{\uparrow}(\lambda, x, x_n; y, y_n) \to 0$ *as* $n \to \infty$ *for all* $\lambda \in (0,1)$, *then* $x_n \xrightarrow{M^{\triangleright}} x$ *and* $y_n \xrightarrow{M^{\triangleright}} y$ *as* $n \to \infty$.

Proof. For any fixed $\lambda \in (0,1)$, using Lemma 1, it follows that there exists $\lambda_0 \in (0,1)$, satisfying:

$$(1-\lambda_0) * (1-\lambda_0) > 1 - \lambda.$$

To prove part (i), we just prove the first case, since the other cases can be similarly obtained. Suppose that $M(x_n, x, t) \to 1$ and $M(y_n, y, t) \to 1$ as $n \to \infty$ for all $t > 0$. Then, given any $t > 0$ and $\delta > 0$, there exists $n_{t,\delta}^{(1)}, n_{t,\delta}^{(2)} \in \mathbb{N}$, satisfying $|M(x_n, x, t) - 1| < \delta$ for $n \geq n_{t,\delta}^{(1)}$ and $|M(y_n, y, t) - 1| < \delta$ for $n \geq n_{t,\delta}^{(2)}$. Given any $\epsilon \in (0, 1)$, there exists $n_\epsilon \in \mathbb{N}$, satisfying

$$\left|M\left(x_n, x, \frac{\epsilon}{2}\right) - 1\right| < \lambda_0 \text{ and } \left|M\left(y_n, y, \frac{\epsilon}{2}\right) - 1\right| < \lambda_0,$$

for $n \geq n_\epsilon$. We also have:

$$M\left(x_n, x, \frac{\epsilon}{2}\right) > 1 - \lambda_0 \text{ and } M\left(y_n, y, \frac{\epsilon}{2}\right) > 1 - \lambda_0,$$

for $n \geq n_\epsilon$. The increasing property of t-norm says that:

$$\zeta\left(x_n, x; y_n, y, \frac{\epsilon}{2}\right) = M\left(x_n, x, \frac{\epsilon}{2}\right) * M\left(y_n, y, \frac{\epsilon}{2}\right) \geq (1 - \lambda_0) * (1 - \lambda_0) > 1 - \lambda.$$

The first result of part (ii) of Proposition 17 says that:

$$\Psi^{\uparrow}(\lambda, x_n, x; y_n, y) \leq \frac{\epsilon}{2} < \epsilon,$$

for $n \geq n_\epsilon$. This shows that $\Psi^{\uparrow}(\lambda, x_n, x; y_n, y) \to 0$ as $n \to \infty$.

To prove the converse, suppose that $\Psi^{\uparrow}(\lambda, x_n, x; y_n, y) \to 0$ as $n \to \infty$ for all $\lambda \in (0, 1)$. Given any $\delta > 0$ and $\lambda \in (0, 1]$, there exists $n_{\delta,\lambda} \in \mathbb{N}$, satisfying $|\Psi^{\uparrow}(\lambda, x_n, x; y_n, y)| < \delta$ for all $n \geq n_{\delta,\lambda}$. For any fixed $t > 0$ and given any $\epsilon \in (0, 1)$, there exists $n_\epsilon \in \mathbb{N}$, satisfying:

$$\Psi^{\uparrow}\left(\epsilon, x_n, x; y_n, y\right) = \left|\Psi^{\uparrow}\left(\epsilon, x_n, x; y_n, y\right)\right| < t, \tag{62}$$

for $n \geq n_\epsilon$, which implies:

$$\zeta(x_n, x; y_n, y, t) > 1 - \epsilon,$$

for $n \geq n_\epsilon$ by the first result of part (i) of Proposition 16, i.e.,

$$M\left(x_n, x, t\right) * M\left(y_n, y, t\right) > 1 - \epsilon,$$

for $n \geq n_\epsilon$. Lemma 2 says that:

$$M\left(x_n, x, t\right) > 1 - \epsilon \text{ and } M\left(y_n, y, t\right) > 1 - \epsilon,$$

for $n \geq n_\epsilon$. This shows that the sequences $\{x_n\}_{n=1}^{\infty}$ and $\{y_n\}_{n=1}^{\infty}$ in X converge to x and y, respectively.

To prove part (ii), the first result to the fourth result can be similarly obtained by the third result of part (ii) of Proposition 17. For proving the fifth result, the fact $\Psi^{\uparrow}(\lambda, x_n, x; y_n, y) \to 0$ implies the inequality (Equation (62)). The third result of part (i) of Proposition 16 says that $\zeta(x, x_n; y, y_n, t) > 1 - \epsilon$, which implies $M\left(x, x_n, x, t\right) > 1 - \epsilon$ and $M\left(y, y_n, y, t\right) > 1 - \epsilon$. In other words, we have $x_n \xrightarrow{M^{\triangleleft}} x$ and $y_n \xrightarrow{M^{\triangleleft}} y$ as $n \to \infty$. The remaining three results can be similarly obtained. This completes the proof. □

Example 7. *From Example 1, we see that:*

$$x_n \xrightarrow{M^{\triangleright}} x \textit{ if and only if } \lim_{n\to\infty} d(x_n, x) = 0,$$

and:

$$x_n \xrightarrow{M^{\triangleleft}} x \textit{ if and only if } \lim_{n\to\infty} d(x, x_n) = 0.$$

From Example 6, we have:

$$\Psi^{\uparrow}(\lambda, x, y; u, v) = \frac{C + \sqrt{C^2 + D}}{2},$$

where:

$$C = \frac{(d(x,y) + d(u,v))(1-\lambda)}{\lambda} \text{ and } D = \frac{d(x,y) \cdot d(u,v) \cdot (1-\lambda)}{\lambda}.$$

It is clear to see that $x_n \stackrel{M^{\triangleright}}{\longrightarrow} x$ and $y_n \stackrel{M^{\triangleright}}{\longrightarrow} y$ as $n \to \infty$ if and only if $\Psi^{\uparrow}(\lambda, x_n, x; y_n, y) \to 0$ as $n \to \infty$ for all $\lambda \in (0,1)$. The other convergence presented in Theorem 5 can be similarly verified.

According to Definition 5, we can similarly define the concepts of four types of the joint Cauchy sequences with respect to $\Psi^{\uparrow}$. We omit the details.

Theorem 6. *Let (X, M) be a fuzzy semi-metric space along with a t-norm $*$. We also assume that M satisfies the rational condition, and that the t-norm $*$ is right-continuous at 0 with respect to the first or second argument. Let $\{x_n\}_{n=1}^{\infty}$ and $\{y_n\}_{n=1}^{\infty}$ be two sequences in X. Then we have the following properties:*

(i) *Suppose that M satisfies the $\circ$-triangle inequality for $\circ \in \{\bowtie, \triangleright, \triangleleft\}$. Then the following statements hold true:*

- *$\{x_n\}_{n=1}^{\infty}$ and $\{y_n\}_{n=1}^{\infty}$ are two $>$-Cauchy sequences in a metric sense if and only if $\{x_n\}_{n=1}^{\infty}$ and $\{y_n\}_{n=1}^{\infty}$ are the joint $(\lambda, >, >)$-Cauchy sequences with respect to $\Psi^{\uparrow}$ for any $\lambda \in (0,1)$.*
- *$\{x_n\}_{n=1}^{\infty}$ is a $>$-Cauchy sequences in a metric sense and $\{y_n\}_{n=1}^{\infty}$ is a $<$-Cauchy sequences in a metric sense if and only if $\{x_n\}_{n=1}^{\infty}$ and $\{y_n\}_{n=1}^{\infty}$ are the joint $(\lambda, >, <)$-Cauchy sequences with respect to $\Psi^{\uparrow}$ for any $\lambda \in (0,1)$.*
- *$\{x_n\}_{n=1}^{\infty}$ is a $<$-Cauchy sequences in a metric sense and $\{y_n\}_{n=1}^{\infty}$ is a $>$-Cauchy sequences in a metric sense if and only if $\{x_n\}_{n=1}^{\infty}$ and $\{y_n\}_{n=1}^{\infty}$ are the joint $(\lambda, >, <)$-Cauchy sequences with respect to $\Psi^{\uparrow}$ for any $\lambda \in (0,1)$.*
- *$\{x_n\}_{n=1}^{\infty}$ and $\{y_n\}_{n=1}^{\infty}$ are two $<$-Cauchy sequences in a metric sense if and only if $\{x_n\}_{n=1}^{\infty}$ and $\{y_n\}_{n=1}^{\infty}$ are the joint $(\lambda, <, <)$-Cauchy sequences with respect to $\Psi^{\uparrow}$ for any $\lambda \in (0,1)$.*

(ii) *Suppose that M satisfies the $\diamond$-triangle inequality. Then the following statements hold true:*

- *Let $\{x_n\}_{n=1}^{\infty}$ and $\{y_n\}_{n=1}^{\infty}$ be two $>$-Cauchy sequences in a metric sense. Given any fixed $\lambda \in (0,1)$, if $\Psi^{\uparrow}(\lambda, x_m, x_n; y_m, y_n) < +\infty$ for all $m > n$, then $\{x_n\}_{n=1}^{\infty}$ and $\{y_n\}_{n=1}^{\infty}$ are the joint $(\lambda, >, >)$-Cauchy sequences with respect to $\Psi^{\uparrow}$ for any $\lambda \in (0,1)$.*
- *Let $\{x_n\}_{n=1}^{\infty}$ be a $>$-Cauchy sequence in a metric sense and let $\{y_n\}_{n=1}^{\infty}$ be a $<$-Cauchy sequence in a metric sense. Given any fixed $\lambda \in (0,1)$, if $\Psi^{\uparrow}(\lambda, x_m, x_n; y_n, y_m) < +\infty$ for all $m > n$, then $\{x_n\}_{n=1}^{\infty}$ and $\{y_n\}_{n=1}^{\infty}$ are the joint $(\lambda, >, <)$-Cauchy sequences with respect to $\Psi^{\uparrow}$ for any $\lambda \in (0,1)$.*
- *Let $\{x_n\}_{n=1}^{\infty}$ be a $<$-Cauchy sequence in a metric sense and let $\{y_n\}_{n=1}^{\infty}$ be a $>$-Cauchy sequence in a metric sense. Given any fixed $\lambda \in (0,1)$, if $\Psi^{\uparrow}(\lambda, x_n, x_m; y_m, y_n) < +\infty$ for all $m > n$, then $\{x_n\}_{n=1}^{\infty}$ and $\{y_n\}_{n=1}^{\infty}$ are the joint $(\lambda, <, >)$-Cauchy sequences with respect to $\Psi^{\uparrow}$ for any $\lambda \in (0,1)$.*
- *Let $\{x_n\}_{n=1}^{\infty}$ and $\{y_n\}_{n=1}^{\infty}$ be two $<$-Cauchy sequences in a metric sense. Given any fixed $\lambda \in (0,1)$, if $\Psi^{\uparrow}(\lambda, x_n, x_m; y_n, y_m) < +\infty$ for all $m > n$, then $\{x_n\}_{n=1}^{\infty}$ and $\{y_n\}_{n=1}^{\infty}$ are the joint $(\lambda, <, <)$-Cauchy sequences with respect to $\Psi^{\uparrow}$ for any $\lambda \in (0,1)$.*
- *Suppose that $\{x_n\}_{n=1}^{\infty}$ and $\{y_n\}_{n=1}^{\infty}$ are the joint $(\lambda, >, >)$-Cauchy sequences with respect to $\Psi^{\uparrow}$ for any $\lambda \in (0,1)$ Then $\{x_n\}_{n=1}^{\infty}$ and $\{y_n\}_{n=1}^{\infty}$ are two $>$-Cauchy sequences in a metric sense.*
- *Suppose that $\{x_n\}_{n=1}^{\infty}$ and $\{y_n\}_{n=1}^{\infty}$ are the joint $(\lambda, >, <)$-Cauchy sequences with respect to $\Psi^{\uparrow}$ for any $\lambda \in (0,1)$. Then $\{x_n\}_{n=1}^{\infty}$ is a $<$-Cauchy sequences in a metric sense and $\{y_n\}_{n=1}^{\infty}$ is a $<$-Cauchy sequences in a metric sense.*

- *Suppose that $\{x_n\}_{n=1}^{\infty}$ and $\{y_n\}_{n=1}^{\infty}$ are the joint $(\lambda, <, >)$-Cauchy sequences with respect to $\Psi^{\uparrow}$ for any $\lambda \in (0,1)$ Then $\{x_n\}_{n=1}^{\infty}$ is a $<$-Cauchy sequences in a metric sense and $\{y_n\}_{n=1}^{\infty}$ is a $>$-Cauchy sequences in a metric sense.*
- *Suppose that $\{x_n\}_{n=1}^{\infty}$ and $\{y_n\}_{n=1}^{\infty}$ are the joint $(\lambda, <, <)$-Cauchy sequences with respect to $\Psi^{\uparrow}$ for any $\lambda \in (0,1)$. Then $\{x_n\}_{n=1}^{\infty}$ and $\{y_n\}_{n=1}^{\infty}$ are two $<$-Cauchy sequences in a metric sense.*

Proof. For any fixed $\lambda \in (0,1)$, using Lemma 1, it follows that there exists $\lambda_0 \in (0,1)$, satisfying:

$$(1-\lambda_0) * (1-\lambda_0) > 1-\lambda.$$

To prove part (i), we just prove the first case, since the other cases can be similarly obtained. Suppose that $\{x_n\}_{n=1}^{\infty}$ and $\{y_n\}_{n=1}^{\infty}$ are $>$-Cauchy sequences in a metric sense. Therefore, given any $t > 0$ and $\delta > 0$, there exists $n_{t,\delta} \in \mathbb{N}$ such that $m > n \geq n_{t,\delta}$ implies $M(x_m, x_n, t) > 1-\delta$ and $M(y_m, y_n, t) > 1-\delta$. Now, given any $\epsilon \in (0,1)$, there exists $n_\epsilon \in \mathbb{N}$ such that $m > n \geq n_\epsilon$ implies:

$$M\left(x_m, x_n, \frac{\epsilon}{2}\right) > 1-\lambda_0 \text{ and } M\left(y_m, y_n, \frac{\epsilon}{2}\right) > 1-\lambda_0.$$

The increasing property of t-norm says that:

$$\begin{aligned} \zeta\left(x_m, x_n; y_m, y_n, \frac{\epsilon}{2}\right) &= M\left(x_m, x_n, \frac{\epsilon}{2}\right) * M\left(y_m, y_n, \frac{\epsilon}{2}\right) \\ &\geq (1-\lambda_0) * (1-\lambda_0) > 1-\lambda. \end{aligned}$$

Further, the first result of part (ii) of Proposition 17 says that:

$$\Psi^{\uparrow}(\lambda, x_m, x_n; y_m, y_n) \leq \frac{\epsilon}{2} < \epsilon,$$

for $m > n \geq n_\epsilon$.

To prove the converse, from the assumption, we see that for any fixed $t > 0$ and given any $\epsilon \in (0,1)$, there exists $n_\epsilon \in \mathbb{N}$ such that $m > n \geq n_\epsilon$ implies $\Psi^{\uparrow}(\epsilon, x_m, x_n; y_m, y_n) < t$. Therefore, using the first result of part (i) of Proposition 16, we obtain $\zeta(x_m, x_n; y_m, y_n, t) > 1-\epsilon$ for $m > n \geq n_\epsilon$, i.e.,

$$M(x_m, x_n, t) * M(y_m, y_n, t) > 1-\epsilon,$$

for $m > n \geq n_\epsilon$. Lemma 2 says that:

$$M(x_m, x_n, t) > 1-\epsilon \text{ and } M(y_m, y_n, t) > 1-\epsilon,$$

for $m > n \geq n_\epsilon$. This shows that $\{x_n\}_{n=1}^{\infty}$ and $\{y_n\}_{n=1}^{\infty}$ are $>$-Cauchy sequences in a metric sense.

To prove part (ii), the first result to the fourth result can be similarly obtained by the third result of part (ii) of Proposition 17. For proving the fifth result, using the assumption, we see that for any fixed $t > 0$ and given any $\epsilon \in (0,1)$, there exists $n_\epsilon \in \mathbb{N}$ such that $m > n \geq n_\epsilon$ implies $\Psi^{\uparrow}(\epsilon, x_m, x_n; y_m, y_n) < t$. The third result of part (i) of Proposition 16 says that $\zeta(x_n, x_m; y_n, y_m, t) > 1-\epsilon$ for $m > n \geq n_\epsilon$. Therefore, we obtain:

$$M(x_n, x_m, t) * M(y_n, y_m, t) > 1-\epsilon,$$

for $m > n \geq n_\epsilon$. This shows that $\{x_n\}_{n=1}^{\infty}$ and $\{y_n\}_{n=1}^{\infty}$ are $<$-Cauchy sequences in a metric sense. The remaining three results can be similarly obtained. This completes the proof. □

Funding: This research received no external funding.

Conflicts of Interest: The author declares no conflict of interest.

References

1. Kramosil, I.; Michalek, J. Fuzzy metric and statistical metric spaces. *Kybernetika* **1975**, *11*, 336–344.
2. Schweizer, B.; Sklar, A. Statistical metric spaces. *Pac. J. Math.* **1960**, *10*, 313–334. [CrossRef]
3. Schweizer, B.; Sklar, A.; Thorp, E. The metrization of statistical metric spaces. *Pac. J. Math.* **1960**, *10*, 673–675. [CrossRef]
4. Schweizer, B.; Sklar, A. Triangle inequalities in a class of statistical metric spaces. *J. Lond. Math. Soc.* **1963**, *38*, 401–406. [CrossRef]
5. Hadžić, O.; Pap, E. *Fixed Point Theory in Probabilistic Metric Spaces*; Kluwer Academic Publishers: Dordrecht, The Netherlands, 2001.
6. Chang, S.S.; Cho, Y.J.; Kang, S.M. *Nonlinear Operator Theory in Probabilistic Metric Space*; Nova Science Publishers: New York, NY, USA, 2001.
7. Kaleva, O.; Seikkala, S. On fuzzy metric spaces. *Fuzzy Sets Syst.* **1984**, *12*, 215–229. [CrossRef]
8. George, A.; Veeramani, P. On some results in fuzzy metric spaces. *Fuzzy Sets Syst.* **1994**, *64*, 395–399. [CrossRef]
9. George, A.; Veeramani, P. On some results of analysis for fuzzy metric spaces. *Fuzzy Sets Syst.* **1997**, *90*, 365–368. [CrossRef]
10. Gregori, V.; Romaguera, S. Some properties of fuzzy metric spaces. *Fuzzy Sets Syst.* **2002**, *115*, 399–404. [CrossRef]
11. Gregori, V.; Romaguera, S. Fuzzy quasi-metric spaces. *Appl. Gen. Topol.* **2004**, *5*, 129–136. [CrossRef]
12. Gregori, V.; Romaguera, S.; Sapena, A. A note on intuitionistic fuzzy metric spaces. *Chaos Solitons Fractals* **2006**, *28*, 902–905. [CrossRef]
13. Wu, H.-C. Hausdorff topology induced by the fuzzy metric and the fixed point theorems in fuzzy metric spaces. *J. Korean Math. Soc.* **2015**, *52*, 1287–1303. [CrossRef]
14. Wu, H.-C. Common coincidence points and common fixed points in fuzzy semi-metric spaces. *Mathematics* **2018**, *6*, 29. [CrossRef]
15. Wu, H.-C. Fuzzy semi-metric spaces. *Mathematics* **2018**, *6*, 106. [CrossRef]
16. Wu, H.-C. Convergence in fuzzy semi-metric spaces. *Mathematics* **2018**, *6*, 170. [CrossRef]

applied system innovation

MDPI

Article

An Adaptive Neuro-Fuzzy Propagation Model for LoRaWAN

Salaheddin Hosseinzadeh [1], Hadi Larijani [1,*], Krystyna Curtis [1] and Andrew Wixted [2]

[1] School of Computing, Engineering and Built Environment, Glasgow Caledonian University, Glasgow G4 0BA, UK; salaheddin.hosseinzadeh@gcu.ac.uk (S.H.); krystyna.curtis@gcu.ac.uk (K.C.)

[2] School of Engineering and Built Environment, Nathan campus, Griffith University, 170 Kessels Road, Brisbane, QLD 4111, Australia; andrew.wixted@staff.griffithcollege.edu.au

* Correspondence: H.Larijani@gcu.ac.uk; Tel.: +44-0141-331-3190

Received: 12 November 2018; Accepted: 13 March 2019; Published: 18 March 2019

Abstract: This article proposes an adaptive-network-based fuzzy inference system (ANFIS) model for accurate estimation of signal propagation using LoRaWAN. By using ANFIS, the basic knowledge of propagation is embedded into the proposed model. This reduces the training complexity of artificial neural network (ANN)-based models. Therefore, the size of the training dataset is reduced by 70% compared to an ANN model. The proposed model consists of an efficient clustering method to identify the optimum number of the fuzzy nodes to avoid overfitting, and a hybrid training algorithm to train and optimize the ANFIS parameters. Finally, the proposed model is benchmarked with extensive practical data, where superior accuracy is achieved compared to deterministic models, and better generalization is attained compared to ANN models. The proposed model outperforms the nondeterministic models in terms of accuracy, has the flexibility to account for new modeling parameters, is easier to use as it does not require a model for propagation environment, is resistant to data collection inaccuracies and uncertain environmental information, has excellent generalization capability, and features a knowledge-based implementation that alleviates the training process. This work will facilitate network planning and propagation prediction in complex scenarios.

Keywords: propagation modeling; adaptive-network-based fuzzy system; LoRa & LoRaWAN; radio wave propagation; artificial neural networks; subtracting clustering

1. Introduction

In recent years, the exponentially increasing number of wireless devices has made the maintenance and efficient planning of wireless networks crucial. Therefore, it is essential to gain a better understanding of propagation and be able to readily predict it in various scenarios. The area of radio propagation has been studied over many decades, during which the physics of the electromagnetic wave and its propagation has been unchanged. Hence, the same free space path loss (FSPL) formula holds to this date. However, telecommunication devices, network requirements and the propagation environment have undergone many developments. Perhaps, adapting to these rapid changes and lack of efficiency and generalization in current models has been the main research incentive in this field. It is clear that establishing an adaptive, efficient and accurate modeling of modern wireless technologies, that is, LoRaWAN, is of great importance. The main contributions of this paper are:

(a) A novel method of propagation modeling is proposed by using an adaptive-network-based fuzzy inference system (ANFIS). This knowledge-based system enables an adaptive model that is capable of incorporating new modeling parameters, depending on the wireless technology, propagation environment and an expert's knowledge.

(b) A robust method of determining the network size is used to avoid overfitting.

(c) Results from the models are critically analyzed with real-world measurements using a relatively new wireless technology.

This novel implementation has several advantages, however, for the purpose of comparison, firstly, a brief review of current models would be required. The rest of the paper is organized as follows: Section 2 provides a critical review of the conventional propagation models. Section 3 explains the implementation details and advantages of the proposed model. In Section 4, the results of the practical and comparative analysis are demonstrated, and Section 5 provides the findings of this research along with final conclusions.

2. Brief Review of Common Models

There are numerous practical propagation reports and models for various scenarios, such as indoor, outdoor, urban, sea, foliage, undergrounds and even tunnels. Reviewing all of these models would be out of the scope of this research. Instead, only some of the most influential and well-established models, based on the appearance in published literatures, are reviewed. A more comprehensive review of the propagation models is provided in [1].

2.1. Okumura and Hata Models

Okumura's signal propagation investigation has been a cornerstone in this field. Okumura's model [2] was built upon a practical data collection in Tokyo, Japan, within the frequency range of 150–2000 MHz. In his model, the path loss estimation depends on the FSPL, antenna gain factor, propagation gains due to mobile/base-station antennae heights and collective correction gain. The latter, collective correction gain, is to compensate for the type of environment, average slope of terrain and finally land/sea parameters. This model is somewhat a more comprehensive version of the log-distance model. Like other nondeterministic methods, Okumura's model lacks the higher accuracy of deterministic models. In addition, independent parameters of this model, such as frequency, and mobile and base station antenna heights, were limited in their range. Okumura's findings then became the basis of the Hata model [3], also known as the Okumura–Hata model. This model introduces many more independent parameters, such as reflection, diffraction and scattering factors. It suggests new correction factors for suburban and rural environments, and extends the range of the parameters of Okumura's model, such as mobile and base station antenna heights. The International Telecommunication Union (ITU) model [4] was inspired by Okumura's and Hata's models.

2.2. Walfisch, Ikegami and Ray Tracing Models

Walfisch and Bertoni [5] proposed a semi-deterministic model that added new modeling parameters into Okumura's model. These parameters were added to account for the multiple-screen diffraction, caused by the buildings and structures. Ikegami et al. [6] took a new approach by using a simplified ray-tracing method. They assumed multiple-reflected/diffracted waves were highly attenuated and, therefore, only accounted for first-reflected/diffracted rays. The other compromise of this deterministic model was perhaps due to the limited computational power, since the reflection/diffraction losses were approximated with constant values. There are several 2D and 3D ray-tracing algorithms, based on the geometrical optics, uniform theory of diffraction and geometrical theory of diffraction. These models have complemented Ikegami's efforts. For instance, it is possible to consider two- and three-fold reflections/diffraction, and reflection/diffraction factors depend on the angle of incidence and permittivity of materials. Further detailed summaries of these models with their implementations are provided in [7–10]. Nevertheless, there are the following drawbacks to these deterministic models:

(a) Having precise building data is a prerequisite, meaning that, ideally, a 3D topographical model of the environment and structures is required [6,11]. However, such a database may not be readily available, especially for outdoor environments.

(b) Computational complexity of the ray tracing is extremely cumbersome even in a relatively small indoor environment and the accuracy of the model depends on the accuracy of the structural model [12]. Therefore, it is not really a practical solution for outdoor large-scale modeling.
(c) Finally, ray tracing incorporates the "small-scale" fading which varies with the order of wavelength [13], and small-scale fading is not predictable for ranges beyond half-wavelengths [14].

These drawbacks are even more aggravated when the radio frequency is increased.

2.3. COST Action 231 Model

The European Cooperation in Science and Technology (COST) Action 231 [15] proposed a new propagation model, plus an extension to the previous Hata model. The COST–Hata PCS Extension simply extended the original Hata model to be applicable to frequencies up to 2 GHz. The COST–Walfisch–Ikegami (COST W.I.) model, however, provided a revised version of the Walfisch and Ikegami models, except that it does not require a 2/3D model of the environment. Eliminating this requirement is perhaps the main advantage of COST W.I., making it the most commonly used outdoor model. COST W.I. has been recommended by ITU and the European Telecommunication Standards Institute (ETSI) [16]. The model still requires the height of the buildings, width of the streets and some other environmental information, and therefore, it is not a fully deterministic model [17]. Although COST W.I. (or models similar to it) may seem to be an acceptable compromise between accuracy and computational complexity, it is rather a granulated set of formulas which has the following disadvantages:

(a) Some of the model parameters lack clear physical or practical interpretation. For instance, the angle of arrival of the beam relative to the axis of the road (ϕ) and road width (w_s) are vague at a crossroad, junction, riverside or a wide highway with partial line of sight (LoS), and undefined where there is no road. These conditions and many others that may occur are not considered in the COST W.I. formulas.
(b) The constant factors that control the loss due to intensity of multiscreen diffraction are discontinuous, resulting in abrupt changes in the result. For instance, scatter loss $L_{rts}(\phi)$ has sharp transitions as it has been approximated with only three piecewise linear functions. Therefore, a small variation in ϕ can affect the estimation drastically.
(c) There are several different permutations and possibilities to calculate the total path loss. The model first differentiates between LoS and NLoS (non-line of sight) conditions. Next, there are two ways of deriving the path loss depending on the d_{LoS}. The NLoS has been further subdivided to scatter and diffraction losses. Next, there are three different conditional statements to calculate the scatter loss depending on the ϕ. Finally, there are another 11 further subdivisions (depending on BS and MS heights and environment type) to calculate the multiscreen diffraction loss. However, in reality, even differentiating between LoS and NLoS is difficult, resulting in the creation of near-LoS or partial-LoS conditions [18].

2.4. Hybrid and Artificial Neural Network-Based Models

The next generation of the propagation models relies on artificial neural networks (ANNs). These models are based on the training of an ANN with empirically collected data. Usually this involves training the ANN from scratch, where the ANN has to learn to derive the very basic mechanisms of radio propagation. For instance, the ANN even needs to learn the trivial fact that increasing the link distance has a negative impact on the signal strength. It is the ANN's responsibility to learn the FSPL from the distance and frequency of transmission or understand the attenuating impact of obstacles on the LoS. A review of the ANN-based models is provided in [19]. The drawbacks of this approach include:

(a) the time-consuming and exhaustive data collection process that is required for the training of ANNs. As an instance, in [20], authors collected 600,000 data samples in a relatively small indoor environment and trained the network for several hours. Not only is such data collection in a small environment tedious but it also defeats the purpose of estimation [19].

(b) ANN requires considerable training time as it contains numerous neurons in each layer; furthermore, an overly complex ANN may lead to data overfitting and hence failing to reach a generalized solution [21,22].

To address these challenges, a hybrid model for a simple indoor environment was proposed [19]. The model comprised an optimized multiwall model (MWM), whose estimations were monitored by an ANN. Therefore, the ANN only had to learn and compensate the deficiencies of the MWM. This strategy drastically reduced the required training data samples and improved the accuracy of the estimation. A relatively similar strategy was employed in [23,24] using the COST W.I. There are, however, two drawbacks of the latter implementations:

(a) Selecting an appropriate and flexible model for a complicated outdoor environment where many propagation mechanisms are involved is challenging. In fact, Hosseinzadeh et al. [19] had to slightly adjust the MWM according to their requirement.

(b) Conventional models are not tailored or flexible enough to represent the novel characteristics of communication devices. For instance, to the best of our knowledge, there is no propagation model that takes the spreading factor (s_m) of a chirp spread spectrum modulation (CSS) [25] into consideration. Theoretically, a higher s_m provides a higher processing gain and, therefore, improves the range. Including such technology-dependent parameters not only increases the modeling accuracy and makes the models more compatible with new devices, but it also helps modeling the signal propagation of the whole system rather than estimation of the path loss only.

3. Proposed Model Description

To address the explained shortfalls and challenges, an adaptive neuro-fuzzy model is proposed. This allows the user to initiate the system with incorporated linguistic knowledge of propagation and then train the system further to achieve a higher accuracy. Next, the essential fuzzy/linguistic input–output parameters of a propagation model are identified. Furthermore, to avoid overfitting and achieve a better generalization, an efficient clustering method to determine the optimum size of the nodes is used. Finally, the system is trained with a hybrid algorithm that tunes the input and output parameters of the fuzzy system. This section explains the implementation details.

3.1. Adaptive Neuro-Fuzzy Inference System for Propagation Estimation

Fuzzy systems are universal approximators of nonlinear dynamic systems [26,27]. The idea of fuzzy sets, fuzzy logics and consequently fuzzy inference systems was first proposed by Zadeh [28]. As stated by Zadeh, fuzzy systems "provide an approximate and yet effective means of describing the behavior of systems which are too complex or too ill-defined to admit of precise mathematical analysis" [29]. The humanistic nature of the fuzzy systems allows us to define a complex system with fuzzy/linguistic variables using a human-like reasoning instead of using conventional mathematical tools or precise quantitative analysis. Fuzzy systems provide some degree of resistance to handle vague, ambiguous, imprecise, noisy, missing and uncertain information [30–33]. This should provide the level of flexibility required to deal with data that is hard or rather impossible to accurately infuse into the model, such as the ϕ and w_s. This resistance also relaxes the inevitable inaccuracies in data collection. This is mainly due to the fuzzification of continuous variables. The fuzzification process transforms the crisp value of the inputs (x) to degrees of membership $\mu(x)$ using a membership function (μ). Next, these membership functions μ are tuned using the gradient-descent algorithm to optimize the output. Changes in the μ are therefore affecting the degrees of membership $\mu(x)$ of the inputs (x).

The proposed ANFIS architecture comprises first-order Tagaki-Sugeno (T-S)-type fuzzy systems [34], where the output membership functions are first-order polynomials. Therefore, a hybrid training allows a linear least-squares estimation to be used for the identification of the consequent parameters, and a gradient descent optimization is used to identify the premise parameters [30,35]. Compared to most neural networks, ANFIS also has fewer parameters, many of which can be tuned with linear least-squares. These features give ANFIS the advantage of fast training and computational speed; furthermore, since there are fewer tunable parameters, the pitfall of overfitting the data would be avoided [36].

The T-S fuzzy implication (if–then rule) is analogous to that of defining a nonlinear input–output mapping. The process can be interpreted as the decomposition of a system into a finite number of subsystems and then approximating each subsystem. The output of the T-S is determined by the aggregation of the implications. Considering a number of implications R^i, with antecedents (premise) A^i and consequences y^i, implication ith $(i = 1, \ldots, n)$ is of the format of Equation (1),

$$\begin{gathered} R^i : \text{if } A^i \text{ then } y^i \\ A^i\text{: If } x_1 \text{ is } \mu_1 \text{ and } \ldots \text{ and } x_k \text{ is } \mu_k \equiv \mu_1(x_1)\Lambda \ldots \Lambda\mu_k(x_k) \\ y^i = p_0 + x_1 p_1 + \underset{\rightarrow}{} + x_k p_k \\ \overline{A}^i = \frac{A^i}{\sum_i A^i}, \end{gathered} \tag{1}$$

where x is the input vector (premise variable), μ_k contains the membership functions of the kth input, p_k is the consequence parameters vector, and $\overline{A}^i$ is the normalized firing strength, or truth value, of the implication R^i.

3.2. Model Input

A relatively similar set of inputs that are defined in the COST231 model was considered, however, three additional inputs were added based on our knowledge of propagation and common sense. In addition, three of the COST231 inputs (base station height, mobile station height and their height difference) were combined into one. Many of these modeling inputs were acquired from Google maps to further facilitate the modeling. The only output of the system is the received signal strength indicator ($RSSI$). These inputs are explained as follows:

(a) spreading factors (s_m) of LoRa's CSS modulation ($7 \leq s_m \leq 12$);
(b) height difference ($\Delta h = h_{bs} - h_{ms}$) between the base station (h_{bs}) and mobile station (h_{ms}), where h_{ms} is the altitude of earth at the location of measurement;
(c) free space path loss ($l_{fs} = 20 \log \frac{\lambda}{4\pi d_{los}}$) to include the effect of frequency, λ is the wavelength of transmission and d_{los} is the LoS distance (regardless of obstruction) between transceivers;
(d) clutter ratio (c_{los}) in the LoS, total number of buildings and structures in LoS, regardless of their heights;
(e) acute angle between the LoS and the axis of the road (ϕ);
(f) relative width of the street (w_s);
(g) d_w defined as the length of LoS that is on the water divided by the d_{los};

where $h_{bs}, h_{ms}, \phi, w_s$ and d_w are acquired from Google map images and, therefore, may have some inaccuracies. For certain scenarios, some of these parameters may not be very important, or may not exist, and therefore, would not apply at all.

3.3. Model Identification

Various membership functions including triangular, trapezoidal and sigmoidal were examined for the fuzzification; however, a normalized Gaussian membership function, with the general form of Equation (2), yielded the best result, where σ is the standard deviation and determines the spread of μ, and τ is the mean, which determines the center of the μ.

$$\mu_{c,\sigma}(x) = e^{-(x-\tau)^2/2\sigma^2} \tag{2}$$

Two approaches were considered for the identification of the premise structure. The first approach was to define fuzzy if–then rules using all the possible permutations of all or some of the fuzzified inputs. For instance, using common sense knowledge of propagation one could define the following implication:

- "If d_{los} is short and s_m is high and c_{los} is clear then $RSSI$ is good."

This states that "if the transmission was done over a short distance, with a high spreading factor, and the LoS was relatively clear of clutter, then reception should be good, regardless of other input parameters". However, this approach can be prone to combinatorial explosion of rules, especially for complex systems. Considering that there was a total of seven inputs, each with two membership functions, then the total number of rules is 2^7.

The second approach was to use a clustering method [36]. A subtractive clustering [37] was chosen for the identification of the rules, since subtractive clustering does not require an initial estimate of the center or the number of clusters [38]. Other clustering algorithms could be used, where eventually, each cluster center forms a fuzzy rule.

A first-order T-S model was selected, as it provided a higher accuracy compared to zero-order T-S. Hence, output membership functions were of the form y in Equation (1), where the output linear functions (p_k) were identified by linear least-squares optimization.

4. Analysis and Results

About 5000 data samples were collected over a relatively large area (4.25 km $\times$ 2.7 km) in the commercial area of Glasgow, Scotland. Data was collected from three base stations at different locations, with 1931, 1820 and 1256 samples being collected from BS1, BS2 and BS3, respectively. Figure 1 shows the area of the investigation, where some of the measurement locations are pinpointed with markers, and gateways are labeled as BS1, BS2 and BS3. Base stations are equipped with the same antennae, mounted relatively at the same height from the ground. Data was analyzed to provide an insight of the performance of the proposed model.

To check the goodness of fit and benchmarking with other models, the most commonly used measures in the literature were reported, RMSE (root mean square error in dB), E_σ (error standard deviation dB) and E_m (mean error dB). Unfortunately, the first two measures depend on the range or scale of data and E_m ignores the error sign. Therefore, to address these issues, the Nash-Sutcliffe efficiency (NSE) coefficient is used as a measure of the goodness of the fit. Having a universal measure of performance benchmarking is especially important, as various wireless technologies have different sensitivities. This difference impacts the dynamic range of measured data and, therefore, its RMSE scale; however, the NSE is less prone to the dynamic range. NSE ranges from $-\infty$ to 1, where 1 would indicate a perfect match between the model predictions and measurements [39].

In addition, to investigate the model's generalization capability, instead of training with one BS at a time, data from all the three BSs was used to train and validate the model. For the purpose of comparison, an ANN model was also used to model the propagation. A feedforward ANN was chosen with three hidden layers of size seven, 14 and four neurons for each layer, respectively. The best ANN structure was chosen heuristically after trying ANNs with two to five hidden layers of various neuron sizes. Results in Table 1 demonstrate the average of a 10-fold cross-validation analysis; 90% of data was used for training.

Figure 1. Area of practical investigation 4.25 km × 2.7 km, and base station locations (BS1, BS2, BS3). A few measurement locations are pinpointed with markers.

Table 1. ANFIS and ANN model performance with 10-fold cross-validation.

		ANFIS			
Base Station	**RMSE**	E_σ	E_m	**MAE**	**NSE**
BS1	5.783	5.788	0.022	4.582	0.58
BS2	6.631	6.621	0.050	5.196	0.56
BS3	6.290	6.285	0.026	5.017	0.38
BS1,2,3	7.060	7.056	0.014	5.495	0.48
		ANN			
BS1	6. 233	6.231	0.002	4.905	0.51
BS2	8.249	8.257	−0.103	6.457	0.44
BS3	7.045	7.037	0.029	5.617	0.30
BS1,2,3	8.128	8.127	0.167	6.388	0.38

The results of the COST W.I. model are tabulated in Table 2 to make a comparison with other practical investigations conducted in [15].

Table 2. COST W.I. model performance with optimization.

Base Station	**RMSE**	E_σ	E_m	**MAE**	**NSE**
BS1	12.37	10.61	6.367	10.54	0.039
BS2	18.94	14.12	12.63	16.25	−0.27
BS3	13.33	12.46	4.752	11.04	−0.30
BS1,2,3	18.14	12.02	13.58	15.86	−0.37

Figure 2 compares the measurements and estimation for BS1, and Figure 3 compares the overall estimation of all base stations with measurements.

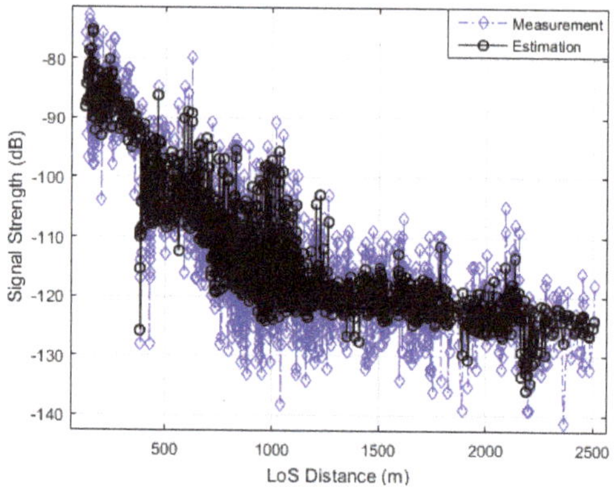

Figure 2. ANFIS estimations vs practical measurements at BS1.

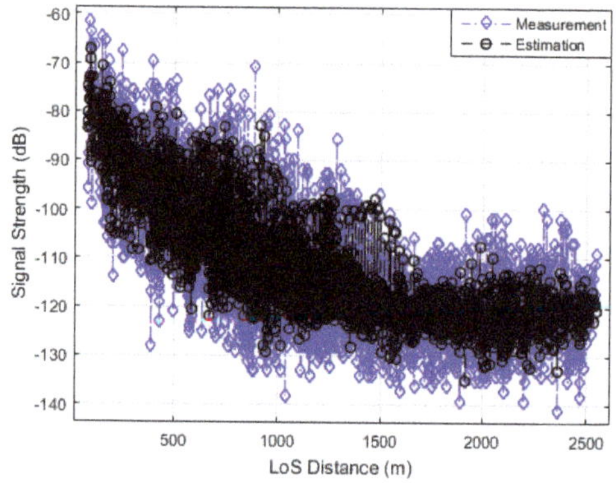

Figure 3. ANFIS estimation vs practical measurements at all BSs.

A series of models were benchmarked against similar practical measurements in COST Action 231 [15]. In this comparison, 970, 335 and 1031 samples were collected from three different base stations in Munich, Germany. Propagation models were then used to estimate the signal strength only for each individual site. Since the examined models were either deterministic or semi-deterministic, 2D or 3D building layout and building height information were provided to the models in the study conducted by COST, whereas the COST231 in this article only used the collected environmental information that was explained earlier (Section 3.2). In Table 3, only the range of measures (maximum and minimum of E_σ and E_m) of COST performance analysis report are included. A detailed performance of each model can be found in Table 4.5.2 of COST Action 231 Chapter 4 [15]. Unfortunately, other performance metrics such and RMSE, NSE and MAE are not stated in this report. Therefore, the RMSE in Table 3 is extracted for instances with $E_m \approx 0$ (if $E_m = 0$ then RMSE $= E_\sigma$). Table 3 is provided as a measure of overall modelling accuracy that can be achieved given the availability of 2D or 3D environmental information.

Table 3. Range of standard deviation and mean in COST W.I. measurements [15].

E_σ		E_m		RMSE	
min	max	min	max	min	max
6.0	21.6	−0.1	21.8	7	13.8

To further observe the generalization capability of the ANFSI model, only 20% of the data was used for training. These results are tabulated in Table 4.

Table 4. ANFIS model performance with 20% training data.

Base Station	RMSE	E_σ	E_m	MAE	NSE
BS1	6.193	6.191	−0.037	4.887	0.51
BS2	8.035	8.014	−0.017	6.233	0.46
BS3	7.451	7.445	0.153	5.837	0.27
BS1,2,3	8.0917	8.0891	−0.008	6.317	0.38

5. Discussion and Conclusions

The decomposition of a propagation system into smaller subsystems has been the ultimate goal of the Okumura, Hata and COST models. In fact, the suggestion of the breaking point-distance phenomenon in ITU recommendation P.1411 [40] follows the same idea. These attempts used crisp or Boolean logic to differentiate between a limited set of propagation conditions or scenarios. In contrast to sudden transitions, fuzzy logics make it possible to have smoother transitions, while mitigating the uncertainties within the data. ANFIS further allowed for the implementation of an expert's knowledge into the system, which addressed part of the challenges of the ANN models.

Comparison of the models used in this investigation indicated that the ANFIS and ANN models resulted in a remarkably better performance in terms of estimation, compared to the COST W.I. model. The ANFIS model resulted in a better performance compared to the ANN model. E_σ and NSE were consistently improved by about 1 dB and 10%, respectively. ANFIS was found as a better generalization candidate. In this study, the performance of ANN in Table 1 was almost identical to the ANFIS results in Table 4. This is while the ANN was trained with 90% of the data, whereas ANFIS achieved the same results with only 20% of the data.

Furthermore, two new parameters were added into the model without having to formulate them. s_m was required due to the wireless technology of choice, and d_w was added due to the features of the propagation environment. Inclusion of these parameters in the modeling reduced the RMSE and NSE of the ANFIS model by 0.55 dB and 7%, respectively. These two parameters, however, did not make a significant change to the ANN model results. This might be due to the limited number of measurements (380 samples) that had d_w. This is the most likely reason, given that ANFIS, with a better generalization, could benefit from this parameter.

In this investigation, the proposed ANFIS model was used for outdoor environments. However, it can be easily adopted for indoor propagation as well. This is as simple as providing the impacting propagation parameters for the system and roughly describing their effect using fuzzy linguistic reasoning. For instance, in an indoor environment, the effect of a higher number of walls, windows or doors on LoS can increase the loss.

Author Contributions: S.H. and A.W. conceived and designed and performed the experiments; S.H. analyzed the data; H.L. and K.C. contributed materials/analysis tools; S.H., K.C. and H.L. wrote the paper reported.

Funding: This research received no external funding.

Acknowledgments: The authors would like to thank Glasgow Caledonian University for funding this research. Also, Stream Technologies for facilitating the data collection and measurements, and Innovate UK (KTP).

Conflicts of Interest: The authors declare no conflict of interest.

References

1. Sarkar, T.K.; Ji, Z.; Kim, K.; Medouri, A.; Salazar-Palma, M. A survey of various propagation models for mobile communication. *IEEE Antennas Propag. Mag.* **2003**, *45*, 51–82. [CrossRef]
2. Okumura, Y.; Ohmori, E.; Kawano, T.; Fukuda, K. Field strength and its variability in VHF and UHF land-mobile radio service. *Rev. Electr. Commun. Lab.* **1968**, *16*, 825–873.

3. Hata, M. Empirical formula for propagation loss in land mobile radio services. *IEEE Trans. Veh. Technol.* **1980**, *29*, 317–325. [CrossRef]
4. *Propagation Data and Prediction Methods for the Planning of Indoor Radio Communication Systems and the Radio Local Area Networks in the Frequency Range 900 MHz to 100 GHz;* ITU-R Recommendations Series, P. International Telecommunication Union: Geneva, Switzerland, 2015.
5. Walfisch, J.; Bertoni, H.L. A theoretical model of UHF propagation in urban environments. *IEEE Trans. Antennas Propag.* **1988**, *36*, 1788–1796. [CrossRef]
6. Ikegami, F.; Takeuchi, T.; Yoshida, S. Theoretical prediction of mean field strength for urban mobile radio. *IEEE Trans. Antennas Propag.* **1991**, *39*, 299–302. [CrossRef]
7. Hosseinzadeh, S.; Larijani, H.; Curtis, K. An enhanced modified multi wall propagation model. In Proceedings of the Global Internet of Things Summit (GIoTS), Geneva, Switzerland, 6–9 June 2017.
8. Hosseinzadeh, S. Multi-wall Signal Propagation Model. 2016. Available online: http://www.mathworks.com/matlabcentral/fileexchange/61340-multi-wall--cost231----free-space-signal-propagation-models (accessed on 10 November 2018).
9. Hosseinzadeh, S. 3D Ray Tracing For Indoor Radio Propagation. 2017. Available online: https://uk.mathworks.com/matlabcentral/fileexchange/64695-3d-ray-tracing-for-indoor-radio-propagation (accessed on 10 November 2018).
10. Hosseinzadeh, S.; Larijani, H.; Curtis, K.; Wixted, A.; Amini, A. Empirical propagation performance evaluation of LoRa for indoor environment. In Proceedings of the 2017 IEEE 15th International Conference on Industrial Informatics (INDIN), Emden, Germany, 24–26 July 2017.
11. Damosso, E. *Digital Mobile Radio towards Future Generation Systems: COST Action 231;* European Commission: Luxembourg, Belgium, 1999.
12. Yuan, D.; Shen, D. Analysis of the Bertoni-Walfisch propagation model for mobile radio. In Proceedings of the 2011 Second International Conference on Mechanic Automation and Control Engineering (MACE), Hohhot, China, 15–17 July 2011.
13. Qiu, L.; Jiang, D.; Hanlen, L. Neural network prediction of radio propagation. In Proceedings of the 6th Australian Communications Theory Workshop, Brisbane, Australia, 2–4 February 2005.
14. Teal, P.D.; Kennedy, R.A. Bounds on extrapolation of field knowledge for long-range prediction of mobile signals. *IEEE Trans. Wirel. Commun.* **2004**, *3*, 672–676. [CrossRef]
15. *COST, Final Report for COST Action 231;* Chapter 4; COST: Luxembourg, Belgium, 1999.
16. Correia, L.M. A view of the COST 231-Bertoni-Ikegami model. In Proceedings of the 3rd European Conf. Antennas and Propagation, Berlin, Germany, 23–27 March 2009.
17. Hamim, S.F.; Jamlos, M.F. An overview of outdoor propagation prediction models. In Proceedings of the 2014 IEEE 2nd International Symposium on Telecommunication Technologies (ISTT), Langkawi, Malaysia, 24–26 November 2014.
18. Sorrentino, A.; Nunziata, F.; Ferrara, G.; Migliaccio, M. An effective indicator for NLOS, nLOS, LOS propagation channels conditions. In Proceedings of the 2012 6th European Conference on Antennas and Propagation (EUCAP), Prague, Czech Republic, 26–30 March 2012.
19. Hosseinzadeh, S.; Almoathen, M.; Larijani, H.; Curtis, K. A Neural Network Propagation Model for LoRaWAN and Critical Analysis with Real-World Measurements. *Big Data Cogn. Comput.* **2017**, *1*, 7. [CrossRef]
20. Neskovic, A.; Neskovic, N. Microcell electric field strength prediction model based upon artificial neural networks. *AEU-Int. J. Electron. Commun.* **2010**, *64*, 733–738. [CrossRef]
21. Parsons, J.D. *The Mobile Radio Propagation Channel;* Wiley & Sons: Chichester, UK, 1992.
22. Ostlin, E.; Zepernick, H.-J.; Suzuki, H. Macrocell path-loss prediction using artificial neural networks. *IEEE Trans. Veh. Technol.* **2010**, *59*, 2735–2747. [CrossRef]
23. Popescu, I.; Nikitopoulos, D.; Constantinou, P.; Nafornita, I. ANN prediction models for outdoor environment. In Proceedings of the 2006 IEEE 17th International Symposium on Personal, Indoor and Mobile Radio Communications, Helsinki, Finland, 11–14 September 2006.
24. Gschwendtner, B.E.; Landstorfer, F.M. Adaptive propagation modelling using a hybrid neural network technique. *Electron. Lett.* **1996**, *32*, 162–164. [CrossRef]
25. Semtech Corporation, LoRa™ Modulation Basics. 2015. Available online: https://www.semtech.com/uploads/documents/an1200.22.pdf (accessed on 10 November 2018).

26. Castro, J.L. Fuzzy logic controllers are universal approximators. *IEEE Trans. Syst. Man Cybern.* **1995**, *25*, 629–635. [CrossRef]
27. Kosko, B. Fuzzy systems as universal approximators. *IEEE Trans. Comput.* **1994**, *43*, 1329–1333. [CrossRef]
28. Zadeh, L.A. Fuzzy sets. In *Fuzzy Sets, Fuzzy Logic, And Fuzzy Systems: Selected Papers by Lotfi A Zadeh*; World Scientific: Singapore, 1996; pp. 394–432.
29. Zadeh, L.A. Fuzzy logic. *Computer* **1988**, *21*, 83–93. [CrossRef]
30. Hosseinzadeh, S. A Fuzzy Inference System for Unsupervised Deblurring of Motion Blur in Electron Beam Calibration. *Appl. Syst. Innov.* **2018**, *1*, 48. [CrossRef]
31. Jang, J.-S.R. ANFIS: Adaptive-network-based fuzzy inference system. *IEEE Trans. Syst. Man Cybern.* **1993**, *23*, 665–685. [CrossRef]
32. Priyono, A.; Ridwan, M.; Alias, A.J.; Atiq, R.; Rahmat, O.K.; Hassan, A.; Ali, M. Generation of fuzzy rules with subtractive clustering. *J. Teknol.* **2005**, *43*, 143–153. [CrossRef]
33. Hosseinzadeh, S. Unsupervised spatial-resolution enhancement of electron beam measurement using deconvolution. *Vacuum* **2016**, *123*, 179–186. [CrossRef]
34. Takagi, T.; Sugeno, M. Fuzzy identification of systems and its applications to modeling and control. *IEEE Trans. Syst. Man Cybern.* **1985**, *SMC-15*, 116–132. [CrossRef]
35. De Mingo López, L.F.; Blas, N.G.; Arteta, A. The optimal combination: Grammatical swarm, particle swarm optimization and neural networks. *J. Comput. Sci.* **2012**, *3*, 46–55. [CrossRef]
36. Chiu, S.L. Fuzzy model identification based on cluster estimation. *J. Intell. Fuzzy Syst.* **1994**, *2*, 267–278.
37. Surmann, H.A. Selenschtschikow and others, Automatic generation of fuzzy logic rule bases: Examples I. In Proceedings of the NF2002: First International ICSC Conference on Neuro-Fuzzy Technologies CUBA, Havana, Cuba, 16–19 January 2002.
38. Yager, R.R.; Filev, D.P. Generation of fuzzy rules by mountain clustering. *J. Intell. Fuzzy Syst.* **1994**, *2*, 209–219.
39. Ritter, A.; Muñoz-Carpena, R. Performance evaluation of hydrological models: Statistical significance for reducing subjectivity in goodness-of-fit assessments. *J. Hydrol.* **2013**, *480*, 33–45. [CrossRef]
40. *Propagation Data and Prediction Methods for the Planning of Short-range Outdoor Radio Communication Systems and Radio Local Area Networks in the Frequency Range 300 MHz to 100 GHz*; ITU-R Recommendations Series, P.; International Telecommunication Union: Geneva, Switzerland, 2017.

applied system innovation

MDPI

Article

New Fuzzy Numerical Methods for Solving Cauchy Problems

Hussein ALKasasbeh [1,*], Irina Perfilieva [2], Muhammad Zaini Ahmad [1] and Zainor Ridzuan Yahya [1]

1 Institute of Engineering Mathematics, Universiti Malaysia Perlis, Kampus Tetap Pauh Putra, Arau 02600, Perlis, Malaysia; mzaini@unimap.edu.my (M.Z.A.); zainoryahya@unimap.edu.my (Z.R.Y.)
2 Institute for Research and Applications of Fuzzy Modelling, University of Ostrava, NSC IT4Innovations, 30. dubna 22, 701 03 Ostrava, Czech Republic; Irina.Perfilieva@osu.cz
* Correspondence: hussein.ahmad.alkasasbeh@gmail.com

Received: 7 April 2018; Accepted: 3 May 2018; Published: 11 May 2018

Abstract: In this paper, new fuzzy numerical methods based on the fuzzy transform (F-transform or FT) for solving the Cauchy problem are introduced and discussed. In accordance with existing methods such as trapezoidal rule, Adams Moulton methods are improved using FT. We propose three new fuzzy methods where the technique of FT is combined with one-step, two-step, and three-step numerical methods. Moreover, the FT with respect to generalized uniform fuzzy partition is able to reduce error. Thus, new representations formulas for generalized uniform fuzzy partition of FT are introduced. As an application, all these schemes are used to solve Cauchy problems. Further, the error analysis of the new fuzzy methods is discussed. Finally, numerical examples are presented to illustrate these methods and compared with the existing methods. It is observed that the new fuzzy numerical methods yield more accurate results than the existing methods.

Keywords: fuzzy partition; fuzzy transform; new iterative method; Cauchy problems

---

1. Introduction

In fact, most mathematical models in engineering and science requires the solution of ordinary differential equations (ODEs). Generally, it is difficult to obtain the closed form solutions for ODEs, especially, for nonlinear and nonhomogeneous cases. Many models often lead to ordinary differential equations which consist of Cauchy problems are an important branch of modern mathematics that arises naturally in different areas of applied sciences, physics, and engineering. Thus, many researchers start developing methods for solving Cauchy problems are of particular importance [1–3].

FT was coined by Perfilieva as a new mathematical method was developed [4]. The core idea of FT is a fuzzy partition of a universe into fuzzy subsets. The technique of FT has been successfully applied into other mathematical problems as well including image processing, analysis of time series and elsewhere [5–7]. This idea has been applied to Cauchy problems was first published as well as other numerical classical methods [8], by proposing generalized Euler and Euler- Cauchy methods, so that the Mid-point FT method was demonstrated in [9]. The success of these applications is due in part to the fact that FT is capable to accurately approximate any continuous function. Thus, we will propose new fuzzy numerical methods for Cauchy problems with help of the FT and new iterative method.

The motivation of the proposed study comes from the papers [3,8,10]. Numeric Solution to the Cauchy problem was considered and the authors showed that the error can be reduced by using FT with uniform fuzzy partitions [8,9]. At the same time, [10,11], the concept of generalized fuzzy partition was proposed. Besides others, a necessary and sufficient condition making it possible to design easily the generalized fuzzy partition was provided [12]. This is important for various practical applications of FT. Further [3], the authors have proposed modifications trapezoidal rule and Adams-Moulton methods (2 and 3-step) to solve ODEs based on the new iterative method was introduced [2].

In this paper, we discuss the problem that considered in [8,9]. The triangular and raised cosine generating function was replaced by new representations formulas for generalized uniform fuzzy partition of FT such as power of the triangular and raised cosine generating function. We study approximation properties of the FT based on powers of triangular and raised cosine generalized uniform fuzzy partition can be constructed in such way that the FT can reduce error. Also, we propose modifications in the FT introduced by I. Perfilieva [4] with respect to new representations formulas for generalized uniform fuzzy partition of FT and then the technique of FT is combined with traditional methods based on the new iterative method [2,3] to solve Cauchy problems. It is observed that the new methods proposed are more accurate results than the fuzzy approximation method [8,9].

This paper is organized as follows. In Section 2, we introduce the basic concepts and results of the FT with respect to the generalized uniform fuzzy partition needed throughout this paper. The main part of this paper is Sections 3 and 4, new representations for basic functions of FT, followed by the modified one step, 2-step , and 3-step based on new representations formulas for generalized uniform fuzzy partition of FT. In Section 5, numeric examples are discussed. Concluding remarks are presented in Section 6.

Throughout the paper, we denote by $\mathbb{N}$, $\mathbb{N}^+$, $\mathbb{Z}$, $\mathbb{R}$, and $\mathbb{R}^+$ the sets of natural (including zero), positive natural, integer, real , and positive real numbers, respectively.

2. Basic Concepts

In this section, we give some definitions and introduce the necessary notation in [10], which will be used throughout the paper. Throughout this section, we deal with an interval $[a,b] \subset \mathbb{R}$ of real numbers.

Definition 1. *(generalized uniform fuzzy partition) Let $x_i \in [a,b]$, $i = 1,\ldots,n$, be fixed nodes such that $a = x_1 < \ldots < x_n = b$, $n \geq 2$. We say that the fuzzy sets $A_i : [a,b] \to [0,1]$ constitute a generalized fuzzy partition of $[a,b]$ if for every $i = 1,\ldots,n$ there exists $h > 0$ such that $x_0 = x_1$, $x_n = x_{n+1}$, $[x_i - h,\ x_i + h] \subseteq [a,b]$ and the following conditions are fulfilled:*

1. *(positivity and locality) – $A_i(x) > 0$ if $x \in (x_{i-1}, x_{i+1})$ and $A_i(x) = 0$ if $x \in [a,b] \setminus (x_{i-1}, x_{i+1})$;*
2. *(continuity) – A_i is continuous on $[x_{i-1}, x_{i+1}]$;*
3. *(covering) – for $x \in [a,b]$, $\sum_{i=1}^{n} A_i(x) > 0$.*

Fuzzy sets $A_1,\ldots,\ A_n$ are called basic functions. It is important to remark that by conditions of locality and continuity, $\int_a^b A_i(x)dx > 0$. A generalized of uniform fuzzy partition of $[a,b]$ is defined for equidistant nodes, i.e., for all $i = 1,\ldots,n-1$, $x_i = x_{i+1} + h$, where $h = (b-a)\,/\,(n-1)$ and two additional properties are satisfied,

4. *$A_i(x_i - x) = A_i(x_i + x)$ for all $x \in [0,h]$, $i = 2,\ldots,n-1$;*
5. *$A_i(x) = A_{i-1}(x-h)$ and $A_{i+1}(x) = A_i(x-h)$ for all $x \in [x_i,\ x_{i+1}]$, $i = 2,\ldots,n-1$;*

then the fuzzy partition is called h-uniform generalized fuzzy partition. Throughout this paper, we will write generalized uniform fuzzy partition instead of h-uniform generalized fuzzy partition.

Definition 2. *(generating function) A function $K : [-1,1] \to [0,1]$ is called a generating function if it is assumed to be even, continuous and $K(x) > 0$ if $x \in (-1,1)$. The function $K : [-1,1] \to \mathbb{R}$ is even if for all $x \in [0,1]$, $K(-x) = K(x)$.*

The following definition recall the concept of generalized fuzzy partition which can be easily extended to the interval $[a,b]$. We assume that $[a,b]$ is partitioned by $A_1,\ldots,\ A_n$, according to Definition 1.

Definition 3. *A generalized uniform fuzzy partition of interval* $[a, b]$*, determined by the triplet* (K, h, a)*, can be defined using generating function K (Definition 2). Then, basic functions of a generalized uniform fuzzy partition are shifted copies of K defined by*

$$A_i(x) = K\left(\frac{x - x_i}{h}\right), \quad x \in [x_i - h, x_i + h],$$

for all $i = 1, \ldots, n$*. The parameter h is called the bandwidth or the shift of the fuzzy partition and the nodes* $x_i = a + ih$ *are called the central point of the fuzzy sets* $A_1, \ldots, A_n$*.*

Remark 1. *A fuzzy partition is called Ruspini if the following condition*

$$A_i(x) + A_{i+1}(x) = 1, \ i = 1, \ldots, n-1, \tag{1}$$

holds for any $x \in [x_i, x_{i+1}]$*. This condition is often called Ruspini condition.*

3. New Representations of Basic Functions for Particular Cases

In this section, we propose two subsection, new representations of basic functions constitute a generalized uniform fuzzy partition of interval $[a, b]$ and then FT technique based on new representations of basic functions.

3.1. Power of the Triangular and Raised Cosine Generalized Uniform Fuzzy Partition

Two types of basic functions, triangular and sinusoidal shaped membership functions, were proposed by [4,8]. Later [13], the authors considered different shapes for the basic functions of fuzzy partition. Furthermore, a generalized fuzzy partition appeared in connection with the notion of a higher-degree F-transform [11]. Its even weaker version was implicitly introduced to satisfy the requirements of image compression [14]. Recently, the different conditions for generalized uniform fuzzy partitions was proposed by [10,12]. Table 1 provides the definition two types of generating function, triangular and raised cosine generating functions [7,10–12,15].

Table 1. Generating functions of strong uniform fuzzy partition.

Triangular Generating Function	Raised Cosine Generating Function
$\max\{1 - \lvert x\rvert, 0\}$	$\frac{1}{2}(1 + \cos(\pi x))_{\lvert[-1,1]}$

In the following, we present new representations for generating function. In particular, we present three new representations, based on the triangular and raised cosine generating functions: two generating function based on the triangular generating functions and one generating function based on the raised cosine generating function.

Definition 4. *(natural order triangular generating function) Let* $K_{T_i^m} : \mathbb{R} \to [0, 1]$*,* $i = 1, 2$*, be defined by*

$$1.\ K_{T_1^m}(x) = \begin{cases} (1 - |x|)^m, & |x| \le 1, \\ 0, & otherwise \end{cases} = \min\left((1 - |x|)^m, 1\right), \tag{2}$$

$$2.\ K_{T_2^m}(x) = \begin{cases} 1 - (|x|)^m, & |x| \le 1, \\ 0, & otherwise \end{cases} = \min\left(1 - (|x|)^m, 1\right), \tag{3}$$

are called power of the triangular (shaped) generating functions, when $m \in \mathbb{N}^+$*.*

Definition 5. *(odd natural order raised cosine generating function) Let* $K_{C^m} : \mathbb{R} \to [0,1]$ *be defined by*

$$K_{C^m}(x) = \begin{cases} \frac{1}{2}\left(1+\cos^m(\pi x)\right), & |x| \le 1; \\ 0, & \textit{otherwise.} \end{cases} \tag{4}$$

is called power of the raised cosine generating function , when m is an odd natural number (i.e., $m = 2k-1,\ k \in \mathbb{N}^+$*).*

Remark 2. *Particularly, we can check the validity of Equation (4) using the following relation*

$$\begin{aligned} K_{C^m}(x) &= \begin{cases} \frac{1}{2}\left(1+\cos^m(\pi x)\right), & |x| \le 1, \\ 0, & \textit{otherwise.} \end{cases} \\ &= \begin{cases} \frac{1}{2}\left(1+\sin^m\left(\frac{\pi}{2}(2x+1)\right)\right), & |x| \le 1, \\ 0, & \textit{otherwise.} \end{cases} \end{aligned}$$

Lemma 1. *If* $K_{T_i^n}(x),\ i = 1,2,\ (K_{C^m}(x))$ *determines power of the triangular (raised cosine) generating functions, then*

1. $\int_{-1}^{1} K_{T_1^n}(x)\,dx = \frac{2}{n+1}$, 2. $\int_{-1}^{1} K_{T_2^n}(x)\,dx = \frac{2n}{n+1}$, 3. $\int_{-1}^{1} K_{C^m}(x)\,dx = 1$,

or equivalent

1. $\int_{-h}^{h} K_{T_1^n}\left(\frac{t}{h}\right)dt = \frac{2h}{n+1}$, 2. $\int_{-h}^{h} K_{T_2^n}\left(\frac{t}{h}\right)dt = \frac{2nh}{n+1}$, 3. $\int_{-h}^{h} K_{C^m}\left(\frac{t}{h}\right)dt = h$,

where $0 \le \left|\frac{2}{n+1}\right| \le 1,\ 1 \le \left|\frac{2n}{n+1}\right| \le 2$ *, h be positive real numbers, m is an odd natural number and* $n \in \mathbb{N}^+$*.*

Proof. The proof can be easily obtained by using integration methods within the boundaries and then substitution $x = t/h$. □

On the basis of Definitions 4 and 5, Lemma 1, and according to Definition 3, we can also be defined using generating function αK for $\alpha > 0$ (in general, not necessarily satisfy Ruspini condition). Thus, basic functions of a generalized uniform fuzzy partition are shifted copies of αK defined by

$$A_k(x, x_0) = \alpha K\left(\frac{x-x_0}{h} - k\right),\ x \in [x_{i-1}, x_{i+1}]. \tag{5}$$

In particular, let $K_{T_1^m}$, $K_{T_2^m}$, (and K_{C^m}) be power of the triangular (and raised cosine) generating function defined above. We will say that a generalized uniform fuzzy partition is power of a triangular (or of raised cosine) generalized uniform fuzzy partition if its generating function K belongs to $\alpha K_{T_1^m}, \alpha K_{T_2^m}$, (or αK_{C^m}) whenever $\alpha = 1/\left(\int_{-1}^{1} K(t)dt\right)$. Indeed, the equality α immediately follows from $\int_{-1}^{1} \alpha K_{T_1^m}(t)\,dt = 1 \Rightarrow \alpha = 1/\left(\int_{-1}^{1} K_{T_1^m}(t)\,dt\right)$. In the following, we modified the definition a triangular and raised cosine generalized uniform fuzzy partition by propose that power of the triangular and raised cosine generalized uniform fuzzy partitions can be simply using the equality $\alpha = 1/\left(\int_{-1}^{1} K(t)dt\right)$.

Definition 6. *Let* $m \in \mathbb{N}^+$*. A system of fuzzy sets* $\{A_k \mid k \in \mathbb{Z}\}$ *defined by*

$$1.\ A_k(x, x_0) = \alpha K_{T_1^m}\left(\tfrac{x-x_0}{h} - k\right),\ \alpha = \tfrac{m+1}{2}, \tag{6}$$

$$2.\ A_k(x, x_0) = \alpha K_{T_2^m}\left(\tfrac{x-x_0}{h} - k\right),\ \alpha = \tfrac{m+1}{2m}, \tag{7}$$

is called power of the triangular generalized uniform fuzzy partition of the real line determined by the triplet $(K_{T_i^m}, h, x_0)$, $i = 1, 2$. *Further, let m is an odd natural number. A system of fuzzy sets* $\{A_k \mid k \in \mathbb{Z}\}$ *defined by*

$$3.\ A_k(x, x_0) = \alpha K_{C^m}\left(\tfrac{x-x_0}{h} - k\right), \quad \alpha = 1, \tag{8}$$

is called power of the raised cosine generalized uniform fuzzy partition of the real line determined by the triplet $(K_{C^m}, h,, x_0)$. *The parameter h is bandwidth of the fuzzy partition and* $x_0 + kh = x_k$.

Definition 7. *Let* $x_1 < \ldots < x_n$ *be fixed nodes within* $[a, b] \subset \mathbb{R}$, *such that* $x_1 = a$, $x_n = b$ *and* $n \geq 2$. *We consider nodes* $x_1, \ldots, x_n$ *are equidistant, with distance (shift)* $h = (b - a) / (n - 1)$. *A system of fuzzy sets* $B_1, \ldots, B_n : [a, b] \to [0, 1]$ *be power of a triangular and raised cosine generalized uniform fuzzy partitions of* $[a, b]$ *if it is defined by*

$$B_k(x) = \begin{cases} A_k(x, a), & x \in [a, b], \\ 0, & \text{otherwise.} \end{cases} \quad \text{or equivalent} \quad B_k(x) = \begin{cases} \alpha K\left(\frac{x - x_k}{h}\right), & x \in [a, b], \\ 0, & \text{otherwise.} \end{cases} \tag{9}$$

where $x_k = a + kh$. *In the sequel, we denote K for a generating function determined by the Formulas (2)–(4). Further,* α, $A_k(x, a)$, $k = 1, \ldots, n$, *are determined by the Formulas (6)–(8).*

Lemma 2. *If* $B_k(x)$ *determines power of the raised cosine generalized uniform fuzzy partition of* $[a, b]$, *then* $B_k(x)$ *satisfied Ruspini condition (1) when m (see (4)) is an odd natural number.*

Proof. Indeed, if $x \in [a, b]$, there exists $k \in \{1, \ldots, n-1\}$ such that $x \in [x_k, x_{k+1}]$. By (4) and (8), and Remark 1, we get

$$\begin{aligned} B_k(x) + B_{k+1}(x) = A_k(x, a) + A_{k+1}(x, a) &= \alpha K_{C^m}\left(\frac{x - x_k}{h}\right) + \alpha K_{C^m}\left(\frac{x - x_{k+1}}{h}\right), \\ &= \frac{1}{2}\left(1 + \cos^m\left(\pi\left(\frac{x - x_k}{h}\right)\right)\right) + \frac{1}{2}\left(1 + \cos^m\left(\pi\left(\frac{x - x_{k+1}}{h}\right)\right)\right), \\ &= 1 + \frac{1}{2}\left(\cos^m\left(\frac{\pi}{h}(x - x_k)\right) + \cos^m\left(\frac{\pi}{h}(x - x_{k+1})\right)\right). \end{aligned}$$

By the properties of trigonometric functions, notice thatcos $(\theta + \pi) = -\cos(\theta)$, it is easy to see that

$$\cos^m\left(\pi\left(\frac{x - x_k}{h}\right)\right) + \cos^m\left(\pi\left(\frac{x - x_{k+1}}{h}\right)\right) = \cos^m\left(\pi\left(\frac{x - x_{k+1}}{h}\right) + \pi\right) + \cos^m\left(\pi\left(\frac{x - x_{k+1}}{h}\right)\right).$$

Thus, if m is an odd natural number, the result is 0. □

In the following, if K is a normal generating function (i.e., $K(0) = 1$, not necessarily satisfy Ruspini condition), we use generating function αK for $\alpha > 0$, where $(\alpha K)(x) = \alpha \cdot K(x)$.

Lemma 3. *If basic functions* B_k, $k = 1, \ldots, n$, *of a generalized uniform fuzzy partition are shifted copies of* αK, $\alpha > 0$, *defined by the Formula (5) and moreover, K is normal as an additional condition. Then, for each* $k = 1, \ldots, n$, $B_k(x_k) = \alpha$, $x_k \in [x_k - h, x_k + h]$.

Proof. A generating function K is said to be normal if $K(0) = 1$. By the Formula (5) and a generating function K is normal, we get $B_k(x_k) = \alpha K\left(\frac{x_k - x_k}{h}\right) = \alpha K(0) = \alpha > 0$. □

Corollary 1. *Let the assumptions of Lemma 3 be fulfilled, but fuzzy sets* B_k, $k = 1, \ldots, n$, $n \geq 2$, *determined by Definition 7. Then, for each* $k = 1, \ldots, n$, $B_k(x_k) = \alpha$, $x_k \in [x_k - h, x_k + h]$, *where* α *is defined by Definition 7.*

Proof. Indeed, the proof immediately follows from Definition 7 and Lemma 3. □

Corollary 2. *Let the assumptions of Lemma 3 be fulfilled, but fuzzy sets B_k, $k = 1, \ldots, n$, $n \geq 2$, determined by Definition 3. Then, for each $k = 1, \ldots, n$, $B_k(x_k) = 1$, $x_k \in [x_k - h, x_k + h]$.*

3.2. New FT Based Power of the Triangular and Raised Cosine Generalized Uniform Fuzzy Partition

In this subsection, we present the main principles of F-transform detailed in [8,10,11] that are modified with respect to power of the triangular and raised cosine generalized uniform fuzzy partition. Further, we will show that FT components with respect to power of the triangular and raised cosine generalized uniform fuzzy partition can be simplified and approximated of an original function, say f.

Definition 8. *Let f be a continuous function on $[a, b]$ and $B_k(t)$, $k = 1, \ldots, n$, be power of the triangular and raised cosine generalized uniform fuzzy partition of $[a, b]$, $n \geq 2$. A vector of real numbers $F[f] = (F_1, F_2, \ldots, F_n)$ given by*

$$F_k = \frac{\int_a^b f(t)\, B_k(t)\, dt}{\int_a^b B_k(t)\, dt}, \tag{10}$$

for $k = 1, \ldots, n$ is called the direct FT of f with respect to power of the triangular and raised cosine generalized uniform fuzzy partition B_k.

In the following, we assume a generating function K in the Formulas (2)–(4). We will simplify the representation (10).

Lemma 4. *Let $f \in C([a, b])$ and according to Definition 7, fuzzy sets B_k, $k = 1, \ldots, n$, $n \geq 2$, be power of a triangular and raised cosine generalized uniform fuzzy partition of $[a, b]$ with a generating function K, then representation (10) of direct FT can be simplified as follows for $k = 1, \ldots, n$*

$$F_k = \frac{\int_{-1}^{1} f(th + t_k)\, K(t)\, dt}{\int_{-1}^{1} K(t)\, dt} = \frac{\int_{-h}^{h} f(t + t_k)\, K(\frac{t}{h})\, dt}{\int_{-h}^{h} K(\frac{t}{h})\, dt}.$$

Proof. In this proof, we will write a generating function K instead of (2)–(4). By Definition 7, we get

$$B_k(t) = \alpha K\left(\frac{t - t_k}{h}\right), \quad t \in [t_k - h, t_k + h],$$

for $k = 1, \ldots, n$, $t_0 = t_1$, $t_{n+1} = t_n$, and substituting $u = \frac{t - t_k}{h}$ and then substituting $t = s/h$. Thus, we get

$$\int_{t_{k-1}}^{t_{k+1}} f(t)\, B_k(t)\, dt = \alpha h \int_{-1}^{1} f(th + t_k)\, K(t)\, dt = \alpha \int_{-h}^{h} f(t + t_k)\, K(\frac{t}{h})\, dt$$

$$\int_{t_{k-1}}^{t_{k+1}} B_k(t)\, dt = \alpha h \int_{-1}^{1} K(t)\, dt = \alpha \int_{-h}^{h} K(\frac{t}{h})\, dt$$

and its corresponding results with representation (10). □

Indeed, the previous lemma holds for every fuzzy partition generated by a kernel. Now, we will simplify the above given expressions for the coefficients $F[f] = (F_1, F_2, \ldots, F_n)$ in the representation (10) even more. This fact is very important for applications which are more flexible and consequently easier to use.

Lemma 5. *Let the assumptions of Lemma 4 be fulfilled. Then, the coefficients* $F[f] = (F_1, F_2, \ldots, F_n)$ *in the expression (10) of the FT component* F_k *of* f *as follows:*

$$F_k = \frac{1}{h}\int_a^b f(t)\, B_k(t)\, dt = \frac{\alpha}{h}\int_a^b f(t)\, K\left(\frac{t-t_k}{h}\right) dt, \tag{11}$$

for $k = 1, \ldots, n$, *where interval* $[a, b]$ *is partitioned by power of the triangular and raised cosine generalized uniform fuzzy partition* $B_1, \ldots, B_n$ *and* α *is defined by Definition 7.*

Proof. Let $k \in \{1, \ldots, n\}$ and consider set of fuzzy sets $B_k(x)$ from power of the triangular and raised cosine generalized uniform fuzzy partition of $[a, b]$ in (9). We will prove the equality $\int_{t_{k-1}}^{t_{k+1}} B_k(t)\, dt = h$. We get by virtue of Lemmas 1 and 4, and (6):

$$\int_{t_{k-1}}^{t_{k+1}} B_k(t)\, dt = \int_{t_{k-1}}^{t_{k+1}} A_k(t, a), dt = \int_{t_k-h}^{t_k+h} \left(\frac{m+1}{2}\right) K_{T_1^m}\left(\frac{t-t_k}{h}\right) dt = h\int_{-1}^{1} \left(\frac{m+1}{2}\right) K_{T_1^m}(t)\, dt = h,$$

where h is the bandwidth of the fuzzy partition and $t_k = a + kh$. Similarly, the other Formulas (7) and (8) will be proved and then its corresponding in the expression (10). □

Lemma 6. *Let* $f \in C[a, b]$. *Then for any* $\varepsilon > 0$ *there exist* $n_\varepsilon \in \mathbb{N}$ *and* $B_1, \ldots, B_{n_\varepsilon}$ *be basic functions form power of the triangular and raised cosine generalized uniform fuzzy partition of* $[a, b]$. *Let* F_k, $k = 1\ldots, n$, *be the integral FT components of* f *with respect to* $B_1, \ldots, B_{n_\varepsilon}$. *Then for each* $k = 1\ldots, n_\varepsilon - 1$ *the following estimations hold:* $|f(t) - F_i| \leq \varepsilon$ *for each* $t \in [a, b] \cap [t_k, t_{k+1}]$ *and* $i = k, k+1$.

Proof. see [4]. □

Corollary 3. *Let the conditions of Lemma 6 be fulfilled. Then for each* $k = 1\ldots, n_\varepsilon - 1$ *the following estimations hold:* $|F_k - F_{k+1}| < \varepsilon$.

Proof. According to [4,16], let $t \in [a, b] \cap [t_k, t_{k+1}]$. Then by Lemma 6, for any $k = 1, \ldots, n-1$ we obtain $|f(t) - F_k| < \varepsilon/2$ and $|f(t) - F_{k+1}| < \varepsilon/2$. Thus, $|F_k - F_{k+1}| \leq |f(t) - F_k| + |f(t) - F_{k+1}| < \frac{\varepsilon}{2} + \frac{\varepsilon}{2} = \varepsilon$. □

The following theorem estimates the difference between the original function and its direct FT with respect to power of the triangular and raised cosine generalized uniform fuzzy partition.

Theorem 1. *Let* $f(t) \in C^2[a, b]$ *and the conditions of Lemma 5 be fulfilled. Then for* $k = 1, \ldots, n$

$$F_k = \alpha f(t_k) + \mathcal{O}(h^2), \tag{12}$$

where $\alpha > 0$ *or* α *is defined by Definition 7.*

Proof. By locality condition for Definition 1, Lemmas 3 and 5, and according to the proof of Lemma 9.3 [8], using the trapezoid formula with nodes t_{k-1}, t_k, t_{k+1} to the numerical computation of the integral, we get for $\alpha > 0$

$$\begin{aligned}
F_k &= \frac{1}{h}\int_{t_{k-1}}^{t_{k+1}} f(t)\, B_k(t)\, dt, \\
&= \frac{1}{h} \cdot \frac{h}{2}\left(f(t_{k-1})\, B_k(t_{k-1}) + 2f(t_k)\, B_k(t_k) + f(t_{k+1})\, B_k(t_{k+1})\right) + \mathcal{O}\left(h^2\right), \\
&= f(t_k)\, B_k(t_k) + \mathcal{O}\left(h^2\right) = f(t_k)\, A_k(t_k, a) + \mathcal{O}\left(h^2\right), \\
&= f(t_k)\, \alpha K(0) + \mathcal{O}\left(h^2\right), \\
&= \alpha f(t_k) + \mathcal{O}\left(h^2\right).
\end{aligned}$$

□

Definition 9. *Let $F[f] = (F_1, F_2, \ldots, F_n)$ be direct FT of a function $f \in C[a,b]$ with respect to the fuzzy partition $B_k(t)$, $k = 1, \ldots, n$ of $[a,b]$. Then, the function $\hat{f}$ defined on $[a,b]$*

$$\hat{f}(t) = \frac{\sum_{k=1}^{n} F_k B_k(t)}{\sum_{k=1}^{n} B_k(t)}, \tag{13}$$

is called the inverse FT of f.

Corollary 4. *Let the assumptions of Lemma 2 and moreover, Let $\hat{f}(t)$ be the inverse FT of f with respect to power of the raised cosine generating function . Then, for all $t \in [a,b]$ the following holds: $\hat{f}(t) = \sum_{k=1}^{n} F_k B_k(t)$.*

Proof. This proof immediately follows from Defintion 9, Lemma 2 and then using $\sum_{k=1}^{n} B_k(t) = 1$. □

The following lemma estimates the difference between the original function and its inverse FT.

Lemma 7. *Let the assumptions of Theorem 1 and let $\hat{f}(t)$ be the inverse FT of f with respect to the fuzzy partition of $[a,b]$ is given by Definition 7. Then, for all $t \in [a,b]$ the following estimation holds:*

$$\hat{f}(t) = \alpha f(t_k) + \mathcal{O}\left(h^2\right). \tag{14}$$

Proof. Let $t \in [a,b]$ so that $x \in [t_k, t_{k+1}]$ for some $k = 1, \ldots, n$. By Theorem 1,

$$\begin{aligned} \hat{f}(t) - \alpha f(t_k) &= \frac{\sum_{k=1}^{n} F_k B_k(t)}{\sum_{k=1}^{n} B_k(t)} - \alpha f(t) = \frac{\sum_{k=1}^{n} F_k B_k(t)}{\sum_{k=1}^{n} B_k(t)} - \frac{\sum_{k=1}^{n} \alpha f(t_k) B_k(t)}{\sum_{k=1}^{n} B_k(t)} \\ &= \frac{\sum_{k=1}^{n} (F_k - \alpha f(t_k)) B_k(t)}{\sum_{k=1}^{n} B_k(t)} - \mathcal{O}\left(h^2\right). \end{aligned}$$

□

Corollary 5. *Let the assumptions of Lemma 7, then $\left|\hat{f}(t) - f(t)\right| < \varepsilon$.*

Proof. The proof easily follows from the proof of Lemma 7 and then using Lemma 6 as follows:

$$\left|\hat{f}(t) - f(t)\right| = \frac{\sum_{k=1}^{n} |F_k - f(t)| B_k(t)}{\sum_{k=1}^{n} B_k(t)} < \varepsilon.$$

□

Remark 3. *According to the Definitions 1 and 2, if the normality is considered to be an additional condition for generating function (i.e., $K(0) = 1$) and generalized uniform fuzzy partition of $[a,b]$ satisfies $A_k(x_k) = \alpha$, $\alpha > 0$, then it is easy to see that the inverse FT $\hat{f}(t_k) = F_k$ for all $k = 1, \ldots, n$. This is true for Definition 7. Moreover, if orthogonality condition (Ruspini condition (1)) is replaced by covering condition in Definition 1 and generalized uniform fuzzy partition of $[a,b]$ satisfies $A_k(x_k) = \alpha = 1$, then it is easy to also see that the inverse FT $\hat{f}(t_k) = F_k$ for all $k = 1, \ldots, n$. This is true for Formula (8) only.*

Important property of the direct FT as well as inverse FT is their linearity, namely, given $f, g \in C[a,b]$ and $\alpha, \beta \in R$, if $h = \alpha f + \beta g$, then $F[h] = \alpha F[f] + \beta F[g]$ and $\hat{h} = \alpha\hat{f} + \beta\hat{g}$. In the next section, we present new fuzzy numerical methods based on the FT and a new iterative method to numeric solution of the Cauchy problem.

4. New Fuzzy Numerical Methods for Cauchy Problem

Consider the initial value problem (IVP) for the Cauchy problem:

$$y' = f(t,y), \quad y(t_1) = y_1, \quad a = t_1 \leq t \leq t_n = b. \tag{15}$$

where $y_1 \in \mathbb{R}$ and f is continuous function on $[a,b] \times \mathbb{R}$ and satisfies Lipschitz condition. In fact, the analytical solution of problem (15) is often difficult and sometimes impossible to obtain. Instead, numerical analysis is interested with obtaining approximate solutions with errors within reasonable bounds. Thus, a usage of fuzzy numerical methods seems to be suitable.

In [8,9], the authors have presented Euler method and Mid-point rule, based on FT to numeric solution of Cauchy problem (15). A new iterative method (NIM) has been proposed for solving linear (nonlinear) functional equations, ordinary differential equations and delay differential equations [2,3].

In this section, we present three new schemes to solve Cauchy problem (15), that use the FT and NIM. Our motivation stems from the classical approach, trapezoidal rule (1-step) and Adams Moulton methods (2 and 3-step). For the rest of this paper, suppose that we are given the Cauchy problem (15), where the function f on $[a,b]$ are sufficiently smooth and we assume that all necessary requirements for constructing the FT of the solution of Cauchy problem (15) are fulfilled. Now, we present numerical Scheme I, II, and III. The first scheme uses 1-step method, while the second one uses 2-step method, and the third uses 3-step method.

4.1. Numeric Scheme I: Modified Trapezoidal Rule Based on FT and NIM for Cauchy Problem

In the present subsection, we will construct a numeric scheme of the more advanced method known as the Trapezoidal Rule. Recall that it is a one-step method with second-order accuracy, which can be considered as a Runge–Kutta method. We propose modification of trapezoidal rule based on FT and NIM for solving Cauchy problem. Modification of the trapezoidal rule can be improved by the FT to solve Cauchy problem (15). We contributed to numeric methods of Cauchy problem (15) by scheme provides formulas for the FT components, Y_k, $k = 2, \ldots, n-1$, of the unknown function $y(t)$ with respect to choose some power of the triangular (or raised cosine) generalized uniform fuzzy partition, $B_1, \ldots, B_n$, of interval $[a,b]$ with parameter h to approximate solution of Cauchy problem (15). The first, choose the number $n \geq 2$ and compute $h = (b-a)\,/\,(n-1)$, then construct the generalized uniform fuzzy partition of $[a,b]$ using Definition 7. Note that each function B_k spans over three nodes t_{k-1}, t_k, t_{k+1}, $k = 2, \ldots, n-1$. Nevertheless, $B_k(t_{k-1}) = B_k(t_{k+1}) = 0$ and $B_k(t_k) = 1$. Now, we apply the FT and NIM to Cauchy problem (15) and obtain the numeric Scheme I for $k = 1, \ldots, n-1$ as follows (see [3,8] for technical details):

$$\begin{aligned} Y_1 &= y_1, \\ Y^{*}_{k+1} &= Y_k + hF_k/2, \\ Y^{**}_{k+1} &= Y^{*}_{k+1} + hF^{*}_{k+1}/2, \\ Y_{k+1} &= Y_k + h\left(F_k + F^{**}_{k+1}\right)/2, \end{aligned} \tag{16}$$

where

$$F_k = \frac{\int_a^b f(t,Y_k)B_k(t)dt}{\int_a^b B_k(t)dt}, \quad F^{*}_{k+1} = \frac{\int_a^b f(t,Y^{*}_{k+1})B_{k+1}(t)dt}{\int_a^b B_{k+1}(t)dt}, \quad F^{**}_{k+1} = \frac{\int_a^b f(t,Y^{**}_{k+1})B_{k+1}(t)dt}{\int_a^b B_{k+1}(t)dt}. \tag{17}$$

In the sequel, the approximate solution of Cauchy problem (15) can be obtained using the inverse FT as follows:

$$y_n(t) = \sum_{k=1}^{n} Y_k B_k(t). \tag{18}$$

4.2. Numeric Scheme II: Modified 2-Step Adams Moulton Method Based on FT and NIM for Cauchy Problem

The Scheme I uses 1-step method for solving Cauchy problem (15). In this subsection, we improve 2-step Adams Moulton method using FT and NIM for solving Cauchy problem (15). The 2-step Adams Moulton method can be improved to effectively approximate the solution of (15) by the FT components, Y_k, $k = 2, \ldots, n-1$, of the unknown function $y(t)$ with respect to choose some power of the triangular (or raised cosine) generalized uniform fuzzy partition (9). Let $Y_1 = y_1$ and $Y_2 = y_2$ if possible; otherwise, we can compute FT component Y_2 from numeric Scheme I. Analogously to [3,8], we apply the FT and NIM to Cauchy problem (15) and obtain the numeric Scheme II in the following form for $k = 2, \ldots, n-1$:

$$\begin{aligned} Y^*_{k+1} &= Y_k + h\left(8F_k - F_{k-1}\right)/12, \\ Y^{**}_{k+1} &= Y^*_{k+1} + 5hF^*_{k+1}/12, \\ Y_{k+1} &= Y_k + h\left(8F_k - F_{k-1} + 5F^{**}_{k+1}\right)/12, \end{aligned} \tag{19}$$

where

$$F_{k-1} = \frac{\int_a^b f(t, Y_{k-1})B_{k-1}(t)dt}{\int_a^b B_{k-1}(t)dt}, \quad F_k = \frac{\int_a^b f(t, Y_k)B_k(t)dt}{\int_a^b B_k(t)dt},$$

$$F^*_{k+1} = \frac{\int_a^b f(t, Y^*_{k+1})B_{k+1}(t)dt}{\int_a^b B_{k+1}(t)dt}, \text{ and } F^{**}_{k+1} = \frac{\int_a^b f(t, Y^{**}_{k+1})B_{k+1}(t)dt}{\int_a^b B_{k+1}(t)dt}.$$

Then, obtain the desired approximation for y by the inverse FT (18) applied to $[Y_1, \ldots, Y_n]$.

4.3. Numeric Scheme III: Modified 3-Step Adams Moulton Method Based on FT and NIM for Cauchy Problem

In this subsection, we improve 3-step Adams Moulton method using FT and NIM for solving Cauchy problem (15). The 3-step Adams Moulton method can be improved to effectively approximate the solution of (15) by the FT components, Y_k, $k = 2, \ldots, n-1$, of the unknown function $y(t)$ with respect to choose some power of the triangular (or raised cosine) generalized uniform fuzzy partition (see Definition 7), $B_1, \ldots, B_n$, of interval $[a, b]$ with parameter $h = (b-a)/(n-1)$, $n \geq 2$. Let $Y_1 = y_1$, $Y_2 = y_2$ and $Y_3 = y_3$ if possible; otherwise, we can compute FT components Y_2 and Y_3 from numeric Scheme I. Now, we apply the FT and NIM to Cauchy problem (15) and obtain the following numeric Scheme III for $k = 3, \ldots, n-1$ (see [3,8] for technical details):

$$\begin{aligned} Y^*_{k+1} &= Y_k + h\left(19F_k - 5F_{k-1} + F_{k-2}\right)/24, \\ Y^{**}_{k+1} &= Y^*_{k+1} + 9hF^*_{k+1}/24, \\ Y_{k+1} &= Y_k + h\left(19F_k - 5F_{k-1} + F_{k-2} + 9F^{**}_{k+1}\right)/24, \end{aligned} \tag{20}$$

where

$$F_{k-2} = \frac{\int_a^b f(t, Y_{k-2})A_{k-2}(t)dt}{\int_a^b A_{k-2}(t)dt}, \quad F_{k-1} = \frac{\int_a^b f(t, Y_{k-1})A_{k-1}(t)dt}{\int_a^b A_{k-1}(t)dt}, \quad F_k = \frac{\int_a^b f(t, Y_k)A_k(t)dt}{\int_a^b A_k(t)dt},$$

$$F^*_{k+1} = \frac{\int_a^b f(t, Y^*_{k+1})A_{k+1}(t)dt}{\int_a^b A_{k+1}(t)dt}, \text{ and } F^{**}_{k+1} = \frac{\int_a^b f(t, Y^{**}_{k+1})A_{k+1}(t)dt}{\int_a^b A_{k+1}(t)dt}.$$

In the sequel, the inverse FT (18) approximates the solution $y(t)$ of the Cauchy problem (15).

4.4. Error Analysis of Fuzzy Numeric Method for Cauchy Problem

In this subsection, we present error analysis for numeric scheme I only, because the technique of error analysis for rest numeric schemes (Schemes II and III) can be obtained analogously. Consider the Formula (16). If $y(t_k) = y_k$ and Y_k denote the exact solution and the numerical solution and substituting the exact solution in the Formula (16), we get

$$\begin{aligned} y^*_{k+1} &= y_k + hF^e_k/2, \\ y^{**}_{k+1} &= y^*_{k+1} + hF^{e*}_{k+1}/2, \\ y_{k+1} &= y_k + h\left(F^e_k + F^{e**}_{k+1}\right)/2, \end{aligned} \tag{21}$$

where

$$F_k^e = \frac{\int_a^b f(t,y_k)B_k(t)dt}{\int_a^b B_k(t)dt}, \quad F_{k+1}^{e*} = \frac{\int_a^b f(t,y_{k+1}^*)B_{k+1}(t)dt}{\int_a^b B_{k+1}(t)dt}, \quad F_{k+1}^{e**} = \frac{\int_a^b f(t,y_{k+1}^{**})B_{k+1}(t)dt}{\int_a^b B_{k+1}(t)dt}, \tag{22}$$

and the truncation error T_k of the Scheme I is given by

$$T_k = \frac{y_{k+1} - y_k}{h} - \frac{1}{2}\left(F_k^e + F_{k+1}^{e**}\right). \tag{23}$$

Rearranging (16), we get

$$0 = \frac{Y_{k+1} - Y_k}{h} - \frac{1}{2}\left(F_k + F_{k+1}^{**}\right). \tag{24}$$

If we denote the error $e_{k+1} = Y_{k+1} - y_{k+1}$ and subtracting (24) from (23), so:

$$T_k h = e_{k+1} - e_k - \frac{h}{2}\left(F_k - F_k^e\right) - \frac{h}{2}\left(F_{k+1}^{**} - F_{k+1}^{e**}\right). \tag{25}$$

Lemma 8. *Let f is assumed to be sufficiently smooth function of its arguments on $[a,b]$ and satisfies the Lipschitz condition with the constant L with respect to y, then we get for $k = 1,\ldots,n$,*

$$|e_{k+1}| \leq \ |e_k|\,(1+c) + Th \quad \text{and} \quad \left|F_k^e - F_{k+1}^{e**}\right| \leq \ LhM_2$$

where $c = hL + \frac{h^2L^2}{2} + \frac{h^3L^3}{8}$, $T = \max_{1\leq k\leq n} |T_k|$, M_2 is upper bound for f, and F_k^e, F_{k+1}^{e**} are determined by Formula (22).

Proof. By hypothesis, f satisfies the Lipschitz condition and using Lemma 5, Formulas (16), (17), (21) and (22), we get

$$\begin{aligned}
|F_k - F_k^e| &\leq \frac{1}{h}\left|\int_a^b f(t,Y_k)B_k(t)dt - \int_a^b f(t,y_k)B_k(t)dt\right| \leq L\,|e_k| \\
|F_{k+1}^* - F_{k+1}^{e*}| &\leq \frac{1}{h}\left|\int_a^b f(t,Y_{k+1}^*)B_{k+1}(t)dt - \int_a^b f(t,y_{k+1}^*)B_{k+1}(t)dt\right| \leq L\,|Y_{k+1}^* - y_{k+1}^*| \\
|F_{k+1}^{**} - F_{k+1}^{e**}| &\leq \frac{1}{h}\left|\int_a^b f(t,Y_{k+1}^{**})B_{k+1}(t)dt - \int_a^b f(t,y_{k+1}^{**})B_{k+1}(t)dt\right| \leq L\,|Y_{k+1}^{**} - y_{k+1}^{**}| \\
|Y_{k+1}^* - y_{k+1}^*| &\leq |(Y_k + hF_k/2) - (y_k + hF_k^e/2)| \leq |e_k|\left(1 + \frac{hL}{2}\right) \\
|Y_{k+1}^{**} - y_{k+1}^{**}| &\leq \left|(Y_{k+1}^* + hF_{k+1}^*/2) - (y_{k+1}^* + hF_{k+1}^{e*}/2)\right| \leq |e_k|\left(1 + \frac{hL}{2}\right)^2 \\
|e_{k+1}| &\leq |e_k| + \frac{hL}{2}\,|e_k| + \frac{hL}{2}\,|y_{k+1}^{**} - Y_{k+1}^{**}| + Th \\
&\leq |e_k| + \frac{hL}{2}\,|e_k| + \frac{hL}{2}\,|e_k|\left(1 + \frac{hL}{2}\right)^2 + Th \\
&= |e_k|\left(1 + hL + \frac{h^2L^2}{2} + \frac{h^3L^3}{8}\right) + Th
\end{aligned}$$

Furthermore, by using $|f(t,y(t))| \leq M_2$, we get

$$\begin{aligned}
|F_k^e - F_{k+1}^{e**}| &= \left|\frac{1}{h}\left(\int_a^b \left(f(t,y_k) - f(t+h,y_{k+1}^{**})\right) B_k(t)dt\right)\right| \\
&= \left|\frac{1}{h}\left(\int_a^b \left(f(t,y_k) - f(t,y_{k+1}^{**}) + f(t,y_{k+1}^{**}) - f(t+h,y_{k+1}^{**})\right) B_k(t)dt\right)\right| \\
&\leq L\left|y_k - y_{k+1}^{**}\right| \\
&= \frac{Lh}{2}\left|-F_k^e - F_{k+1}^{e*}\right| \\
&= \frac{Lh}{2}\left|\frac{1}{h}\int_a^b \left(f(t,y_k) + f(t,y_{k+1}^{*})\right) B_k(t)dt\right| \\
&\leq LhM_2
\end{aligned}$$

This completes the proof. □

Theorem 2. *Consider the the numeric Scheme I (16), where $f \in C^2[a,b]$ and satisfies the Lipschitz condition with the constant L with respect to y. Then, the solution Y_k, $k = 1,\ldots, n$, obtained by the numeric scheme I (16) for solving Cauchy problem (15) satisfies*

$$|e_k| = |Y_k - y_k| \leq \frac{hM}{2L}e^{kc}, \tag{26}$$

where $c = hL + \frac{h^2L^2}{2} + \frac{h^3L^3}{8}$, M_1, M_2 are upper bound for f', f, respectively, on $[a,b]$, and $M_1 + M_2L = M$.

Proof. By hypothesis, y'' exists and bounded on $[a,b]$ with $\max_{a\leq t\leq b}|y''(t)| = M_1$ by assuming that $f \in C^2[a,b]$. Then, using Lemma 8, (23) and Taylor's theorem for $k = 1,\ldots,n-1$, we get

$$\begin{aligned}
T_k &= \frac{y_{k+1} - y_k}{h} - \frac{1}{2}\left(F_k^e + F_{k+1}^{e**}\right) \\
&= \frac{1}{2}hy''(\xi_k) + f(t_k,y_k) - \frac{1}{2}\left(F_k^e + F_{k+1}^{e**}\right) \\
&= \frac{1}{2}hy''(\xi_k) + f(t_k,y_k) - F_k^e + \frac{1}{2}F_k^e - \frac{1}{2}F_{k+1}^{e**} \\
&= \frac{1}{2}hy''(\xi_k) + \frac{1}{2}\left(F_k^e - F_{k+1}^{e**}\right)
\end{aligned}$$

where $\xi_k \in [t_k, t_{k+}]$. Now, using Lemma 8

$$\begin{aligned}
T = \max_{1\leq k\leq n}|T_k| &\leq \frac{1}{2}h\left|y''(\xi_k)\right| + \frac{LhM_2}{2} \\
&\leq \frac{h}{2}(M_1 + LM_2) = \frac{hM}{2}
\end{aligned}$$

Now, by virtue of Lemma 8 and we have used $e_1 = 0$, $(1+c)^k \leq e^{kc}$, we get for $k = 1,\ldots,n$

$$\begin{aligned}
|e_k| &\leq \frac{(1+c)^k - 1}{c}Th \leq \frac{(1+c)^k}{L + \frac{hL^2}{2} + \frac{h^2L^3}{8}}T \\
&\leq \frac{T}{L}e^{kc} \leq \frac{hM}{2L}e^{kc}
\end{aligned}$$

where $c = hL + \frac{h^2L^2}{2} + \frac{h^3L^3}{8}$. Thus, if the step length $h \to 0$, then for all k, the error, $|e_k|$ converges to zero. So the method is convergent. This completes the proof. □

5. Numerical Examples

In this section, we present examples of the Cauchy problem (15).

Example 1. *Consider the following initial value problem with initial conditions* $y(0) = 1$ *and with a smooth right-hand function*

$$y'(t) = t^2 - y, \quad t \in [0,2]. \tag{27}$$

Example 2. *Consider the Cauchy problem (15) with oscillating right-hand function. We take* $f(t,y) = 1 + 2y\cos(t^2) + \sin(2t^2)$, $t(\frac{\pi}{2}) = 2.1951$, $a = \frac{\pi}{2}$ *and* $b = \frac{3\pi}{2}$.

The results are listed in Tables 2–4 by fuzzy numerical methods proposed in this paper with respect to case $K_{T_1^{201}}$ and Table 5 by fuzzy numerical methods proposed in this paper with respect to case $K_{T_1^1}$, $K_{T_1^3}$, $K_{T_1^{201}}$, K_{C^1}. The Euclidean distance is given by Norm ℓ_2 defined as $\|Y - y(t)\|_2 = \sqrt{\sum_k (Y_k - y(t_k))^2}$ and mean square error (MSE) defined as MSE $= \frac{1}{n}\left(\|Y_k - y(t_k)\|_2\right)^2$. This is an easily computable quantity for a particular sample. Concluding remarks are summarized as follows:

- In view of Table 2, a comparison between the Euler method (Euler-FT) [8], the Mid-point rule (Mid-FT), Scheme I and II [9] and three new schemes (16), (19) and (20) in this paper for Example 1. We can easily observe from Table 2, the better results (in comparison with the Euler-FT method [8]) are obtained by the three new schemes in this paper and the best result (in comparison with the Scheme I, II and II) is obtained by the Scheme III. Also, the better results (in comparison with the Mid-point rule (Mid-FT), Scheme I and II [9]) are obtained by the Scheme II (19) and Scheme III (20) in this paper where all fuzzy numerical methods used the FT components and the best approximation is shown by the Scheme III (20) with FT components.

Table 2. Comparison of numeric results for Example 1. The columns contain the exact and seven approximate solutions of the Cauchy problem (27) with a smooth right-hand function: the first three approximate solution is obtained by the three new schems ((16), (19) and (20)), the fourth approximate solution by the Euler-FT [8] with FT components and the last three by the schemes are proposed in [9]. The best approximation is shown by the Scheme III proposed above (20) with FT components.

t_i	Solution $y(t)$	Proposed Scheme I	Proposed Scheme II	Proposed Scheme III	Euler-FT in [8]	Mid-FT in [9]	Scheme I in [9]	Scheme II in [9]
0	1	1	1	1	1	1	1	1
0.1	0.905163	0.905350	0.905163	0.905163	0.900166	0.905162	0.904392	0.904297
0.2	0.821269	0.821605	0.821322	0.821269	0.811316	0.8213	0.819722	0.819741
0.3	0.749182	0.749630	0.749274	0.749221	0.734351	0.749235	0.746860	0.747182
0.4	0.689680	0.690208	0.689798	0.689742	0.670083	0.689786	0.686592	0.687391
0.5	0.643469	0.644047	0.643602	0.643546	0.619241	0.643611	0.639629	0.641061
0.6	0.611188	0.611788	0.611324	0.611271	0.582484	0.611397	0.606615	0.608821
0.7	0.593415	0.594012	0.593543	0.593495	0.560402	0.593665	0.588129	0.591239
0.8	0.590671	0.591243	0.590781	0.590741	0.553528	0.590998	0.584697	0.588828
0.9	0.603430	0.603956	0.603513	0.603483	0.562342	0.603799	0.596795	0.602053
1	0.632121	0.632581	0.632168	0.632149	0.587274	0.632571	0.624851	0.631332
1.1	0.677129	0.677507	0.677132	0.677127	0.628714	0.677618	0.669253	0.677045
1.2	0.738806	0.739085	0.738757	0.738768	0.687009	0.739381	0.730353	0.739535
1.3	0.817468	0.817635	0.817360	0.817388	0.762475	0.818075	0.808466	0.819111
1.4	0.913403	0.913443	0.913229	0.913276	0.855394	0.914099	0.903881	0.916053
1.5	1.026870	1.026772	1.026624	1.026692	0.966021	1.027588	1.016856	1.030615
1.6	1.158103	1.157857	1.157779	1.157869	1.094586	1.158915	1.147625	1.163024
1.7	1.307316	1.306911	1.306909	1.307022	1.241294	1.308138	1.296400	1.313489
1.8	1.474701	1.474127	1.474205	1.474343	1.406331	1.47562	1.463372	1.482195
1.9	1.660431	1.659681	1.659842	1.660006	1.589864	1.661347	1.648715	1.669312
2	1.864665	1.863636	1.863899	1.864097	1.779378	1.865684	1.852585	1.874993

- In Tabel 3, a comparison of MSE and a comparison of Norm ℓ_2 for Examples 1 and 2. We can easily observe, the best results are obtained by the three new schemes in this paper and the better results (in comparison with the other numerical classical methods) are obtained by all fuzzy numerical methods used the FT components except Euler-FT [8] for these examples.

Table 3. The values of MSE and Norm ℓ_2 for Example 1– 2.

Method	Ex.1		Ex.2	
	Norm ℓ_2	MSE	Norm ℓ_2	MSE
Proposed Scheme I	2.21945×10^{-03}	2.34569×10^{-07}	3.42892×10^{-01}	5.59882×10^{-03}
Proposed Scheme II	1.28684×10^{-03}	7.88551×10^{-08}	3.76033×10^{-01}	6.73336×10^{-03}
Proposed Scheme III	9.28253×10^{-04}	4.10311×10^{-08}	2.15401×10^{-01}	2.20942×10^{-03}
Euler-FT [8]	2.20790×10^{-01}	2.32134×10^{-03}	$3.74484 \times 10^{+00}$	6.67801×10^{-01}
Mid-FT [9]	2.56525×10^{-03}	3.13357×10^{-07}	6.73731×10^{-01}	2.16149×10^{-02}
Scheme I [9]	3.54973×10^{-02}	6.00026×10^{-05}	8.42893×10^{-01}	3.38319×10^{-02}
Scheme II [9]	1.90439×10^{-02}	1.72701×10^{-05}	5.90233×10^{-01}	1.65893×10^{-02}
Trapezoidal Rule	4.30423×10^{-02}	8.82208×10^{-05}	$1.93095 \times 10^{+00}$	1.77551×10^{-01}
2-Step Adams Moulton	3.49968×10^{-02}	5.83228×10^{-05}	$1.85289 \times 10^{+00}$	1.63485×10^{-01}
3-Step Adams Moulton	3.14968×10^{-02}	4.72405×10^{-05}	$1.57237 \times 10^{+00}$	1.17732×10^{-01}

- In view of Table 4, a comparison between the three new schemes (16), (19) and (20) in this paper and the Trapezoidal Rule, 2-Step Adams Moulton Method and 3-Step Adams Moulton Method based on Euler method for Example 2. We can easily observe from Table 4, the better results are obtained by the three new schemes in this paper and the best result (in comparison with the Scheme I, II and II) is obtained by the Scheme III.

Table 4. Comparison of numeric results for Example 2. The columns contain the exact and six approximate solutions of the Cauchy problem (27) with oscillating right-hand function: the first three approximate solution is obtained by the three new schems ((16), (19), and (20)), the last three approximate solution by the Trapezoidal Rule, 2-Step Adams Moulton Method and 3-Step Adams Moulton Method. The best approximation is shown by the Scheme III proposed above (20) with FT components.

t_i	Solution $y(t)$	Proposed Scheme I	Proposed Scheme II	Proposed Scheme III	Trap [1]	2-Step Adams [2]	3-Step Adams [3]
1.570796327	2.195062	2.195062	2.195062	2.195062	2.195062	2.195062	2.195062
1.727875959	1.883281	1.894259	1.883281	1.883281	1.860613	1.883281	1.883281
1.884955592	1.485003	1.511046	1.490853	1.485003	1.454428	1.463813	1.485003
2.042035225	1.185605	1.224868	1.191621	1.184830	1.172418	1.163839	1.177378
2.199114858	1.206758	1.256721	1.208147	1.202292	1.205264	1.180648	1.194336
2.35619449	1.688183	1.733796	1.675538	1.676788	1.638071	1.613025	1.638504
2.513274123	2.546629	2.558069	2.508798	2.525411	2.371288	2.370415	2.421052
2.670353756	3.420051	3.381690	3.362740	3.396118	3.110292	3.151241	3.228492
2.827433388	3.817594	3.751660	3.766365	3.802239	3.459038	3.534435	3.617396
2.984513021	3.479187	3.451288	3.463039	3.476930	3.153857	3.226956	3.285059
3.141592654	2.711291	2.760842	2.722280	2.707811	2.480046	2.498676	2.521585
3.298672286	2.305201	2.404556	2.301686	2.280860	2.168197	2.117053	2.127181
3.455751919	2.871345	2.942818	2.810863	2.818639	2.622265	2.558398	2.599754
3.612831552	4.080085	4.035446	3.952587	4.015230	3.555034	3.556356	3.660448
3.769911184	4.767095	4.645081	4.647076	4.733825	4.104830	4.188576	4.317767
3.926990817	4.209785	4.184589	4.183879	4.213375	3.643127	3.728492	3.801548
4.08407045	3.258243	3.383482	3.263962	3.224157	2.935940	2.895967	2.895039
4.241150082	3.481873	3.609332	3.386338	3.370921	3.111499	2.989980	3.008814
4.398229715	4.873146	4.799588	4.642440	4.733200	4.094280	4.055938	4.179701
4.555309348	5.501192	5.331327	5.311775	5.444657	4.561691	4.652699	4.813484
4.71238898	4.498591	4.551357	4.485128	4.493916	3.817903	3.867167	3.912209

[1] Trapezoidal Rule; [2] 2-Step Adams Moulton Method; [3] 3-Step Adams Moulton Method.

- In Tabel 5, a comparison between computation errors for three schemes based on the FT with respect to the power of the triangular and raised cosine generalized uniform fuzzy partition determined by Formula (9), where the advantage of the $K_{T_1^m}$ for Examples 1 and 2 is evident.

Table 5. The values of MSE and Norm ℓ_2 for Examples 1 and 2 by the three schemes with respect to the power of the triangular and raised cosine generalized uniform fuzzy partition are proposed in this paper. The best approximation is shown by using $K_{T_1^{201}}$.

Proposed Scheme	Case	Ex.1		Ex.2	
		Norm ℓ_2	MSE	Norm ℓ_2	MSE
I	$K_{T_1^1}$	7.81857×10^{-03}	2.91095×10^{-06}	6.38151×10^{-01}	1.93922×10^{-02}
	$K_{T_1^3}$	5.06528×10^{-03}	1.22176×10^{-06}	4.65538×10^{-01}	1.03203×10^{-02}
	$K_{T_1^{201}}$	2.21945×10^{-03}	2.34569×10^{-07}	3.42892×10^{-01}	5.59882×10^{-03}
	K_{C^1}	6.96371×10^{-03}	2.30920×10^{-06}	5.79002×10^{-01}	1.59640×10^{-02}
II	$K_{T_1^1}$	5.92425×10^{-03}	1.67127×10^{-06}	5.45616×10^{-01}	1.41761×10^{-02}
	$K_{T_1^3}$	3.58895×10^{-03}	6.13360×10^{-07}	4.29959×10^{-01}	8.80307×10^{-03}
	$K_{T_1^{201}}$	1.28684×10^{-03}	7.88551×10^{-08}	3.76033×10^{-01}	6.73336×10^{-03}
	K_{C^1}	5.18129×10^{-03}	1.27837×10^{-06}	5.01710×10^{-01}	1.19864×10^{-02}
III	$K_{T_1^1}$	5.31047×10^{-03}	1.34291×10^{-06}	4.42442×10^{-01}	9.32167×10^{-03}
	$K_{T_1^3}$	3.09350×10^{-03}	4.55702×10^{-07}	2.88083×10^{-01}	3.95199×10^{-03}
	$K_{T_1^{201}}$	9.28253×10^{-04}	4.10311×10^{-08}	2.15401×10^{-01}	2.20942×10^{-03}
	K_{C^1}	4.59684×10^{-03}	1.00624×10^{-06}	3.84860×10^{-01}	7.05320×10^{-03}

This constitutes an important improvement to previous methods which do not provide such information except in the methods such as Euler-FT proposed in [8] and Mid-FT , Scheme I, and Scheme II [9] for Cauchy problems by the more efficient way of computation approximate solutions. Thus, this study will be of particular importance.

6. Conclusions

We extended applicability of fuzzy numeric methods to the initial value problem (the Cauchy problem). We proposed three new numeric methods based on the FT and NIM and then analyzed their suitability. We considered in the case of the generalized uniform fuzzy partition is power of the triangular (raised cosine) generalized uniform fuzzy partition and showed that the newly proposed schemes outperform the Euler-FT [8] and Mid-FT , Scheme I, and Scheme II [9] especially on examples where the generalized uniform fuzzy partition is power of the triangular generalized uniform fuzzy partition by using generating function (2). Alos, the newly proposed schemes in this paper outperform the Trapezoidal Rule, 2-Step Adams Moulton Method and 3-Step Adams Moulton Method. Moreover, we proved that the Scheme I determines an approximate solution which converges to the exact solution. This constitutes an important improvement to previous results were coined by I. Perfilieva [8]. To conclude previous sections, the proposed schemes are more accurate and stable. In particular, these schemes can be used for solving initial value problem.

Author Contributions: H. A. ALKasasbeh and M. Z. Ahmad proposed and designed the numerical methods. H. A. ALKasasbeh performed the numerical experiments. I. Perfilieva evaluated the results and supported this work. M. Z. Ahmad project administration. Z. R. Yahya provided software and data curation.

Acknowledgments: This work of Irina Perfilieva has been supported by the project "LQ1602 IT4Innovations excellence in science" and by the Grant Agency of the Czech Republic (project No. 16-09541S). Also, many thanks given to Universiti Malaysia Perlis for providing all facilities until this work completed successfully.

Conflicts of Interest: The authors declare no conflicts of interest.

References

1. Ahmad, M.Z.; Hasan, M.K.; Baets, B.D. Analytical and numerical solutions of fuzzy differential equations. *Inf. Sci.* **2013**, *236*, 156–167. [CrossRef]
2. Daftardar-Gejji, V.; Jafari, H. An iterative method for solving nonlinear functional equations. *J. Math. Anal. Appl.* **2006**, *316*, 753–763. [CrossRef]
3. Sukale, Y.; Daftardar-Gejji, V. New Numerical Methods for Solving Differential Equations. *Int. J. Appl. Comput. Math.* **2017**, *3*, 1639–1660. [CrossRef]
4. Perfilieva, I. Fuzzy transforms: Theory and applications. *Fuzzy Sets Syst.* **2006**, *157*, 993–1023. [CrossRef]
5. Perfilieva, I.; Baets, B.D. Fuzzy transforms of monotone functions with application to image compression. *Inf. Sci.* **2010**, *180*, 3304–3315. [CrossRef]
6. Nguyen, L.; Novák, V. Filtering out high frequencies in time series using F-transform with respect to raised cosine generalized uniform fuzzy partition. In Proceedings of the 2015 IEEE International Conference on Fuzzy Systems (FUZZ-IEEE), Istanbul, Turkey, 2–5 August 2015; pp. 1–8.
7. Perfilieva, I.; Hodáková, P.; Hurtík, P. Differentiation by the F-transform and application to edge detection. *Fuzzy Sets Syst.* **2016**, *288*, 96–114. [CrossRef]
8. Perfilieva, I. Fuzzy transform: Application to the Reef growth problem. In *Fuzzy Logic in Geology*; Demicco, R.V., Klir, G.J., Eds.; Academic Press: Amsterdam, The Netherlands, 2003; Chapter 9, pp. 275–300.
9. Khastan, A.; Perfilieva, I.; Alijani, Z. A new fuzzy approximation method to Cauchy problems by fuzzy transform. *Fuzzy Sets Syst.* **2016**, *288*, 75–95. [CrossRef]
10. Perfilieva, I. F-Transform. In *Handbook of Computational Intelligence*; Kacprzyk, J., Pedrycz, W., Eds.; Springer: Berlin/Heidelberg, Germany, 2015; pp. 113–130.
11. Perfilieva, I.; Daňková, M.; Bede, B. Towards a higher degree F-transform. *Fuzzy Sets Syst.* **2011**, *180*, 3–19. [CrossRef]
12. Holčapek, M.; Perfilieva, I.; Novák, V.; Kreinovich, V. Necessary and sufficient conditions for generalized uniform fuzzy partitions. *Fuzzy Sets Syst.* **2015**, *277*, 97–121. [CrossRef]
13. Bede, B.; Rudas, I.J. Approximation properties of fuzzy transforms. *Fuzzy Sets Syst.* **2011**, *180*, 20–40. [CrossRef]
14. Hurtik, P.; Perfilieva, I. Image Compression Methodology Based on Fuzzy Transform. In *International Joint Conference CISIS'12-ICEUTE '12-SOCO '12 Special Sessions*; Springer: Berlin/Heidelberg, Germany, 2013; pp. 525–532.
15. Loquin, K.; Strauss, O. Histogram density estimators based upon a fuzzy partition. *Stat. Probab. Lett.* **2008**, *78*, 1863–1868. [CrossRef]
16. Jahedi, S.; Mehdipour, M.; Rafizadeh, R. Approximation of integrable function based on ϕ—transform. *Soft Comput.* **2013**, *18*, 2015–2022. [CrossRef]

applied system innovation

MDPI

Article

New Approximation Methods Based on Fuzzy Transform for Solving SODEs: I

Hussein ALKasasbeh [1,*], Irina Perfilieva [2], Muhammad Zaini Ahmad [1] and Zainor Ridzuan Yahya [1]

1 Institute of Engineering Mathematics, Universiti Malaysia Perlis, Kampus Tetap Pauh Putra, Arau 02600, Perlis, Malaysia; mzaini@unimap.edu.my (M.Z.A.); zainoryahya@unimap.edu.my (Z.R.Y.)
2 Institute for Research and Applications of Fuzzy Modelling, University of Ostrava, NSC IT4Innovations, 30. dubna 22, 701 03 Ostrava, Czech Republic; Irina.Perfilieva@osu.cz
* Correspondence: hussein.ahmad.alka@gmail.com

Received: 17 June 2018; Accepted: 14 August 2018; Published: 23 August 2018

Abstract: In this paper, new approximation methods for solving systems of ordinary differential equations (SODEs) by fuzzy transform (FzT) are introduced and discussed. In particular, we propose two modified numerical schemes to solve SODEs where the technique of FzT is combined with one-stage and two-stage numerical methods. Moreover, the error analysis of the new approximation methods is discussed. Finally, numerical examples of the proposed approach are confirmed, and applications are presented.

Keywords: fuzzy partition; fuzzy transform; numerical methods; systems of ordinary differential equations

1. Introduction

Differential equations have great potential to model and understand real-world problems in science and engineering. In many cases, differential equations cannot be solved analytically, so that numerical methods are required. Therefore, numerical methods have been elaborated frequently in scientific research for solving differential equations, for example [1–4]. In this connection, fuzzy approaches successfully cope with solving differential equations. One of fuzzy approaches that has been proposed in the literature is fuzzy transform (FzT).

FzT is a general mathematical technique coined and developed by Perfilieva [5]. The study of FzT is rapidly expanding as a new branch of approximation method based on fuzzy sets. The main idea of FzT is usually forming a fuzzy partition of a universe into fuzzy subsets. Two shapes for the basic functions of fuzzy partition, triangular- and sinusoidal-shaped membership functions, were considered by [6]. Applications of the FzT can be used in the construction of approximate models, the approximation of functions and the solution of differential equations. FzT has been generalized from the case of constant components to the case of polynomial components [7]. Later, FzT was successfully used by [8] for a second order initial value problem. From this idea, FzT was proposed for numerical solutions of two point boundary value problems by the more efficient way in comparison with the similar ones obtained by the finite difference method [9].

Recently, in [10], a new numerical method based on FzT to solve a class of delay differential equations by means of the Picard-like numerical scheme was presented. The author demonstrated the stability of the method, and the obtained results have good agreement with existing methods. Furthermore, in some cases, a better approximation was achieved through sinusoidal-shaped basic functions, while the Bernstein basis polynomials allow better results in the other examples. On the other hand, a new approach to the fuzzy boundary value problem in the form of the fuzzy relation was investigated by [11]. Another approach for the second order linear differential equation with constant coefficients and Dirichlet boundary conditions was introduced by [12], where the ability of FzT was demonstrated to deal with boundary value problems affected by noise, and the results were compared with the finite difference method. To confirm again the ability of FzT with respect to noisy and non-noisy source functions, FzT based on the shooting method was introduced by [13] for solving a nonlinear second-order ODE, and the obtained results were better than the classical method, namely the second order Runge–Kutta. Further, in [14], FzT to approximate the solution of boundary value problems by minimizing the integral squared error in the two-norm was considered. The trigonometric function based on higher degree FzT and the accuracy of the resulting approximation increase with the increase in the degree of FzT were presented by [15], while the weighted transform method was discussed by [16]. The conditions for quasi-consensus in a multi-agent system with sampled data based on FzT were proposed by [17]. In [18], the dynamical properties of a two-neuron system with respect to FzT and a single delay have been investigated. Multi-step FzT was first studied by [19] for solving ODEs. From this perspective, FzT was considered for solving a class of ODEs.

In this regard, many approximation methods have been studied for solving ODEs, for example using neural networks [20], embedded three and two pairs of nonlinear methods [21], electrical analogy [22], multi-general purpose graphical processing units [23], the differential transform method [24] and a Galerkin finite element method [25]. A numerical method based on the trapezoidal rule for the Cauchy–Smoluchowski problem was discussed by [26]. In [27,28], the authors studied ODEs with the initial value as the triangular fuzzy number. A new fuzzification of the classical Euler method and then incorporating an unconstrained optimization technique were proposed by [1]. Furthermore, most real-life problems involve systems of ODEs, for example the Lotka–Volterra prey predator model based on an autonomous model [29], a non-autonomous model [30,31] and fuzzy initial populations [4].

In this study, our aim is to extend the applicability of the FzT to general coupled Systems of ODEs (SODEs) where this method works better than its classical counterpart. The motivation of the present research stems from the fuzzy approach as follows. The first application of the FzT for solving ODEs had been proposed by [6] where a generalization of the Euler method for an ordinary Cauchy problem was developed and its potential in comparison with classical methods (Euler method) was demonstrated. The same approach has been successfully used by [32] to solve the Cauchy problem for a more accurate comparison with the classical method (the second-order Runge–Kutta method) and with the generalization of Euler method based on FzT, as proposed by [6]. Further, in [19], new fuzzy methods based on FzT for solving the Cauchy problem were presented, and the authors compared the results with existing numerical results in [6,32] and with classical methods, including one, two and three steps. All these fuzzy approximation methods performed better than the classical trapezoidal rule (one step) and the classical Adam–Moulton method (two and three steps) and outperformed the previous fuzzy methods in [6,32].

In this contribution, two new approximation methods are presented in detail to solve SODEs where the technique of FzT is combined with one-stage and two-stage numerical methods. The first approximation method improves the Euler method (one-stage), and the other approximation method improves trapezoidal rule (two-stage). The primary focus of this contribution is to demonstrate the applicability of the FzT for functions of two variables based on the uniform fuzzy partition. The error analysis is discussed in the context of the uniform fuzzy partition. Algorithms inspired by the FzT are shown for solving SODEs. Two new approximation methods are applied to the Lotka–Volterra prey-predator model. This contribution is an important modification relative to classical methods, the Euler method and the trapezoidal rule. Thus, these methods are compared with the Euler method and trapezoidal rule. Both approximation methods with the help of FzT provide better numerical solutions than the classical Euler method and the classical trapezoidal rule.

The paper is organized as follows. In Section 2, several related concepts and results associated with the FzT are reviewed. In Section 3, we construct procedures to obtain an approximate solution for SODEs by using the FzT method. Applications are discussed in Section 4. Finally, conclusions are given in Section 5.

2. Basic Concepts

Throughout this section, we deal with an interval $[a,b] \subset \mathbb{R}$ of real numbers. Let $[a,b]$ be an interval on the real line $\mathbb{R}$. Fuzzy sets on $[a,b]$ will be identified with their membership functions mapping from $[a,b]$ into $[0,1]$. We will assume an interval $[a,b]$ as a real domain. In this section, we remind about the definitions and claims that were introduced and proven by [5].

Definition 1. *(Fuzzy partition) Let $x_1 < \cdots < x_n$ be fixed nodes within $[a,b]$ such that $x_1 = a,\ x_n = b$ and $n \geq 2$. The fuzzy sets $A_1, \ldots, A_n$ are often called basic functions. We say that fuzzy sets $A_1, \ldots, A_n \subset [a,b]$ establish a fuzzy partition of $[a,b]$ if they fulfill the following conditions for $k = 1, \ldots, n$ (for the uniformity of notation, we set $x_0 = a$ and $x_{n+1} = b$):*

1. *$A_k(x) : [a, b] \to [0,1]$ is continuous with $A_k(x_k) = 1$, $A_k(x) > 0$ if $x \in (x_{k-1}, x_{k+1})$ and $A_k(x) = 0$ if $x \notin (x_{k-1}, x_{k+1})$;*
2. *$A_k(x)$, $k = 2, \ldots, n$, strictly increases on $[x_{k-1}, x_k]$, and $A_k(x)$, $k = 1, \ldots, n-1$, strictly decreases on $[x_k, x_{k+1}]$;*
3. *For all $x \in [a, b]$, $\sum_{k=1}^{n} A_k(x) = 1$. This is called the Ruspini condition.*

We say that a fuzzy partition of $[a,b]$ is h-uniform if its nodes $x_1, \ldots, x_n$, where $n \geq 2$ are equidistant. This means that $x_k = a + h(k-1)$, $k = 1, \ldots, n$, where $h = \frac{b-a}{n-1}$, $n \geq 2$, and the two additional properties are fulfilled:

- $A_k(x_k - x) = A_k(x_k + x)$, $k = 2, \ldots, n-1$, for all $x \in [0,h]$ and:
- $A_k(x) = A_{k-1}(x-h)$ and $A_{k+1}(x) = A_k(x-h)$, $k = 2, \ldots, n-1$, for all $x \in [x_k, x_{k+1}]$.

Two uniform fuzzy partitions with triangular- and sinusoidal-shaped basic functions can be found in [5,6]. Throughout this paper, we will write uniform fuzzy partition instead of h-uniform fuzzy partition.

Definition 2. *Let f be a continuous function on $[a,b]$ and $A_k(x)$, $k = 1, \ldots, n$, be a uniform fuzzy partition of $[a,b]$, $n \geq 2$. A vector of real numbers $F[f] = (F_1, F_2, \ldots, F_n)$ given by (to complete this notation, we set $x_1 = a$ and $x_{n+1} = b$):*

$$F_k[f] = \frac{\int_a^b f(x)\, A_k(x)\, dx}{\int_a^b A_k(x)\, dx} = \frac{\int_{x_{k-1}}^{x_{k+1}} f(x)\, A_k(x) dx}{\int_{x_{k-1}}^{x_{k+1}} A_k(x) dx},\ k = 1, \ldots, n, \tag{1}$$

is called the direct FzT of f with respect to $A_1, \ldots, A_n$.

Remark 1. *The elements $F_1[f], \ldots, F_n[f]$ are called components of the FzT. If $A_1, \ldots, A_n$ forms a uniform fuzzy partition, then the expression (1) can be simplified as follows:*

$$F_1[f] = \frac{2}{h} \int_{x_1}^{x_2} f(x)\, A_1(x) dx,\ F_n[f] = \frac{2}{h} \int_{x_{n-1}}^{x_n} f(x)\, A_n(x) dx,$$

$$F_k[f] = \frac{1}{h} \int_{x_{k-1}}^{x_{k+1}} f(x)\, A_k(x) dx,\ k = 2, \ldots, n-1. \tag{2}$$

Definition 3. *Let $F[f] = (F_1, F_2, \ldots, F_n)$ be the direct FzT of $f \in C[a,b]$ with respect to $A_k(x)$, $k = 1, \ldots, n$. Then, the inverse FzT of f, $\hat{f} : [a,b] \to R$, given by:*

$$\hat{f}(x) = \sum_{k=1}^{n} F_k\, A_k(x). \tag{3}$$

Lemma 1. *[6] Let $f(x)$ be continuous on $[a,b]$ and twice continuously differentiable in (a,b), and let basic functions form a uniform fuzzy partition of $[a,b]$. Then, for each $k = 1,\ldots,n$:*

$$F_k = f(x_k) + \mathcal{O}(h^2). \tag{4}$$

Remark 2. *An important property of the direct FzT, as well as inverse FzT is their linearity, namely, given $f,\ g \in C[a,b]$ and $\alpha,\ \beta \in \mathbb{R}$, if $h = \alpha f + \beta g$, then $F[h] = \alpha F[f] + \beta F[g]$ and $\hat{h} = \alpha\hat{f} + \beta\hat{g}$.*

3. FzT for Solving SODEs

In this section, we present methodological remarks and numerical schemes for solving SODEs.

3.1. Methodological Remarks to Applications of the FzT

Consider the Initial Value Problem (IVP) for the SODEs:

$$\begin{cases} x'(t) = f\left(t, x\left(t\right), y\left(t\right)\right), & x\left(t_1\right) = \alpha, t_1 \le t \le t_n, \\ y'(t) = g\left(t, x\left(t\right), y\left(t\right)\right), & y\left(t_1\right) = \beta, \end{cases} \tag{5}$$

where $\alpha,\ \beta \in \mathbb{R}$, f and g are continuous functions on $[t_1,\ t_n] \times \mathbb{R} \times \mathbb{R}$ and satisfy the Lipschitz condition. Unfortunately, the analytical solution $(x(t),\ y(t))$ of Problem (5) is often difficult and sometimes impossible to obtain. Thus, approximate solutions by means of FzT are extremely important for solving (5). A numerical method for (5) is an algorithm that computes FzT components $X_k \approx x(t_k)$ and $Y_k \approx y(t_k)$, for each $k = 2,\ldots,n$ (to complete this notation, we set $\alpha = X_1 = x(t_1)$ and $\beta = Y_1 = y(t_1)$).

Below, we extend the main principles of FzT detailed in Formulas (6) that are needed later.

Definition 4. *Let $f\ (g)$ be a continuous function on $[t_1,\ t_n]$ and $A_1,\ldots,A_n$ be the fuzzy partition of $[t_1,\ t_n]$. A vector of real numbers $F_k[f] = (F_1[f],\ldots,F_n[f])$ $(G_k[g] = (G_1[g],\ldots,G_n[g]))$ given by:*

$$F_k[f] = \frac{\int_{t_1}^{t_n} f(t, x\left(t\right), y\left(t\right))\, A_k(t)\, dt}{\int_{t_1}^{t_n} A_k(t)\, dt} \quad \left(G_k[g] = \frac{\int_{t_1}^{t_n} g(t, x\left(t\right), y\left(t\right))\, A_k(t)\, dt}{\int_{t_1}^{t_n} A_k(t)\, dt} \right), \tag{6}$$

is called the direct FzT of $f\ (g)$ that is extended with independent variable t and two dependent variables x and y.

Remark 3. *We need a way to approximate the direct FzT components 6. This is discussed in Corollary 1.*

In the following, the necessary steps of the FzT are given.

1. Construction of the fuzzy partition:
   (a) Specify the number n of components, and compute the step $h = (t_n - t_1)\ /\ (n-1)$.
   (b) Construct the nodes $t_1 < \ldots < t_n$, where $t_k = t_1 + h(k-1)$.
   (c) Select the shape of basic functions. We mostly use triangular- or sinusoidal-shaped basic functions. Recall that the shape of the basic functions determines the course of $\hat{f}$, that is whether it is piecewise linear or nonlinear.
   (d) Construct a uniform fuzzy partition of $[t_1, t_n]$ by triangular- or sinusoidal-shaped basic functions [5].
2. Computation of FzT: We replace $x'\ (t)$ and $y'\ (t)$ by their approximations based on the Taylor expansion as new functions with respect to the fuzzy partition $A_1,\ldots,A_n$ by Step 1. In this way, similarly to [6], we transfer the original SODEs to the space of fuzzy units, solve them in the new

space and then transfer them back by the inverse FzT. Compute the approximation for x and y by the inverse FzT applied to $[X_1, \ldots, X_n]$ and $[Y_1, \ldots, Y_n]$. In the next subsections, the schemes provide formulas for the computation of components of FzT.

3.2. Numerical Scheme I for SODEs

In this subsection, we present a modified scheme to solve SODEs using the FzT. Suppose that the functions f and g on $[t_1, t_n]$ are sufficiently smooth in (5). For solving SODEs (5) on $[t_1, t_n]$, the interval is divided into $n-1$ subintervals. Let us choose some uniform fuzzy partition of interval $[t_1, t_n]$ with parameter $h = (t_n - t_1)/(n-1)$, $n \geq 2$, and basic functions $A_1, \ldots, A_n$. In view of the methodological remarks in Subsection 3.1, we describe the complete sequence of steps, which leads to the approximation solution of SODEs (5) (see [6] for technical details). Before we apply the direct FzT to both parts of the differential equation, we will use the Taylor expansion and replace the first derivatives of the left-hand sides in (5) by their approximations, i.e.,

$$\begin{cases} x(t+h) = x(t) + hx'(t) + \mathcal{O}(h^2), \\ y(t+h) = y(t) + hy'(t) + \mathcal{O}(h^2). \end{cases} \tag{7}$$

Denote $\begin{cases} x^+(t) = x(t+h), \\ y^+(t) = y(t+h), \end{cases}$ as new functions and then apply the direct FzT components $\begin{cases} F_n \\ G_n \end{cases}$ to both parts of Equation (7).

$$\begin{cases} F_n[x'(t)] = \frac{1}{h}(F_n[x^+] - F_n[x]) + \mathcal{O}(h^2), \\ G_n[y'(t)] = \frac{1}{h}(G_n[y^+] - G_n[y]) + \mathcal{O}(h^2), \end{cases}$$

where $\begin{cases} F_n[x'] = [X'_1, \ldots, X'_{n-1}], \\ G_n[y'] = [Y'_1, \ldots, Y'_{n-1}], \end{cases}$ $\begin{cases} F_n[x] = [X_1, \ldots, X_{n-1}], \\ G_n[y] = [Y_1, \ldots, Y_{n-1}], \end{cases}$ and $\begin{cases} F_n[x^+] = [X^+_1, \ldots, X^+_{n-1}], \\ G_n[y^+] = [Y^+_1, \ldots, Y^+_{n-1}]. \end{cases}$

Now, prove that:

$$\begin{cases} X^+_1 = X_2 + \mathcal{O}(h^2), \\ X^+_k = X_{k+1},\ k = 2, \ldots, n-2, \end{cases} \quad \text{and} \quad \begin{cases} Y^+_1 = Y_2 + \mathcal{O}(h^2), \\ Y^+_k = Y_{k+1},\ k = 2, \ldots, n-2. \end{cases}$$

For the values $k = 1$ and $k = n-1$ by Lemma 1, we have $\begin{cases} X^+_k = X_{k+1} + \mathcal{O}(h^2), \\ Y^+_k = Y_{k+1} + \mathcal{O}(h^2), \end{cases}$, and for $k = 2, \ldots, n-2$, we have:

$$\begin{cases} X^+_k = \frac{1}{h}\int_{t_{k-1}}^{t_{k+1}} x(t+h).A_k(t)dt = \frac{1}{h}\int_{t_k}^{t_{k+2}} x(s).A_{k+1}(s)ds = X_{k+1}, \\ Y^+_k = \frac{1}{h}\int_{t_{k-1}}^{t_{k+1}} y(t+h).A_k(t)dt = \frac{1}{h}\int_{t_k}^{t_{k+2}} y(s).A_{k+1}(s)ds = Y_{k+1}. \end{cases}$$

Then, we get:

$$\begin{cases} hX'_k = (X_{k+1} - X_k) + \mathcal{O}(h^2), \\ hY'_k = (Y_{k+1} - Y_k) + \mathcal{O}(h^2). \end{cases}, k = 1, \ldots, n-1. \tag{8}$$

Therefore, we can introduce the $(n-1) \times n$ matrix:

$$D = \frac{1}{h}\begin{pmatrix} -1 & 1 & 0 & \cdots & 0 \\ 0 & -1 & 1 & \cdots & 0 \\ \vdots & & & \ddots & \vdots \\ 0 & 0 & \cdots & -1 & 1 \end{pmatrix}.$$

Thus, according to (5), the equality (8) can be rewritten (up to $\mathcal{O}\left(h^2\right)$) as matrix equality:

$$\begin{cases} F_n[x'] = DF_n[x], \\ G_n[y'] = DG_n[y], \end{cases} \tag{9}$$

where $\begin{cases} F_n[x'] = [X'_1, \dots, X'_{n-1}]^T, \\ G_n[y'] = [Y'_1, \dots, Y'_{n-1}]^T, \end{cases}$ and $\begin{cases} F_n[x] = [X_1, \dots, X_{n-1}]^T, \\ G_n[y] = [Y_1, \dots, Y_{n-1}]^T. \end{cases}$

Now, let us return to the problem (5) and apply the FzT to both sides of the differential equation. Based on the linearity of FzT and Formula (9), we obtain the following system with respect to the unknown $F_n[x]$ and $G_n[y]$:

$$\begin{cases} DF_n[x] = F_n[f], \\ DG_n[y] = G_n[g], \end{cases} \tag{10}$$

where $\begin{cases} F_n[f] = [F_1, \dots, F_{n-1}]^T, \\ G_n[g] = [G_1, \dots, G_{n-1}]^T, \end{cases}$ are the FzT of $f(t, x(t), y(t))$ $(g(t, x(t), y(t)))$ as a function of t w.r.t. the chosen basic functions $A_1, \dots, A_n$. Note that the last components F_n and G_n are not presented in $F_n[f]$ and $G_n[g]$, respectively, due to the dimensionality limitation, and (10) does not include two initial conditions of (5). Thus, we complete the matrix D by adding the first row as the initial value as follows:

$$D^* = \frac{1}{h}\begin{pmatrix} 1 & 0 & 0 & \cdots & 0 \\ -1 & 1 & 0 & \cdots & 0 \\ 0 & -1 & 1 & \cdots & 0 \\ \vdots & & & \ddots & \vdots \\ 0 & 0 & \cdots & -1 & 1 \end{pmatrix},$$

so that D^* is an $n \times n$ nonsingular matrix. Based on the initial conditions and the matrix D^*, we also expand $F_n[f]$ and $G_n[g]$ by adding the first element, as follows:

$$\begin{cases} F_n^*[f] = [\frac{x_1}{h}, F_1, \dots, F_{n-1}]^T, \\ G_n^*[g] = [\frac{y_1}{h}, G_1, \dots, G_{n-1}]^T. \end{cases}$$

Then, the problem (5) can be completely represented by the following expression with respect to the unknown $F_n[x]$ and $G_n[y]$:

$$\begin{cases} D^*.F_n[x] = F_n^*[f], \\ D^*.G_n[y] = G_n^*[g]. \end{cases} \tag{11}$$

The solution of (11) can be obtained by the following formula:

$$\begin{cases} F_n[x] = (D^*)^{-1}.F_n^*[f], \\ G_n[y] = (D^*)^{-1}.G_n^*[g]. \end{cases} \tag{12}$$

In fact, to obtain the solution of (12), we should compute the inverse matrix of D^*. Therefore, we have:

$$(D^*)^{-1} = h \begin{pmatrix} 1 & 0 & 0 & \cdots & 0 \\ 1 & 1 & 0 & \cdots & 0 \\ 0 & 1 & 1 & \cdots & 0 \\ \vdots & & & \ddots & \vdots \\ 1 & 1 & \cdots & 1 & 1 \end{pmatrix},$$

and by (12), we get:

$$X_{k+1} = X_k + hF_k, \qquad X_1 = \alpha, \mid Y_{k+1} = Y_k + hG_k, \qquad Y_1 = \beta,\ k = 1, \ldots, n-1, \tag{13}$$

where F_k (G_k) is given by Formula (6). Formula (13) can be applied to the computation of $X_2, \ldots, X_n$ and $Y_2, \ldots, Y_n$. However, it cannot be applied directly by using the function $f(t,x,y)$ or $g(t,x,y)$, because it uses unknown functions x and y. Therefore, we will use the same trick as in [6] and replace functions by their FzT components:

$$\hat{F}_k[f] = \frac{\int_{t_1}^{t_n} f(t, X_k, Y_k)\, A_k(t)\, dt}{\int_{t_1}^{t_n} A_k(t)\, dt}, \ \hat{G}_k[g] = \frac{\int_{t_1}^{t_n} g(t, X_k, Y_k)\, A_k(t)\, dt}{\int_{t_1}^{t_n} A_k(t)\, dt}, \ k = 1, \ldots, n-1. \tag{14}$$

Thus, the components of FzT of x and y can be approximated from the following Scheme I:

$$X_{k+1} = X_k + h\hat{F}_k, \qquad X_1 = \alpha, \ \mid \ Y_{k+1} = Y_k + h\hat{G}_k, \qquad Y_1 = \beta,\ k = 1, \ldots, n-1. \tag{15}$$

Finally, the approximate solution of (5) can be obtained using the inverse FzT as follows:

$$x_n(t) = \sum_{k=1}^{n} X_k\, A_k(t), \ y_n(t) = \sum_{k=1}^{n} Y_k\, A_k(t), \tag{16}$$

where $A_k(t),\ k = 1, 2, \ldots, n$ are given basic functions. The proposed method is similar to the well-known Euler method and under similar assumptions. It has the same degree of accuracy. In the next theorem, we obtain an error estimate in the context of a fuzzy partition and error analysis of numerical Scheme I.

Theorem 1. *Let $f, g : [t_1, t_n] \to \mathbb{R}$ be twice continuously differentiable on $[t_1, t_n]$. Let, moreover, $f, g : [t_1, t_n] \times \mathbb{R} \times \mathbb{R} \to \mathbb{R}$ be Lipschitz continuous with respect to x and y, i.e. there exists a constant $L \in \mathbb{R}$, such that for all $t \in [t_1,\ t_n]$ and $x, x', y, y' \in \mathbb{R}$,*

$$\begin{aligned} |f(t,x,y) - f(t,x',y')| &\le L(|x - x'| + |y - y'|), \\ |g(t,x,y) - g(t,x',y')| &\le L(|x - x'| + |y - y'|). \end{aligned} \tag{17}$$

Assume that $\{A_k \mid k = 1, \ldots, n\}$, $n \ge 2$, is a uniform fuzzy partition of $[t_1, t_n]$. Then, the local (global) error of Scheme I (14)–(15) is of the order h^2 (h).

Proof. Let us choose and fix some k, where $2 \le k \le n$, and assume that X_k (Y_k) is the k-th FzT component of x (y). We consider the SODEs (5) and their FzT representation by the system of equations (11). We start with the following easy consequences from the Taylor expansions:

$$\begin{aligned} x(t_{k+1}) &= x(t_k) + hf(t_k, x_k, y_k) + \mathcal{O}\left(h^2\right), \\ y(t_{k+1}) &= y(t_k) + hg(t_k, x_k, y_k) + \mathcal{O}\left(h^2\right). \end{aligned}$$

Let $e_k = x_k - X_k$, $d_k = y_k - Y_k$, $x(t_k) = x_k$ and $y(t_k) = y_k$, then according to (14)–(15), we get:

$$e_{k+1} = e_k + h\left(f(t_k, x_k, y_k) - \hat{F}_k\right) + \mathcal{O}\left(h^2\right), \tag{18}$$

$$d_{k+1} = d_k + h\left(g(t_k, x_k, y_k) - \hat{G}_k\right) + \mathcal{O}\left(h^2\right). \tag{19}$$

By the assumption that f and g have continuous second order derivatives on $[t_1, t_n]$ and are Lipschitz continuous with respect to x and y, therefore, using the trapezoid rule and Remark 1, we get:

$$\begin{aligned} |f(t_k, x_k, y_k) - \hat{F}_k| &= |f(t_k, x_k, y_k) - \frac{1}{h}\int_{t_{k-1}}^{t_{k+1}} f(t, X_k, Y_k)A_k(t)dt|, \\ &= |f(t_k, x_k, y_k) - \frac{1}{h}\frac{h}{2}2\, f(t_k, X_k, Y_k) + \mathcal{O}(h^2)|, \\ &= |f(t_k, x_k, y_k) - f(t_k, X_k, Y_k) + \mathcal{O}(h^2)|, \\ &\leq L(|e_k| + |d_k|) + \mathcal{O}(h^2). \end{aligned} \tag{20}$$

Similarly,

$$|g(t_k, x_k, y_k) - \hat{G}_k| \leq L(|e_k| + |d_k|) + \mathcal{O}(h^2).$$

Therefore,

$$|e_{k+1}| \leq |e_k| + hL(|e_k| + |d_k|) + \mathcal{O}(h^2),$$
$$|d_{k+1}| \leq |d_k| + hL(|e_k| + |d_k|) + \mathcal{O}(h^2).$$

Denoting $\delta_k = \max(|e_k|, |d_k|)$ and using the obvious equality $\max(a + b, c + b) = \max(a, c) + b$, we obtain from the above :

$$\delta_{k+1} \leq \delta_k + 2hL(|e_k| + |d_k|) + \mathcal{O}(h^2), \tag{21}$$

and finally,

$$\delta_{k+1} \leq \delta_k(1 + 2hL) + \mathcal{O}(h^2).$$

By iteration, we come to:

$$\delta_{k+1} \leq \delta_1(1+C)^k + \mathcal{O}(h^2)(1 + (1+C) + \ldots + (1+C)^{k-1}) = \delta_1(1+C)^k + \mathcal{O}(h^2)\frac{(1+C)^k - 1}{C},$$

where $C = 2hL$. For $k = n - 1$, we have:

$$\delta_n \lesssim e^{2L}\delta_1 + \mathcal{O}(h), \tag{22}$$

where we made use of the following fact: $h = (t_n - t_1)/(n - 1)$, and the following asymptotic equalities: $(1 + 1/n)^n \sim e$, $(1 + a)^n \sim 1 + na$.

By (21) and (22), we conclude that Scheme I has the local order h^2 and the global order h. □

Remark 4. *Theorem 1 extends Theorem 9.1 of [6] to the SODEs.*

Corollary 1. *Let the assumptions of Theorem 1 be fulfilled. Then, for each $k = 1, \ldots, n$:*

$$F_k - \hat{F}_k = \mathcal{O}(h^2), \text{ and } G_k - \hat{G}_k = \mathcal{O}(h^2),$$

where F_k, G_k are defined by (6) and $\hat{F}_k$, $\hat{G}_k$ are defined by (14).

Proof. By Lemma 1, $x_k - X_k = \mathcal{O}(h^2)$ and $y_k - Y_k = \mathcal{O}(h^2)$ and (20), we get:

$$\hat{F}_k - f(t_k, x_k, y_k) = \mathcal{O}(h^2),$$

and using the trapezoid rule and Remark 1, we obtain:

$$\begin{aligned} F_k &= \frac{1}{h}\int_{t_{k-1}}^{t_{k+1}} f(t, x_k, y_k)A_k(t)dt, \\ &= \frac{1}{h}\frac{h}{2}2\, f(t_k, x_k, y_k) + \mathcal{O}(h^2), \\ &= f(t_k, x_k, y_k) + \mathcal{O}(h^2). \end{aligned}$$

which together with the previous estimation proves that:

$$F_k - \hat{F}_k = \mathcal{O}(h^2).$$

Similarly,

$$g(t_k, x_k, y_k) - \hat{G}_k = \mathcal{O}(h^2) \text{ and } G_k - \hat{G}_k = \mathcal{O}(h^2). \quad \square$$

Corollary 2. *The approximation method for (5) is given by Scheme I (14)–(15) with the local error $\mathcal{O}(h^2)$. The approximate solution to (5) can be found by taking the inverse FzT (16) where $A_1, \ldots, A_n$ are fixed basic functions.*

Theorem 2. *Let f, $g \in C^2[t_1, t_n]$ and bounded on $I = [t_1, t_n]$. Let, moreover, basic functions $\{A_k \mid k = 1, \ldots, n\}$, $n \geq 2$, form a uniform fuzzy partition of I. Assume that F_k (G_k), $k = 1\ldots, n$, and F'_k (G'_k), $k = 1\ldots, n$, are the FzT components of f (g) and f' (g') with respect to $\{A_k \mid k = 1, \ldots, n\}$, respectively. Then, for $k = 1, \ldots, n-1$:*

$$|F_{k+1} - F_k| \leq h\,|F'_k| + \frac{h^2}{2}M_f,\ |G_{k+1} - G_k| \leq h\,|G'_k| + \frac{h^2}{2}M_g, \tag{23}$$

where $M_f = \max_{t \in I} |f''(t, x(t), y(t))|$ and $M_g = \max_{t \in I} |g''(t, x(t), y(t))|$.

Proof. Let us write the following result from Taylor's theorem:

$$\begin{aligned} f(t+h, x(t), y(t)) =& f(t, x(t), y(t)) + hf'(t, x(t), y(t)) + \frac{h^2}{2}f''(\varepsilon, x(\varepsilon), y(\varepsilon)), \\ g(t+h, x(t), y(t)) =& g(t, x(t), y(t)) + hg'(t, x(t), y(t)) + \frac{h^2}{2}g''(\xi, x(\xi), y(\xi)). \end{aligned}$$

where $t_k < \varepsilon$, $\xi < t_{k+1}$. Using the linearity of the FzT by Remark 2 and the properties of the uniform fuzzy partition by Definition 1, we get:

$$\begin{aligned} F_k[f(t+h, x, y)] &= \frac{1}{h}\int_{t_{k-1}}^{t_{k+1}} f(t+h, x, y)A_k(t)dt, \\ &= \frac{1}{h}\int_{t_k}^{t_{k+2}} f(t, x, y)A_{k+1}(t)dt, \\ &= F_{k+1}[f(t, x, y)],\ k = 2, 3, \ldots, n-1, \end{aligned}$$

and:

$$F_{k+1} = F_k + hF'_k + \mathcal{O}\left(h^2\right),\ G_{k+1} = G_k + hG'_k + \mathcal{O}\left(h^2\right),$$
$$F_{k+1} \leq F_k + hF'_k + \frac{h^2}{2}M_f,\ G_{k+1} \leq G_k + hG'_k + \frac{h^2}{2}M_g,$$

which easily leads to (23). □

Lemma 2. *Consider Scheme I (14)–(15). If the set of fuzzy sets* $\{A_k \mid k = 1, \ldots, n-1\}$, $n \geq 2$, *is a uniform fuzzy partition of* $[t_1, t_n]$, *then we have for fixed* $k = 1, \ldots, n-1$:

$$X_{k+1} - X_k = \begin{cases} 2\int_{t_1}^{t_n} f(t, X_k, Y_k)A_1(t)dt & \text{if} \quad k = 1, \\ \int_{t_1}^{t_n} f(t, X_k, Y_k)A_k(t)dt & \text{if} \quad k = 2, \ldots, n-1, \end{cases}$$
$$Y_{k+1} - Y_k = \begin{cases} 2\int_{t_1}^{t_n} g(t, X_k, Y_k)A_1(t)dt & \text{if} \quad k = 1, \\ \int_{t_1}^{t_n} g(t, X_k, Y_k)A_k(t)dt & \text{if} \quad k = 2, \ldots, n-1, \end{cases} \tag{24}$$

Proof. The proof can be obtained from Remark 1; in particular, by the properties of the uniform fuzzy partition $\int_{t_1}^{t_n} A_k(t)dt = h/2$ for $k = 1$ and $\int_{t_1}^{t_n} A_k(t)dt = h$ for $k = 2, \ldots, n-1$ and after substituting this into (14)–(15). □

Lemma 3. *Suppose that* f, g *are continuous and bounded on* $I = [t_1, t_n]$, *and consider that Scheme I (14)–(15) with respect to the basic functions forms a uniform fuzzy partition of* I. *Then, we have for fixed* $k = 1, \ldots, n-1$:

$$|X_{k+1} - X_k| \leq M_1 h,\ \ |Y_{k+1} - Y_k| \leq M_2 h,$$

where $M_1 = \max_{t \in [t_1, t_n]} |f(t, x, y)|$ *and* $M_2 = \max_{t \in [t_1, t_n]} |g(t, x, y)|$.

Proof. Let us choose a value of k in the range $1 \leq k \leq n-1$. By using Lemma 2, we get:

$$\begin{aligned} |X_2 - X_1| &= \left|2\int_{t_1}^{t_2} f(t, X_k, Y_k)A_1(t)dt\right|, \\ &\leq 2\int_{t_1}^{t_2} |f(t, X_k, Y_k)A_1(t)|\, dt \ \leq 2M_1 \int_{t_{k-1}}^{t_{k+1}} A_1(t)dt = M_1 h, \\ |X_{k+1} - X_k| &= \left|\int_{t_{k-1}}^{t_{k+1}} f(t, X_k, Y_k)A_k(t)dt\right|, \\ &\leq \int_{t_{k-1}}^{t_{k+1}} |f(t, X_k, Y_k)A_k(t)|\, dt \ \leq M_1 \int_{t_{k-1}}^{t_{k+1}} A_k(t)dt = M_1 h. \end{aligned}$$

Similarly, $|Y_2 - Y_1| \leq M_2 h$ and $|Y_{k+1} - Y_k| \leq M_2 h$. □

Throughout the assumptions of Theorem 3, we consider (t, x_1, x_2) instead (t, x, y).

Theorem 3. *Suppose that* $f(t, x_1, x_2), g(t, x_1, x_2) \in C^2[t_1, t_2]$. *Let* $\left|\frac{\partial f}{\partial x_i}\right| \leq L_f \left(\left|\frac{\partial g}{\partial x_i}\right| \leq L_g\right)$, $i = 1, 2$, *and* $|f(t, x_1, x_2)| \leq M_1$ $(|g(t, x_1, x_2)| \leq M_2)$. *Consider Scheme I (14)–(15) for some positive integer* k, *and* $\{A_k \mid k = 1, \ldots, n-1\}$, $n \geq 2$, *is a uniform fuzzy partition of* $[t_1, t_n]$, *then the following hold true:*

1. *for a value of k in the range $1 \le k \le n-1$:*

$$\left|\hat{F}_k - \hat{F}_{k-1}\right| \le L_f h U_{k,k-1}, \quad \left|\hat{G}_k - \hat{G}_{k-1}\right| \le L_g h U_{k,k-1},$$

where $U_{k,k-1} = |X_k - X_{k-1}| + |Y_k - Y_{k-1}|$.

2. *for all $k = 1, \dots, n-1$:*

$$|X_{n+1} - X_n| \le \frac{Mh}{2} e^{n(2Lh^2)}, \quad |Y_{n+1} - Y_n| \le \frac{Mh}{2} e^{n(2Lh^2)},$$

where $L = L_f + L_g$ and $M = \sum_i^2 M_i$.

Proof. 1. Using (6), we can get for each $k = 2, \dots, n-1$ and $t \in I \cap [t_k, t_{k+1}]$:

$$\left|\hat{F}_k - \hat{F}_{k-1}\right| = \left| \frac{\int_{t_{k-1}}^{t_{k+1}} f(t, X_k, Y_k) A_k(t) dt}{\int_{t_{k-1}}^{t_{k+1}} A_k(t) dt} - \frac{\int_{t_{k-2}}^{t_k} f(t, X_{k-1}, Y_{k-1}) A_{k-1}(t) dt}{\int_{t_{k-2}}^{t_k} A_{k-1}(t) dt} \right|.$$

Based on Remark 1 and Definition 1, the properties of the uniform fuzzy partition, we replace t by $t-h$ and then $A_{k-1}(t-h)$ by $A_k(t)$. Thus,

$$\begin{aligned}
\left|\hat{F}_k - \hat{F}_{k-1}\right| &= \left| \frac{\int_{t_{k-1}}^{t_{k+1}} f(t, X_k, Y_k) A_k(t) dt}{\int_{t_{k-1}}^{t_{k+1}} A_k(t) dt} - \frac{\int_{t_{k-1}}^{t_{k+1}} f(t-h, X_{k-1}, Y_{k-1}) A_k(t) dt}{\int_{t_{k-1}}^{t_{k+1}} A_k(t) dt} \right|, \\
&\le L_f \left(|X_k - X_{k-1}| + |Y_k - Y_{k-1}|\right) \int_{t_{k-1}}^{t_{k+1}} A_k(t) dt, \\
&= L_f h U_{k,k-1}.
\end{aligned} \tag{25}$$

In a similar way, $\left|\hat{G}_k - \hat{G}_{k-1}\right| \le L_g h U_{k,k-1}$.

2. We first prove the estimate for $k = 1$. Then, we show that for all $k = 2, \dots, n-1$, by using Lemma 2, for $k = 1$,

$$\begin{aligned}
|X_2 - X_1| &= \left| \int_{t_1}^{t_2} f(t, X_k, Y_k) A_k(t) dt \right|, \\
&\le \int_{t_1}^{t_2} |f(t, X_k, Y_k) A_k(t)| \, dt \quad \le M_1 \int_{t_1}^{t_2} A_k(t) dt \le \frac{Mh}{2},
\end{aligned}$$

where $M = \sum_i^2 M_i$. By (15), we get:

$$X_{k+1} - X_k = X_k - X_{k-1} + h\left(\hat{F}_k - \hat{F}_{k-1}\right), \quad Y_{k+1} - Y_k = Y_k - Y_{k-1} + h\left(\hat{G}_k - \hat{G}_{k-1}\right).$$

By using (25), we get:

$$\begin{aligned}
|X_{k+1} - X_k| &\le |X_k - X_{k-1}| + Lh^2 \left(|X_k - X_{k-1}| + |Y_k - Y_{k-1}|\right), \\
&\le |X_k - X_{k-1}| + |Y_k - Y_{k-1}| + 2Lh^2 \left(|X_k - X_{k-1}| + |Y_k - Y_{k-1}|\right), \\
&\le \left[1 + 2Lh^2\right] \left(|X_k - X_{k-1}| + |Y_k - Y_{k-1}|\right), \\
&\le \left(1 + 2Lh^2\right)^k U_{2,1} \le \frac{Mh}{2} \left(1 + 2Lh^2\right)^k,
\end{aligned}$$

where $U_{2,1} = |X_2 - X_1| + |Y_2 - Y_1|$, $L = L_f + L_g$ and $M = \sum_i^2 M_i$. In particular,

$$|X_{n+1} - X_n| \leq \frac{Mh}{2}\left(1 + 2Lh^2\right)^n \leq e^{n(2Lh^2)} \frac{Mh}{2},$$

where we have used inequality $(1 + 2h^2L)^n \leq e^{n(2h^2L)}$, $n \geq 0$. Analogously, $|Y_{n+1} - Y_n| \leq e^{n(2Lh^2)} \frac{Mh}{2}$. which concludes the proof. □

Remark 5. *The following estimations are used in Theorem 4 for $k = 1, \ldots, n-1$. Let f (g) satisfy the Lipschitz condition in the second and third arguments; we get:*

$$\begin{aligned} |f(t_k, x_k, y_k) - f(t_k, X_k, Y_k)| &\leq L_1 |x_k - X_k| + L_2 |y_k - Y_k|, \\ |g(t_k, x_k, y_k) - g(t_k, X_k, Y_k)| &\leq L_3 |x_k - X_k| + L_4 |y_k - Y_k|. \end{aligned}$$

From (20), we get:

$$\begin{aligned} f(t_k, x_k, y_k) - \hat{F}_k &\leq f(t_k, x_k, y_k) - f(t_k, X_k, Y_k) + \tfrac{h^2}{12} 2.M_2, \\ |f(t_k, x_k, y_k) - \hat{F}_k| &\leq L_1 |x_k - X_k| + L_2 |y_k - Y_k| + \tfrac{h^2}{6} M_2, \\ |g(t_k, x_k, y_k) - \hat{G}_k| &\leq L_3 |x_k - X_k| + L_4 |y_k - Y_k| + \tfrac{h^2}{6} M_2. \end{aligned}$$

Thus,

$$|f(t_k, X_k, Y_k) - \hat{F}_k| \leq \frac{h^2}{6} M_2, \text{ and } |g(t_k, X_k, Y_k) - \hat{G}_k| \leq \frac{h^2}{6} M_2,$$

where $M_2 = M_{2f} + M_{2g}$, $M_{2f} = \max_{t \in [t_1, t_n]} |f''(t, x, y)|$ and $M_{2g} = \max_{t \in [t_1, t_n]} |g''(t, x, y)|$.

Now, we show that the proposed Scheme I is convergent.

Theorem 4. *Let the assumptions of Theorem (3) be fulfilled, and further assume that f (g) satisfies a Lipschitz condition in the second and third arguments. Consider Scheme I (14)–(15) for some positive integer k, and $\{A_k \mid k = 1, \ldots, n-1\}$, $n \geq 2$, is a uniform fuzzy partition of $[t_1, t_n]$. Thus, if a sequence of $h = \{h_1, \ldots, h_m\} \to 0$, $m > 0$, and with each h, we compute the $X_{k,h}, Y_{k,h}$ component, then $|x(t_k) - X_{k,h}|$, $|y(t_k) - Y_{k,h}|$ converges to zero for each $k = 1, \ldots, n-1$.*

Proof. Let us drop the h subscript in the errors, writing $|x(t_k) - X_k|$ and $|y(t_k) - Y_k|$. Now, when $k = 1$, the result is clearly true, since $x(t_1) = X_1 = x_1$, $y(t_1) = Y_1 = y_1$. By Taylor's theorem, we have:

$$\begin{aligned} x(t_{k+1}) &= x(t_k) + hf(t_k, x_k, y_k) + \frac{h^2}{2} f'(\varepsilon_k, x(\varepsilon_k), y(\varepsilon_k)), \\ y(t_{k+1}) &= y(t_k) + hg(t_k, x_k, y_k) + \frac{h^2}{2} g'(\xi_k, x(\xi_k), y(\xi_k)), \end{aligned}$$

where $t_k < \varepsilon_k$, $\xi_k < t_{k+1}$. Denote $e1_k = x_k - X_k$, $e2_k = y_k - Y_k$, $x_k = x(t_k)$ and $y_k = y(t_k)$. Then, by (15), we get:

$$\begin{aligned} e1_{k+1} &= e1_k + h\left(f(t_k, x_k, y_k) - f(t_k, X_k, Y_k) + f(t_k, X_k, Y_k) - \hat{F}_k\right) + \frac{h^2}{2} x''(\varepsilon_k), \\ e2_{k+1} &= e2_k + h\left(g(t_k, x_k, y_k) - g(t_k, X_k, Y_k) + g(t_k, X_k, Y_k) - \hat{G}_k\right) + \frac{h^2}{2} y''(\xi_k). \end{aligned}$$

By virtue of Remark 5, we get:

$$|e1_{k+1}| \leq |e1_k| + hL_1 |e1_k| + hL_2 |e2_k| + \frac{h^2}{2}\left(M_{1f} + \frac{h}{3}M_2\right),$$
$$|e2_{k+1}| \leq |e2_k| + hL_4 |e2_k| + hL_3 |e1_k| + \frac{h^2}{2}\left(M_{1g} + \frac{h}{3}M_2\right),$$

where $M_{1f} = \max_{t\in[t_1,t_n]} |x''(t)|$ and $M_{1g} = \max_{t\in[t_1,t_n]} |y''(t)|$. Therefore,

Case 1. If $c = M_{1f} + M_{1g} + \frac{2h}{3}M_2$, $L = \sum_{i=1}^{4} L_i$, we get:

$$\begin{aligned}|e1_{k+1}| &\leq |e1_{k+1}| + |e2_{k+1}| \leq (1+2hL)\left(|e1_k| + |e2_k|\right) + h^2c/2,\\ &\leq (1+2hL)\,U_k + \frac{h^2}{2}c \leq (1+2hL)^k\,U_1 + \left(\sum_{j=0}^{k-1}(1+2hL)^j\right)\frac{h^2}{2}c,\\ &= (1+2hL)^k\,U_1 + \left(\frac{(1+2hL)^k - 1}{2hL}\right)\frac{h^2}{2}c \leq e^{k(2hL)}\left(U_1 + \frac{hc}{4L}\right) - \frac{hc}{4L},\end{aligned}$$

where $U_k = |e1_k| + |e2_k|$. Indeed, we have used inequality $(1+2hL)^k \leq e^{k(2hL)}$, $k \geq 0$, and the quantity $\sum_{j=0}^{k-1}(1+2hL)^j$ is a finite geometric series; these can be calculated by:

$$2Lh\left(\sum_{j=0}^{k-1}(1+2hL)^j\right) = (1+2Lh)\left(\sum_{j=0}^{k-1}(1+2hL)^j\right) - \left(\sum_{j=0}^{k-1}(1+2hL)^j\right) = (1+2hL)^k - 1.$$

In particular, when $U_1 = 0$, this implies that:

$$|x_n - X_n| \leq \frac{hc}{4L}\left(e^{2L(t_n-t_1)} - 1\right),\ |y_n - Y_n| \leq \frac{hc}{4L}\left(e^{2L(t_n-t_1)} - 1\right). \tag{26}$$

Case 2. In view of Remark 5, let $L = L_i, i = 1, \ldots, 4$

$$\left|f(t_k, x_k, y_k) - \hat{F}_k\right| \leq \frac{h^2}{6}M_2 + 2L\max\{|x_k - X_k|, |y_k - Y_k|\},$$
$$\left|g(t_k, x_k, y_k) - \hat{G}_k\right| \leq \frac{h^2}{6}M_2 + 2L\max\{|x_k - X_k|, |y_k - Y_k|\}.$$

Thus, $e1_{k+1} = x_k - X_k$, $e2_k = y_k - Y_k$,

$$|e1_{k+1}| \leq |e1_k| + 2Lh\max\{|e1_k|, |e2_k|\} + \frac{h^2}{2}c_1,$$
$$|e2_{k+1}| \leq |e2_k| + 2Lh\max\{|e1_k|, |e2_k|\} + \frac{h^2}{2}c_2,$$

where $c_1 = M_{1f} + \frac{h}{3}M_2$ and $c_2 = M_{1g} + \frac{h}{3}M_2$. Consequently,

$$|e1_k| \leq (1+4Lh)^k\,|U_1| + h^2c_1\frac{(1+4Lh)^k - 1}{4Lh},$$
$$|e2_k| \leq (1+4Lh)^k\,|U_1| + h^2c_2\frac{(1+4Lh)^k - 1}{4Lh},$$

where $U_1 = |e1_1| + |e2_1|$. In particular, when $U_1 = 0$, this implies that:

$$|x_n - X_n| \leq hc_1 \frac{e^{4L(t_n - t_1)} - 1}{4L}, \; |y_n - Y_n| \leq hc_2 \frac{e^{4L(t_n - t_1)} - 1}{4L}, \tag{27}$$

and if a sequence of $h \to 0$, we get $|e1_{n,h}| \to 0$, $|e2_{n,h}| \to 0$, which concludes the proof. □

3.3. Numerical Scheme II for SODEs

In this subsection, we will construct numerical Scheme II, a more advanced method than that of Scheme I. The components of FzT of x and y can be approximated by the average of the two methods, Scheme I (14)–(15) and the backward Scheme I (or implicit Scheme I), then the FzT is given by:

$$\left.\begin{array}{rlr|rlr} X_p = & X_k + h\hat{F}_k, & X_1 = \alpha, & Y_p = & Y_k + h\hat{G}_k, & Y_1 = \beta, \\ X_c = & X_k + h\hat{F}_{k+1}, & X_1 = \alpha, & Y_c = & Y_k + h\hat{G}_{k+1}, & Y_1 = \beta, \\ X_{k+1} = & \frac{1}{2}\left(X_p + X_c\right), & & Y_{k+1} = & \frac{1}{2}\left(Y_p + Y_c\right), & \end{array}\right\} \tag{28}$$

for $k = 1, \ldots, n-1$.

The problem with the previous scheme (28) is that the unknown quantities $\hat{F}_{k+1}$ and $\hat{G}_{k+1}$ appear on both sides (an implicit method). Therefore, one solution to this problem would be to use an explicit method such as another fuzzy approach. The following Scheme II for $k = 1, \ldots, n-1$, $X_1 = \alpha$, $Y_1 = \beta$:

$$\left.\begin{array}{rcl|rcl} X^*_{k+1} & = & X_k + h\hat{F}_k, & Y^*_{k+1} & = & Y_k + h\hat{G}_k, \\ X_{k+1} & = & X_k + \frac{h}{2}\left(\hat{F}_k + \hat{F}^*_{k+1}\right), & Y_{k+1} & = & Y_k + \frac{h}{2}\left(\hat{G}_k + \hat{G}^*_{k+1}\right), \end{array}\right\} \tag{29}$$

where:

$$\left.\begin{array}{rcl|rcl} \hat{F}_k[f] & = & \frac{\int_{t_1}^{t_n} f(t,X_k,Y_k)A_k(t)dt}{\int_{t_1}^{t_n} A_k(x)dx}, & \hat{G}_k[g] & = & \frac{\int_{t_1}^{t_n} g(t,X_k,Y_k)A_k(t)dt}{\int_{t_1}^{t_n} A_k(x)dx}, \\ \hat{F}^*_{k+1}[f] & = & \frac{\int_{t_1}^{t_n} f(t,X^*_{k+1},Y^*_{k+1})A_{k+1}(t)dt}{\int_{t_1}^{t_n} A_{k+1}(x)dx}, & \hat{G}^*_{k+1}[g] & = & \frac{\int_{t_1}^{t_n} g(t,X^*_{k+1},Y^*_{k+1})A_{k+1}(t)dt}{\int_{t_1}^{t_n} A_{k+1}(x)dx}. \end{array}\right\} \tag{30}$$

This method computes the approximate coordinates $[X_1, \ldots, X_n]$ and $[Y_1, \ldots, Y_n]$ of the direct FzT of the functions $x(t)$ and $y(t)$, respectively. In the sequel, the inverse FzT (16) approximates the solution $x(t)$ $(y(t))$ of the SODEs (5).

In the next theorem, we obtain an error estimate in the context of a fuzzy partition and error analysis of approximation Scheme II.

Theorem 5. *Suppose that $f(t,x_1,x_2)$, $g(t,x_1,x_2) \in C^2[t_1,t_2]$. Let $\left|\frac{\partial f}{\partial x_i}\right| \leq L_f$ $\left(\left|\frac{\partial g}{\partial x_i}\right| \leq L_g\right)$, $i = 1,2$, and $|f(t,x_1,x_2)| \leq M_1$ $(|g(t,x_1,x_2)| \leq M_2)$. Consider Scheme II (29)–(30) for some positive integer k and $\{A_k \mid k = 1, \ldots, n-1\}$, $n \geq 2$, to be a uniform fuzzy partition of $[t_1, t_n]$, then the following hold true:*

1. *for a value of k in the range $1 \leq k \leq n-1$:*

$$|\hat{F}^*_{k+1} - \hat{F}^*_k| \leq Lh\left(1 + 2Lh^2\right)U_{k,k-1}, \qquad |\hat{G}^*_{k+1} - \hat{G}^*_k| \leq Lh\left(1 + 2Lh^2\right)U_{k,k-1},$$

where $U_{k,k-1} = |X_k - X_{k-1}| + |Y_k - Y_{k-1}|$.

2. *for all $k = 1, \ldots, n-1$*

$$|X_{n+1} - X_n| \leq \frac{Mh}{2}e^{n(2Lh^2)}, \; |Y_{n+1} - Y_n| \leq \frac{Mh}{2}e^{n(2Lh^2)},$$

where $L = L_f + L_g$ and $M = \sum_i^2 M_i$.

Proof. The proof is similar to the proof of Theorem 3, so we just write out the procedure. The proofs of Part 1 is as follows.

$$\begin{aligned} |X^*_{k+1} - X^*_k| + |Y^*_{k+1} - Y^*_k| &\leq |X_k - X_{k-1}| + h\,|\hat{F}_k - \hat{F}_{k-1}| + |Y_k - Y_{k-1}| + h\,|\hat{G}_k - \hat{G}_{k-1}|, \\ |\hat{F}^*_{k+1} - \hat{F}^*_k| &\leq L_f h \left(|X^*_{k+1} - X^*_k| + |Y^*_{k+1} - Y^*_k|\right), \\ |\hat{F}_k - \hat{F}_{k-1}| &\leq L_f h U_{k,k-1}, \text{ and } |\hat{G}_k - \hat{G}_{k-1}| \leq L_g h U_{k,k-1}. \end{aligned}$$

The proof of Part 2, using (29), gives:

$$\begin{aligned} |X_{k+1} - X_k| &\leq |X_k - X_{k-1}| + Lh^2\left(1 + Lh^2\right) U_{k,k-1}, \\ |X_{k+1} - X_k| &\leq |X_{k+1} - X_k| + |Y_{k+1} - Y_k| \leq \left(1 + 2Lh^2\left(1 + Lh^2\right)\right)^k U_{2,1}, \\ &\leq \frac{Mh}{2}\left(1 + 2Lh^2\left(1 + Lh^2\right)\right)^k. \end{aligned}$$

where $U_{k,k-1} = |X_k - X_{k-1}| + |Y_k - Y_{k-1}|$ and $M = M_1 + M_2$. In particular:

$$|X_{n+1} - X_n| \leq \exp\left(n\left(2Lh^2\right)\left(1 + Lh^2\right)\right)\frac{Mh}{2}, \quad |Y_{n+1} - Y_n| \leq \exp\left(n\left(2Lh^2\right)\left(1 + Lh^2\right)\right)\frac{Mh}{2},$$

which concludes the proof. □

Lemma 4. *Let f and g have continuous second order derivatives on $t \in [t_1, t_n]$, and f (g) satisfies a Lipschitz condition in the second and third arguments. Then, a local error of Scheme II (29)–(30) is of the order h^3.*

Proof. We consider the SODEs (5). We start with the Taylor expansion and the forward divided difference approximation of the second derivative (please see Appendix A for more details), i.e.,

$$\begin{aligned} x(t_{k+1}) &= x(t_k) + hx'(t_k) + \frac{h^2}{2}\left(\frac{x'(t_{k+1}) - x'(t_k)}{h} - \frac{h}{2}x'''(\varepsilon_{2k})\right) + \frac{h^3}{6}x'''(\varepsilon_{1k}), \\ x(t_{k+1}) &= x(t_k) + \frac{h}{2}x'(t_k) + \frac{h}{2}x'(t_{k+1}) + h^3\left[\frac{1}{6}x'''(\varepsilon_{1k}) - \frac{1}{4}x'''(\varepsilon_{2k})\right], \end{aligned}$$

where $\varepsilon_{ik} \in (t_k, t_{k+1})$, $i = 1, 2$. The first derivative can be replaced by the right-hand side of the differential Equation (5). The Taylor expansion becomes:

$$x(t_{k+1}) = x(t_k) + \frac{h}{2}\left(f(t_k, x_k, y_k) + f\left(t_{k+1}, x_k + hf(t_k, x_k, y_k),\ y_k + hg(t_k, x_k, y_k)\right)\right) + e_f h^3,$$

where $e_f = \left[\frac{1}{6}x'''(\varepsilon_{1k}) - \frac{1}{4}x'''(\varepsilon_{2k}) + \frac{1}{4}f_2\right]$ and $f_2 = \frac{\partial}{\partial x}f(\varepsilon_{3k}, x(\varepsilon_{3k}), y(\varepsilon_{3k})) + \frac{\partial}{\partial y}f(\varepsilon_{3k}, x(\varepsilon_{3k}), y(\varepsilon_{3k}))$. We can write this as:

$$\begin{aligned} x_{k+1} &= x_k + \frac{h}{2}\left(K_0 + K_f\right) + e_f h^3, \\ y_{k+1} &= y_k + \frac{h}{2}\left(K_1 + K_g\right) + e_g h^3, \end{aligned}$$

where:

$$K_0 = f(t_k, x_k, y_k),\ K_1 = g(t_k, x_k, y_k), K_f = f(t_{k+1},\ x_k + hK_0,\ y_k + hK_1),\ K_g = g(t_{k+1},\ x_k + hK_0,\ y_k + hK_1),$$

$$e_g = \left[\frac{1}{6}y'''(\varepsilon_{1k}) - \frac{1}{4}y'''(\varepsilon_{2k}) + \frac{1}{4}g_2 y''(\varepsilon_{3k})\right] \text{ and } g_2 = \frac{\partial}{\partial x}g(\varepsilon_{3k}, x(\varepsilon_{3k}), y(\varepsilon_{3k})) + \frac{\partial}{\partial y}g(\varepsilon_{3k}, x(\varepsilon_{3k}), y(\varepsilon_{3k})).$$

Now, let f (g) satisfy a Lipschitz condition in the second and third arguments. By Lemma 1 and Remark 5, we have:

$$|f(t_k, x_k, y_k) - f(t, X_k, Y_k)| \leq L\left(|x_k - X_k| + |y_k - Y_k|\right) \leq \alpha_f h^2 \leq \alpha h^2, |g(t_k, x_k, y_k) - g(t, X_k, Y_k)| \leq \alpha_g \leq \alpha h^2,$$

where $\alpha = \alpha_f + \alpha_g$ is a positive constant. Once again, using Remark 5 and according to (29)–(30), we obtain for fixed $k = 1, \ldots, n-1$:

$$\begin{aligned} x_{k+1} - X_{k+1} &= x_k - X_k + \frac{h}{2}\left(K_0 - \hat{F}_k\right) + \frac{h}{2}\left(K_f - \hat{F}^*_{k+1}\right) + e_f h^3, \\ K_0 - \hat{F}_k &\leq \left|f(t_k, x_k, y_k) - f(t_k, X_k, Y_k) + f(t_k, X_k, Y_k) - \hat{F}_k\right| \leq \alpha h^2 + \frac{1}{6}M_2 h^2, \\ K_1 - \hat{G}_k &= \left|g(t_k, x_k, y_k) - g(t_k, X_k, Y_k) + g(t_k, X_k, Y_k) - \hat{F}_k\right| \leq \alpha h^2 + \frac{1}{6}M_2 h^2. \end{aligned}$$

By the trapezium formula, we have:

$$f(t_{k+1}, X^*_{k+1}, Y^*_{k+1}) - \hat{F}^*_{k+1} \leq f(t_{k+1}, X^*_{k+1}, Y^*_{k+1}) - f(t_{k+1}, X^*_{k+1}, Y^*_{k+1}) + \frac{1}{6}M_{2f}h^2 = \frac{1}{6}M_{2f}h^2 \leq \frac{1}{6}M_2 h^2,$$

Note that:

$$f(t_{k+1}, x_k + hK_0, y_k + hK_1) - f(t_{k+1}, X^*_{k+1}, Y^*_{k+1}) \leq L\left[|x_k - X_k| + h\left|K_0 - \hat{F}_k\right| + |y_k - Y_k| + h\left|K_1 - \hat{G}_k\right|\right].$$

Next,

$$\begin{aligned} K_f - \hat{F}^*_{k+1} &= f(t_{k+1}, x_k + hK_0, y_k + hK_1) - f(t_{k+1}, X^*_{k+1}, Y^*_{k+1}) + f(t_{k+1}, X^*_{k+1}, Y^*_{k+1}) - \hat{F}^*_{k+1}, \\ &\leq \left[\alpha h^2 + \frac{1}{6}M_2 h^2\right](2Lh + 1). \end{aligned}$$

This leads to:

$$|x_{k+1} - X_{k+1}| \leq |x_k - X_k| + \tfrac{h}{2}\left(\alpha h^2 + \tfrac{1}{6}M_2 h^2\right) + \tfrac{h}{2}\left(\left[\alpha h^2 + \tfrac{1}{6}M_2 h^2\right](2Lh + 1)\right) + e_f h^3 = |x_k - X_k| + E_f h^3. \tag{31}$$

Similarly, $|y_{k+1} - Y_{k+1}| \leq |y_k - Y_k| + E_g h^3$, where E_f and E_g are appropriate constants that depend on f and g, respectively. Therefore, the error of this method is $O(h^3)$. □

To see that Scheme II is globally a second-order method, we need to establish its convergence.

Theorem 6. *Let the assumptions of Lemma 4 be fulfilled. Consider Scheme II (29)–(30) for some positive integer k, and $\{A_k \mid k = 1, \ldots, n-1\}$, $n \geq 2$, is a uniform fuzzy partition of $[t_1, t_n]$. Thus, if a sequence of $h \to 0$, and with each h, we compute the $X_{k,h}, Y_{k,h}$ component, then $\left|x(t_k) - X_{k,h}\right|$, $\left|y(t_k) - Y_{k,h}\right|$ converges to zero for each $k = 1, \ldots, n-1$.*

Proof. The proof is similar to the proof of Theorem 4, so we just write out the procedure. According to Remark 5 and (31), we get:

$$\begin{aligned} x_{k+1} - X_{k+1} &= x_k - X_k + \frac{h}{2}\left(K_0 - \hat{F}_k\right) + \frac{h}{2}\left(K_f - \hat{F}^*_{k+1}\right) + e_f h^3, \\ K_0 - \hat{F}_k &\leq L_1|x_k - X_k| + L_2|y_k - Y_k| + \frac{1}{6}M_2 h^2, \\ K_1 - \hat{G}_k &\leq L_3|x_k - X_k| + L_4|y_k - Y_k| + \frac{1}{6}M_2 h^2, \\ f(t_{k+1}, X^*_{k+1}, Y^*_{k+1}) - \hat{F}^*_{k+1} &= \frac{1}{6}M_{2f}h^2 \leq \frac{1}{6}M_2 h^2, \end{aligned}$$

$$f(t_{k+1}, x_k + hK_0, y_k + hK_1) - f(t_{k+1}, X^*_{k+1}, Y^*_{k+1}) \leq L_1\left(|x_k - X_k| + h\left|K_0 - \hat{F}_k\right|\right) + L_2\left(|y_k - Y_k| + h\left|K_1 - \hat{G}_k\right|\right),$$

$$\begin{aligned} K_f - \hat{F}^*_{k+1} &= f(t_{k+1}, x_k + hK_0, y_k + hK_1) - f(t_{k+1}, X^*_{k+1}, Y^*_{k+1}) + f(t_{k+1}, X^*_{k+1}, Y^*_{k+1}) - \hat{F}^*_{k+1}, \\ &\leq L_1|x_k - X_k| + hL_1\left|K_0 - \hat{F}_k\right| + L_2|y_k - Y_k| + hL_2\left|K_1 - \hat{G}_k\right| + \frac{1}{6}M_2h^2. \end{aligned}$$

Once again, using Remark 5 gives:

$$\begin{aligned} |x_{k+1} - X_{k+1}| \leq\ & |x_k - X_k| + \frac{h}{2}\left(L_1|x_k - X_k| + L_2|y_k - Y_k| + \frac{1}{6}M_2h^2\right) + \\ & \frac{h}{2}\left(L_1|x_k - X_k| + hL_1\left|K_0 - \hat{F}_k\right| + L_2|y_k - Y_k| + hL_2\left|K_1 - \hat{G}_k\right| + \frac{1}{6}M_2h^2\right) + \\ & \left(\frac{1}{4}SM_1 - \frac{1}{12}M_2\right)h^3, \end{aligned}$$

where

$e_f \leq \frac{1}{6}M_{2f} - \frac{1}{4}M_{2f} + \frac{1}{4}SM_{1f} = \frac{1}{4}SM_{1f} - \frac{1}{12}M_{2f}$,

$M_1 = M_{1f} + M_{1g}$, $M_{1f} = \max\limits_{t\in[t_1,t_n]} |f'(t,x,y)|$, $M_{1g} = \max\limits_{t\in[t_1,t_n]} |g'(t,x,y)|$,

$M_2 = M_{2f} + M_{2g}$, $M_{2f} = \max\limits_{t\in[t_1,t_n]} |f''(t,x,y)|$, $M_{2g} = \max\limits_{t\in[t_1,t_n]} |g''(t,x,y)|$,

$S = S_f + S_g$ is the upper bound of $\frac{\partial f}{\partial x_i}\left(\frac{\partial g}{\partial x_i}\right)$, $i = 1, 2$, $x = x_1$ and $y = x_2$.

Simplifying, then:

$$\begin{aligned} |x_{k+1} - X_{k+1}| \leq\ & |x_k - X_k| + \\ & + \frac{h}{2}(2L_1 + hL_1L_1 + hL_2L_3)|x_k - X_k| + \frac{h}{2}(2L_2 + hL_1L_2 + hL_2L_4)|y_k - Y_k| \\ & + \left(\frac{1}{4}SM_1 + \frac{1}{12}[hL_1 + hL_2 + 1]M_2\right)h^3. \end{aligned}$$

Similarly,

$$\begin{aligned} |y_{k+1} - Y_{k+1}| \leq\ & |y_k - Y_k| \\ & + \frac{h}{2}(2L_4 + hL_3L_2 + hL_4L_4)|y_k - Y_k| + \frac{h}{2}(2L_3 + hL_3L_1 + hL_4L_3)|x_k - X_k| \\ & + \left(\frac{1}{4}SM_1 + \frac{1}{12}[hL_3 + hL_4 + 1]M_2\right)h^3. \end{aligned}$$

Therefore,

Case 1. If $L = \sum_{i=1}^{4} L_i$ and $c = \frac{1}{2}SM_1 + \frac{1}{6}[2hL + 1]M_2$, we get

$$|x_{k+1} - X_{k+1}| \leq |x_{k+1} - X_{k+1}| + |y_{k+1} - Y_{k+1}| \leq [1 + 2hL(1 + hL)](|x_k - X_k| + |y_k - Y_k|) + \frac{h^3}{2}c,$$

$$|y_{k+1} - Y_{k+1}| \leq |x_{k+1} - X_{k+1}| + |y_{k+1} - Y_{k+1}| \leq [1 + 2hL(1 + hL)](|x_k - X_k| + |y_k - Y_k|) + \frac{h^3}{2}c.$$

By using the proof of Theorem 4, when $U_1 = 0$, this implies that:

$$|x_n - X_n| \leq \frac{h^2 c}{4L\left(1+\frac{L(t_n-t_1)}{n}\right)}\left[\exp\left(2L(t_n-t_1)\left(1+\frac{L(t_n-t_1)}{n}\right)\right)-1\right]. \quad (32)$$

In a similar way,

$$|y_n - Y_n| \leq \frac{h^2 c}{4L\left(1+\frac{L(t_n-t_1)}{n}\right)}\left[\exp\left(2L(t_n-t_1)\left(1+\frac{L(t_n-t_1)}{n}\right)\right)-1\right]. \quad (33)$$

Case 2. If $L = L_i, i = 1, \ldots, 4$ and:

$$\begin{aligned} \left|f(t_k, x_k, y_k) - \hat{F}_k\right| &\leq \tfrac{h^2}{6} M_2 + 2L \max\{|x_k - X_k|, |y_k - Y_k|\}, \\ \left| g(t_k, x_k, y_k) - \hat{G}_k \right| &\leq \tfrac{h^2}{6} M_2 + 2L \max\{|x_k - X_k|, |y_k - Y_k|\}. \end{aligned}$$

Thus,

$$\begin{cases} |x_{k+1} - X_{k+1}| \leq |x_k - X_k| + 2Lh(1+hL) \max\{|x_k - X_k|, |y_k - Y_k|\} + \frac{h^3}{2}c, \\ |y_{k+1} - Y_{k+1}| \leq |y_k - Y_k| + 2Lh(1+hL) \max\{|x_k - X_k|, |y_k - Y_k|\} + \frac{h^3}{2}c, \end{cases}$$

where $c = \frac{1}{2}SM_1 + \frac{1}{6}[2hL+1]M_2$. Consequently,

$$\begin{cases} |x_k - X_k| \leq (1+4Lh(1+hL))^k |U_1| + h^3 c \frac{(1+4Lh(1+hL))^k - 1}{4Lh(1+hL)}, \\ |y_k - Y_k| \leq (1+4Lh(1+hL))^k |U_1| + h^3 c \frac{(1+4Lh(1+hL))^k - 1}{4Lh(1+hL)}, \end{cases}$$

where $U_k = |x_k - X_k| + |y_k - Y_k|$. In particular,
when $U_1 = 0$, we get:

$$\begin{cases} |x_n - X_n| \leq h^2 c \frac{\left[\exp\left(4L(t_n-t_1)\left(1+\frac{L(t_n-t_1)}{n}\right)\right)-1\right]}{4L\left(1+\frac{L(t_n-t_1)}{n}\right)}, \\ |y_n - Y_n| \leq h^2 c \frac{\left[\exp\left(4L(t_n-t_1)\left(1+\frac{L(t_n-t_1)}{n}\right)\right)-1\right]}{4L\left(1+\frac{L(t_n-t_1)}{n}\right)}, \end{cases} \quad (34)$$

and if $h = \{h_1, \ldots, h_m\} \to 0$, $m > 0$ in (32), (33) and (34), we get $\left|x_n - X_{n,h}\right| \to 0$, $\left|y_n - Y_{n,h}\right| \to 0$, which concludes the proof. □

4. Applications

One of the main problems of mathematics appears with variable coefficients when $\alpha(t), \beta(t), \delta(t), \gamma(t)$ are analytic functions and added to the model. The new differential equations are represented by non-autonomous SODEs. In this model, time varying values for the growth rate of the prey, the efficiency of the predator, being the ability to capture prey, the death rate of the predator and the growth rate of the predator are considered. It is important to remark that since in this problem, the coefficients are time varying, careful attention must be paid in order to obtain the correct recurrence equation system of the model. The model, incorporating the above functions, is as follows [30,33,34]:

$$\begin{aligned} \frac{dx}{dt} &= \alpha(t)x(t) - \beta(t)x(t)y(t), & x(0) = x_1 \\ \frac{dy}{dt} &= \delta(t)x(t)y(t) - \gamma(t)y(t), & y(0) = y_1 \end{aligned} \quad (35)$$

Three examples are discussed in order to prove the results obtained by Scheme I (14)–(15) and Scheme II (29)–(30), two examples for the numerical solution of the model (35) and one example for the linear case.

Example 1. *Consider the problem of the Lotka–Volterra prey-predator model (35). We take*

$\alpha(t) = 4 + \tan(t)$, $\beta(t) = \exp(2t)$, $\gamma(t) = -2$, $\delta(t) = \cos(t)$, $x(0) = -4$ *and* $y(0) = 4$.

The exact solution for these coefficients is $x(t) = \frac{-4}{\cos(t)}$, $y(t) = 4\exp(-2t)$, *as proposed by [30,33,34].*

Example 2. *Consider the problem of the Lotka–Volterra prey-predator model (35) with*

$\alpha(t) = -t$, $\beta(t) = -t$, $\gamma(t) = t$, $\delta(t) = t$, $x(0) = 2$ *and* $y(0) = 2$.

The exact solution for these coefficients is $x(t) = \frac{2}{2-\exp(t^2/2)}$, $y(t) = \frac{2}{2-\exp(t^2/2)}$, *as proposed by [30,33].*

Example 3. *Consider the following non-autonomous SODEs with initial values (5):*

$$\begin{cases} x'(t) = x(t) - y(t) + 2t - t^2 - t^3 & , x(0) = 1 \ , t \in [0,1] \\ y'(t) = x(t) + y(t) - 4t^2 + t^3 & , y(0) = 0. \end{cases} \tag{36}$$

The exact solution of (36) is given by $x(t) = e^t \cos(t) + t^2$ *and* $y(t) = e^t \sin(t) - t^3$.

The results are listed in Tables 1–7 by the proposed fuzzy approximation methods with respect to the raised cosine generating function and Table 8 by the proposed fuzzy approximation methods with respect to the triangular generating function and raised cosine generating function. The proposed fuzzy approximation methods are generated by Algorithms A1 and A2 (please see Appendix B). The mean square error (MSE) is defined as MSE $= \frac{1}{n}\left(\|Y_k - y(t_k)\|_2\right)^2$. This is an easily computable quantity for a particular sample. From the numerical tests, the results are summarized as follows:

1. In view of Tables 2–7, a comparison is made between the two new proposed schemes (15), (29), the Euler method and the trapezoidal rule based on the Euler method for Examples 1–3.
2. Moreover, a comparison of MSE for Examples 1–3 is shown in Table 1. It is observed that the new fuzzy approximation methods yield more accurate results in comparison with the classical Euler and classical trapezoidal rule (one-step). The best result (in comparison with the Schemes I and II) is obtained by Scheme II.
3. In Table 8, a comparison is given between the errors for two proposed schemes based on the FzT with respect to fuzzy partitions determined by [6,19].

The better results (in comparison with the non-linear case) are obtained by the linear case and non-autonomous SODEs in Example 3. Further, the results obtained using proposed fuzzy approximation methods for Examples 1–3 are shown in Figure 1 by using the raised cosine generating function. In view of Figure 1, the graphical results of Examples 1–3 show a comparison between numerical schemes (I and II) and the exact solution. Furthermore, in view of Figure 1, a comparison is given between the numerical results of Examples 1 and 2 and exact solutions for $h = 0.01$, while a comparison is given between the numerical results of Example 3 and exact solutions for $h = 0.1$. All the graphs are plotted using MATLAB software. This constitutes an important improvement to the previous fuzzy approach, which did not provide such information for SODEs. Thus, this study will be particularly important.

Remark 6. *We compare new results based on FzT with the conventional numerical methods. For a discussion of the conventional numerical methods, the Euler method and trapezoidal rule to solve SODEs, see for example [35,36].*

Table 1. The values of MSE for Example 1–3.

Method	Example 1		Example 2		Example 3	
	$x(t)$	$y(t)$	$x(t)$	$y(t)$	$x(t)$	$y(t)$
Scheme I	2.91443×10^{-1}	6.67431×10^{-3}	1.12399×10^{-1}	1.12399×10^{-1}	2.92534×10^{-4}	1.58256×10^{-3}
Scheme II	2.24139×10^{-2}	3.77476×10^{-4}	3.04846×10^{-4}	3.04846×10^{-4}	1.72082×10^{-5}	4.10161×10^{-5}
Euler	6.99731×10^{-1}	1.19826×10^{-2}	1.14890×10^{-1}	1.14890×10^{-1}	5.43867×10^{-4}	1.68059×10^{-3}
Trapezoidal	5.99915×10^{-1}	1.40165×10^{-3}	2.75574×10^{-2}	2.75574×10^{-2}	3.19103×10^{-5}	5.21595×10^{-4}

Table 2. Comparison of numerical results of $x(t)$ for Example 3.

t_i	Solution $x(t)$	Proposed Scheme I	Proposed Scheme II	Euler	Trapezoidal
0.00	1.00000	1.00000	1.00000	1.00000	1.00000
0.10	1.10965	1.10581	1.11259	1.10000	1.10945
0.20	1.23706	1.22517	1.24000	1.21890	1.23671
0.30	1.37957	1.36112	1.38257	1.35438	1.37924
0.40	1.53406	1.51087	1.53717	1.50368	1.53402
0.50	1.69689	1.67119	1.70015	1.66355	1.69755
0.60	1.86386	1.83827	1.86733	1.83022	1.86575
0.70	2.03020	2.00777	2.03392	1.99935	2.03396
0.80	2.19055	2.17473	2.19456	2.16601	2.19692
0.90	2.33891	2.33356	2.34322	2.32461	2.34870
1.00	2.46869	2.47798	2.47776	2.46891	2.48270

Table 3. Comparison of numerical results of $y(t)$ for Example 3.

t_i	Solution $y(t)$	Proposed Scheme I	Proposed Scheme II	Euler	Trapezoidal
0.00	0.00000	0.00000	0.00000	0.00000	0.00000
0.10	0.10933	0.09948	0.10781	0.10000	0.10805
0.20	0.23466	0.21563	0.23168	0.21610	0.23113
0.30	0.37191	0.34407	0.36755	0.34440	0.36521
0.40	0.51694	0.48088	0.51126	0.48098	0.50621
0.50	0.66544	0.62209	0.65855	0.62184	0.64994
0.60	0.81285	0.76359	0.80488	0.76288	0.79202
0.70	0.95430	0.90109	0.94543	0.89979	0.92782
0.80	1.08451	1.03002	1.07494	1.02801	1.05236
0.90	1.19767	1.14549	1.18766	1.14261	1.16023
1.00	1.28736	1.24213	1.28463	1.23823	1.24549

Table 4. Comparison of numerical results of $x(t)$ for Example 1.

t_i	Solution $x(t)$	Proposed Scheme I	Proposed Scheme II	Euler	Trapezoidal
0.00	−4.00000	−4.00000	−4.00000	−4.00000	−4.00000
0.05	−4.00501	−3.97865	−3.99616	−4.00000	−4.00714
0.10	−4.02008	−3.99232	−4.01193	−4.01429	−4.02627
0.15	−4.04543	−4.01914	−4.03732	−4.04276	−4.05752
0.20	−4.08136	−4.05937	−4.07264	−4.08574	−4.10134
0.25	−4.12834	−4.11358	−4.11833	−4.14389	−4.15850
0.30	−4.18701	−4.18268	−4.17490	−4.21830	−4.23015
0.35	−4.25816	−4.26794	−4.24305	−4.31052	−4.31783
0.40	−4.34282	−4.37105	−4.32359	−4.42260	−4.42361
0.45	−4.44224	−4.49423	−4.41752	−4.55722	−4.55014
0.50	−4.55798	−4.64028	−4.52605	−4.71784	−4.70086
0.55	−4.69195	−4.81278	−4.65062	−4.90891	−4.88023
0.60	−4.84651	−5.01628	−4.79297	−5.13613	−5.09399
0.65	−5.02460	−5.25662	−4.95520	−5.40685	−5.34964
0.70	−5.22984	−5.54127	−5.13983	−5.73061	−5.65700
0.75	−5.46680	−5.87990	−5.34992	−6.11990	−6.02912
0.80	−5.74130	−6.28518	−5.58925	−6.59122	−6.48349
0.85	−6.06076	−6.77381	−5.86249	−7.16663	−7.04390
0.90	−6.43490	−7.36820	−6.17551	−7.87602	−7.74310
0.95	−6.87660	−8.09874	−6.53582	−8.76042	−8.62699
1.00	−7.40326	−9.00740	−6.96630	−9.87710	−9.76103

Table 5. Comparison of numerical results of $y(t)$ for Example 1.

t_i	Solution $y(t)$	Proposed Scheme I	Proposed Scheme II	Euler	Trapezoidal
0.00	4.00000	4.00000	4.00000	4.00000	4.00000
0.05	3.61935	3.60013	3.62249	3.60000	3.62045
0.10	3.27492	3.24497	3.28040	3.24090	3.27766
0.15	2.96327	2.92506	2.97057	2.91774	2.96757
0.20	2.68128	2.63645	2.69004	2.62635	2.68671
0.25	2.42612	2.37573	2.43612	2.36315	2.43201
0.30	2.19525	2.13994	2.20634	2.12506	2.20079
0.35	1.98634	1.92646	1.99847	1.90938	1.99067
0.40	1.79732	1.73299	1.81047	1.71374	1.79951
0.45	1.62628	1.55749	1.64050	1.53607	1.62541
0.50	1.47152	1.39815	1.48689	1.37451	1.46664
0.55	1.33148	1.25333	1.34811	1.22742	1.32166
0.60	1.20478	1.12159	1.22279	1.09333	1.18908
0.65	1.09013	1.00161	1.10969	0.97093	1.06762
0.70	0.98639	0.89223	1.00768	0.85906	0.95615
0.75	0.89252	0.79241	0.91576	0.75670	0.85366
0.80	0.80759	0.70122	0.83301	0.66295	0.75923
0.85	0.73073	0.61784	0.75862	0.57703	0.67208
0.90	0.66120	0.54154	0.69184	0.49827	0.59150
0.95	0.59827	0.47170	0.63202	0.42612	0.51692
1.00	0.54134	0.40778	0.57734	0.36016	0.44787

Table 6. Comparison of numerical results of $x(t)$ for Example 2.

t_i	Solution $x(t)$	Proposed Scheme I	Proposed Scheme II	Euler	Trapezoidal
0.00	2.00000	2.00000	2.00000	2.00000	2.00000
0.05	2.00250	2.00149	2.00325	2.00000	2.00250
0.10	2.01008	2.00650	2.01082	2.00500	2.01005
0.15	2.02289	2.01660	2.02364	2.01508	2.02279
0.20	2.04124	2.03197	2.04201	2.03042	2.04101
0.25	2.06557	2.05294	2.06636	2.05134	2.06511
0.30	2.09650	2.07996	2.09731	2.07830	2.09567
0.35	2.13485	2.11365	2.13568	2.11191	2.13344
0.40	2.18171	2.15485	2.18256	2.15301	2.17942
0.45	2.23852	2.20462	2.23939	2.20265	2.23493
0.50	2.30720	2.26437	2.30808	2.26226	2.30167
0.55	2.39031	2.33595	2.39116	2.33365	2.38192
0.60	2.49133	2.42177	2.49211	2.41923	2.47868
0.65	2.61513	2.52506	2.61574	2.52224	2.59605
0.70	2.76863	2.65022	2.76888	2.64702	2.73967
0.75	2.96202	2.80329	2.96157	2.79961	2.91754
0.80	3.21093	2.99285	3.20907	2.98854	3.14130
0.85	3.54059	3.23143	3.53586	3.22625	3.42845
0.90	3.99443	3.53788	3.98346	3.53151	3.80653
0.95	4.65413	3.94192	4.62841	3.93381	4.32100
1.00	5.69348	4.49277	5.61875	4.48201	5.05197

Table 7. Comparison of numerical results of $y(t)$ for Example 2.

t_i	Solution $y(t)$	Proposed Scheme I	Proposed Scheme II	Euler	Trapezoidal
0.00	2.00000	2.00000	2.00000	2.00000	2.00000
0.05	2.00250	2.00149	2.00325	2.00000	2.00250
0.10	2.01008	2.00650	2.01082	2.00500	2.01005
0.15	2.02289	2.01660	2.02364	2.01508	2.02279
0.20	2.04124	2.03197	2.04201	2.03042	2.04101
0.25	2.06557	2.05294	2.06636	2.05134	2.06511
0.30	2.09650	2.07996	2.09731	2.07830	2.09567
0.35	2.13485	2.11365	2.13568	2.11191	2.13344
0.40	2.18171	2.15485	2.18256	2.15301	2.17942
0.45	2.23852	2.20462	2.23939	2.20265	2.23493
0.50	2.30720	2.26437	2.30808	2.26226	2.30167
0.55	2.39031	2.33595	2.39116	2.33365	2.38192
0.60	2.49133	2.42177	2.49211	2.41923	2.47868
0.65	2.61513	2.52506	2.61574	2.52224	2.59605
0.70	2.76863	2.65022	2.76888	2.64702	2.73967
0.75	2.96202	2.80329	2.96157	2.79961	2.91754
0.80	3.21093	2.99285	3.20907	2.98854	3.14130
0.85	3.54059	3.23143	3.53586	3.22625	3.42845
0.90	3.99443	3.53788	3.98346	3.53151	3.80653
0.95	4.65413	3.94192	4.62841	3.93381	4.32100
1.00	5.69348	4.49277	5.61875	4.48201	5.05197

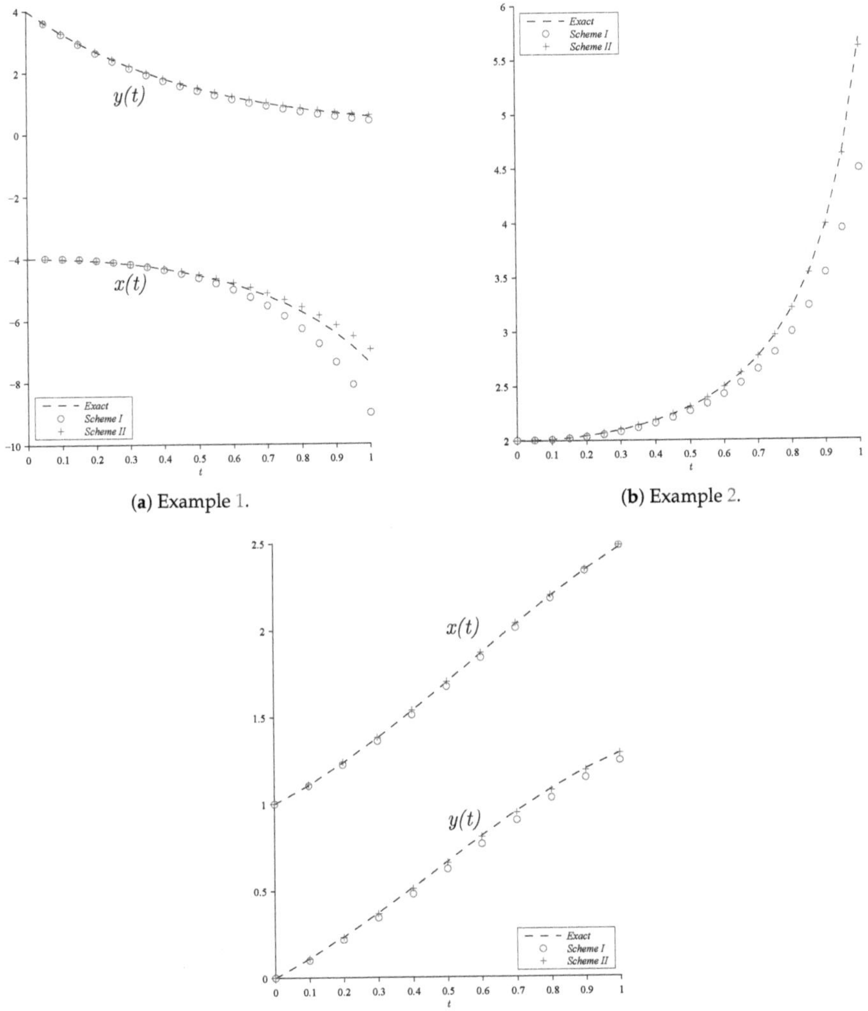

(**a**) Example 1.

(**b**) Example 2.

(**c**) Example 3.

Figure 1. A comparison between three fuzzy numerical methods and the exact solution for three examples.

Table 8. The values of MSE for Examples 1–3 by the different types of fuzzy partitions.

	Case	Proposed Scheme for $x(t)$		Proposed Scheme for $y(t)$	
		I	II	I	II
Ex.1	T [1]	2.48353×10^{-1}	3.37890×10^{-2}	6.03282×10^{-3}	5.38734×10^{-4}
	C [2]	2.91443×10^{-1}	2.24139×10^{-2}	6.67431×10^{-3}	3.77476×10^{-4}
Ex.2	T	1.12099×10^{-1}	3.01900×10^{-4}	1.12099×10^{-1}	3.01900×10^{-4}
	C	1.12399×10^{-1}	3.04846×10^{-4}	1.12399×10^{-1}	3.04846×10^{-4}
Ex.3	T	2.71905×10^{-4}	2.08807×10^{-5}	1.61509×10^{-3}	4.84176×10^{-5}
	C	2.92534×10^{-4}	1.72082×10^{-5}	1.58256×10^{-3}	4.10161×10^{-5}

[1] Triangular generating function; [2] Raised cosine generating function.

5. Conclusions

We extended the applicability of fuzzy-based numerical methods to the problems of conventional mathematics. In particular, we contributed to approximation methods of the SODEs. Two approximation methods based on the FzT were proposed and their error estimate analyzed. Moreover, we proved that two approximation methods, namely Schemes I and II, determine an approximate solution, which converges to the exact solution, and the local truncation error of the Scheme I (Scheme II) is $\mathcal{O}(h^2)$ $(\mathcal{O}(h^3))$. As an application, a system of nonlinear differential equations is solved by using Schemes I and II. From the numerical results, it is observed that the new fuzzy approximation methods yield more accurate results in comparison with the classical Euler method (one-stage) and classical trapezoidal rule (two-stage). Hence, the new fuzzy approximation methods provided alternative techniques for solving differential equations with better results, and the objective of this research was achieved and tested.

As a consequence, it should be noted that the numerical solutions depend on the types of uniform fuzzy partitions. For cases $\left(1 - \left|\frac{x-x_k}{h}\right|\right)$ and $\frac{1}{2}\left(1 + \cos\left(\pi\left(\frac{x-x_k}{h}\right)\right)\right)$, the shape of the basic functions determines the form of representation (linear or non-linear) of the numerical solution. This agrees with the results proposed by [5,6] using uniform fuzzy partitions. It is also worth pointing out that the results in this research are better in comparison with the classical numerical methods using uniform fuzzy partitions for linear and nonlinear cases. Thus, the proposed method is very much suitable for solving SODEs (5) in a linear or nonlinear case under the assumption of f and g satisfying the Lipschitz condition. If we want to obtain the best approximation of f and g as possible, then the number n of components should be large. It should be stressed that the application of the FzT can be used for removing noise from the given data. This is especially important for various practical applications of FzT. The proposed methods can also be applied to the n-dimensional system of first-order coupled differential equations in the case of a non-noisy or noisy right-hand side. The discussion will continue in [37] to give more details about the fuzzy partition and the modification of multiple steps.

Author Contributions: Conceptualization and performed the numerical experiments, H.A. ALKasasbeh; Evaluated the results and supported this work, I. Perfilieva; Project administration and designed the numerical methods, M. Z. Ahmad; Software and data curation, Z. R. Yahyaa.

Funding: The work of Irina Perfilieva has been supported by the project "IT4Innovations excellence in science, LQ1602" and by the Grant Agency of the Czech Republic (Project No. 16-09541S).

Acknowledgments: The authors would like to express their thanks to the editors and the anonymous referees for their valuable comments and suggestions that contributed to the paper. Many thanks are given to Universiti Malaysia Perlis for providing all facilities until this work was completed successfully.

Conflicts of Interest: The authors declare no conflicts of interest.

Appendix A. Taylor Series

A Taylor series is given that:

$$\begin{aligned}
x(t_{k+1}) &= x(t_k) + hx'(t_k) + \frac{h^2}{2}x''(t_k) + \frac{h^3}{6}x'''(\varepsilon_{1k}), \\
&= x(t_k) + hx'(t_k) + \frac{h^2}{2}\left(\frac{x'(t_{k+1}) - x'(t_k)}{h} - \frac{h}{2}x'''(\varepsilon_{2k})\right) + \frac{h^3}{6}x'''(\varepsilon_{1k}), \\
&= x(t_k) + \frac{h}{2}x'(t_k) + \frac{h}{2}x'(t_{k+1}) + h^3\left[\frac{1}{6}x'''(\varepsilon_{1k}) - \frac{1}{4}x'''(\varepsilon_{2k})\right],
\end{aligned} \tag{A1}$$

where $x''(t) = \frac{x'(t_{k+1})-x'(t_k)}{h} - \frac{h}{2}x'''(\varepsilon_{2k})$. Calculus can be used to derive that:

$$\begin{aligned}
x'(t_{k+1}) &= f\left(t_{k+1}, x(t_{k+1}), y(t_{k+1})\right) \\
&= f\left(t_{k+1}, x(t_k) + hf(t_k, x(t_k), y(t_k)), y(t_k) + hg(t_k, x(t_k), y(t_k))\right) \\
&\quad + \frac{h^2}{2}f_2(\varepsilon_{3k}, x(\varepsilon_{3k}), y(\varepsilon_{3k}))x''(\varepsilon_{3k}),
\end{aligned}$$

where $f_2(\varepsilon_{3k}, x(\varepsilon_{3k}), y(\varepsilon_{3k})) = \frac{\partial}{\partial x}f(\varepsilon_{3k}, x(\varepsilon_{3k}), y(\varepsilon_{3k})) + \frac{\partial}{\partial y}f(\varepsilon_{3k}, x(\varepsilon_{3k}), y(\varepsilon_{3k}))$.

Substituting Equation (A1), it is given that:

$$\begin{aligned}
x(t_{k+1}) &= x(t_k) + \frac{h}{2}x'(t_k) \\
&\quad + \frac{h}{2}f\left(t_{k+1}, x(t_k) + hf(t_k, x(t_k), y(t_k)), y(t_k) + hg(t_k, x(t_k), y(t_k))\right) \\
&\quad + \frac{h}{2}\frac{h^2}{2}f_2(\varepsilon_{3k}, x(\varepsilon_{3k}), y(\varepsilon_{3k}))x''(\varepsilon_{3k}) \\
&\quad + h^3\left[\frac{1}{6}x'''(\varepsilon_{1k}) - \frac{1}{4}x'''(\varepsilon_{2k})\right], \\
x(t_{k+1}) &= x(t_k) + \\
&\quad \frac{h}{2}\left(x'(t_k) + f\left(t_{k+1}, x(t_k) + hf(t_k, x(t_k), y(t_k)), y(t_k) + hg(t_k, x(t_k), y(t_k))\right)\right) \\
&\quad + h^3\left[\frac{1}{6}x'''(\varepsilon_{1k}) - \frac{1}{4}x'''(\varepsilon_{2k}) + \frac{1}{4}f_2(\varepsilon_{3k}, x(\varepsilon_{3k}), y(\varepsilon_{3k}))x''(\varepsilon_{3k})\right].
\end{aligned}$$

It can be rewritten as:

$$x(t_{k+1}) = x(t_k) + \frac{h}{2}\left(K_0 + K_f\right) + e_f h^3,$$

where:

$K_0 = x'(t) = f(t_k, x(t_k), y(t_k))$, $K_1 = y'(t) = g(t_k, x(t_k), y(t_k))$, $K_f = f(t_{k+1}, x(t_k) + hK_0, y(t_k) + hK_1)$,
$e_f = \frac{1}{6}x'''(\varepsilon_{1k}) - \frac{1}{4}x'''(\varepsilon_{2k}) + \frac{1}{4}f_2(\varepsilon_{3k}, x(\varepsilon_{3k}), y(\varepsilon_{3k}))x''(\varepsilon_{3k})$, and $t_k < \varepsilon_{1k}, \varepsilon_{2k}, \varepsilon_{3k} < t_{k+1}$.

Similarly,

$$y(t_{k+1}) = y(t_k) + \frac{h}{2}\left(K_1 + K_g\right) + e_g h^3,$$

where:

$$K_g = g(t_{k+1}, x(t_k) + hK_0, y(t_k) + hK_1),\ e_g = \frac{1}{6}y'''(\xi_{1k}) - \frac{1}{4}y'''(\xi_{2k}) + \frac{1}{4}g_2(\xi_{3k}, x(\xi_{3k}), y(\xi_{3k}))y''(\xi_{3k}),$$

$$g_2(\varepsilon_{3k}, x(\varepsilon_{3k}), y(\varepsilon_{3k})) = \frac{\partial}{\partial x}g(\varepsilon_{3k}, x(\varepsilon_{3k}), y(\varepsilon_{3k})) + \frac{\partial}{\partial y}g(\varepsilon_{3k}, x(\varepsilon_{3k}), y(\varepsilon_{3k})),$$

and $t_k < \xi_{1k}, \xi_{2k}, \xi_{3k} < t_{k+1}$.

Appendix B. Algorithms

In this Appendix, the algorithms of the approximation methods based on FzT for Sections 3.2 and 3.3 are explained in detail. Pseudocode is used to describe the algorithms and a simplified code that is easy to read. This pseudocode specifies the form of the input to be supplied and the form of the desired output. As a consequence, a stopping technique independent of the numerical technique is incorporated into each algorithm to avoid infinite loops. Two punctuation symbols are used in the algorithms: a period (.) indicates the termination of a step, and a semicolon (;) separates tasks within a step. The integral symbol (integral(function,upper limits,lower limits)) is used to denote a definite integral. The steps in the algorithms follow the rules of structured program construction. They have been arranged so that there should be minimal difficulty translating pseudocode into any programming language suitable for scientific applications. We approximate the solution of SODEs (5) at $(N+1)$ equally-spaced numbers in the interval $[a, b]$ as follows.

Algorithm A1. One-stage (modified Euler) algorithm for the system of ODEs.

INPUT: $f(t,x,y)$ and $g(t,x,y)$ in Equation (5); endpoints a, b; integer N; initial condition y_1.
Step 1 Set $h = (b-a)/N$; $X_1 = x_1$; $Y_1 = y_1$; $t_1 = a$; $k = 1, \ldots, N+1$; $t_k = a + (k-1)h$.
Step 2 Define the generalized uniform fuzzy partitions as $A_k(t) = \frac{1}{2}\left(1 + \cos\left(\pi\left(\frac{t-t(k)}{h}\right)\right)\right)$.
Step 3 For $k = 1$ to N, do Steps 4–7.
  Step 4 $F(k) =$integral$(f(t, X(k), Y(k))A_k(t), t(k-1), t(k+1))/$integral$(A_k(t), t(k-1), t(k+1))$.
  Step 5 $G(k) =$integral$(g(t, X(k), Y(k))A_k(t), t(k-1), t(k+1))/$integral$(A_k(t), t(k-1), t(k+1))$.
  Step 6 $X(k+1) = X(k) + hF(k)$.
  Step 7 $Y(k+1) = Y(k) + hG(k)$.
end.
OUTPUT: Approximation X and Y to x and y, respectively, at the $(N+1)$ values of t.

Algorithm A2. Two-stage (modified trapezoidal rule) algorithm for the system of ODEs.

INPUT: $f(t,x,y)$; $g(t,x,y)$; endpoints a, b; integer N; initial condition y_1.
Step 1 Set $h = (b-a)/N$; $X_1 = x_1$; $Y_1 = y_1$; $t_1 = a$; $k = 1, \ldots, N+1$; $t_k = a + (k-1)h$.
Step 2 Define the generalized uniform fuzzy partitions as $A_k(t) = \frac{1}{2}\left(1 + \cos\left(\pi\left(\frac{t-t(k)}{h}\right)\right)\right)$.
Step 3 For $k = 1$ to N, do Steps 4–11.
  Step 04 $F(k) =$integral$(f(t, X(k), Y(k))A_k(t), t(k-1), t(k+1))/$integral$(A_k(t), t(k-1), t(k+1))$.
  Step 05 $G(k) =$integral$(g(t, X(k), Y(k))A_k(t), t(k-1), t(k+1))/$integral$(A_k(t), t(k-1), t(k+1))$.
  Step 06 $Xstar(k+1) = X(k) + hF(k)$.
  Step 07 $Ystar(k+1) = Y(k) + hG(k)$.
  Step 08 $Fstar(k+1) =$integral$(f(t, Xstar(k+1), Ystar(k+1))A_{k+1}(t), t(k), t(k+2))/$integral$(A_{k+1}(t), t(k), t(k+2))$.
  Step 09 $Gstar(k+1) =$integral$(g(t, Xstar(k+1), Ystar(k+1))A_{k+1}(t), t(k), t(k+2))/$integral$(A_{k+1}(t), t(k), t(k+2))$.
  Step 10 $X(k+1) = X(k) + h\,(F(k) + Fstar(k+1))\,/2$.
  Step 11 $Y(k+1) = Y(k) + h\,(G(k) + Gstar(k+1))\,/2$.
end.
OUTPUT: Approximation X and Y to x and y, respectively, at the $(N+1)$ values of t.

References

1. Ahmad, M.Z.; Hasan, M.K.; Baets, B.D. Analytical and numerical solutions of fuzzy differential equations. *Inf. Sci.* **2013**, *236*, 156–167. [CrossRef]
2. Shawagfeh, N.; Kaya, D. Comparing numerical methods for the solutions of systems of ordinary differential equations. *Appl. Math. Lett.* **2004**, *17*, 323–328. [CrossRef]
3. Atkinson, K.; Han, W.; Stewart, D. *Numerical Solution of Ordinary Differential Equations*; Wiley: New York, NY, USA, 2009.
4. Ahmad, M.Z.; De Baets, B. A Predator-Prey Model with Fuzzy Initial Populations. In Proceeding of the Joint 13th IFSA World Congress and 6th EUSFLAT Conference, Lisbon, Portugal, 20–24 July 2009; pp. 1311–1314.
5. Perfilieva, I. Fuzzy transforms: Theory and applications. *Fuzzy Sets Syst.* **2006**, *157*, 993–1023. [CrossRef]
6. Perfilieva, I. Fuzzy transform: Application to the Reef growth problem. In *Fuzzy Logic in Geology*; Demicco, R.V., Klir, G.J., Eds.; Academic Press: Amsterdam, The Netherlands, 2003; Chapter 9, pp. 275–300.
7. Perfilieva, I.; Daňková, M.; Bede, B. Towards a higher degree F-transform. *Fuzzy Sets Syst.* **2011**, *180*, 3–19. [CrossRef]
8. Chen, W.; Shen, Y. Approximate solution for a class of second-order ordinary differential equations by the fuzzy transform. *J. Intell. Fuzzy Syst.* **2014**, *27*, 73–82.
9. Alireza, K.; Zahra, A.; Irina, P. Fuzzy transform to approximate solution of two-point boundary value problems. *Math. Meth. Appl. Sci.* **2017**, *40*, 6147–6154.
10. Tomasiello, S. An alternative use of fuzzy transform with application to a class of delay differential equations. *Int. J. Comput. Math.* **2017**, *94*, 1719–1726. [CrossRef]
11. Hodakova, P.; Perfilieva, I.; Valasek, R. A new approach to fuzzy boundary value problem. In *Uncertainty Modelling in Knowledge Engineering and Decision Making*; World Scientific Proceedings Series on Computer Engineering and Information Science; World Scientific Publishing Company: Singapore, 2016; Volume 10, pp. 276–281; ISBN: 978-981-3146-96-9.
12. Perfilieva, I.; Števuliáková, P.; Valášek, R. F-transform for numerical solution of two-point boundary value problem. *Iran. J. Fuzzy Syst.* **2017**, *14*, 1–13.
13. Perfilieva, I.; Števuliáková, P.; Valášek, R. F-transform-based shooting method for nonlinear boundary value problems. *Soft Comput.* **2017**, *21*, 3493–3502. [CrossRef]
14. Alijani, Z.; Khastan, A.; Khattri, S.K.; Tomasiello, S. Fuzzy Transform to Approximate Solution of Boundary Value Problems via Optimal Coefficients. In Proceedings of the 2017 International Conference on High Performance Computing Simulation (HPCS), Genoa, Italy, 17–21 July 2017; pp. 466–471.
15. Alikhani, R.; Zeinali, M.; Bahrami, F.; Shahmorad, S.; Perfilieva, I. Trigonometric F^m-transform and its approximative properties. *Soft Comput.* **2017**, *21*, 3567–3577. [CrossRef]
16. Jahedi, S.; Javadi, F.; Mehdipour, M.J. Weighted transform and approximation of some functions on unbounded sets. *Soft Comput.* **2017**, *21*, 3579–3585. [CrossRef]
17. Tomasiello, S.; Gaeta, M.; Loia, V. Quasi–consensus in Second–Order Multi–agent Systems with Sampled Data Through Fuzzy Transform. *J. Uncertain Syst.* **2016**, *10*, 243–250.
18. Tomasiello, S. A First Investigation on the Dynamics of Two Delayed Neurons through Fuzzy Transform Approximation. In Proceedings of the 2017 International Conference on High Performance Computing Simulation (HPCS), Genoa, Italy, 17–21 July 2017; pp. 460–465.
19. Alkasasbeh, H.A.; Perfilieva, I.; Ahmad, M.Z.; Yahya, Z.R. New fuzzy numerical methods for solving Cauchy problems. *Appl. Syst. Innov.* **2018**, *1*, 15. [CrossRef]
20. Parapari, H.F.; Menhaj, M.B. Solving nonlinear ordinary differential equations using neural networks. In Proceedings of the 2016 4th International Conference on Control, Instrumentation, and Automation (ICCIA), Qazvin, Iran, 27–28 January 2016; pp. 351–355.
21. Ramos, H.; Singh, G.; Kanwar, V.; Bhatia, S. An embedded 3 (2) pair of nonlinear methods for solving first order initial-value ordinary differential systems. *Numer. Algorithms* **2017**, *75*, 509–529. [CrossRef]
22. Perez, J.F.S.; Conesa, M.; Alhama, I. Solving ordinary differential equations by electrical analogy: A multidisciplinary teaching tool. *Eur. J. Phys.* **2016**, *37*, 065703. [CrossRef]
23. Al-Omari, A.; Arnold, J.; Taha, T.; Schüttler, H.B. Solving Large Nonlinear Systems of First-Order Ordinary Differential Equations with Hierarchical Structure Using Multi-GPGPUs and an Adaptive Runge Kutta ODE Solver. *IEEE Access* **2013**, *1*, 770–777. [CrossRef]

24. Opanuga, A.; Edeki, S.; Okagbue, H.; Akinlabi, G.; Osheku, A.; Ajayi, B. On numerical solutions of systems of ordinary differential equations by numerical-analytical method. *Appl. Math. Sci.* **2014**, *8*, 8199–8207. [CrossRef]
25. Al-Omari, A.; Schuttler, H.B.; Arnold, J.; Taha, T. Solving Nonlinear Systems of First Order Ordinary Differential Equations Using a Galerkin Finite Element Method. *IEEE Access* **2013**, *1*, 408–417. [CrossRef]
26. Matveev, S.A.; Smirnov, A.P.; Tyrtyshnikov, E. A fast numerical method for the Cauchy problem for the Smoluchowski equation. *J. Comput. Phys.* **2015**, *282*, 23–32. [CrossRef]
27. Mondal, S.P.; Roy, T.K. First order homogeneous ordinary differential equation with initial value as triangular intuitionistic fuzzy number. *J. Uncertain. Math. Sci.* **2014**, *2014*, 1–17. [CrossRef]
28. Mondal, S.P.; Roy, T.K. System of Differential Equation with Initial Value as Triangular Intuitionistic Fuzzy Number and its Application. *Int. J. Appl. Comput. Math.* **2015**, *1*, 449–474. [CrossRef]
29. Paul, S.; Mondal, S.P.; Bhattacharya, P. Numerical solution of Lotka Volterra prey predator model by using Runge-Kutta-Fehlberg method and Laplace Adomian decomposition method. *Alex. Eng. J.* **2016**, *55*, 613–617. [CrossRef]
30. Yusufoğlu, E.; Erbaş, B. He's variational iteration method applied to the solution of the prey and predator problem with variable coefficients. *Phys. Lett. A* **2008**, *372*, 3829–3835. [CrossRef]
31. Li, J.; Zhao, A. Stability analysis of a non-autonomous Lotka-Volterra competition model with seasonal succession. *Appl. Math. Modell.* **2016**, *40*, 763–781. [CrossRef]
32. Khastan, A.; Perfilieva, I.; Alijani, Z. A new fuzzy approximation method to Cauchy problems by fuzzy transform. *Fuzzy Sets Syst.* **2016**, *288*, 75–95. [CrossRef]
33. Bougoffa, L. Solvability of the predator and prey system with variable coefficients and comparison of the results with modified decomposition. *Appl. Math. Comput.* **2006**, *182*, 383 –387. [CrossRef]
34. González-Parra, G.C.; Arenas, A.J.; Cogollo, M.R. Numerical-analytical solutions of predator-prey models. *WSEAS Trans. Biol. Biomed.* **2013**, *10*, 79–87.
35. Butcher, J.C. *Numerical Methods for Ordinary Differential Equations*, 3rd ed.; John Wiley & Sons, Ltd.: Chichester, UK, 2016.
36. Burden, R.L.; Faires, J.D. *Numerical Analysis*, 9th ed.; Brooks/Cole Cengage Learning: Boston, MA, USA, 2010; ISBN: 978-0-538-73351-9.
37. Alkasasbeh, H.A.; Perfilieva, I.; Ahmad, M.Z.; Yahya, Z.R. New approximation methods based on fuzzy transform for solving SODEs: II. *Appl. Syst. Innov.* **2018**, *1*, 30.

applied system innovation

MDPI

Article

New Approximation Methods Based on Fuzzy Transform for Solving SODEs: II

Hussein ALKasasbeh [1,*], Irina Perfilieva [2] and Muhammad Zaini Ahmad [1] and Zainor Ridzuan Yahya [1]

1 Institute of Engineering Mathematics, Universiti Malaysia Perlis, Kampus Tetap Pauh Putra, Arau, Perlis 02600, Malaysia; mzaini@unimap.edu.my (M.Z.A.); zainoryahya@unimap.edu.my (Z.R.Y.)

2 Institute for Research and Applications of Fuzzy Modelling, University of Ostrava, NSC IT4Innovations, 30. Dubna 22, 701 03 Ostrava, Czech Republic; Irina.Perfilieva@osu.cz

* Correspondence: hussein.ahmad.alka@gmail.com

Received: 17 June 2018; Accepted: 14 August 2018; Published: 23 August 2018

Abstract: In this research, three approximation methods are used in the new generalized uniform fuzzy partition to solve the system of differential equations (SODEs) based on fuzzy transform (FzT). New representations of basic functions are proposed based on the new types of a uniform fuzzy partition and a subnormal generating function. The main properties of a new uniform fuzzy partition are examined. Further, the simpler form of the fuzzy transform is given alongside some of its fundamental results. New theorems and lemmas are proved. In accordance with the three conventional numerical methods: Trapezoidal rule (one step) and Adams Moulton method (two and three step modifications), new iterative methods (NIM) based on the fuzzy transform are proposed. These new fuzzy approximation methods yield more accurate results in comparison with the above-mentioned conventional methods.

Keywords: fuzzy partition; fuzzy transform; numerical methods; NIM; systems of ordinary differential equations

1. Introduction

Differential equation is particularly useful for different areas of applied sciences and engineering. Many differential equations have no closed form solutions. Thus, many researchers are developing approximation methods for solving differential equations, for example [1–3]. In this paper, we continue the study of approximation methods based on FzT to solutions of differential equations.

The core idea of FzT is a fuzzy partition of a universe into fuzzy subsets. The first fuzzy partition of FzT with the Ruspini condition was introduced by [4] and was extensively investigated by [5]. This condition implies normality of the fuzzy partition. In addition, the fuzzy partition with the generalized Ruspini condition (fuzzy r-partition) was introduced by [6]. This fuzzy partition was achieved by replacing the partition of unity by fuzzy r-partition. This type of partition was used by [6,7] for smoothing or filtering data based on the inverse FzT. Further, a generalized fuzzy partition appeared in connection with the notion of the FzT, where FzT components are polynomials of degree m [8]. By [9], different types of fuzzy partitions are taken into consideration such as B-splines, Shepard kernels, Bernstein basis polynomials and Favard-Szasz-Mirakjan operators. Later, the higher degree FzT based on B-splines was proposed [10] to improve the quality of the function approximation of two variables.

A generalized fuzzy partition was implicitly introduced by [11] with the purpose of meeting the requirements of image compression. In addition, a generalized fuzzy partition can also be considered in connection with radial membership functions [12]. Further, necessary and sufficient conditions for modeling the generalized fuzzy partition was provided by [13]. Recently, a new representation formula for basic functions of FzT and a new fuzzy numerical method based on block pulse functions for numerical solution of integral equations were presented by [14]. The approximation method based on the FzT with Shepard-type basic functions for linear Fredholm integral equations was discussed by [15]. New representations of the generalized uniform fuzzy partitions with the normal case to obtain better approximation solutions for solving Cauchy problems were presented by [16].

FzT is a soft computing method developed by Perfilieva [5] that has many applications, for example, in differential and integral equations. FzT for solving ordinary Cauchy problems with one variable was initiated by [4]. The generalization of the Euler method has been discussed by [17] for solving ordinary Cauchy problems. The author has applied this technique to reef growth and sea level variations models. Further, FzT has been generalized from the case of constant components to the case of polynomial components by [8]. Later, the first and second degree FzT based mid-point rule for solving the Cauchy problem and the uncertain initial value problem have been proposed by [18]. Furthermore, an algorithm to obtain the approximate solutions of second order initial value problems was constructed by [19]. From this idea, FzT for numerical solutions of two point boundary value problems was proposed by [20].

FzT of two variables based on finite differences method was used by [21] for solving a type of partial differential equations with Dirichlet boundary conditions and initial conditions. In addition, the first degree FzT of two variables was introduced by [22]. By [23], the partial derivatives using the first FzT were approximated and modification of the Canny edge detector was proposed. Furthermore, the uniform stability result for the vibrations of a telegraph equation using FzT of two variables was proposed by [24]. The composition of inverse and direct discrete FzT method was extended to numerical solution of Fredholm integral equations and Volterra Fredholm integral equations [25]. The general form of the higher order FzT was constructed by [26] for solving differential and integral equations using any arbitrary basis functions. The FzT has investigated for solving the Volterra population growth model using the approximation for the Caputo derivative [27]. A new numerical method based FzT was demonstrated to solve a class of delay differential equations by means of the Picard-like numerical scheme [28]. FzT was considered to approximate the solution of boundary value problems by minimizing the integral squared error in 2-norm [29]. In [30], the dynamical properties of a two neuron system with respect to FzT and a single delay have been investigated. The conditions under which quasi-consensus in a multi-agent system with sampled data based on FzT were proposed by [31].

NIM was proposed to solve nonlinear functional equations and the existence of solution for nonlinear Volterra integral equations [2]. At the same time, NIM was introduced for solving nonlinear equations by using a different decomposition technique [32]. From this conception, NIM was considered in terms up to fourth-order in Taylor series for solving nonlinear equations [33]. Sufficiency conditions have been presented for convergence of the NIM [34]. A new predictor-corrector approach was developed based on NIM for fractional differential equations [35]. Classical methods are modified by [3] to derive numerous formulas for solving the differential equations.

The motivation of the proposed study comes from [16,36,37]. In [16], new fuzzy numerical methods to solve the Cauchy problem was considered and the authors showed that the error can be reduced by FzT and NIM with respect to new generalized uniform fuzzy partitions, namely power of the triangular and raised cosine generalized uniform fuzzy partitions, where generating functions are normal (see also [37] for another approach). In addition, two basic approximation methods, modified Euler method and Trapezoidal rule, with help from FzT for solving SODEs are analyzed in detail by [36]. For this purpose, more generally, new generalized uniform fuzzy partitions are proposed in this study, where a generating function is not normal.

The membership functions in underlying fuzzy partitions are often called basic functions. There has been a growing interest in investigating the properties of fuzzy partitions. However, the problem arises on how one can effectively construct the basic function of fuzzy partitions. In this paper, new representations of basic functions are proposed. This is achieved by introducing new generalized uniform fuzzy partitions, where a generating function is not normal. Further, new fuzzy numerical methods based on NIM and FzT for solving SODEs are introduced and discussed. In particular, we consider functions of two variables with initial conditions. In accordance with the existing methods, Trapezoidal rule and Adams Moulton are improved using FzT and NIM. The methods are combined with one-step, two-step and three-step. As an application, all these methods are used to solve a general model of the dynamical system, i.e., Lotka–Volterra equation with derivatives and with variable coefficients. Furthermore, numerical examples are presented. It is observed that the new fuzzy numerical methods yield more accurate results than classical Trapezoidal rule and classical Adams Moulton methods (2 and 3-step).

The paper is organized as follows. The main part of the paper is Sections 3 and 4, which provides new representations for basic functions of FzT, followed by the modified one step, 2-step and 3-step based on NIM and FzT method with respect to new representations formulas for generalized uniform fuzzy partition of FzT. In Section 5, numerical examples are discussed. Finally, conclusions are given in Section 6.

2. Basic Concepts

In this section, we give some definitions and introduce the necessary notation following [38], which will be used throughout the paper. Throughout this section, we deal with an interval $[a,b] \subset \mathbb{R}$ of real numbers.

Definition 1. *(generalized uniform fuzzy partition) Let $t_i \in [a,b]$, $i = 1,\ldots,n$, be fixed nodes such that $a = t_1 < \ldots < t_n = b$, $t_0 = t_1$, $t_n = t_{n+1}$, $n \geq 2$ and $[t_i - h,\ t_i + h] \subseteq [a,b]$. We say that the fuzzy sets $A_i : [a,b] \to [0,1]$ constitute a generalized fuzzy partition of $[a,b]$ if the following conditions are fulfilled:*

1. *(positivity and locality)—$A_i(t) > 0$ if $t \in (t_{i-1}, t_{i+1})$ and $A_i(t) = 0$ if $t \in [a,b] \setminus (t_{i-1}, t_{i+1})$;*
2. *(continuity)—A_i is continuous on $[t_{i-1}, t_{i+1}]$;*
3. *(covering)—for $t \in [a,b]$, $\sum_{i=1}^{n} A_i(t) > 0$.*

Fuzzy sets $A_1,\ldots, A_n$ are called basic functions. It is important to remark that by conditions of locality and continuity, $\int_a^b A_i(t)dt > 0$. A generalized uniform fuzzy partition of $[a,b]$ is defined for equidistant nodes, i.e., for all $i = 1,\ldots,n-1$, $t_i = t_{i+1} + h$, where $h = (b-a)/(n-1)$ and two additional properties are satisfied,

4. *$A_i(t_i - t) = A_i(t_i + t)$ for all $t \in [0,h]$, $i = 2,\ldots,n-1$;*
5. *$A_i(t) = A_{i-1}(t-h)$ and $A_{i+1}(t) = A_i(t-h)$ for all $t \in [t_i,\ t_{i+1}]$, $i = 2,\ldots,n-1$;*

then the fuzzy partition is called h-uniform generalized fuzzy partition.

Definition 2. *(generating function) A function $K : [-1,1] \to [0,1]$ is called a generating function if it is assumed to be even, continuous and $K(t) > 0$ if $t \in (-1,1)$. The function $K : [-1,1] \to \mathbb{R}$ is even if for all $t \in [0,1]$, $K(-t) = K(t)$.*

The following definition recalls the concept of the generalized fuzzy partition which can be easily extended to the interval $[a,b]$. We assume that $[a,b]$ is partitioned by $A_1,\ldots, A_n$, according to Definition 1.

Definition 3. *A h-uniform generalized fuzzy partition of interval $[a,b]$, determined by the triplet $(K,\ h,\ a)$, can be defined using generating function K (Definition 2). Then, basic functions of a h-uniform generalized fuzzy partition are shifted copies of K defined by*

$$A_i(t) = K\left(\frac{t-t_i}{h}\right),\quad t \in [t_i - h,\ t_i + h],$$

for all $i = 1, \dots, n$*. The parameter* h *is called the bandwidth or the shift of the fuzzy partition and the nodes* $t_i = a + ih$ *are called the central point of the fuzzy sets* $A_1, \dots, A_n$.

Remark 1. *A* h*-uniform fuzzy partition is called Ruspini if the following condition*

$$A_i(t) + A_{i+1}(t) = 1,\ i = 1, \dots, n-1, \tag{1}$$

holds for any $t \in [t_i, t_{i+1}]$*. This condition is often called Ruspini condition.*

New Iterative Method

NIM have proposed by [2] for solving linear and nonlinear functional equations of the form

$$u = f_1 + N(u), \tag{2}$$

where f_1 is a known function and N a non linear operator. Solutions obtained by this method are in the form of rapidly converging infinite series which can be effectively approximated by calculating only the first few terms. In this method non linear operator N is decomposed as $N(u) = N(u_0) + \sum_{i=1}^{\infty} \left\{ N\left(\sum_{n=0}^{i} u_n\right) - N\left(\sum_{n=0}^{i-1} u_n\right) \right\}$. In [2], the authors were defined the recurrence relation:

$$\begin{cases} u_0 = f_1, \\ u_1 = N(u_0), \\ u_{m+1} = N(u_0 + \cdots + u_m) - N(u_0 + \cdots + u_{m-1}), \quad m = 1, 2, \dots \end{cases} \tag{3}$$

Then$(u_1 + \cdots + u_{m+1}) = N(u_0 + \cdots + u_m)$, $m = 1, 2, 3, \dots$, and $u = f_1 + \sum_{i=1}^{\infty} u_i = f_1 + N(u_0) + [N(u_0 + u_1) - N(u_0)] + \cdots = f_1 + N(u)$. Hence u satisfies the functional (2).

3. New Representations for Basic Functions of FzT

Let us recall the basic facts of an FzT of a continuous real function f as presented by [5,17]. The first step in the definition of the FzT of f involves the selection of a fuzzy partition of the domain $[a, b]$ by a finite number $n \geq 2$ of fuzzy sets $B_k(t)$, $k = 1, \dots, n$. In those papers, five axioms specified $B_k(t)$, $k = 1, \dots, n$, in the fuzzy partition: normality, locality, continuity, unimodality (monotonicity) and orthogonality (Ruspini condition). A fuzzy partition is called uniform if the fuzzy sets $B_k(t)$, $k = 2, \dots, n-1$, are shifted copies of symmetrized B_1 (more details can be found in [17]). The membership functions $B_k(t)$, $k = 1, \dots, n$, in a fuzzy partition are called basic functions. Later, a generalized fuzzy partition appeared in connection with the notion of a higher-degree FzT [8]. Furthermore, summarize both these notions in [38]. Three axioms specify $B_k(t)$, $k = 1, \dots, n$, in the fuzzy partition: positivity and locality, continuity and covering. Recently, the different conditions for generalized uniform fuzzy partitions was proposed [13,38] while another approach was demonstrated by [37] where a function can be reconstructed from its F-transform components. In the following, we modify the definition h-uniform generalized fuzzy partition.

3.1. Generalized Uniform Fuzzy Partitions with the Generalized Normal Case

Let us recall the h-uniform generalized fuzzy partition of real line can be defined using generating function K. Then, basic functions of the h-uniform generalized fuzzy partition are shifted copies of K. On the basis of Definition 1 can be also defined using a generating function $\lambda\beta K(t)$ where $\beta = 1/K(0)$, $K(0) \neq 0$, $\beta > 0$ and $\lambda > 0$ (in general, not necessarily satisfying normal and Ruspini condition) which is that $K(t)$ assumed to be even, continuous and $K(t) > 0$ if $t \in (-1, 1)$. Therefore, we will modify the basic functions of the h-uniform generalized fuzzy partition so that they are shifted copies of $\lambda\beta K$ defined by

$$A_k(t,t_0) = \lambda\beta K\left(\frac{t-t_0}{h} - k\right),\ t \in [t_k - h, t_k + h],\ k \in \mathbb{Z}. \tag{4}$$

The parameter h is bandwidth of the fuzzy partition and $t_0 + kh = t_k$. The concept of the h-uniform generalized fuzzy partition can be easily extended to the interval $[a,b]$ as follows.

Definition 4. *Let $t_1 < \ldots < t_n$ be fixed nodes within $[a,b] \subset \mathbb{R}$, such that $t_1 = a$, $t_n = b$ and $n \geq 2$. We consider nodes $t_1, \ldots, t_n$ are equidistant, with distance (shift) $h = (b-a)\,/\,(n-1)$. A system of fuzzy sets $B_1, \ldots, B_n$ $[a,b] \to [0,1]$ be a generalized uniform fuzzy partitions of $[a,b]$ if it is defined by*

$$B_k(t) \quad = \quad \begin{cases} A_k(t,a), & t \in [a,b], \\ 0, & otherwise. \end{cases} \quad = \quad \begin{cases} \lambda\beta K\left(\frac{t-t_k}{h}\right), & t \in [a,b], \\ 0, & otherwise. \end{cases} \tag{5}$$

where $t_k = a + (k-1)h$, $\beta = 1/K(0)$, $K(0) \neq 0$, $\beta > 0$ and $\lambda > 0$. In the sequel, a generating function denote by K and basic functions of FzT denote by B_k, $k = 1, \ldots, n$.

Lemma 1. *If basic functions B_k, $k = 1, \ldots, n$, of a h-uniform generalized fuzzy partition are shifted copies of $\lambda\beta K$ defined by (5). Then, for each $k = 1, \ldots, n$, $B_k(t_k) = \lambda$, $t_k \in [t_k - h, t_k + h]$.*

Proof. By (5), we get $B_k(t_k) = \lambda\beta K\left(\frac{t_k - t_k}{h}\right) = \lambda$. □

3.2. Simpler Form of F-Transform Components Based on Generalized Uniform Fuzzy Partitions with the Generalized Normal Case

In this subsection, we present the main principles of FzT with respect to new representations of h-uniform generalized fuzzy partition. Further, we will show that FzT components with respect to new representations of h-uniform generalized fuzzy partition can be simplified and approximated of an original function, say f.

Definition 5. *Let f be a continuous function on $[a,b]$ and $B_k(t)$, $k = 1, \ldots, n$, be h-uniform generalized fuzzy partition of $[a,b]$, $n \geq 2$. A vector of real numbers $F[f] = (F_1, F_2, \ldots, F_n)$ given by*

$$F_k = \frac{\int_a^b f(t)\, B_k(t)\, dt}{\int_a^b B_k(t)\, dt}, \tag{6}$$

for $k = 1, \ldots, n$ is called the direct FzT of f with respect to B_k.

In the following, we will simplify the representation (6).

Lemma 2. *Let $f \in C([a,b])$ and according to Definition 4, fuzzy sets B_k, $k = 1, \ldots, n$, $n \geq 2$, be a h-uniform generalized fuzzy partition of $[a,b]$ with a generating function K, then representation (6) of direct FzT can be simplified for $k = 1, \ldots, n$ as follows*

$$F_k = \frac{\int_{-1}^{1} f(th + t_k)\, K(t)\, dt}{\int_{-1}^{1} K(t)\, dt} = \frac{\int_{-h}^{h} f(t + t_k)\, K(\frac{t}{h})\, dt}{\int_{-h}^{h} K(\frac{t}{h})\, dt}. \tag{7}$$

Proof. By Definition 4, we get

$$B_k(t) = \lambda\beta K\left(\frac{t - t_k}{h}\right),\quad t \in [t_k - h, t_k + h],$$

for $k = 1, \ldots, n$, $t_0 = t_1$, $t_{n+1} = t_n$, and substituting $u = \frac{t-t_k}{h}$ and then substituting $t = s/h$. Thus, we get

$$\int_{t_{k-1}}^{t_{k+1}} f(t)\, B_k(t)\, dt = \lambda\beta h \int_{-1}^{1} f(th + t_k)\, K(t)\, dt = \lambda\beta \int_{-h}^{h} f(t + t_k)\, K(\frac{t}{h})\, dt,$$

$$\int_{t_{k-1}}^{t_{k+1}} B_k(t)\, dt = \lambda\beta h \int_{-1}^{1} K(t)\, dt = \lambda\beta \int_{-h}^{h} K(\frac{t}{h})\, dt,$$

and its corresponding results with representation (6). □

If $\lambda > 0$, the Lemma 1 still hold by choosing suitable constant λ, satisfying $\lambda = 1/\left(\int_{-1}^{1} \beta K(t)dt\right)$, where $\int_{-1}^{1} \beta K(t)dt > 0$. So, we will restrict ourselves to h-uniform generalized fuzzy partition with $0 < \lambda = 1/\left(\int_{-1}^{1} \beta K(t)dt\right)$, where $\int_{-1}^{1} \beta K(t)dt \neq 0$. In the following, we will simplify the above given expressions for the coefficients $F[f] = (F_1, F_2, \ldots, F_n)$ in the representation (6). This fact is very important for applications which are more flexible and consequently easier to use.

Corollary 1. *Let the assumptions of Lemma 2 be fulfilled and* $0 < \lambda = 1/\left(\int_{-1}^{1} \beta K(t)dt\right)$, *where* $\int_{-1}^{1} \beta K(t)dt \neq 0$. *Then, the coefficients* $F[f] = (F_1, F_2, \ldots, F_n)$ *in the expression (6) of the FzT component* F_k *of* f *as follows:*

$$F_k = \frac{1}{h} \int_{a}^{b} f(t)\, B_k(t)\, dt = \frac{\lambda\beta}{h} \int_{a}^{b} f(t)\, K\left(\frac{t - t_k}{h}\right) dt, \tag{8}$$

for $k = 1, \ldots, n$, *where interval* $[a, b]$ *is partitioned by the h-uniform generalized fuzzy partition* $B_1, \ldots, B_n$.

Proof. Let $k \in \{1, \ldots, n\}$ and consider set of fuzzy sets $B_k(t)$ be the h-uniform generalized fuzzy partition of $[a, b]$ defined by (5). Using the proof of Lemma 2, we get

$$\int_{t_{k-1}}^{t_{k+1}} B_k(t)\, dt = \int_{t_{k-1}}^{t_{k+1}} A_k(t, a), dt = \int_{t_k - h}^{t_k + h} \lambda\beta K\left(\frac{t - t_k}{h}\right) dt = h\lambda \int_{-1}^{1} \beta K(t)\, dt = h, \tag{9}$$

where $0 < \lambda = 1/\left(\int_{-1}^{1} \beta K(t)dt\right)$, $\int_{-1}^{1} \beta K(t)dt \neq 0$, h is bandwidth of the fuzzy partition and $t_k = a + (k-1)h$ and then its corresponding in the expression (6). □

Lemma 3. *Let* $f \in C[a, b]$. *Then for any* $\varepsilon > 0$ *there exist* $n_\varepsilon \in \mathbb{N}$ *and* $B_1, \ldots, B_{n_\varepsilon}$ *be basic functions form the h-uniform generalized fuzzy partition of* $[a, b]$. *Let* F_k, $k = 1 \ldots, n$, *be the integral FzT components of* f *with respect to* $B_1, \ldots, B_{n_\varepsilon}$. *Then for each* $k = 1 \ldots, n_\varepsilon - 1$ *the following estimations hold:* $|f(t) - F_i| \leq \varepsilon$ *for each* $t \in [a, b] \cap [t_k, t_{k+1}]$ *and* $i = k, k+1$.

Proof. see [5]. □

Corollary 2. *Let the conditions of Lemma 3 be fulfilled. Then for each* $k = 1 \ldots, n_\varepsilon - 1$ *the following estimations hold:* $|F_k - F_{k+1}| < \varepsilon$.

Proof. According to [5,39], let $t \in [a, b] \cap [t_k, t_{k+1}]$. Then by Lemma 3, for any $k = 1, \ldots, n-1$ we obtain that $|f(t) - F_k| < \varepsilon/2$ and $|f(t) - F_{k+1}| < \varepsilon/2$. Thus,

$$|F_k - F_{k+1}| \leq |f(t) - F_k| + |f(t) - F_{k+1}| < \frac{\varepsilon}{2} + \frac{\varepsilon}{2} = \varepsilon. \quad \square$$

The following theorem estimates the difference between the original function and its direct FzT with respect to the h-uniform generalized fuzzy partition.

Theorem 1. *Let* $f(t) \in C^2[a, b]$ *and the conditions of Lemma 2 be fulfilled. Then for* $k = 1, \ldots, n$

$$F_k = \lambda f(t_k) + \mathcal{O}(h^2), \tag{10}$$

where $0 < \lambda = 1/\left(\int_{-1}^{1} \beta K(t)dt\right)$ *and* $\int_{-1}^{1} \beta K(t)dt \neq 0$.

Proof. By locality condition of definition of h-uniform generalized fuzzy partition, Corollary 1, Lemma 1, and according to [17], using the trapezoid formula with nodes t_{k-1}, t_k, t_{k+1} to the numerical computation of the integral, we get for $k = 1, \ldots, n$ and $0 < \lambda = 1/\left(\int_{-1}^{1} \beta K(t)dt\right)$

$$\begin{aligned} F_k &= \frac{1}{h}\int_{t_{k-1}}^{t_{k+1}} f(t)\, B_k(t)\, dt, \\ &= \frac{1}{h}\frac{h}{2}\left(f(t_{k-1})\, B_k(t_{k-1}) + 2f(t_k)\, B_k(t_k) + f(t_{k+1})\, B_k(t_{k+1})\right) + \mathcal{O}\left(h^2\right), \\ &= f(t_k)\, B_k(t_k) + \mathcal{O}\left(h^2\right) = \lambda f(t_k) + \mathcal{O}\left(h^2\right). \quad \square \end{aligned} \tag{11}$$

Corollary 3. *Let* $f(t) \in C^2[a,b]$ *and the conditions of Lemma 2 be fulfilled. Let moreover,* f *be Lipschitz continuous with respect to* t*, i.e., there exists a constant* $L \in \mathbb{R}$*, such that for all* $t \in [a,\, b]$ *and* $t, t' \in \mathbb{R}$*,*

$$|f(t) - f(t')| \leq L|t - t'|. \tag{12}$$

Then for $k = 1, \ldots, n$

$$\left|f(t) - \frac{1}{\lambda}F_k\right| \leq Lh + \frac{h^2}{6\lambda}M,$$

where $0 < \lambda = 1/\left(\int_{-1}^{1} \beta K(t)dt\right)$, $\int_{-1}^{1} \beta K(t)dt \neq 0$, $M = \max\limits_{t\in[t_{k-1}, t_{k+1}]} |f''(t)|$ *and* $|t - t_k| < h$ *whenever* $t \in [t_{k-1},\, t_{k+1}]$.

Proof. By the assumption f has continuous second order derivatives on $[a,b]$ and is Lipschitz continuous with respect to t. Therefore, using the trapezoid rule and let us choose a value of k in the range $1 \leq k \leq n$ and $t \in [t_{k-1},\, t_{k+1}]$, we get for $0 < \lambda = 1/\left(\int_{-1}^{1} \beta K(t)dt\right)$

$$\begin{aligned} \left|f(t) - \frac{1}{\lambda}F_k\right| &= \left|f(t) - \frac{1}{h\lambda}\int_{t_{k-1}}^{t_{k+1}} f(t)\, B_k(t)\, dt\right|, \\ &= \left|f(t) - \frac{1}{h\lambda}\left[h\lambda f(t_k) - \frac{h^3}{12}\left(f''(\xi_{k-1}) + f''(\xi_{k+1})\right)\right]\right|, \\ &\leq |f(t) - f(t_k)| + \frac{h^2}{12\lambda}2M, \\ &\leq L\,|t - t_k| + \frac{h^2}{6\lambda}M \ \leq Lh + \frac{h^2}{6\lambda}M, \end{aligned} \tag{13}$$

where $\xi_{k-1} \in (t_{k-1}, t_k)$, $\xi_{k+1} \in (t_k, t_{k+1})$ and $M = \max\limits_{t\in[t_{k-1}, t_{k+1}]} |f''(t)|$. $\square$

Remark 2. *In view of (13), if* $0 < \lambda \leq 1$*. Then,* $\left|f(t) - \frac{1}{\lambda}F_k\right| \leq Lh + \frac{h^2}{6}M$.

Definition 6. *Let* $F[f] = (F_1,\, F_2, \ldots,\, F_n)$ *be direct FzT of a function* $f \in C[a,b]$ *with respect to the fuzzy partition* $B_k(t)$, $k = 1, \ldots, n$ *of* $[a,b]$*. Then, the function* $\hat{f}$ *defined on* $[a,b]$

$$\hat{f}(t) = \frac{\sum_{k=1}^{n} F_k B_k(t)}{\sum_{k=1}^{n} B_k(t)}, \tag{14}$$

is called the inverse FzT of f.

The following lemma estimates the difference between the original function and its inverse FzT.

Lemma 4. *Let the assumptions of Theorem 1 and let $\hat{f}(t)$ be the inverse FzT of f with respect to the fuzzy partition of $[a,b]$ is given by Definition 4. Then, the following estimation holds for $t \in [a,b]$ and $k = 1, \ldots, n$*

$$\hat{f}(t) = \lambda f(t_k) + \mathcal{O}\left(h^2\right), \tag{15}$$

where $0 < \lambda = 1/\left(\int_{-1}^{1} \beta K(t)dt\right)$ *and* $\int_{-1}^{1} \beta K(t)dt \neq 0$.

Proof. Let $t \in [a,b]$ so that $t \in [t_k, t_{k+1}]$ for some $k = 1, \ldots, n$. By Theorem 1,

$$\begin{aligned}\hat{f}(t) - \lambda f(t_k) &= \frac{\sum_{k=1}^{n} F_k B_k(t)}{\sum_{k=1}^{n} B_k(t)} - \lambda f(t_k) = \frac{\sum_{k=1}^{n} F_k B_k(t)}{\sum_{k=1}^{n} B_k(t)} - \frac{\sum_{k=1}^{n} \lambda f(t_k) B_k(t)}{\sum_{k=1}^{n} B_k(t)}, \\ &= \frac{\sum_{k=1}^{n} (F_k - \lambda f(t_k)) B_k(t)}{\sum_{k=1}^{n} B_k(t)} = \mathcal{O}\left(h^2\right). \quad \square\end{aligned}$$

Theorem 2. *Let $f \in C[a,b]$. Thus, for any $\varepsilon > 0$ there exist $n_\varepsilon \in \mathbb{N}$ and $B_1, \ldots, B_{n_\varepsilon}$ be the h-uniform generalized fuzzy partition of $[a,b]$ defined by (5). Then, the following estimations hold $\left|\hat{f}(t) - f(t)\right| < \varepsilon$ for each $t \in [a,b] \cap [t_k, t_{k+1}]$.*

Proof. From the proof of Lemma 4 and then using Lemma (3) in the sense that for all $k = 1, \ldots, n$,

$$\left|\hat{f}(t) - f(t)\right| = \frac{\sum_{k=1}^{n} |F_k - f(t)| B_k(t)}{\sum_{k=1}^{n} B_k(t)} < \varepsilon. \quad \square$$

Remark 3. *According to Definition (4), it is easy to see that the inverse FzT $\hat{f}(t_k) = F_k$ for all $k = 1, \ldots, n$.*

On the basis of Definition 4, necessary steps of a new method to construct generalized uniform fuzzy partitions of $[-1,1]$ for solve case K is not normal in the following.

1. Select the generating function K which is assumed to be even, continuous and $K(t) > 0$ if $t \in (-1,1)$.
2. Specify the value $\beta = 1/K(0)$, where $K(0) \neq 0$ to get the normal generating function K and then compute the value $\lambda = 1/\left(\int_{-1}^{1} \beta K(t)dt\right)$, where $\int_{-1}^{1} \beta K(t)dt \neq 0$.
3. If conditions $\beta > 0$ and $\lambda > 0$ holds, then construct generalized uniform fuzzy partitions of $[-1,1]$ by $\lambda\beta K(t)$.

Example 1. *Let $K : \mathbb{R} \to [0,1]$ be defined by*

$$K(t) = (1 + \cos(\pi t))^m.$$

One can see in Table 1 the h-uniform generalized fuzzy partition of $[a,b]$ determined by Definition 4.

Table 1. Example 1.

$K_{C_2^m}(t)$	β	λ	$B_k = \lambda\beta K\left(\frac{t-t_k}{h}\right)$
$(1+\cos(\pi t))^m$	$\frac{1}{2^m}$	$\frac{\sqrt{\pi}\Gamma(m+1)}{2\Gamma\left(m+\frac{1}{2}\right)}$	$\left(\frac{\sqrt{\pi}\Gamma(m+1)}{2\Gamma\left(m+\frac{1}{2}\right)}\right)\frac{1}{2^m}\left(1+\cos\left(\pi\frac{t-t_k}{h}\right)\right)^m$

The following remark is for modified Trapezoidal rule based on FzT and NIM to solve SODEs.

Remark 4. *In view of Equation (9), $\int_{t_{k-1}}^{t_{k+1}} B_k(t)\,dt = h$. This means that $\int_{t_k}^{t_{k+1}} B_k(t)\,dt = \frac{h}{2}$.*

Important property of the direct FzT as well as inverse FzT is their linearity, namely, given $f,\ g \in C[a,b]$ and $\alpha,\ \beta \in R$, if $h = \alpha f + \beta g$, then $F[h] = \alpha F[f] + \beta F[g]$ and $\hat{h} = \alpha\hat{f} + \beta\hat{g}$.

4. New Fuzzy Numerical Methods for Solving SODEs

Consider the initial value problem (IVP) for the SODEs:

$$\begin{cases} x'(t) = f(t,x,y), & x(t_1) = x_1,\ a = t_1 \le t \le t_n = b \\ y'(t) = g(t,x,y), & y(t_1) = y_1, \end{cases} \tag{16}$$

where $x_1,\ y_1 \in R$ and f, g are continuous function on $D = [a,b] \times R \times R$. If $f\ (g)$ satisfies a Lipschitz condition on D in the variable $x\ (y)$, then the initial-value problem (16) has a unique solution $x(t)\ (y(t))$ for $a \le t \le b$. In many cases, the problem (16) cannot be solved analytically so that numerical solutions are required. In [16], new representations of basic function based on the FzT are constructed for solving generalized Cauchy problems with help of NIM, FzT and classical methods (one-step, two-step and three-step) have presented while Euler method and Mid-point rule, based on FzT to solve Cauchy problem proposed by [17,18]. Further, NIM has been proposed for solving ODEs and delay differential equations [3]. Moreover, Adams-Bashforth methods and Adams-Moulton methods are noted as two families of multistep methods in literature. Multistep methods refer to using several previous values from the previous steps. The Adams-Bashforth methods were presented by John Couch Adams to solve a differential equation modelling capillary action due to Francis Bashforth and it follows that the Adams-Moulton method was developed improved multistep methods for solving ballistic equations by Forest Ray Moulton. In particular, the Adams-Moulton method is similar to the Adams-Bashforth method and the Adams-Moulton method was used Newton's method to solve the implicit equation. Clearly, Adams-Bashforth methods are explicit methods and the Adams–Moulton methods are implicit methods, for example, see ([40], p. 111).

Necessary steps of construction of the generalized uniform fuzzy partitions can be summarized as follows.

1. Specify the number n of components and compute the step $h = (t_n - t_1)\ /\ (n-1)$. If we want to obtain as best approximation of f as possible, then n should be large.
2. Construct the nodes $t_1 < \ldots < t_n$, where $t_k = t_1 + h(k-1)$.
3. Select the shape of basic functions. This is achieved by selecting the shape of generating function.
4. Construct a h-uniform generalized fuzzy partition of $[t_1, t_n]$ by new representations of basic functions are defined by Definition 4.

To begin the derivation of a modified Trapezoidal rule (1-step) and Adams Moulton method (2 and 3-step), integrate (16) on the interval $[t_k, t_{k+1}]$, $k = 1, \ldots, n-1$ to obtain=

$$\begin{aligned} x(t_{k+1}) &= x(t_k) + \int_{t_k}^{t_{k+1}} f(s, x(s), y(s))\, ds, \\ y(t_{k+1}) &= y(t_k) + \int_{t_k}^{t_{k+1}} g(s, x(s), y(s))\, ds. \end{aligned} \tag{17}$$

Consider the following integral

$$\begin{aligned} I_f &= \int_{t_k}^{t_{k+1}} f(s, x(s), y(s))\, ds, \\ I_g &= \int_{t_k}^{t_{k+1}} g(s, x(s), y(s))\, ds. \end{aligned} \tag{18}$$

However, we cannot integrate $f(s, x(s), y(s))$ and $g(s, x(s), y(s))$ without knowing $x(s)$ and $y(s)$. So, the above integral (18) can be approximated by the following approach

$$I_f \approx \int_{t_k}^{t_{k+1}} f_1\left(s, x(s), y\left(s\right)\right) ds,$$
$$I_g \approx \int_{t_k}^{t_{k+1}} g_1\left(s, x(s), y\left(s\right)\right) ds, \tag{19}$$

where f_1 and g_1 are the approximation of f and g on the interval $[t_k, t_{k+1}]$. Choosing different f_1 (g_1) leads to different schemes. In particular, we choose f_1 (g_1) which contributes to the one, two and three-step methods based on FzT. Later, modification of these methods based on FzT and NIM.

In this section, we present three new schemes to solve SODEs (16) that use the F-transform and NIM and suppose that the functions f and g on $[a, b]$ are sufficiently smooth. The first scheme uses 1-step method while the second one uses the 2-step method and the last uses the 3-step method.

4.1. Numerical Scheme I: Modified Trapezoidal Rule Based on FzT and NIM for SODEs

According to necessary steps of construction of the generalized uniform fuzzy partitions in Section 4, we contributed to approximation methods of SODEs (16) by scheme provides formulas for the FzT components, X_k (Y_k), $k = 2, \dots, n-1$, of the unknown function $x(t)$ $(y(t))$ with respect to choose some of the h-uniform generalized fuzzy partition, $B_1, \dots, B_n$, of interval $[a, b]$ with parameter h to approximate solution of SODEs (16). As initial step, choose the number $n \geq 2$ and compute $h = (b-a)/(n-1)$, then construct the h-uniform generalized fuzzy partition of $[a, b]$ using Definition 4. Let $X_1 = x_1$ and $Y_1 = y_1$. In the following, we apply the FzT and NIM to the SODEs (16) for obtaining the numerical Scheme I, where $k = 1, \dots, n-1$.

First, let f_1 (g_1) in the Equation (19) is chosen as

$$f_1 = B_k F_k + B_{k+1} F_{k+1},$$
$$g_1 = B_k G_k + B_{k+1} G_{k+1}, \tag{20}$$

where

$$F_k = \frac{\int_a^b f(t, X_k, Y_k)\, B_k(t)\, dt}{\int_a^b B_k(t)\, dt}, \quad G_k = \frac{\int_a^b g(t, X_k, Y_k)\, B_k(t)\, dt}{\int_a^b B_k(t)\, dt}, \tag{21}$$

and B_k represents the generalized uniform fuzzy partitions that are defined by Definition 4. Then, substituting (20) into (19) for $k = 1, \dots, n-1$

$$I_f \approx \int_{t_k}^{t_{k+1}} B_k F_k ds + \int_{t_k}^{t_{k+1}} B_{k+1} F_{k+1} ds,$$

$$I_g \approx \int_{t_k}^{t_{k+1}} B_k G_k ds + \int_{t_k}^{t_{k+1}} B_{k+1} G_{k+1} ds.$$

By Remark 4 in the interval $[t_k, t_{k+1}]$, we have

$$I_f \approx \frac{h}{2}\left(F_k + F_{k+1}\right), \; I_g \approx \frac{h}{2}\left(G_k + G_{k+1}\right).$$

Hence, the one step method based on FzT for (17) is derived as follows, where $k = 1, \dots, n-1$.

$$X_{k+1} = X_k + \frac{h}{2}\left(F_k + F_{k+1}\right),$$
$$Y_{k+1} = Y_k + \frac{h}{2}\left(G_k + G_{k+1}\right), \tag{22}$$

where F_k and G_k are defined by (21).

This method computes the approximate coordinates $[X_1, \ldots, X_n]$ and $[Y_1, \ldots, Y_n]$ of the FzT for the functions $x(t)$ and $y(t)$. The problem with the previous scheme (22) is that the unknown quantities F_{k+1} and G_{k+1} which means that $X_{k+1}(Y_{k+1})$ appears on both sides and an implicit method. Therefore, one solution to this problem would be to use an explicit method such as another fuzzy approach, namely Scheme I. For this purpose, the scheme (22) is of the form

$$X_{k+1} = f_x + N(X_{k+1}), \quad \Big| \quad Y_{k+1} = f_y + N(Y_{k+1}),$$

and can be solved by NIM (2), where

$$f_x = X_k + \frac{h}{2}F_k, \text{ and } N(X_{k+1}) = \frac{h}{2}F_{k+1}.$$

$$f_y = Y_k + \frac{h}{2}G_k, \text{ and } N(Y_{k+1}) = \frac{h}{2}G_{k+1}.$$

The three term approximation of the NIM (3) gives the following formulas for solving SODEs (16):

$$\left.\begin{array}{lll|lll} u_{x0} &=& X_k + \frac{h}{2}F_k, & u_{y0} &=& Y_k + \frac{h}{2}G_k, \\ u_{x1} &=& N(u_{x0}), & u_{y1} &=& N(u_{y0}), \\ u_{x2} &=& N(u_{x0}+u_{x1}) - N(u_{x0}), & u_{y2} &=& N(u_{y0}+u_{y1}) - N(u_{y0}), \end{array}\right\}$$

Hence, the three term approximate solution is

$$u_x = u_{x0} + u_{x1} + u_{x2} = u_{x0} + N(u_{x0} + u_{x1})$$

and

$$u_y = u_{y0} + N(u_{y0} + u_{y1}),$$

which leads to the following formulas.

$$\left.\begin{array}{lll|lll} X^*_{k+1} &=& X_k + hF_k/2, & Y^*_{k+1} &=& Y_k + hG_k/2, \\ X^{**}_{k+1} &=& X^*_{k+1} + hF^*_{k+1}/2, & Y^{**}_{k+1} &=& Y^*_{k+1} + hG^*_{k+1}/2, \\ X_{k+1} &=& X_k + h\left(F_k + F^{**}_{k+1}\right)/2, & Y_{k+1} &=& Y_k + h\left(G_k + G^{**}_{k+1}\right)/2, \end{array}\right\} \tag{23}$$

where

$$\left.\begin{array}{lll|lll} F_k &=& \dfrac{\int_a^b f(t, X_k, Y_k) B_k(t)\, dt}{\int_a^b B_k(t)\, dt}, & G_k &=& \dfrac{\int_a^b g(t, X_k, Y_k) B_k(t)\, dt}{\int_a^b B_k(t)\, dt}, \\ F^*_{k+1} &=& \dfrac{\int_a^b f(t, X^*_{k+1}, Y^*_{k+1}) B_{k+1}(t)\, dt}{\int_a^b B_{k+1}(t)\, dt}, & G^*_{k+1} &=& \dfrac{\int_a^b g(t, X^*_{k+1}, Y^*_{k+1}) B_{k+1}(t)\, dt}{\int_a^b B_{k+1}(t)\, dt}, \\ F^{**}_{k+1} &=& \dfrac{\int_a^b f(t, X^{**}_{k+1}, Y^{**}_{k+1}) B_{k+1}(t)\, dt}{\int_a^b B_{k+1}(t)\, dt}, & G^{**}_{k+1} &=& \dfrac{\int_a^b g(t, X^{**}_{k+1}, Y^{**}_{k+1}) B_{k+1}(t)\, dt}{\int_a^b B_{k+1}(t)\, dt}. \end{array}\right\} \tag{24}$$

In the sequel, the approximate solution of SODEs (16) can be obtained using the inverse FzT as follows:

$$x_n(t) = \sum_{k=1}^{n} X_k B_k(t), \; y_n(t) = \sum_{k=1}^{n} Y_k B_k(t). \tag{25}$$

4.2. Numerical Scheme II: Modified 2-Step Adams Moulton Method Based on FzT and NIM for SODEs

The Scheme I uses 1-step method for solving SODEs (16). In this subsection, we improve 2-step Adams Moulton method using FzT and NIM for solving SODEs (16). Let us recall that the modified 2-step Adams Moulton method proposed by [16]. From this idea, the modified 2-step Adams Moulton method can be extended to approximate the solution of (16) by necessary steps of construction of the generalized uniform fuzzy partitions in Section 4. It is worth noting that three terms of NIM were used in [16], while four terms of NIM are used in this study. Let FzT components, X_k $(Y_k), k = 2, \ldots, n-1$, of the unknown function $x(t)$ $(y(t))$ with respect to choose some of the h-uniform generalized fuzzy partition (5) and let $X_1 = x_1, Y_1 = y_1, X_2 = x_2$ and $Y_2 = y_2$ if possible; otherwise, we can compute FzT components X_2 and Y_2 from the numerical Scheme I. In the following, we apply the F-transform and NIM to the SODEs (16) for obtaining the numerical Scheme II, where $k = 2, \ldots, n-1$. First, if f_1 in the Equation (19) is approximated by

$$f_1 = (p_0 + p_1) F_{k+1} + p_2 F_k + p_3 F_{k-1}, \tag{26}$$

where

$$F_k = \frac{\int_a^b f(t, X_k, Y_k)\, B_k(t)\, dt}{\int_a^b B_k(t)\, dt},$$

$p_k = (-1)^k \int_0^1 \binom{-s+1}{k} ds$. Substituting (26) into (19), then for $k = 1, \ldots, n-1$

$$I_f \approx \frac{h}{12}\left(5F_{k+1} + 8F_k - F_{k-1}\right).$$

Similarity,

$$I_g \approx \frac{h}{12}\left(5G_{k+1} + 8G_k - G_{k-1}\right).$$

Thus, the two step method based FzT for (17) is given for $k = 1, \ldots, n-1$ as

$$X_{k+1} = X_k + h\left(8F_k - F_{k-1} + 5F_{k+1}\right)/12, \quad \Big| \quad Y_{k+1} = Y_k + h\left(8G_k - G_{k-1} + 5G_{k+1}\right)/12, \tag{27}$$

where

$$F_k = \frac{\int_a^b f(t, X_k, Y_k) B_k(t) dt}{\int_a^b A_k(t) dt}, \quad \Big| \quad G_k = \frac{\int_a^b g(t, X_k, Y_k) B_k(t) dt}{\int_a^b B_k(t) dt},$$

The problem with the previous scheme (27) is that the unknown quantities F_{k+1} and G_{k+1}. Therefore, one solution to this problem would be to use an explicit method. For this purpose, the scheme (27) is of the form

$$X_{k+1} = f_x + N(X_{k+1}), \quad \Big| \quad Y_{k+1} = f_y + N(Y_{k+1}),$$

and can be solved by NIM (2), where

$$f_x = X_k + \frac{h}{12}\left(8F_k - F_{k-1}\right), \text{ and } N(X_{k+1}) = \frac{5h}{12} F_{k+1}.$$

$$f_y = Y_k + \frac{h}{12}\left(8G_k - G_{k-1}\right), \text{ and } N(Y_{k+1}) = \frac{5h}{12} G_{k+1}.$$

The four term approximation of the NIM (3) gives the following formulas for solving SODEs (16):

$$
\left.\begin{array}{lll|lll}
u_{x0} & = & X_k + \frac{h}{12}\left(8F_k - F_{k-1}\right), & u_{y0} & = & Y_k + \frac{h}{12}\left(8G_k - G_{k-1}\right), \\
u_{x1} & = & N\left(u_{x0}\right), & u_{y1} & = & N\left(u_{y0}\right), \\
u_{x2} & = & N\left(u_{x0}+u_{x1}\right) - N\left(u_{x0}\right), & u_{y2} & = & N\left(u_{y0}+u_{y1}\right) - N\left(u_{y0}\right), \\
u_{x3} & = & N\left(u_{x0}+u_{x1}+u_{x2}\right) - N\left(u_{x0}+u_{x1}\right). & u_{y3} & = & N\left(u_{y0}+u_{y1}+u_{y2}\right) - N\left(u_{y0}+u_{y1}\right).
\end{array}\right\}
$$

Hence, the four term approximate solution is

$$u_x = u_{x0} + u_{x1} + u_{x2} + u_{x3} = u_{x0} + N\left(u_{x0} + N\left(u_{x0} + u_{x1}\right)\right)$$

and

$$u_y = u_{y0} + N\left(u_{y0} + N\left(u_{y0} + u_{y1}\right)\right),$$

which leads to the following formulas.

$$
\left.\begin{array}{lll|lll}
X^{*}_{k+1} & = & X_k + h\left(8F_k - F_{k-1}\right)/12, & Y^{*}_{k+1} & = & Y_k + h\left(8G_k - G_{k-1}\right)/12, \\
X^{**}_{k+1} & = & X^{*}_{k+1} + 5hF^{*}_{k+1}/12, & Y^{**}_{k+1} & = & Y^{*}_{k+1} + 5hG^{*}_{k+1}/12, \\
X^{***}_{k+1} & = & X^{*}_{k+1} + 5hF^{**}_{k+1}/12, & Y^{***}_{k+1} & = & Y^{*}_{k+1} + 5hG^{**}_{k+1}/12, \\
X_{k+1} & = & X_k + h\left(8F_k - F_{k-1} + 5F^{***}_{k+1}\right)/12, & Y_{k+1} & = & Y_k + h\left(8G_k - G_{k-1} + 5G^{***}_{k+1}\right)/12,
\end{array}\right\} \quad (28)
$$

where

$$
\begin{array}{lll|lll}
F_{k-1} & = & \dfrac{\int_a^b f(t, X_{k-1}, Y_{k-1}) B_{k-1}(t)dt}{\int_a^b A_{k-1}(t)dt}, & G_{k-1} & = & \dfrac{\int_a^b g(t, X_{k-1}, Y_{k-1})) B_{k-1}(t)dt}{\int_a^b B_{k-1}(t)dt}, \\
F_k & = & \dfrac{\int_a^b f(t, X_k, Y_k) B_k(t)dt}{\int_a^b A_k(t)dt}, & G_k & = & \dfrac{\int_a^b g(t, X_k, Y_k) B_k(t)dt}{\int_a^b B_k(t)dt}, \\
F^{*}_{k+1} & = & \dfrac{\int_a^b f(t, X^{*}_{k+1}, Y^{*}_{k+1}) B_{k+1}(t)dt}{\int_a^b B_{k+1}(t)dt}, & G^{*}_{k+1} & = & \dfrac{\int_a^b g(t, X^{*}_{k+1}, Y^{*}_{k+1}) B_{k+1}(t)dt}{\int_a^b B_{k+1}(t)dt}, \\
F^{**}_{k+1} & = & \dfrac{\int_a^b f(t, X^{**}_{k+1}, Y^{**}_{k+1}) B_{k+1}(t)dt}{\int_a^b B_{k+1}(t)dt}, & G^{**}_{k+1} & = & \dfrac{\int_a^b g(t, X^{**}_{k+1}, Y^{**}_{k+1}) B_{k+1}(t)dt}{\int_a^b B_{k+1}(t)dt}, \\
F^{***}_{k+1} & = & \dfrac{\int_a^b f(t, X^{***}_{k+1}, Y^{***}_{k+1}) B_{k+1}(t)dt}{\int_a^b B_{k+1}(t)dt}, & G^{***}_{k+1} & = & \dfrac{\int_a^b g(t, X^{***}_{k+1}, Y^{***}_{k+1}) B_{k+1}(t)dt}{\int_a^b B_{k+1}(t)dt}.
\end{array}
$$

Then, obtain the desired approximation for x and y by the inverse FzT (25) applied to $[X_1, \ldots, X_n]$ and $[Y_1, \ldots, Y_n]$.

4.3. Numerical Scheme III: Modified 3-Step Adams Moulton Method Based on FzT and NIM for SODEs

In this subsection, we improve 3-step Adams Moulton method using FzT and NIM for solving SODEs (16). The modified 3-step Adams Moulton method proposed by [16] for solving Cauchy problems. From this idea, we can propose to approximate the solution of (16) by NIM and FzT components, X_k (Y_k), $k = 2, \ldots, n-1$, of the unknown function $x(t)$ $(y(t))$ with respect to choose some of the h-uniform generalized fuzzy partition (see Definition 4), $B_1, \ldots, B_n$, of interval $[a, b]$ with parameter $h = (b-a)/(n-1)$, $n \geq 2$. Let $X_1 = x_1$, $Y_1 = y_1$, $X_2 = x_2$, $Y_2 = y_2$, $X_3 = x_3$, and $Y_3 = y_3$ if possible; otherwise, we can compute FzT components X_2, Y_2, X_3 and Y_3 from the numerical

Scheme I. Now, we apply the F -transform and NIM to the SODEs (16) and obtain the following numerical Scheme III for $k = 3, \ldots, n-1$:

According to steps of deriving Equation (27) and then steps of NIM in previous Subsection 4.2, we get the four term approximation of the NIM as follows.

$$\left.\begin{array}{lll|lll} u_{x0} &=& X_k + \frac{h}{24}\left(19F_k - 5F_{k-1} + F_{k-2}\right), & u_{y0} &=& Y_k + \frac{h}{24}\left(19G_k - 5G_{k-1} + G_{k-2}\right), \\ u_{x1} &=& N\left(u_{x0}\right), & u_{y1} &=& N\left(u_{y0}\right), \\ u_{x2} &=& N\left(u_{x0} + u_{x1}\right) - N\left(u_{x0}\right), & u_{y2} &=& N\left(u_{y0} + u_{y1}\right) - N\left(u_{y0}\right), \\ u_{x3} &=& N\left(u_{x0} + u_{x1} + u_{x2}\right) - N\left(u_{x0} + u_{x1}\right). & u_{y3} &=& N\left(u_{y0} + u_{y1} + u_{y2}\right) - N\left(u_{y0} + u_{y1}\right). \end{array}\right\}$$

Hence, the four term approximate solution is

$$u_x = u_{x0} + u_{x1} + u_{x2} + u_{x3} = u_{x0} + N\left(u_{x0} + N\left(u_{x0} + u_{x1}\right)\right)$$

and

$$u_y = u_{y0} + N\left(u_{y0} + N\left(u_{y0} + u_{y1}\right)\right),$$

which leads to the following formulas.

$$\left.\begin{array}{lll|lll} X^*_{k+1} &=& X_k + \frac{h}{24}\left(19F_k - 5F_{k-1} + F_{k-2}\right), & Y^*_{k+1} &=& Y_k + \frac{h}{24}\left(19G_k - 5G_{k-1} + G_{k-2}\right), \\ X^{**}_{k+1} &=& X^*_{k+1} + \frac{9h}{24}F^*_{k+1}, & Y^{**}_{k+1} &=& Y^*_{k+1} + \frac{9h}{24}G^*_{k+1}, \\ X^{***}_{k+1} &=& X^*_{k+1} + \frac{9h}{24}F^{**}_{k+1}, & Y^{***}_{k+1} &=& Y^*_{k+1} + \frac{9h}{24}G^{**}_{k+1}, \\ X_{k+1} &=& X_k, & Y_{k+1} &=& Y_k, \\ &+& \frac{h}{24}\left(19F_k - 5F_{k-1} + F_{k-2} + 9F^{***}_{k+1}\right), & &+& \frac{h}{24}\left(19G_k - 5G_{k-1} + G_{k-2} + 9G^{***}_{k+1}\right), \end{array}\right\} \quad (29)$$

where

$$\begin{array}{lll|lll} F_{k-2} &=& \dfrac{\int_a^b f(t, X_{k-2}, Y_{k-2})B_{k-2}(t)dt}{\int_a^b B_{k-2}(t)dt}, & G_{k-2} &=& \dfrac{\int_a^b g(t, X_{k-2}, Y_{k-2}))B_{k-2}(t)dt}{\int_a^b B_{k-2}(t)dt}, \\ F_{k-1} &=& \dfrac{\int_a^b f(t, X_{k-1}, Y_{k-1})B_{k-1}(t)dt}{\int_a^b B_{k-1}(t)dt}, & G_{k-1} &=& \dfrac{\int_a^b g(t, X_{k-1}, Y_{k-1}))B_{k-1}(t)dt}{\int_a^b B_{k-1}(t)dt}, \\ F_k &=& \dfrac{\int_a^b f(t, X_k, Y_k)B_k(t)dt}{\int_a^b B_k(t)dt}, & G_k &=& \dfrac{\int_a^b g(t, X_k, Y_k)B_k(t)dt}{\int_a^b B_k(t)dt}, \\ F^*_{k+1} &=& \dfrac{\int_a^b f(t, X^*_{k+1}, Y^*_{k+1})B_{k+1}(t)dt}{\int_a^b B_{k+1}(t)dt}, & G^*_{k+1} &=& \dfrac{\int_a^b g(t, X^*_{k+1}, Y^*_{k+1})B_{k+1}(t)dt}{\int_a^b B_{k+1}(t)dt}, \\ F^{**}_{k+1} &=& \dfrac{\int_a^b f(t, X^{**}_{k+1}, Y^{**}_{k+1})B_{k+1}(t)dt}{\int_a^b B_{k+1}(t)dt}, & G^{**}_{k+1} &=& \dfrac{\int_a^b g(t, X^{**}_{k+1}, Y^{**}_{k+1})B_{k+1}(t)dt}{\int_a^b B_{k+1}(t)dt}, \\ F^{***}_{k+1} &=& \dfrac{\int_a^b f(t, X^{***}_{k+1}, Y^{***}_{k+1})B_{k+1}(t)dt}{\int_a^b B_{k+1}(t)dt}, & G^{***}_{k+1} &=& \dfrac{\int_a^b g(t, X^{***}_{k+1}, Y^{***}_{k+1})B_{k+1}(t)dt}{\int_a^b B_{k+1}(t)dt}. \end{array}$$

In the sequel, the inverse FzT (25) approximates the solution $x(t)$ $(y(t))$ of the problem (16).

4.4. Error Analysis of Numerical Scheme I for SODEs

In this subsection, we present error analysis for numerical Scheme I and consider the Formula (23). If $x(t_k) = x_k$ and $y(t_k) = y_k$ denote the exact solution and X_k, Y_k denote the numerical solution. Then, substituting the exact solution in the Formula (23), we get

$$\left.\begin{array}{rcl|rcl} x_{k+1}^{*} &=& x_k + hF_k^e/2, & y_{k+1}^{*} &=& y_k + hG_k^e/2, \\ x_{k+1}^{**} &=& x_{k+1}^{*} + hF_{k+1}^{e*}/2, & y_{k+1}^{**} &=& y_{k+1}^{*} + hG_{k+1}^{e*}/2, \\ x_{k+1} &=& x_k + h\left(F_k^e + F_{k+1}^{e**}\right)/2, & y_{k+1} &=& y_k + h\left(G_k^e + G_{k+1}^{e**}\right)/2, \end{array}\right\} \tag{30}$$

where

$$\left.\begin{array}{rcl|rcl} F_k^e &=& \dfrac{\int_a^b f(t, x_k, y_k) B_k(t)\,dt}{\int_a^b B_k(t)\,dt}, & G_k^e &=& \dfrac{\int_a^b g(t, x_k, y_k) B_k(t)\,dt}{\int_a^b B_k(t)\,dt}, \\ F_{k+1}^{e*} &=& \dfrac{\int_a^b f(t, x_{k+1}^{*}, y_{k+1}^{*}) B_{k+1}(t)\,dt}{\int_a^b B_{k+1}(t)\,dt}, & G_{k+1}^{e*} &=& \dfrac{\int_a^b g(t, x_{k+1}^{*}, y_{k+1}^{*}) B_{k+1}(t)\,dt}{\int_a^b B_{k+1}(t)\,dt}, \\ F_{k+1}^{e**} &=& \dfrac{\int_a^b f(t, x_{k+1}^{**}, y_{k+1}^{**}) B_{k+1}(t)\,dt}{\int_a^b B_{k+1}(t)\,dt}, & G_{k+1}^{e**} &=& \dfrac{\int_a^b g(t, x_{k+1}^{**}, y_{k+1}^{**}) B_{k+1}(t)\,dt}{\int_a^b B_{k+1}(t)\,dt}. \end{array}\right\} \tag{31}$$

and the truncation error Tx_k and Ty_k of the scheme I are given by

$$Tx_k = \tfrac{x_{k+1}-x_k}{h} - \tfrac{1}{2}\left(F_k^e + F_{k+1}^{e**}\right), \;\Big|\; Ty_k = \tfrac{y_{k+1}-y_k}{h} - \tfrac{1}{2}\left(G_k^e + G_{k+1}^{e**}\right). \tag{32}$$

Rearranging (23), we get

$$0 = \tfrac{X_{k+1}-X_k}{h} - \tfrac{1}{2}\left(F_k + F_{k+1}^{**}\right), \;\Big|\; 0 = \tfrac{Y_{k+1}-Y_k}{h} - \tfrac{1}{2}\left(G_k + G_{k+1}^{**}\right). \tag{33}$$

Let $e_{k+1} = X_{k+1} - x_{k+1}$ and $d_{k+1} = Y_{k+1} - y_{k+1}$, then subtracting (33) from (32), we get

$$\begin{aligned} Tx_k h &= e_{k+1} - e_k - \frac{h}{2}\left(F_k - F_k^e\right) - \frac{h}{2}\left(F_{k+1}^{**} - F_{k+1}^{e**}\right), \\ Ty_k h &= d_{k+1} - d_k - \frac{h}{2}\left(G_k - G_k^e\right) - \frac{h}{2}\left(G_{k+1}^{**} - G_{k+1}^{e**}\right), \end{aligned} \tag{34}$$

Similarly to Lemma 8 and Theorem 2 by [16], we have the following results.

Lemma 5. *Let f, g are assumed to be sufficiently smooth functions of its arguments on $[a, b]$ and be Lipschitz continuous with respect to x and y, i.e., there exists a constant $L \in \mathbb{R}$, such that for all $t \in [a, b]$ and $x, x', y, y' \in \mathbb{R}$,*

$$\begin{aligned} |f(t, x, y) - f(t, x', y')| &\le L(|x - x'| + |y - y'|), \\ |g(t, x, y) - g(t, x', y')| &\le L(|x - x'| + |y - y'|). \end{aligned} \tag{35}$$

Assume that $\{B_k \mid k = 1, \ldots, n\}$, $n \ge 2$, is a h-uniform generalized fuzzy partition of $[a, b]$. Then we get for $k = 1, \ldots, n$,

$$\begin{array}{rlcrl} |e_{k+1}| \le & |e_k|(1 + c) + Th & \text{and} & \left|F_k^e - F_{k+1}^{e**}\right| \le & LhM_2, \\ |d_{k+1}| \le & |d_k|(1 + c) + Th & \text{and} & \left|G_k^e - G_{k+1}^{e**}\right| \le & LhM_3, \end{array}$$

where $c = hL + \frac{h^2L^2}{2} + \frac{h^3L^3}{8}$, $T = \max_{1 \le k \le n} |Tx_k, Ty_k|$, F_k^e, F_{k+1}^{e**}, G_k^e, G_{k+1}^{e**} *are determined by Formula (31) and* M_2, M_3 *are upper bound for* f *and* g *respectively on* $[a, b]$.

Theorem 3. *Let* $f, g : [a, b] \to \mathbb{R}$ *be twice continuously differentiable on* $[a, b]$. *Let moreover,* $f, g : [a, b] \times \mathbb{R} \times \mathbb{R} \to \mathbb{R}$ *be Lipschitz continuous with respect to x and y. Assume that* $\{B_k \mid k = 1, \ldots, n\}$, $n \ge 2$, *is a h-uniform generalized fuzzy partition of* $[a, b]$. *Then the scheme I (23) is convergent.*

The technique of error analysis for rest schemes can be obtained analogously to numerical Scheme I.

5. Applications

A general model for the dynamical system may be written as $\frac{dx}{dt} = xg(x, y)$, $\frac{dy}{dt} = yh(x, y)$, where g and h are arbitrary functions of the prey and predator species whose populations are $x(t)$ and $y(t)$ at time t. However, the following problem of Lotka–Volterra equation with derivatives and with variable coefficients $\alpha(t)$, $\beta(t)$, $\delta(t)$, $\gamma(t)$ as functions of time t have not yet been solved by any fuzzy numerical method. The new differential equations are represented by a non-autonomous ordinary differential equation system. The model, incorporating the above functions is as follows [41–43]:

$$\begin{aligned} \frac{dx}{dt} &= \alpha(t)\, x(t) - \beta(t)\, x(t)\, y(t), & x(0) &= x_1 \\ \frac{dy}{dt} &= \delta(t)\, x(t)\, y(t) - \gamma(t)\, y(t), & y(0) &= y_1 \end{aligned} \tag{36}$$

Two examples are discussed in order to prove that the results obtained by Scheme I (23), II (28) and III (29) for the numerical solution of the model (36).

Example 2. *Consider the problem of Lotka-Volterra-prey- predator model (36). We take* $\alpha(t) = 4 + \tan(t)$, $\beta(t) = \exp(2t)$, $\gamma(t) = -2$, $\delta(t) = \cos(t)$, $x(0) = -4$ *and* $y(0) = 4$.

The exact solution for these coefficients is $x(t) = \frac{-4}{\cos(t)}$, $y(t) = 4\exp(-2t)$ *proposed by [41–43].*

Example 3. *Consider the problem of Lotka-Volterra-prey-predator model (36) with* $\alpha(t) = -t$, $\beta(t) = -t$, $\gamma(t) = t$, $\delta(t) = t$, $x(0) = 2$ *and* $y(0) = 2$.

The exact solution for these coefficients is $x(t) = \frac{2}{2-\exp(t^2/2)}$, $y(t) = \frac{2}{2-\exp(t^2/2)}$ *proposed by [41,42].*

The results are listed in Tables 2–6 by the proposed fuzzy approximation methods with respect to case $K_{C_2^{201}}$ is defined by Example 1. The proposed fuzzy approximation methods are generated by Algorithms A1–A3 (Appendix A). The mean square error (MSE) defined as MSE $= \frac{1}{n}(\|Y_k - y(t_k)\|_2)^2$. This is an easily computable quantity for a particular sample. From the numerical tests, the results are summarized as follows:

1. In view of Tables 2–5, a comparison between the three new proposed schemes ((23), (28) and (29)) and the classical Trapezoidal rule (1-step), classical 2-step Adams Moulton Method and classical 3-step Adams Moulton Method based on Euler method for Examples 2 and 3.
2. Moreover, comparison of MSE for Examples 2 and 3 shown in Table 6. It is observed that the new fuzzy approximation methods yield more accurate results in comparison with the classical Trapezoidal rule (one step) and classical Adams Moulton method (two and three steps). Hence, the new fuzzy approximation methods provide alternative techniques for solving SODEs with better results.

Table 2. Comparison of numerical results of $x(t)$ for Example 2.

t_i	Solution $x(t)$	Proposed Scheme I	Proposed Scheme II	Proposed Scheme III	Trap [1]	2-Step Adams [2]	3-Step Adams [3]
0.00	−4.00000	−4.00000	−4.00000	−4.00000	−4.00000	−4.00000	−4.00000
0.05	−4.00501	−4.00506	−4.00501	−4.00501	−4.00714	−4.00501	−4.00501
0.10	−4.02008	−4.02059	−4.02011	−4.02008	−4.02627	−4.02187	−4.02008
0.15	−4.04543	−4.04589	−4.04536	−4.04536	−4.05752	−4.04875	−4.04703
0.20	−4.08136	−4.08128	−4.08121	−4.08119	−4.10134	−4.08610	−4.08434
0.25	−4.12834	−4.12721	−4.12813	−4.12811	−4.15850	−4.13442	−4.13261
0.30	−4.18701	−4.18423	−4.18673	−4.18670	−4.23015	−4.19440	−4.19250
0.35	−4.25816	−4.25304	−4.25783	−4.25779	−4.31783	−4.26691	−4.26487
0.40	−4.34282	−4.33453	−4.34243	−4.34236	−4.42361	−4.35300	−4.35077
0.45	−4.44224	−4.42976	−4.44178	−4.44170	−4.55014	−4.45399	−4.45151
0.50	−4.55798	−4.54001	−4.55745	−4.55733	−4.70086	−4.57151	−4.56871
0.55	−4.69195	−4.66686	−4.69134	−4.69117	−4.88023	−4.70755	−4.70433
0.60	−4.84651	−4.81220	−4.84581	−4.84558	−5.09399	−4.86455	−4.86079
0.65	−5.02460	−4.97833	−5.02378	−5.02346	−5.34964	−5.04556	−5.04112
0.70	−5.22984	−5.16805	−5.22887	−5.22845	−5.65700	−5.25435	−5.24903
0.75	−5.46680	−5.38480	−5.46565	−5.46508	−6.02912	−5.49569	−5.48924
0.80	−5.74130	−5.63286	−5.73990	−5.73914	−6.48349	−5.77559	−5.76768
0.85	−6.06076	−5.91759	−6.05906	−6.05803	−7.04390	−6.10182	−6.09202
0.90	−6.43490	−6.24582	−6.43281	−6.43141	−7.74310	−6.48450	−6.47222
0.95	−6.87660	−6.62636	−6.87400	−6.87209	−8.62699	−6.93706	−6.92148
1.00	−7.40326	−7.07221	−7.40106	−7.39831	−9.76103	−7.47766	−7.45769

[1] Trapezoidal rule; [2] 2-Step Adams Moulton Method; [3] 3-Step Adams Moulton Method.

Table 3. Comparison of numerical results of $y(t)$ for Example 2.

t_i	Solution $y(t)$	Proposed Scheme I	Proposed Scheme II	Proposed Scheme III	Trap [1]	2-Step Adams [2]	3-Step Adams [3]
0.00	4.00000	4.00000	4.00000	4.00000	4.00000	4.00000	4.00000
0.05	3.61935	3.62135	3.61935	3.61935	3.62045	3.61935	3.61935
0.10	3.27492	3.27848	3.27485	3.27492	3.27766	3.27601	3.27492
0.15	2.96327	2.96802	2.96314	2.96327	2.96757	2.96496	2.96415
0.20	2.68128	2.68698	2.68112	2.68124	2.68671	2.68326	2.68261
0.25	2.42612	2.43262	2.42595	2.42607	2.43201	2.42819	2.42769
0.30	2.19525	2.20247	2.19508	2.19520	2.20079	2.19724	2.19688
0.35	1.98634	1.99425	1.98619	1.98630	1.99067	1.98815	1.98792
0.40	1.79732	1.80593	1.79718	1.79729	1.79951	1.79888	1.79875
0.45	1.62628	1.63564	1.62616	1.62627	1.62541	1.62754	1.62752
0.50	1.47152	1.48170	1.47143	1.47153	1.46664	1.47245	1.47253
0.55	1.33148	1.34257	1.33142	1.33153	1.32166	1.33207	1.33224
0.60	1.20478	1.21688	1.20474	1.20485	1.18908	1.20500	1.20526
0.65	1.09013	1.10337	1.09012	1.09023	1.06762	1.08998	1.09033
0.70	0.98639	1.00091	0.98641	0.98652	0.95615	0.98587	0.98632
0.75	0.89252	0.90848	0.89257	0.89269	0.85366	0.89163	0.89217
0.80	0.80759	0.82515	0.80766	0.80779	0.75923	0.80632	0.80696
0.85	0.73073	0.75011	0.73084	0.73097	0.67208	0.72910	0.72984
0.90	0.66120	0.68260	0.66133	0.66148	0.59150	0.65919	0.66004
0.95	0.59827	0.62195	0.59844	0.59860	0.51692	0.59590	0.59686
1.00	0.54134	0.56745	0.54145	0.54163	0.44787	0.53860	0.53969

[1] Trapezoidal rule; [2] 2-Step Adams Moulton Method; [3] 3-Step Adams Moulton Method.

Table 4. Comparison of numerical results of $x(t)$ for Example 3.

t_i	Solution $x(t)$	Proposed Scheme I	Proposed Scheme II	Proposed Scheme III	Trap [1]	2-Step Adams [2]	3-Step Adams [3]
0.00	2.00000	2.00000	2.00000	2.00000	2.00000	2.00000	2.00000
0.05	2.00250	2.00257	2.00250	2.00250	2.00250	2.00250	2.00250
0.10	2.01008	2.01016	2.01007	2.01008	2.01005	2.01006	2.01008
0.15	2.02289	2.02299	2.02289	2.02289	2.02279	2.02285	2.02286
0.20	2.04124	2.04138	2.04124	2.04124	2.04101	2.04117	2.04118
0.25	2.06557	2.06577	2.06558	2.06558	2.06511	2.06546	2.06547
0.30	2.09650	2.09676	2.09652	2.09651	2.09567	2.09633	2.09633
0.35	2.13485	2.13520	2.13488	2.13486	2.13344	2.13459	2.13459
0.40	2.18171	2.18218	2.18176	2.18172	2.17942	2.18132	2.18132
0.45	2.23852	2.23915	2.23860	2.23854	2.23493	2.23795	2.23796
0.50	2.30720	2.30803	2.30731	2.30722	2.30167	2.30637	2.30638
0.55	2.39031	2.39140	2.39047	2.39034	2.38192	2.38910	2.38911
0.60	2.49133	2.49279	2.49157	2.49138	2.47868	2.48956	2.48957
0.65	2.61513	2.61708	2.61549	2.61520	2.59605	2.61251	2.61251
0.70	2.76863	2.77126	2.76917	2.76873	2.73967	2.76465	2.76466
0.75	2.96202	2.96563	2.96288	2.96218	2.91754	2.95585	2.95584
0.80	3.21093	3.21600	3.21235	3.21120	3.14130	3.20103	3.20099
0.85	3.54059	3.54791	3.54308	3.54108	3.42845	3.52398	3.52386
0.90	3.99443	4.00539	3.99914	3.99541	3.80653	3.96482	3.96449
0.95	4.65413	4.67130	4.66407	4.65640	4.32100	4.59671	4.59578
1.00	5.69348	5.72071	5.71719	5.69904	5.05197	5.56751	5.56473

[1] Trapezoidal rule; [2] 2-Step Adams Moulton Method; [3] 3-Step Adams Moulton Method.

Table 5. Comparison of numerical results of $y(t)$ for Example 3.

t_i	Solution $y(t)$	Proposed Scheme I	Proposed Scheme II	Proposed Scheme III	Trap [1]	2-Step Adams [2]	3-Step Adams [3]
0.00	2.00000	2.00000	2.00000	2.00000	2.00000	2.00000	2.00000
0.05	2.00250	2.00257	2.00250	2.00250	2.00250	2.00250	2.00250
0.10	2.01008	2.01016	2.01007	2.01008	2.01005	2.01006	2.01008
0.15	2.02289	2.02299	2.02289	2.02289	2.02279	2.02285	2.02286
0.20	2.04124	2.04138	2.04124	2.04124	2.04101	2.04117	2.04118
0.25	2.06557	2.06577	2.06558	2.06558	2.06511	2.06546	2.06547
0.30	2.09650	2.09676	2.09652	2.09651	2.09567	2.09633	2.09633
0.35	2.13485	2.13520	2.13488	2.13486	2.13344	2.13459	2.13459
0.40	2.18171	2.18218	2.18176	2.18172	2.17942	2.18132	2.18132
0.45	2.23852	2.23915	2.23860	2.23854	2.23493	2.23795	2.23796
0.50	2.30720	2.30803	2.30731	2.30722	2.30167	2.30637	2.30638
0.55	2.39031	2.39140	2.39047	2.39034	2.38192	2.38910	2.38911
0.60	2.49133	2.49279	2.49157	2.49138	2.47868	2.48956	2.48957
0.65	2.61513	2.61708	2.61549	2.61520	2.59605	2.61251	2.61251
0.70	2.76863	2.77126	2.76917	2.76873	2.73967	2.76465	2.76466
0.75	2.96202	2.96563	2.96288	2.96218	2.91754	2.95585	2.95584
0.80	3.21093	3.21600	3.21235	3.21120	3.14130	3.20103	3.20099
0.85	3.54059	3.54791	3.54308	3.54108	3.42845	3.52398	3.52386
0.90	3.99443	4.00539	3.99914	3.99541	3.80653	3.96482	3.96449
0.95	4.65413	4.67130	4.66407	4.65640	4.32100	4.59671	4.59578
1.00	5.69348	5.72071	5.71719	5.69904	5.05197	5.56751	5.56473

[1] Trapezoidal rule; [2] 2-Step Adams Moulton Method; [3] 3-Step Adams Moulton Method.

Table 6. The values of MSE for Examples 2 and 3.

Case		Proposed Scheme for $x(t)$			Classical Method for $x(t)$		
(a) The Values of MSE of $x(t)$ for SODEs							
		I	**II**	**III**	**Trapezoidal Rule**	**2-Step Adams Moulton**	**3-Step Adams Moulton**
Ex.1	$K_{C_2^{201}}$	1.21569×10^{-2}	1.21112×10^{-6}	3.71739×10^{-6}	5.99915×10^{-1}	8.37463×10^{-4}	4.72086×10^{-4}
Ex.2	$K_{C_2^{201}}$	6.02153×10^{-5}	3.29699×10^{-5}	1.77640×10^{-6}	2.75574×10^{-2}	9.75470×10^{-4}	1.01541×10^{-3}
(b) The Values of MSE of $y(t)$ for SODEs							
Case		**Proposed Scheme for $y(t)$**			**Classical Method for $y(t)$**		
		I	**II**	**III**	**Trapezoidal Rule**	**2-Step Adams Moulton**	**3-Step Adams Moulton**
Ex.1	$K_{C_2^{201}}$	1.80417×10^{-4}	1.20902×10^{-8}	2.08739×10^{-8}	1.40165×10^{-3}	2.25155×10^{-6}	1.09674×10^{-6}
Ex.2	$K_{C_2^{201}}$	6.02153×10^{-5}	3.29699×10^{-5}	1.77640×10^{-6}	2.75574×10^{-2}	9.75470×10^{-4}	1.01541×10^{-3}

Further, the results obtained using proposed fuzzy approximation methods for Examples 2 and 3 are shown in Figures 1–3 by using $K_{C_2^{201}}$. In view of Figures 2 and 3, the graphical results of Examples 2 and 3 show a comparison between numerical Schemes (I, II and III) and exact solutions are shown separated from each other for clarity while a comparison between three proposed fuzzy numerical methods (Schemes I, II and III) and exact solution are shown in Figure 1. Furthermore, in view of Figures 2 and 3, a comparison between the numerical results and exact solutions for $h = 0.01$. All the graphs are plotted using MATLAB software.

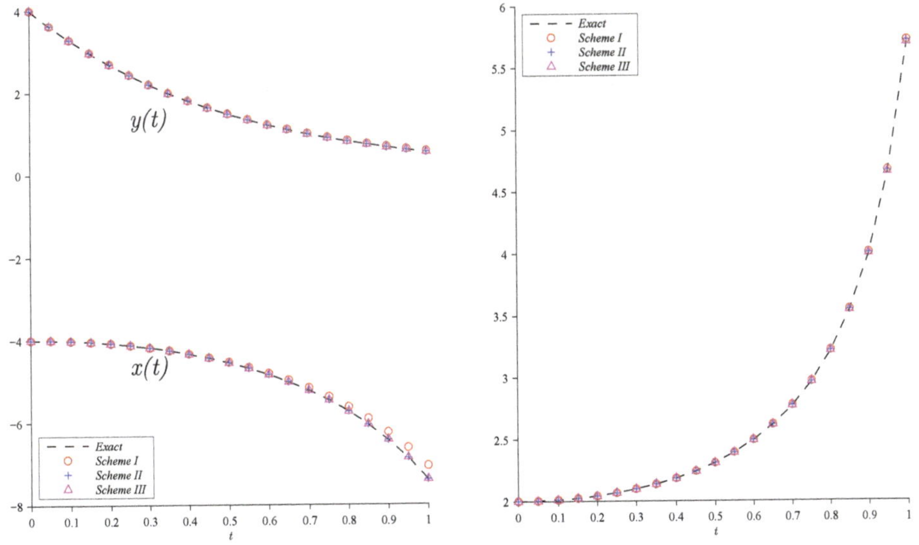

(a) (b)

Figure 1. A comparison between three fuzzy numerical methods and exact solution for two examples. (**a**) Example 2; (**b**) Example 3.

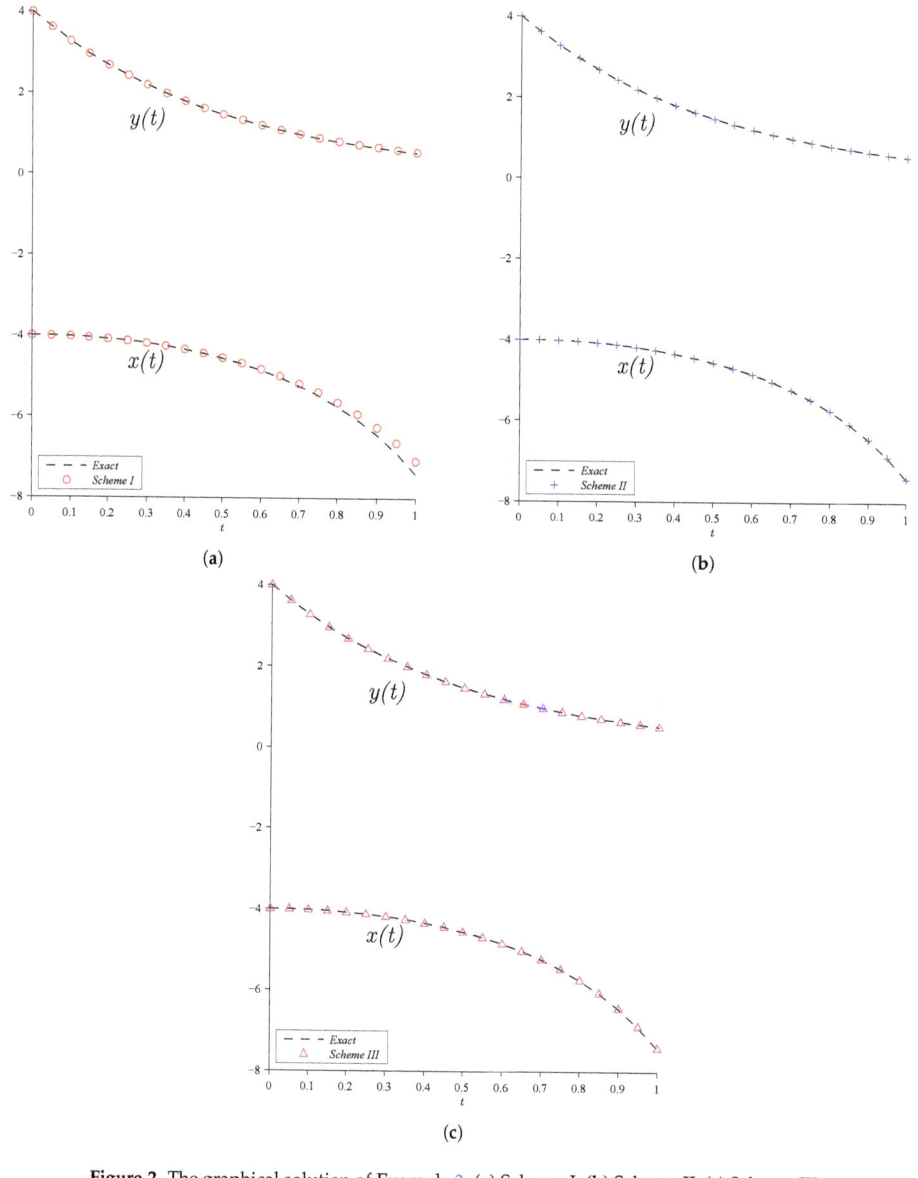

Figure 2. The graphical solution of Example 2. (**a**) Scheme I; (**b**) Scheme II; (**c**) Scheme III.

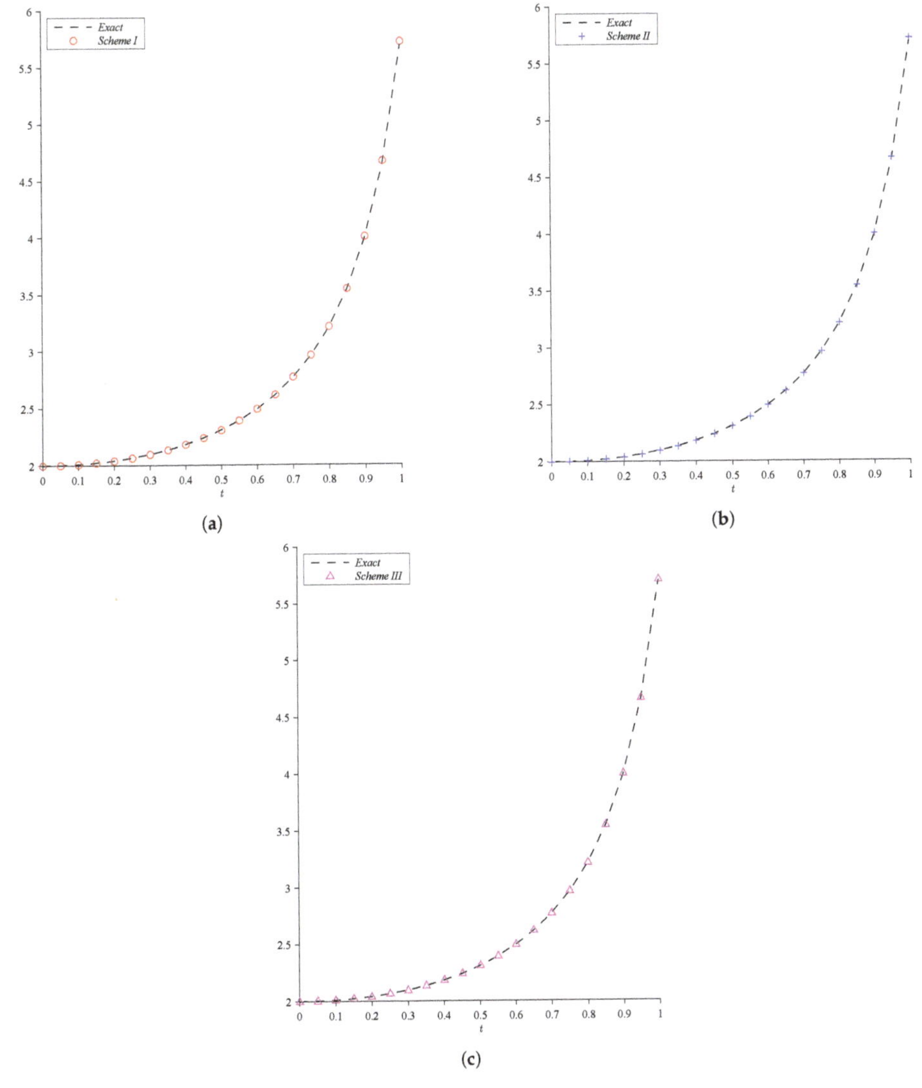

Figure 3. The graphical solution of Example 3. (**a**) Scheme I; (**b**) Scheme II; (**c**) Scheme III.

6. Conclusions

Three approximation methods are used the new generalized uniform fuzzy partitions for solving SODEs. In accordance with the three approximation methods for Cauchy problem by [16], Trapezoidal rule (one step) and Adams Moulton method (two and three steps) are improved using FzT and NIM. The results proved that the first approximation method converged to the exact solution. As an application, a predator-prey model is solved by using three proposed approximation methods. From the numerical results, it is observed that the new fuzzy approximation methods yield more accurate results in comparison with the classical Trapezoidal rule (one step) and classical Adams Moulton method (two and three steps). So, it is recommended to use the proposed methods to solve differential equations.

In this regard, it is well-known that FzT has a certain advantage to cope with problems affected by noise. This is because the FzT components of original and noisy functions are very similar to each other. In addition, we can reduce a higher-order differential equation into a system of first-order differential equations by relabeling the variables. Thus, the proposed methods can also be applied to a higher-order differential equation in the case of non-noisy or noisy right-hand side. From Algorithms A1–A3 (Appendix A), it is observed that the new fuzzy approximation methods are more time consuming in comparison with the considered Trapezoidal rule and the Adams Moulton methods. In the future research, we plan to give more details about running time of proposed methods. Further, we plan to solve a boundary value problem for a second order ordinary differential equation with fuzzy boundary conditions, see preliminary results in [44].

Author Contributions: Conceptualization and performed the numerical experiments, H.A.A.; Evaluated the results and supported this work, I.P.; Project administration and designed the numerical methods, M.Z.A.; Software and data curation, Z.R.Y.

Funding: This work of Irina Perfilieva has been supported by the project "LQ1602 IT4Innovations excellence in science" and by the Grant Agency of the Czech Republic (project No. 16-09541S).

Acknowledgments: The authors would like to express their deep gratitude to the editors and the anonymous referees for their valuable comments and criticism towards the improvement of the paper. Also, many thanks given to Universiti Malaysia Perlis for providing all facilities until this work was completed successfully.

Conflicts of Interest: The authors declare no conflicts of interest.

Appendix A. Algorithms

In this appendix, algorithms of approximation methods based on FzT and NIM for Sections 4.1–4.3 are explained with details. A pseudocode is used to describe the algorithms and simplified code that is easy to read. This pseudocode specifies the form of the input to be supplied and the form of the desired output. As a consequence, a stopping technique independent of the numerical technique is incorporated into each algorithm to avoid infinite loops. Two punctuation symbols are used in the algorithms, a period (.) indicates the termination of a step and a semicolon (;) separates tasks within a step. In algorithms with help of MATLAB software, the definite integral is specified by integral (function,upper limits, lower limits). The steps in the algorithms follow the rules of structured program construction. They have been arranged so that there should be minimal difficulty translating pseudocode into any programming language suitable for scientific applications. To approximate the solution of SODEs (16) at $(N+1)$ equally spaced numbers in the interval $[a, b]$, proceed as follows.

Algorithm A1. One-step algorithm for system of ODEs.

INPUT: $f(t,x,y)$; $g(t,x,y)$; endpoints a, b; integer N; initial condition y_1; m.

Step 1 Set $h = (b-a)/N$; $X_1 = x_1$; $Y_1 = y_1$; $t_1 = a$; $k = 1, \ldots, N+1$; $t_k = a + (k-1)h$.

Step 2 Define the generalized uniform fuzzy partitions as $B_k(t) = \left(\frac{\sqrt{\pi}\Gamma(m+1)}{2\Gamma(m+\frac{1}{2})}\right)\frac{1}{2^m}\left(1+\cos\left(\pi\frac{t-t(k)}{h}\right)\right)^m$.

Step 3 for $k = 1$ to N do Steps 04–15.

Step 04 $F(k) = \text{integral}(f(t, X(k), Y(k))B_k(t), t(k-1), t(k+1))/\text{integral}(B_k(t), t(k-1), t(k+1))$.

Step 05 $G(k) = \text{integral}(g(t, X(k), Y(k))B_k(t), t(k-1), t(k+1))/\text{integral}(B_k(t), t(k-1), t(k+1))$.

Step 06 $Xstar(k+1) = X(k) + hF(k)/2$.

Step 07 $Ystar(k+1) = Y(k) + hG(k)/2$.

Step 08 $Fstar(k+1) = \text{integral}(f(t, Xstar(k+1), Ystar(k+1))B_{k+1}(t), t(k-1), t(k+1))/\text{integral}(B_{k+1}(t), t(k-1), t(k+1))$.

Step 09 $Gstar(k+1) = \text{integral}(g(t, Xstar(k+1), Ystar(k+1))B_{k+1}(t), t(k-1), t(k+1))/\text{integral}(B_{k+1}(t), t(k-1), t(k+1))$.

Step 10 $Xstar2(k+1) = Xstar(k+1) + hFstar(k+1)/2$.

Step 11 $Ystar2(k+1) = Ystar(k+1) + hGstar(k+1)/2$.

Step 12 $Fstar2(k+1) = \text{integral}(f(t, Xstar2(k+1), Ystar2(k+1))B_{k+1}(t), t(k-1), t(k+1))/\text{integral}(B_{k+1}(t), t(k-1), t(k+1))$.

Step 13 $Gstar2(k+1) = \text{integral}(g(t, Xstar2(k+1), Ystar2(k+1))B_{k+1}(t), t(k-1), t(k+1))/\text{integral}(B_{k+1}(t), t(k-1), t(k+1))$.

Step 14 $X(k+1) = X(k) + h\left(F(k) + Fstar2(k+1)\right)/2$.

Step 15 $Y(k+1) = Y(k) + h\left(G(k) + Gstar2(k+1)\right)/2$.

end.

OUTPUT: Approximation X and Y to x and y, respectively at the $(N+1)$ values of t.

Algorithm A2. Two-step algorithm for system of ODEs.

INPUT: $f(t,x,y)$; $g(t,x,y)$; endpoints a, b; integer N; initial condition y_1; m.
Step 1 Set $h = (b-a)/N$; $X_1 = x_1$; $Y_1 = y_1$; $t_1 = a$; $k = 1, \ldots, N+1$; $t_k = a + (k-1)h$.
Step 2 Define the generalized uniform fuzzy partitions as $B_k(t) = \left(\frac{\sqrt{\pi}\Gamma(m+1)}{2\Gamma(m+\frac{1}{2})}\right)\frac{1}{2^m}\left(1+\cos\left(\pi\frac{t-t(k)}{h}\right)\right)^m$.
Step 3 Set $X_2 = x_2$; $Y_2 = y_2$. (In the case of no exact solutions, compute X_2 and Y_2 using Algorithm 1.)
Step 4 for $k = 2$ to N do Steps 05–18.
Step 05 $F(k-1) =$integral$(f(t,X(k-1),Y(k-1))B_{k-1}(t),t(k-1),t(k+1))/$integral$(B_{k-1}(t),t(k-1),t(k+1))$.
Step 06 $G(k-1) =$integral$(g(t,X(k-1),Y(k-1))B_{k-1}(t),t(k-1),t(k+1))/$integral$(B_{k-1}(t),t(k-1),t(k+1))$.
Step 07 $F(k) =$integral$(f(t,X(k),Y(k))B_k(t),t(k-1),t(k+1))/$integral$(B_k(t),t(k-1),t(k+1))$.
Step 08 $G(k) =$integral$(g(t,X(k),Y(k))B_k(t),t(k-1),t(k+1))/$integral$(B_k(t),t(k-1),t(k+1))$.
Step 09 $Xstar(k+1) = X(k) + h(8F(k) - F(k-1))/12$.
Step 10 $Ystar(k+1) = Y(k) + h(8G(k) - G(k-1))/12$.
Step 11 $Fstar(k+1) =$integral$(f(t,Xstar(k+1),Ystar(k+1))B_{k+1}(t),t(k-1),t(k+1))/$integral$(B_{k+1}(t),t(k-1),t(k+1))$.
Step 12 $Gstar(k+1) =$integral$(g(t,Xstar(k+1),Ystar(k+1))B_{k+1}(t),t(k-1),t(k+1))/$integral$(B_{k+1}(t),t(k-1),t(k+1))$.
Step 13 $Xstar2(k+1) = Xstar(k+1) + 5hFstar(k+1)/12$.
Step 14 $Ystar2(k+1) = Ystar(k+1) + 5hGstar(k+1)/12$.
Step 15 $Fstar2(k+1) =$integral$(f(t,Xstar2(k+1),Ystar2(k+1))B_{k+1}(t),t(k-1),t(k+1))/$integral$(B_{k+1}(t),t(k-1),t(k+1))$.
Step 16 $Gstar2(k+1) =$integral$(g(t,Xstar2(k+1),Ystar2(k+1))B_{k+1}(t),t(k-1),t(k+1))/$integral$(B_{k+1}(t),t(k-1),t(k+1))$.
Step 17 $X(k+1) = X(k) + h(8F(k) - F(k-1) + 5Fstar2(k+1))/12$.
Step 18 $Y(k+1) = Y(k) + h(8G(k) - G(k-1) + 5Gstar2(k+1))/12$.
end.
OUTPUT: Approximation X and Y to x and y, respectively at the $(N+1)$ values of t.

Algorithm A3. Three-step algorithm for system of ODEs.

INPUT: $f(t,x,y)$; $g(t,x,y)$; endpoints a, b; integer N; initial condition y_1; m.
Step 1 Set $h = (b-a)/N$; $X_1 = x_1$; $Y_1 = y_1$; $t_1 = a$; $k = 1, \ldots, N+1$; $t_k = a + (k-1)h$.
Step 2 Define the generalized uniform fuzzy partitions as $B_k(t) = \left(\frac{\sqrt{\pi}\Gamma(m+1)}{2\Gamma(m+\frac{1}{2})}\right)\frac{1}{2^m}\left(1+\cos\left(\pi\frac{t-t(k)}{h}\right)\right)^m$.
Step 3 Set $X_2 = x_2$; $Y_2 = y_2$; $X_3 = x_3$; $Y_3 = y_3$. (In the case of no exact solutions, compute X_2, Y_2, X_3 and Y_3 using Algorithm 1 or 2.)
Step 4 for $k = 3$ to N do Steps 05–20.
Step 05 $F(k-2) =$integral$(f(t,X(k-2),Y(k-2))B_{k-2}(t),t(k-1),t(k+1))/$integral$(B_{k-2}(t),t(k-1),t(k+1))$.
Step 06 $G(k-2) =$integral$(g(t,X(k-2),Y(k-2))B_{k-2}(t),t(k-1),t(k+1))/$integral$(B_{k-2}(t),t(k-1),t(k+1))$.
Step 07 $F(k-1) =$integral$(f(t,X(k-1),Y(k-1))B_{k-1}(t),t(k-1),t(k+1))/$integral$(B_{k-1}(t),t(k-1),t(k+1))$.
Step 08 $G(k-1) =$integral$(g(t,X(k-1),Y(k-1))B_{k-1}(t),t(k-1),t(k+1))/$integral$(B_{k-1}(t),t(k-1),t(k+1))$.
Step 09 $F(k) =$integral$(f(t,X(k),Y(k))B_k(t),t(k-1),t(k+1))/$integral$(B_k(t),t(k-1),t(k+1))$.
Step 10 $G(k) =$integral$(g(t,X(k),Y(k))B_k(t),t(k-1),t(k+1))/$integral$(B_k(t),t(k-1),t(k+1))$.
Step 11 $Xstar(k+1) = X(k) + h(19F(k) - 5F(k-1) + F(k-2))/24$.
Step 12 $Ystar(k+1) = Y(k) + h(19G(k) - 5G(k-1) + G(k-2))/24$.
Step 13 $Fstar(k+1) =$integral$(f(t,Xstar(k+1),Ystar(k+1))B_{k+1}(t),t(k-1),t(k+1))/$integral$(B_{k+1}(t),t(k-1),t(k+1))$.
Step 14 $Gstar(k+1) =$integral$(g(t,Xstar(k+1),Ystar(k+1))B_{k+1}(t),t(k-1),t(k+1))/$integral$(B_{k+1}(t),t(k-1),t(k+1))$.
Step 15 $Xstar2(k+1) = Xstar(k+1) + 9hFstar(k+1)/24$.
Step 16 $Ystar2(k+1) = Ystar(k+1) + 9hGstar(k+1)/24$.
Step 17 $Fstar2(k+1) =$integral$(f(t,Xstar2(k+1),Ystar2(k+1))B_{k+1}(t),t(k-1),t(k+1))/$integral$(B_{k+1}(t),t(k-1),t(k+1))$.
Step 18 $Gstar2(k+1) =$integral$(g(t,Xstar2(k+1),Ystar2(k+1))B_{k+1}(t),t(k-1),t(k+1))/$integral$(B_{k+1}(t),t(k-1),t(k+1))$.
Step 19 $X(k+1) = X(k) + h(19F(k) - 5F(k-1) + F(k-2) + 9Fstar2(k+1))/24$.
Step 20 $Y(k+1) = Y(k) + h(19G(k) - 5G(k-1) + G(k-2) + 9Gstar2(k+1))/24$.
end.
OUTPUT: Approximation X and Y to x and y, respectively at the $(N+1)$ values of t.

References

1. Ahmad, M.Z.; Hasan, M.K.; Baets, B.D. Analytical and numerical solutions of fuzzy differential equations. *Inf. Sci.* **2013**, *236*, 156–167. [CrossRef]
2. Daftardar-Gejji, V.; Jafari, H. An iterative method for solving nonlinear functional equations. *J. Math. Anal. Appl.* **2006**, *316*, 753–763. [CrossRef]
3. Sukale, Y.; Daftardar-Gejji, V. New Numerical Methods for Solving Differential Equations. *Int. J. Appl. Comput. Math.* **2017**, *3*, 1639–1660. [CrossRef]
4. Perfilieva, I.; Haldeeva, E. Fuzzy transformation. In Proceedings of the Joint 9th IFSA World Congress and 20th NAFIPS International Conference, Vancouver, BC, Canada, 25–28 July 2001; Volume 4, pp. 1946–1948.
5. Perfilieva, I. Fuzzy transforms: Theory and applications. *Fuzzy Sets Syst.* **2006**, *157*, 993–1023. [CrossRef]

6. Stefanini, L. F-transform with parametric generalized fuzzy partitions. *Fuzzy Sets Syst.* **2011**, *180*, 98–120. [CrossRef]
7. Holčapek, M.; Tichý, T. A smoothing filter based on fuzzy transform. *Fuzzy Sets Syst.* **2011**, *180*, 69–97. [CrossRef]
8. Perfilieva, I.; Daňková, M.; Bede, B. Towards a higher degree F-transform. *Fuzzy Sets Syst.* **2011**, *180*, 3–19. [CrossRef]
9. Bede, B.; Rudas, I.J. Approximation properties of fuzzy transforms. *Fuzzy Sets Syst.* **2011**, *180*, 20–40. [CrossRef]
10. Kokainis, M.; Asmuss, S. Higher Degree F-transforms Based on B-splines of Two Variables. In *Information Processing and Management of Uncertainty in Knowledge-Based Systems;* Carvalho, J.P., Lesot, M.J., Kaymak, U., Vieira, S., Bouchon-Meunier, B., Yager, R.R., Eds.; Springer International Publishing: Cham, Switzerland, 2016; pp. 648–659.
11. Hurtik, P.; Perfilieva, I. Image Compression Methodology Based on Fuzzy Transform. In *International Joint Conference CISIS'12-ICEUTE'12-SOCO'12 Special Sessions;* Springer: Berlin/Heidelberg, Germany, 2013; pp. 525–532.
12. Patané, G. Fuzzy transform and least-squares approximation: Analogies, differences, and generalizations. *Fuzzy Sets Syst.* **2011**, *180*, 41–54. [CrossRef]
13. Holčapek, M.; Perfilieva, I.; Novák, V.; Kreinovich, V. Necessary and sufficient conditions for generalized uniform fuzzy partitions. *Fuzzy Sets Syst.* **2015**, *277*, 97–121. [CrossRef]
14. Khastan, A. A new representation for inverse fuzzy transform and its application. *Soft Comput.* **2017**, *21*, 3503–3512. [CrossRef]
15. Ziari, S.; Perfilieva, I. On the approximation properties of fuzzy transform. *J. Intell. Fuzzy Syst.* **2017**, *33*, 171–180. [CrossRef]
16. Alkasasbeh, H.A.; Perfilieva, I.; Ahmad, M.Z.; Yahya, Z.R. New fuzzy numerical methods for solving Cauchy problems. *Appl. Syst. Innov.* **2018**, *1*, 15. [CrossRef]
17. Perfilieva, I. Fuzzy transform: Application to the Reef growth problem. In *Fuzzy Logic in Geology;* Demicco, R.V., Klir, G.J., Eds.; Academic Press: Amsterdam, The Netherlands, 2003; Chapter 9, pp. 275–300.
18. Khastan, A.; Perfilieva, I.; Alijani, Z. A new fuzzy approximation method to Cauchy problems by fuzzy transform. *Fuzzy Sets Syst.* **2016**, *288*, 75–95. [CrossRef]
19. Chen, W.; Shen, Y. Approximate solution for a class of second-order ordinary differential equations by the fuzzy transform. *J. Intell. Fuzzy Syst.* **2014**, *27*, 73–82.
20. Alireza, K.; Zahra, A.; Irina, P. Fuzzy transform to approximate solution of two-point boundary value problems. *Math. Meth. Appl. Sci.* **2017**, *40*, 6147–6154.
21. Holcapek, M.; Valášek, R. Numerical solution of partial differential equations with the help of fuzzy transform technique. In Proceedings of the 2017 IEEE International Conference on Fuzzy Systems (FUZZ-IEEE), Naples, Italy, 9–12 July 2017; pp. 1–6.
22. Hodáková, P.; Perfilieva, I. F^1-transform of Functions of Two Variables. In *EUSFLAT 2013;* Atlantis Press: Milan, Italy, 2013; pp. 547–553.
23. Perfilieva, I.; Hodáková, P.; Hurtík, P. Differentiation by the F-transform and application to edge detection. *Fuzzy Sets Syst.* **2016**, *288*, 96–114. [CrossRef]
24. Ghosh, R.; Chowdhury, S.; Gorain, G.C.; Kar, S. Uniform stabilization of the telegraph equation with a support by fuzzy transform method. *QSci. Connect* **2014**, *2014*, 19. [CrossRef]
25. Ezzati, R.; Mokhtari, F.; Maghasedi, M. Numerical solution of Volterra-Fredholm integral equations with the help of inverse and direct discrete fuzzy transforms and collocation technique. *Int. J. Ind. Math.* **2012**, *4*, 221–229.
26. Zeinali, M.; Alikhani, R.; Shahmorad, S.; Bahrami, F.; Perfilieva, I. On the structural properties of F^m-transform with applications. *Fuzzy Sets Syst.* **2018**, *342*, 32–52. [CrossRef]
27. Baleanu, D.; Agheli, B.; Adabitabar Firozja, M.; Al Qurashi, M.M. A method for solving nonlinear Volterra's population growth model of noninteger order. *Adv. Differ. Equ.* **2017**, *2017*, 368. [CrossRef]
28. Tomasiello, S. An alternative use of fuzzy transform with application to a class of delay differential equations. *Int. J. Comput. Math.* **2017**, *94*, 1719–1726. [CrossRef]

29. Alijani, Z.; Khastan, A.; Khattri, S.K.; Tomasiello, S. Fuzzy Transform to Approximate Solution of Boundary Value Problems via Optimal Coefficients. In Proceedings of the 2017 International Conference on High Performance Computing Simulation (HPCS), Genoa, Italy, 17–21 July 2017; pp. 466–471.
30. Tomasiello, S. A First Investigation on the Dynamics of Two Delayed Neurons through Fuzzy Transform Approximation. In Proceedings of the 2017 International Conference on High Performance Computing Simulation (HPCS), Genoa, Italy, 17–21 July 2017; pp. 460–465.
31. Tomasiello, S.; Gaeta, M.; Loia, V. Quasi–consensus in Second–Order Multi–agent Systems with Sampled Data Through Fuzzy Transform. *J. Uncertain Syst.* **2016**, *10*, 243–250.
32. Noor, M.A.; Noor, K.I.; Mohyud-Din, S.T.; Shabbir, A. An iterative method with cubic convergence for nonlinear equations. *Appl. Math. Comput.* **2006**, *183*, 1249–1255. [CrossRef]
33. Saeed, R.K.; Aziz, K.M. An iterative method with quartic convergence for solving nonlinear equations. *Appl. Math. Comput.* **2008**, *202*, 435–440. [CrossRef]
34. Bhalekar, S.; Daftardar-Gejji, V. Convergence of the New Iterative Method. *Int. J. Differ. Equ.* **2011**, *2011*, 10. [CrossRef]
35. Daftardar-Gejji, V.; Sukale, Y.; Bhalekar, S. A new predictor-corrector method for fractional differential equations. *Appl. Math. Comput.* **2014**, *244*, 158–182. [CrossRef]
36. Alkasasbeh, H.A.; Perfilieva, I.; Ahmad, M.Z.; Yahya, Z.R. New approximation methods based on fuzzy transform for solving SODEs: I. *Appl. Syst. Innov.* **2018**, *1*, 29.
37. Perfilieva, I.; Holčapek, M.; Kreinovich, V. A new reconstruction from the F-transform components. *Fuzzy Sets Syst.* **2016**, *288*, 3–25. [CrossRef]
38. Perfilieva, I. F-Transform. In *Handbook of Computational Intelligence*; Kacprzyk, J., Pedrycz, W., Eds.; Springer: Berlin/Heidelberg, Germany, 2015; Chapter 7, pp. 113–130.
39. Jahedi, S.; Mehdipour, M.; Rafizadeh, R. Approximation of integrable function based on ø-transform. *Soft Comput.* **2013**, *18*, 2015–2022. [CrossRef]
40. Butcher, J.C. *Numerical Methods for Ordinary Differential Equations*, 3rd ed.; John Wiley & Sons, Ltd.: Hoboken, NJ, USA, 2016.
41. Bougoffa, L. Solvability of the predator and prey system with variable coefficients and comparison of the results with modified decomposition. *Appl. Math. Comput.* **2006**, *182*, 383–387. [CrossRef]
42. Yusufoğlu, E.; Erbaş, B. He's variational iteration method applied to the solution of the prey and predator problem with variable coefficients. *Phys. Lett. A* **2008**, *372*, 3829–3835. [CrossRef]
43. González-Parra, G.C.; Arenas, A.J.; Cogollo, M.R. Numerical-analytical solutions of predator-prey models. *WSEAS Trans. Biol. Biomed.* **2013**, *10*, 79–87.
44. Hodakova, P.; Perfilieva, I.; Valasek, R. A new approach to fuzzy boundary value problem. In *Uncertainty Modelling in Knowledge Engineering and Decision Making*; World Scientific Proceedings Series on Computer Engineering and Information Science; World Scientific: Singapore, 2016; pp. 276–281.

applied system innovation

MDPI

Article

A Performance Study of the Impact of Different Perturbation Methods on the Efficiency of GVNS for Solving TSP

Christos Papalitsas [1,*,†], Panayiotis Karakostas [2,*,†] and Theodore Andronikos [1,†]

1 Department of Informatics, Ionian University, 7 Tsirigoti Square, 49100 Corfu, Greece; andronikos@ionio.gr
2 Department of Applied Informatics, University of Macedonia, GR-546 36 Thessaloniki, Greece
* Correspondence: c14papa@ionio.gr (C.P.); pkarakostas.tm@gmail.com (P.K.); Tel.: +30-266-108-7712 (C.P.)
† These authors contributed equally to this work.

Received: 18 August 2019; Accepted: 18 September 2019; Published: 20 September 2019

Abstract: The purpose of this paper is to assess how three shaking procedures affect the performance of a metaheuristic GVNS algorithm. The first shaking procedure is generally known in the literature as intensified shaking method. The second is a quantum-inspired perturbation method, and the third is a shuffle method. The GVNS schemes are evaluated using a search strategy for both First and Best improvement and a time limit of one and two minutes. The formed GVNS schemes were applied on Traveling Salesman Problem (sTSP, nTSP) benchmark instances from the well-known TSPLib. To examine the potential advantage of any of the three metaheuristic schemes, extensive statistical analysis was performed on the reported results. The experimental data shows that for aTSP instances the first two methods perform roughly equivalently and, in any case, much better than the shuffle approach. In addition, the first method performs better than the other two when using the First Improvement strategy, while the second method gives results quite similar to the third. However, no significant deviations were observed when different methods of perturbation were used for Symmetric TSP instances (sTSP, nTSP).

Keywords: variable neighborhood search; experimental comparison; statistical analysis; traveling salesman problem; soft computing

1. Introduction

Variable Neighborhood Search (VNS) is a metaheuristic approach proposed by Mladenovic and Hansen to solve combinatorial and global optimization problems [1,2]. This framework is primarily designed to systematically modify the neighborhood structure, to reach an optimal (or near-optimal) solution [3]. VNS and its extensions have demonstrated their effectiveness in solving many problems in the combinatorial and global optimization field [4,5].

Each VNS heuristic consists of three parts. The first is a process of shaking (phase of diversification) used to escape from local optimal solutions. The next one is changing the neighborhood, where the next neighborhood structure to be searched will be determined; an approval or rejection criterion will also be applied to the last solution found during this part. The third part is the phase of improvement (intensification) achieved by exploring neighborhood structures by applying various local search moves. This exploration is carried out primarily through one of the following steps to change the neighborhood:

- Cyclic neighborhood change step: Whether there is an improvement in some neighborhood or not, the search continues in the next neighborhood structure in the list.
- Pipe neighborhood change step: If the current solution is improved in some neighborhood, exploration in that neighborhood will continue.

- Skewed neighborhood change step: Accept as new incumbent alternatives that not only improve solutions, but also some that are worse than the current incumbent solution. Such a neighborhood change step is intended to allow valley exploration away from the incumbent solution. A trial solution is evaluated taking into consideration not only the trial's objective values and the incumbent solution, but also their distance.

Variable neighborhood search variants. Many VNS variants have already been developed and used to solve hard optimization problems [6,7]. The most commonly used variants are the Basic VNS (BVNS), the Variable Neighborhood Descent (VND), and the General VNS (GVNS) and the Reduced VNS (RVNS). In the BVNS a method of diversification is alternated with a local search operator. VND consists of an improvement procedure in which neighborhood structures are systematically explored and a neighborhood change step. According to their neighborhood change step, there are different variants of VND. The pipe-VND, which uses the pipe neighborhood change step, appears to be the most efficient way to solve computational problems [6]. General Variable Neighborhood Search (GVNS) is a VNS variant that uses a VND method to improve. In many applications, GVNS has been successfully tested, as several recent works have shown [8,9].

The efficiency of metaheuristics depends on the efficiency of their components. Performance studies are a prerequisite for evaluating different metaheuristics [10] or different components of a metaheuristic algorithm [11]. In this direction and based on the VNS, Huber and Geiger (2017) [12] examined the impact of different order of local search operators in the improvement component of a VNS algorithm. There are similar studies for the impact of the initial solution [13] or the use of different neighborhood change strategies [2] to the overall performance of a VNS algorithm. However, there is a lack of contributions on studying the impact of the shaking components to the overall performance of a VNS algorithm. Papalitsas et al. (2019) [14] attempted an initial study on the impact of diversification methods on the performance of GVNS by focusing on asymmetric TSP instances.

This work is a substantial extension of our recent conference paper [14] in which we investigated the impact of three shaking methods on a GVNS metaheuristic, applied on asymmetric Traveling Salesman Problem (TSP) instances from the TSPLib. In an effort to build a comprehensive view related to that potential impact of diversification methods, the findings of the previous work are integrated with further analysis on the obtained solutions of symmetric and world TSP instances from TSPLib. To examine this potential impact of the different perturbation strategies, the three shaking methods were examined within the same improvement step. Moreover, the resulting GVNS schemes were executed both with First and Best improvement search strategies, and two different time limits were used as the main stopping criteria: 60 s and 120 s. The obtained experimental results were analyzed statistically to establish whether the use of different perturbation methods affects the performance of the GVNS algorithm. Our findings demonstrate that the use of different perturbation strategies clearly affect the solution quality in aTSP instances, while no significant differences were observed for the case of sTSP instances, with the exception of the experiments conducted using Best Improvement and 120 s run time limit. Moreover, to examine the efficiency of the formed methods, a comparison is performed between the obtained results and other recent metaheuristic solution approaches for the TSP in the literature. As it can be confirmed by our experimental results, the proposed GVNS schemes produce better solutions than the other metaheuristics.

Organization

This paper is organized as follows. In Section 2 the proposed GVNS solution methods and their technical components are explained. Section 3 contains the experimental results of our performance analysis, while the statistical tests applied to our numerical results are presented in Section 4. Section 5 provides a comparative study between our algorithms and other metaheuristic solution approaches in the recent literature. Finally, conclusions and ideas for future work are given in Section 6.

2. GVNS Heuristics

The formed GVNS methods use the pipe-VND scheme, which means that the search is taking place in the same neighborhood where the improvement occurs, as their improvement phase.

2.1. Neighborhood Structures

Three local search operators are considered for exploring different solutions:

- **1-0 Relocate**. This move removes node i from its current position in the route and re-inserts it after a selected node b.
- **2-Opt**. The 2-Opt move breaks two arcs in the current solution and reconnects them in a different way.
- **1-1 Exchange**. This move swaps two nodes in the current route.

All three neighborhood structures are incorporated in a pipe-VND scheme, as illustrated in Algorithm 1, where $l_{max} = 3$ denotes the number of neighborhood structures.

Algorithm 1 pipe-VND.

1: **procedure** PVND(N, l_{max})
2:     $l = 1$
3:     **while** $l <= l_{max}$ **do**
4:        **select case**(l)
5:           **case(1)** : $S' \leftarrow$ 1-0 Relocate(S)
6:           **case(2)** : $S' \leftarrow$ 2-Opt(S)
7:           **case(3)** : $S' \leftarrow$ 1-1 Exchange(S)
8:        **end select**
9:        **if** $f(S') < f(S)$ **then**
10:          $S \leftarrow S'$
11:        **else**
12:          $l = l + 1$
13:        **end if**
14:     **end while**
15:     **return** S
16: **end procedure**

2.2. Shaking Methods

To avoid local optimum traps, three different shaking procedures are examined. These perturbation methods are the following:

Shake_1. This diversification method randomly selects one of the predefined neighborhood structures and applies it k times ($1 < k < k_{max}$, where k_{max} is the maximum number of shaking iterations) in the current solution. The method is summarized in Algorithm 2.

Shake_2 [15]. The scientific community seems to tend to revolve around new unconventional computing methods. Overall, unconventional computing is a wide range of proposed new or unusual computing models. Part of these computing models is natural computing [16]. Nature-inspired computing has emerged as an efficient paradigm for designing and simulating innovative computational models inspired by natural phenomena to solve complex nonlinear, dynamic specific problems. Some of the well-known nature-inspired computational systems and algorithms are [17]:

1. Evolutionary, biological-inspired algorithms.
2. Swarm intelligence algorithms inspired by swarm/agent group behavior.
3. Social and cultural algorithms inspired by society's interactions and beliefs.
4. Inspired by quantum physics, Quantum-inspired algorithms.

Algorithm 2 Shake_1.

1: **procedure** SHAKE_1(S, k, l_{max})
2:    $l = random\_integer(1, l_{max})$
3:    **for** $i \leftarrow 1, k$ **do**
4:       select case(l)
5:       case(1)
6:          $S' \leftarrow$ 1-0 Relocate(S)
7:       case(2)
8:          $S' \leftarrow$ 2-Opt(S)
9:       case(3)
10:         $S' \leftarrow$ 1-1 Exchange(S)
11:      end select
12:    **end for**
13:    **return** S'
14: **end procedure**

Quantum Computing Principles

Quantum inspired methods imitate the fundamental principles of quantum computing. Quantum computing, a natural computing subsection and a field recently introduced by Feynman (1980s). Feynman realized that an effective simulation of an actual quantum system using a standard computer is not possible because the simulation of actual quantum processes would be exponentially slowed down [18,19]. Quantum computing is an important addition to the existing standard computing models. A general concept which considers the process as a quantum phenomenon. Quantum computing combines, apart from computer science, definitions, mathematical abstractions, and physics. Mathematics, such as linear algebra, and physics, such as quantum mechanics, are mainly involved.

The *qubit* is the quantum analogue of the classical bit. Similarly, the *quantum register*, which is a collection of qubits, is the quantum analogue of the classical processor register. In each call of this shaking method, a simulated quantum n-qubit register generates a normalized complex n-dimensional unit vector. In this context, normalized means that if $(z_1, \ldots, z_n)$ is the complex vector, then $|z_1|^2 + \ldots + |z_n|^2 = 1$. The dimension n of the complex unit vector is greater than or equal to the dimension of the problem. The complex n-dimensional vector is converted into a real n-dimensional vector, the components of which are real numbers in the interval $[0, 1]$. If z_i and r_i are the ith components of the complex and real vectors respectively, then $r_i = |z_i|^2$, i.e., r_i is equal to the modulus squared of z_i. Moreover, each of the real vector's selected components corresponds to a current solution node. For each node of the incumbent solution, the components are used as a flag. Sorting the first vector affects the order in the solution vector due to the correspondence between components and nodes in a tour and thus drives the exploration effort to another point in the search space. This shaking procedure's pseudocode is given in Algorithm 3.

Algorithm 3 Shake_2.

1: **procedure** SHAKE_2(S, n)
2:    $NQubits \leftarrow$ **QuantumRegister**(n)
3:    Compute the components based on the qubits.
4:    Save the n components in the vector $QCompVector$.
5:    Matching each element in the $QCompVector$ with a node in S.
6:    Descending sorting on $QCompVector$ produces S'.
7:    Recalculate the cost of the new S'.
8:    **return** S'
9: **end procedure**

Shake_3. This shaking method is a shuffle method, where in each iteration the customers are placed in a random order. The method is shown in Algorithm 4.

Algorithm 4 Shake_3.

1: **procedure** SHAKE_3(S)
2:    $S' \leftarrow Shuffle(S)$
3:    **return** S'
4: **end procedure**

2.3. GVNS Schemes

For each perturbation method a GVNS scheme is formed. Specifically, the GVNS_1 contains Shake_1 as its shaking method, GVNS_2 uses Shake_2 to diversify solutions, and GVNS_3 adopts the Shake_3 perturbation method. The initial solution is produced by the Nearest Neighbor heuristic in all GVNS schemes. The pseudocode for three GVNS approaches is given in Algorithms 5–7, respectively.

Algorithm 5 GVNS_1.

1: **procedure** GVNS_1($S, k_{max}, max\_time, l_{max}$)
2:    **while** $time \leq max\_time$ **do**
3:       **for** $k \leftarrow 1, k_{max}$ **do**
4:          $S^* = \text{Shake\_1}(S, k, l_{max})$
5:          $S' = pVND(S^*)$
6:          **if** $f(S') < f(S)$ **then**
7:             $S \leftarrow S'$
8:          **end if**
9:       **end for**
10:    **end while**
11:    **return** S
12: **end procedure**

Algorithm 6 GVNS_2.

1: **procedure** GVNS_2(S, n, max_time)
2:    **while** $time \leq max\_time$ **do**
3:       $S^* = \text{Shake\_2}(S, n)$
4:       $S' = pVND(S^*)$
5:       **if** $f(S') < f(S)$ **then**
6:          $S \leftarrow S'$
7:       **end if**
8:    **end while**
9:    **return** S
10: **end procedure**

Algorithm 7 GVNS_3.

1: **procedure** GVNS_3(S, max_time)
2:    **while** $time \leq max\_time$ **do**
3:       $S^* = \text{Shake\_3}(S)$
4:       $S' = pVND(S^*)$
5:       **if** $f(S') < f(S)$ **then**
6:          $S \leftarrow S'$
7:       **end if**
8:    **end while**
9:    **return** S
10: **end procedure**

It should be mentioned that in all three GVNS methods the neighborhoods are searched with both the First and Best improvement search strategy.

3. Computational Analysis

3.1. Computing Environment & Parameter Settings

The aforementioned methods were implemented in Fortran and were executed in a PC running Windows 64-bit on an Intel Core i7-6700 CPU at 2.6 GHz with 16 GB RAM. The compilation of the code was done using the Intel Fortran 64 compiler XE with the optimization option /O3. The maximum execution time limit was set to $max\_time = 60$ s and $max\_time = 120$ s and the maximum number of shaking iterations in the Shake_1 was experimentally set to $k_{max} = 12$.

3.2. Computational Results

This section presents the computational results of the different perturbation strategies for each class of experiments. The GVNS schemes with the different shaking methods were applied on TSPLIB instances. The TSP is one of the most famous NP-hard combinatorial optimization problems. Solving the TSP means finding the minimum cost route so that the salesman starts from a particular node and returns to that node after passing from all the other nodes once.

All experiments were executed 5 times and the average value of all runs was computed. Tables 1 and 2 contain the aggregated experimental results. Specifically, they show the benchmark name, the optimal value (zOpt), the cost of the three GVNS schemes (GVNS_1, GVNS_2 and GVNS_3) and their corresponding gaps from the optimal value. The results depicted in Table 1 were obtained using the First Improvement search strategy and an execution time limit of 1 min, whereas the results in Table 2 were obtained using the Best Improvement search strategy and the same execution time of 1 min. As mentioned earlier, the cost of each GVNS scheme is the average of 5 runs for each problem. The reported gap is computed as follows: given the outcome x, its gap from the optimal value OV is given by the formula $\frac{100\times(x-OV)}{OV}$.

Table 1. The results shown here were obtained using the First Improvement search strategy and an execution time limit of 1 min [14].

Instance	zOpt	GVNS_1	GVNS_2	GVNS_3	GAP_1 (%)	GAP_2 (%)	GAP_3 (%)
br17.atsp	39	39	39	39	0.00	0.00	0.00
ft53.atsp	6905	7189	7328	7737	4.11	6.13	12.05
ft70.atsp	38673	39782	40691	40537	2.87	5.22	4.82
ftv33.atsp	1286	1318	1339	1450	2.49	4.12	12.75
ftv35.atsp	1473	1484	1499	1596	0.75	1.77	8.35
ftv38.atsp	1530	1546	1585	1579	1.05	3.59	3.20
ftv44.atsp	1613	1651	1760	1797	2.36	9.11	11.41
ftv47.atsp	1778	1821	1992	2101	2.42	12.04	18.17
ftv55.atsp	1608	1666	1985	1912	3.61	23.45	18.91
ftv64.atsp	1839	1961	2382	2395	6.63	29.53	30.23
ftv70.atsp	1950	2136	2557	2484	9.54	31.13	27.38
ftv170.atsp	2755	3487	3923	3923	26.57	42.40	42.40
kro124p.atsp	36230	39024	43187	40259	7.71	19.20	11.12
p43.atsp	5620	5620	5623	5658	0.00	0.05	0.68
rbg323.atsp	1326	1516	1563	1626	14.32	17.87	22.62
rbg358.atsp	1163	1347	1437	1404	15.82	23.55	20.72
rbg403.atsp	2465	2535	2587	2565	9.78	4.42	11.76
rbg443.atsp	2720	2814	2859	2814	3.46	5.11	3.46
ry48p.atsp	14422	14549	14901	14738	0.88	3.32	2.19
Average	**6599.74**	**6920.26**	**7328.37**	**7190.21**	**12.33**	**12.64**	**24.10**

The results in Table 1 indicate a definite pattern, namely that both GVNS_1 and GVNS_2 outperform GVNS_3 in most cases. Recall that GVNS_3 is a shuffle perturbation strategy. For example, consider benchmark ftv47; we can see that the cost of GVNS_1 is 1821, the cost of GVNS_2 is 1992 and of GVNS_3 is 2101. GVNS_1 and GVNS_2 both outperform GVNS_3 and are also relatively close to the optimal value (1778).

Table 2. The results depicted here were obtained using the Best Improvement search strategy and an execution time limit of 1 min [14].

Instance	zOpt	GVNS_1	GVNS_2	GVNS_3	GAP_1 (%)	GAP_2 (%)	GAP_3 (%)
br17.atsp	39	39	39	39	0.00	0.00	0.00
ft53.atsp	6905	7043	7135	7674	2.00	3.33	11.14
ft70.atsp	38673	39507	40206	40539	2.16	3.96	4.83
ftv33.atsp	1286	1289	1286	1379	0.23	0.00	7.23
ftv35.atsp	1473	1476	1473	1533	0.20	0.00	4.07
ftv38.atsp	1530	1538	1541	1599	0.52	0.72	4.51
ftv44.atsp	1613	1632	1644	1728	1.18	1.92	7.13
ftv47.atsp	1778	1792	1816	1940	0.79	2.14	9.11
ftv55.atsp	1608	1642	1665	2012	2.11	3.54	25.12
ftv64.atsp	1839	1908	1986	2193	3.75	7.99	19.25
ftv70.atsp	1950	2110	2157	2346	8.21	10.62	20.31
ftv170.atsp	2755	3341	3852	3923	21.27	39.82	42.40
kro124p.atsp	36230	36501	37076	38195	0.75	2.34	5.42
p43.atsp	5620	5620	5620	5627	0.00	0.00	0.12
rbg323.atsp	1326	1486	1539	1633	12.06	16.06	23.15
rbg358.atsp	1163	1307	1409	1437	12.38	21.15	23.55
rbg403.atsp	2465	2510	2547	2554	11.76	11.76	11.76
rbg443.atsp	2720	2765	2824	2844	1.65	3.16	4.56
ry48p.atsp	14422	14480	14498	14659	0.40	0.12	1.64
Average	**6599.74**	**6736.11**	**6858.58**	**7044.95**	**15.88**	**17.69**	**22.28**

In addition, the provided results in Table 2 lead to the same statement that both GVNS_1 and GVNS_2 produce better results than GVNS_3 in most cases. Table 3 shows the results of the GVNS schemes within a 2 min run time limit and the First Improvement as search strategy. The results of Table 3 mention that GVNS_1 outperform GVNS_2 and GVNS_3 in most cases. However, the main difference from the results of Tables 1 and 2 is that now the behavior of GVNS_2 is closer to that of GVNS_3 solution approach. Table 4 shows the results achieved by the GVNS schemes within a 2 min run time limit and the Best Improvement search strategy. The results of Table 4 corroborate the conclusion of Tables 1 and 2 that both GVNS_1 and GVNS_2 outperform GVNS_3 in most cases.

Table 3. Results using the First Improvement search strategy and an execution time limit of 2 min [14].

Instance	zOpt	GVNS_1	GVNS_2	GVNS_3	GAP_1 (%)	GAP_2 (%)	GAP_3 (%)
br17.atsp	39	39	39	39	0.00	0.00	0.00
ft53.atsp	6905	7024	7498	7752	1.72	8.59	12.27
ft70.atsp	38673	39615	40827	40505	2.44	5.57	4.74
ftv33.atsp	1286	1330	1370	1454	3.42	6.53	13.06
ftv35.atsp	1473	1482	1519	1604	0.61	3.12	8.89
ftv38.atsp	1530	1547	1618	1576	1.11	5.75	3.01
ftv44.atsp	1613	1628	1839	1812	0.93	14.01	12.34
ftv47.atsp	1778	1787	2020	2097	0.51	13.61	17.94
ftv55.atsp	1608	1668	2012	1912	3.73	25.12	18.91
ftv64.atsp	1839	1951	2484	2476	6.09	35.07	34.64
ftv70.atsp	1950	2165	2571	2484	11.03	31.85	27.38
ftv170.atsp	2755	3412	3923	3923	23.85	42.40	42.40
kro124p.atsp	36230	39344	44243	40849	8.60	22.12	12.75
p43.atsp	5620	5620	5628	5657	0.00	0.14	0.66
rbg323.atsp	1326	1499	1576	1586	13.04	18.85	19.60
rbg358.atsp	1163	1329	1410	1406	14.27	21.23	20.89
rbg403.atsp	2465	2509	2586	2547	2.27	4.10	11.76
rbg443.atsp	2720	2808	2849	2811	3.24	4.74	3.35
ry48p.atsp	14422	14475	14936	14708	0.37	3.56	1.98
Average	**6599.74**	**6906.95**	**7418.32**	**7220.95**	**5.29**	**13.96**	**24.78**

Table 4. Results using the Best Improvement search strategy and an execution time limit of 2 min [14].

Instance	zOpt	GVNS_1	GVNS_2	GVNS_3	GAP_1 (%)	GAP_2 (%)	GAP_3 (%)
br17.atsp	39	39	39	39	0.00	0.00	0.00
ft53.atsp	6905	7043	7207	7773	2.00	4.37	12.57
ft70.atsp	38673	39358	40230	40588	1.77	4.03	4.95
ftv33.atsp	1286	1286	1290	1370	0.00	0.31	6.53
ftv35.atsp	1473	1474	1475	1509	0.07	0.14	2.44
ftv38.atsp	1530	1538	1555	1599	0.52	1.63	4.51
ftv44.atsp	1613	1636	1664	1731	1.43	3.16	7.32
ftv47.atsp	1778	1787	1837	1903	0.51	3.32	7.03
ftv55.atsp	1608	1640	1686	2012	1.99	4.85	25.12
ftv64.atsp	1839	1914	2032	2217	4.08	10.49	20.55
ftv70.atsp	1950	2038	2189	2342	4.51	12.26	20.10
ftv170.atsp	2755	3351	3918	3923	21.63	42.21	42.40
kro124p.atsp	36230	36379	37378	37915	0.41	3.17	4.65
p43.atsp	5620	5620	5620	5625	0.00	0.00	0.09
rbg323.atsp	1326	1473	1531	1610	11.08	15.46	21.41
rbg358.atsp	1163	1292	1405	1435	11.09	20.80	23.38
rbg403.atsp	2465	2498	2547	2553	1.30	3.25	11.76
rbg443.atsp	2720	2771	2822	2842	1.88	3.75	4.49
ry48p.atsp	14422	14468	14464	14678	0.32	0.29	1.78
Average	**6599.74**	**6716.05**	**6888.89**	**7034.95**	**3.28**	**7.05**	**22.16**

Tables 5–8 contain the aggregated experimental results for Symmetric TSP instances. Specifically, they contain the benchmark name, the optimal value (zOpt), the cost of the three GVNS variations (GVNS_1, GVNS_2 and GVNS_3) and their corresponding gaps from the optimal value. Table 5 depicts GVNS using the First Improvement search strategy and an execution time limit of 1 min. Table 6 shows GVNS using the Best Improvement as search strategy and an execution time of 1 min. Tables 7 and 8 are executed for 120 s within First and Best improvement search strategy respectively. The reported results show that the GVNS_3 produce better solutions than the other two algorithms, and that GVNS_1 and GVNS_2 do not have significant differences.

Table 5. Results using the First Improvement search strategy and an execution time limit of 1 min.

Instance	zOpt	GVNS_1	GVNS_2	GVNS_3	GAP_1	GAP_2	GAP_3
a280.tsp	2579	2683	2739	2745	4.03	6.20	6.44
att48.tsp	10628	10628	10628	10635	0.00	0.00	0.07
bayg29.tsp	1610	1610	1610	1610	0.00	0.00	0.00
bays29.tsp	2020	2020	2020	2020	0.00	0.00	0.00
bier127.tsp	118282	118636	120066	119966	0.30	1.51	1.42
kroA100.tsp	21282	21296	21375	21398	0.07	0.44	0.55
burma14.tsp	3323	3323	3323	3323	0.00	0.00	0.00
ch130.tsp	6110	6156	6239	6235	0.75	2.11	2.05
ch150.tsp	6528	6583	6720	6723	0.84	2.94	2.99
d493.tsp	35002	36928	37307	37166	5.50	6.59	6.18
kroB100.tsp	22141	22187	22339	22362	0.21	0.89	1.00
kroC100.tsp	20749	20759	20864	20834	0.05	0.55	0.41
kroD100.tsp	21294	21370	21573	21667	0.36	1.31	1.75
kroE100.tsp	22068	22114	22291	22360	0.21	1.01	1.32
kroA150.tsp	26524	26792	27247	27206	1.01	2.73	2.57
kroB150.tsp	26130	26382	26680	26767	0.96	2.10	2.44
kroA200.tsp	29368	29753	30420	30392	1.31	3.58	3.49
kroB200.tsp	29437	30164	30727	30711	2.47	4.38	4.33
d198.tsp	15780	15908	16116	16147	0.81	2.13	2.33
brg180.tsp	1950	1960	2024	2040	0.51	3.79	4.62
berlin52.tsp	7542	7542	7542	7542	0.00	0.00	0.00

Table 5. *Cont.*

Instance	zOpt	GVNS_1	GVNS_2	GVNS_3	GAP_1	GAP_2	GAP_3
dantzig42.tsp	699	699	699	699	0.00	0.00	0.00
eil51.tsp	426	426	426	428	0.00	0.00	0.47
eil76.tsp	538	539	544	545	0.19	1.12	1.30
eil101.tsp	629	630	645	647	0.16	2.54	2.86
fri26.tsp	937	937	937	937	0.00	0.00	0.00
gil262.tsp	2378	2460	2509	2509	3.45	5.51	5.51
gr17.tsp	2085	2085	2085	2085	0.00	0.00	0.00
gr21.tsp	2707	2707	2707	2707	0.00	0.00	0.00
gr24.tsp	1272	1272	1272	1272	0.00	0.00	0.00
gr48.tsp	5046	5046	5046	5048	0.00	0.00	0.04
gr96.tsp	55209	55285	55635	55713	0.14	0.77	0.91
gr120.tsp	6942	6979	7085	7103	0.53	2.06	2.32
gr137.tsp	69853	70207	71158	71330	0.51	1.87	2.11
gr202.tsp	40160	41232	41752	41850	2.67	3.96	4.21
gr229.tsp	134602	137642	139570	140144	2.26	3.69	4.12
gr431.tsp	171414	179950	182365	182884	4.98	6.39	6.69
hk48.tsp	11461	11461	11461	11470	0.00	0.00	0.08
lin105.tsp	14379	14386	14433	14458	0.05	0.38	0.55
lin318.tsp	42029	43641	44175	44183	3.84	5.11	5.13
pcb442.tsp	50778	53301	54176	54486	4.97	6.69	7.30
pr76.tsp	108159	108168	108411	108621	0.01	0.23	0.43
pr107.tsp	44303	44428	44695	44708	0.28	0.88	0.91
pr124.tsp	59030	59045	59169	59222	0.03	0.24	0.33
pr136.tsp	96772	97875	99118	99300	1.14	2.42	2.61
pr144.tsp	58537	58538	58629	58627	0.00	0.16	0.15
pr152.tsp	73682	74016	74379	74299	0.45	0.95	0.84
pr226.tsp	80369	80605	81007	81267	0.29	0.79	1.12
pr264.tsp	49135	50237	50883	50847	2.24	3.56	3.48
pr299.tsp	48191	50331	50917	50883	4.44	5.66	5.59
pr439.tsp	107217	112633	114121	114191	5.05	6.44	6.50
rat99.tsp	1211	1215	1240	1243	0.33	2.39	2.64
rat195.tsp	2323	2371	2456	2453	2.07	5.73	5.60
rd100.tsp	7910	7925	8011	8051	0.19	1.28	1.78
rd400.tsp	15281	15951	16281	16269	4.38	6.54	6.47
si175.tsp	21407	21430	21482	21486	0.11	0.35	0.37
st70.tsp	675	677	675	677	0.30	0.00	0.30
swiss42.tsp	1273	1273	1273	1273	0.00	0.00	0.00
ts225.tsp	126643	126858	128223	128359	0.17	1.25	1.35
tsp225.tsp	3916	4014	4116	4122	2.50	5.11	5.26
u159.tsp	42080	42556	43373	43374	1.13	3.07	3.08
ulysses16.tsp	6859	6859	6859	6859	0.00	0.00	0.00
ulysses22.tsp	7013	7013	7013	7013	0.00	0.00	0.00
ali535.tsp	202339	218486	221688	217544	7.98	9.56	7.51
att532.tsp	27686	29351	29747	28818	6.01	7.44	4.09
brazil58.tsp	25395	33181	25395	25396	30.66	0.00	0.00
brg180.tsp	1950	1959	2040	2158	0.46	4.62	10.67
d657.tsp	48912	52126	53015	51059	6.57	8.39	4.39
d1291.tsp	50801	59103	60214	55243	16.34	18.53	8.74
d1655.tsp	62128	73791	74028	73982	18.77	19.15	19.08
d2103.tsp	80450	86653	86653	86653	7.71	7.71	7.71
dsj1000.tsp	18659688	24056781	24631467	20034159	20.28	23.16	0.17
fl417.tsp	11861	12366	12232	12227	4.26	3.13	3.09
fl1400.tsp	20127	27242	27447	25980	35.35	36.37	29.08
fl1577.tsp	22249	27941	27996	27996	25.58	25.83	25.83
fl3795.tsp	28772	35262	35285	35285	22.56	22.64	22.64
fnl4461.tsp	182566	229963	229963	229963	25.96	25.96	25.96

Table 5. *Cont.*

Instance	zOpt	GVNS_1	GVNS_2	GVNS_3	GAP_1	GAP_2	GAP_3
gr666.tsp	294358	317446	324339	309556	7.84	10.19	5.16
nrw1379.tsp	56638	67679	68964	61769	19.49	21.76	9.06
p654.tsp	34643	36502	36558	35569	5.37	5.53	2.67
pa561.tsp	2763	2928	3053	3003	5.97	10.49	8.69
pcb1173.tsp	56892	70520	71978	61273	23.95	26.52	7.70
pcb3038.tsp	137694	175926	176310	176310	27.77	28.04	28.04
pr1002.tsp	259045	323543	331103	277196	24.90	27.82	7.01
pr2392.tsp	378032	460547	461170	461170	21.83	21.99	21.99
rat575.tsp	6773	7179	7190	7153	5.99	6.15	5.61
rat783.tsp	8806	9634	9610	9341	9.40	9.13	6.08
rl1304.tsp	252948	330540	335779	277603	30.68	32.75	9.75
rl1323.tsp	270199	331586	332103	293133	22.72	22.91	8.49
rl1889.tsp	316536	388695	389270	389270	22.80	22.98	22.98
rl5915.tsp	565530	695602	695602	695602	23.00	23.00	23.00
rl5934.tsp	556045	672412	672412	672412	20.93	20.93	20.93
si535.tsp	48450	48697	48848	48807	0.51	0.82	0.74
si1032.tsp	92650	92883	94571	92909	0.25	2.07	0.28
u574.tsp	36905	40206	40020	39488	8.94	8.44	7.00
u724.tsp	41910	45583	45988	44646	8.76	9.73	6.53
u1060.tsp	224094	297757	308980	242181	32.87	37.88	8.07
u1432.tsp	152970	185839	188807	166714	21.49	23.43	8.98
u1817.tsp	57201	71999	72030	72030	25.87	25.92	25.92
u2152.tsp	64253	78870	79260	79260	22.75	23.36	23.36
u2319.tsp	234256	275453	278765	278765	17.59	19.00	19.00
vm1084.tsp	239297	295088	301477	258248	23.31	25.98	7.92
vm1748.tsp	336556	406536	408102	408102	20.79	21.26	21.26
Average	**266956.86**	**317607.30**	**323886.60**	**276033.63**	**7.31**	**8.06**	**6.03**

Table 6. Results using the Best Improvement search strategy and an execution time limit of 1 min.

Instance	zOpt	GVNS_1	GVNS_2	GVNS_3	GAP_1	GAP_2	GAP_3
a280.tsp	2579	2632	2738	2734	2.06	6.17	6.01
att48.tsp	10628	10628	10628	10631	0.00	0.00	0.03
bayg29.tsp	1610	1610	1610	1610	0.00	0.00	0.00
bays29.tsp	2020	2020	2020	2020	0.00	0.00	0.00
bier127.tsp	118282	118411	119593	119962	0.11	1.11	1.42
kroA100.tsp	21282	21282	21332	21413	0.00	0.23	0.62
burma14.tsp	3323	3323	3323	3323	0.00	0.00	0.00
ch130.tsp	6110	6147	6219	6237	0.61	1.78	2.08
ch150.tsp	6528	6571	6704	6725	0.66	2.70	3.02
d493.tsp	35002	36559	37182	37119	4.45	6.23	6.05
kroB100.tsp	22141	22162	22282	22362	0.09	0.64	1.00
kroC100.tsp	20749	20749	20837	20880	0.00	0.42	0.63
kroD100.tsp	21294	21346	21507	21581	0.24	1.00	1.35
kroE100.tsp	22068	22129	22241	22264	0.28	0.78	0.89
kroA150.tsp	26524	26696	27169	27220	0.65	2.43	2.62
kroB150.tsp	26130	26233	26631	26696	0.39	1.92	2.17
kroA200.tsp	29368	29549	30416	30381	0.62	3.57	3.45
kroB200.tsp	29437	29888	30618	30688	1.53	4.01	4.25
d198.tsp	15780	15845	16077	16090	0.41	1.88	1.96
brg180.tsp	1950	1963	2024	2035	0.67	3.79	4.36
berlin52.tsp	7542	7542	7542	7563	0.00	0.00	0.28
dantzig42.tsp	699	699	699	699	0.00	0.00	0.00
eil51.tsp	426	426	426	428	0.00	0.00	0.47
eil76.tsp	538	538	544	547	0.00	1.12	1.67

Table 6. *Cont.*

Instance	zOpt	GVNS_1	GVNS_2	GVNS_3	GAP_1	GAP_2	GAP_3
eil101.tsp	629	630	645	648	0.16	2.54	3.02
fri26.tsp	937	937	937	937	0.00	0.00	0.00
gil262.tsp	2378	2437	2515	2518	2.48	5.76	5.89
gr17.tsp	2085	2085	2085	2085	0.00	0.00	0.00
gr21.tsp	2707	2707	2707	2707	0.00	0.00	0.00
gr24.tsp	1272	1272	1272	1272	0.00	0.00	0.00
gr48.tsp	5046	5046	5046	5049	0.00	0.00	0.06
gr96.tsp	55209	55293	55521	55582	0.15	0.57	0.68
gr120.tsp	6942	6977	7085	7120	0.50	2.06	2.56
gr137.tsp	69853	69948	70964	71412	0.14	1.59	2.23
gr202.tsp	40160	41079	41693	41720	2.29	3.82	3.88
gr229.tsp	134602	136416	139729	140377	1.35	3.81	4.29
gr431.tsp	171414	177946	181693	182205	3.81	6.00	6.30
hk48.tsp	11461	11461	11461	11471	0.00	0.00	0.09
lin105.tsp	14379	14382	14396	14437	0.02	0.12	0.40
lin318.tsp	42029	43094	44070	44309	2.53	4.86	5.42
pcb442.tsp	50778	52584	54456	54581	3.56	7.24	7.49
pr76.tsp	108159	108159	108278	108513	0.00	0.11	0.33
pr107.tsp	44303	44396	44539	44607	0.21	0.53	0.69
pr124.tsp	59030	59030	59058	59081	0.00	0.05	0.09
pr136.tsp	96772	97262	98966	98879	0.51	2.27	2.18
pr144.tsp	58537	58537	58561	58561	0.00	0.04	0.04
pr152.tsp	73682	73781	74027	73966	0.13	0.47	0.39
pr226.tsp	80369	80462	80861	80834	0.12	0.61	0.58
pr264.tsp	49135	49670	50905	50886	1.09	3.60	3.56
pr299.tsp	48191	49245	50614	50646	2.19	5.03	5.09
pr439.tsp	107217	111621	113347	113038	4.11	5.72	5.43
rat99.tsp	1211	1213	1234	1240	0.17	1.90	2.39
rat195.tsp	2323	2356	2451	2457	1.42	5.51	5.77
rd100.tsp	7910	7927	7963	8022	0.21	0.67	1.42
rd400.tsp	15281	15802	16312	16311	3.41	6.75	6.74
si175.tsp	21407	21420	21472	21476	0.06	0.30	0.32
st70.tsp	675	676	675	676	0.15	0.00	0.15
swiss42.tsp	1273	1273	1273	1273	0.00	0.00	0.00
ts225.tsp	126643	126721	127690	127849	0.06	0.83	0.95
tsp225.tsp	3916	3987	4115	4119	1.81	5.08	5.18
u159.tsp	42080	42329	42969	43024	0.59	2.11	2.24
ulysses16.tsp	6859	6859	6859	6859	0.00	0.00	0.00
ulysses22.tsp	7013	7013	7013	7013	0.00	0.00	0.00
ali535.tsp	202339	213387	218429	218205	5.46	7.95	7.84
att532.tsp	27686	28764	29614	29525	3.89	6.96	6.64
brazil58.tsp	25395	33181	25395	25412	30.66	0.00	0.07
brg180.tsp	1950	1963	2019	2198	0.67	3.54	12.72
d657.tsp	48912	51497	52986	51934	5.29	8.33	6.18
d1291.tsp	50801	54927	57302	55431	8.12	12.80	9.11
d1655.tsp	62128	67236	70399	67683	8.22	13.31	8.94
d2103.tsp	80450	83240	86653	83486	3.47	7.71	3.77
dsj1000.tsp	18659688	20201248	20449409	20063920	8.26	9.59	7.53
fl417.tsp	11861	12023	12119	12161	1.37	2.18	2.53
fl1400.tsp	20127	21244	21198	21166	5.55	5.32	5.16
fl1577.tsp	22249	23721	24355	23736	6.62	9.47	6.68
fl3795.tsp	28772	34663	33535	35214	20.47	16.55	22.39
fnl4461.tsp	182566	217998	204703	199441	19.41	12.13	9.24
gr666.tsp	294358	313338	321656	315223	6.45	9.27	7.09
nrw1379.tsp	56638	60516	62459	60983	6.85	10.28	7.67
p654.tsp	34643	36083	35544	36741	4.16	2.60	6.06
pa561.tsp	2763	2893	2896	3058	4.71	4.81	10.68
pcb1173.tsp	56892	61883	63335	62161	8.77	11.32	9.26
pcb3038.tsp	137694	153475	156692	149788	11.46	13.80	8.78
pr1002.tsp	259045	278408	285203	279922	7.47	10.10	8.06

Table 6. *Cont.*

Instance	zOpt	GVNS_1	GVNS_2	GVNS_3	GAP_1	GAP_2	GAP_3
pr2392.tsp	378032	410784	430379	408360	8.66	13.85	8.02
rat575.tsp	6773	7195	7090	7224	6.23	4.68	6.66
rat783.tsp	8806	9373	9391	9391	6.44	6.64	6.64
rl1304.tsp	252948	282487	282839	274566	11.68	11.82	8.55
rl1323.tsp	270199	293350	300601	285668	8.57	11.25	5.73
rl1889.tsp	316536	344218	356697	342893	8.75	12.69	8.33
rl5915.tsp	565530	680825	695602	695602	20.39	23.00	23.00
rl5934.tsp	556045	664895	672412	661012	19.58	20.93	18.88
si535.tsp	48450	48622	48783	48847	0.36	0.69	0.82
si1032.tsp	92650	92918	93397	92908	0.29	0.81	0.28
u574.tsp	36905	40374	40022	39792	9.40	8.45	7.82
u724.tsp	41910	44662	45765	45273	6.57	9.20	8.02
u1060.tsp	224094	242630	246175	243291	8.27	9.85	8.57
u1432.tsp	152970	165304	170578	165833	8.06	11.51	8.41
u1817.tsp	57201	62782	66243	62050	9.76	15.81	8.48
u2152.tsp	64253	70205	74581	70787	9.26	16.07	10.17
u2319.tsp	234256	243928	249738	245475	4.13	6.61	4.79
vm1084.tsp	239297	256431	262955	255369	7.16	9.89	6.72
vm1748.tsp	336556	362026	373926	360551	7.57	11.10	7.13
Average	**253944.13**	**274974.75**	**278630.04**	**273507.26**	**13.05**	**4.88**	**4.40**

Table 7. Results using the First Improvement search strategy and an execution time limit of 2 min.

Instance	zOpt	GVNS_1	GVNS_2	GVNS_3	GAP_1	GAP_2	GAP_3
a280.tsp	2579	2672	2728	2706	3.61	5.78	4.92
att48.tsp	10628	10628	10628	10724	0.00	0.00	0.90
bayg29.tsp	1610	1610	1610	1615	0.00	0.00	0.31
bays29.tsp	2020	2020	2020	2028	0.00	0.00	0.40
bier127.tsp	118282	118518	119737	119860	0.20	1.23	1.33
kroA100.tsp	21282	21283	21337	21699	0.00	0.26	1.96
burma14.tsp	3323	3323	3323	3323	0.00	0.00	0.00
ch130.tsp	6110	6157	6218	6289	0.77	1.77	2.93
ch150.tsp	6528	6583	6680	6649	0.84	2.33	1.85
d493.tsp	35002	36659	37040	36557	4.73	5.82	4.44
kroB100.tsp	22141	22168	22317	22557	0.12	0.79	1.88
kroC100.tsp	20749	20757	20806	21321	0.04	0.27	2.76
kroD100.tsp	21294	21329	21567	21904	0.16	1.28	2.86
kroE100.tsp	22068	22144	22250	22407	0.34	0.82	1.54
kroA150.tsp	26524	26630	27050	27738	0.40	1.98	4.58
kroB150.tsp	26130	26232	26701	26742	0.39	2.19	2.34
kroA200.tsp	29368	29570	30177	29718	0.69	2.75	1.19
kroB200.tsp	29437	29801	30459	30205	1.24	3.47	2.61
d198.tsp	15780	15871	16077	15964	0.58	1.88	1.17
brg180.tsp	1950	1956	2026	2153	0.31	3.90	10.41
berlin52.tsp	7542	7542	7542	7591	0.00	0.00	0.65
dantzig42.tsp	699	699	699	706	0.00	0.00	1.00
eil51.tsp	426	426	426	430	0.00	0.00	0.94
eil76.tsp	538	538	543	544	0.00	0.93	1.12
eil101.tsp	629	629	642	649	0.00	2.07	3.18
fri26.tsp	937	937	937	948	0.00	0.00	1.17
gil262.tsp	2378	2444	2505	2542	2.78	5.34	6.90
gr17.tsp	2085	2085	2085	2085	0.00	0.00	0.00
gr21.tsp	2707	2707	2707	2707	0.00	0.00	0.00
gr24.tsp	1272	1272	1272	1272	0.00	0.00	0.00
gr48.tsp	5046	5046	5046	5063	0.00	0.00	0.34
gr96.tsp	55209	55247	55549	56109	0.07	0.62	1.63

Table 7. *Cont.*

Instance	zOpt	GVNS_1	GVNS_2	GVNS_3	GAP_1	GAP_2	GAP_3
gr120.tsp	6942	6960	7063	7103	0.26	1.74	2.32
gr137.tsp	69853	70005	71002	71579	0.22	1.64	2.47
gr202.tsp	40160	40973	41598	41672	2.02	3.58	3.76
gr229.tsp	134602	136884	139544	138929	1.70	3.67	3.21
gr431.tsp	171414	178310	181839	180800	4.02	6.08	5.48
hk48.tsp	11461	11461	11461	11664	0.00	0.00	1.77
lin105.tsp	14379	14394	14413	14672	0.10	0.24	2.04
lin318.tsp	42029	43463	44016	44024	3.41	4.73	4.75
pcb442.tsp	50778	52867	53957	52362	4.11	6.26	3.12
pr76.tsp	108159	108159	108373	109446	0.00	0.20	1.19
pr107.tsp	44303	44400	44687	44769	0.22	0.87	1.05
pr124.tsp	59030	59033	59108	59444	0.01	0.13	0.70
pr136.tsp	96772	97158	98762	101877	0.40	2.06	5.28
pr144.tsp	58537	58537	58605	60768	0.00	0.12	3.81
pr152.tsp	73682	73801	74185	75515	0.16	0.68	2.49
pr226.tsp	80369	80533	80840	81472	0.20	0.59	1.37
pr264.tsp	49135	49633	50391	50951	1.01	2.56	3.70
pr299.tsp	48191	49595	50642	50023	2.91	5.09	3.80
pr439.tsp	107217	111921	113644	113269	4.39	5.99	5.64
rat99.tsp	1211	1212	1236	1266	0.08	2.06	4.54
rat195.tsp	2323	2365	2448	2404	1.81	5.38	3.49
rd100.tsp	7910	7925	7998	8122	0.19	1.11	2.68
rd400.tsp	15281	15847	16203	15936	3.70	6.03	4.29
si175.tsp	21407	21428	21475	21514	0.10	0.32	0.50
st70.tsp	675	676	675	691	0.15	0.00	2.37
swiss42.tsp	1273	1273	1273	1273	0.00	0.00	0.00
ts225.tsp	126643	126815	128183	127374	0.14	1.22	0.58
tsp225.tsp	3916	3996	4101	4093	2.04	4.72	4.52
u159.tsp	42080	42341	43196	44203	0.62	2.65	5.05
ulysses16.tsp	6859	6859	6859	6860	0.00	0.00	0.01
ulysses22.tsp	7013	7013	7013	7013	0.00	0.00	0.00
ali535.tsp	202339	217245	219075	216606	7.37	8.27	7.05
att532.tsp	27686	29167	29615	28916	5.35	6.97	4.44
brazil58.tsp	25395	36518	25395	25395	43.80	0.00	0.00
brg180.tsp	1950	1956	2023	2165	0.31	3.74	11.03
d657.tsp	48912	51769	52769	50997	5.84	7.89	4.26
d1291.tsp	50801	56007	60214	55060	10.25	18.53	8.38
d1655.tsp	62128	73181	74028	66384	17.79	19.15	6.85
d2103.tsp	80450	86582	86653	83454	7.62	7.71	3.73
dsj1000.tsp	18659688	22538673	24631467	19905772	20.79	32.00	6.68
fl417.tsp	11861	12118	12233	12190	2.17	3.14	2.77
fl1400.tsp	20127	27026	27447	21220	34.28	36.37	5.43
fl1577.tsp	22249	27461	27996	23500	23.43	25.83	5.62
fl3795.tsp	28772	35731	35285	35285	24.19	22.64	22.64
fnl4461.tsp	182566	229761	229963	229963	25.85	25.96	25.96
gr666.tsp	294358	316182	321701	308908	7.41	9.29	4.94
nrw1379.tsp	56638	67094	68964	60193	18.46	21.76	6.28
p654.tsp	34643	36022	36019	35708	3.98	3.97	3.07
pa561.tsp	2763	2905	3001	3008	5.14	8.61	8.87
pcb1173.tsp	56892	65023	70731	61249	14.29	24.33	7.66
pcb3038.tsp	137694	175799	176310	176310	27.67	28.04	28.04
pr1002.tsp	259045	279694	296142	275098	7.97	14.32	6.20
pr2392.tsp	378032	459687	461170	461170	21.60	21.99	21.99
rat575.tsp	6773	7136	7182	7158	5.36	6.03	5.68
rat783.tsp	8806	9501	9620	9352	7.89	9.24	6.20
rl1304.tsp	252948	322802	335779	274942	27.62	32.75	8.70
rl1323.tsp	270199	316844	332103	291235	17.26	22.91	7.79
rl1889.tsp	316536	388400	389270	389270	22.70	22.98	22.98

Table 7. *Cont.*

Instance	zOpt	GVNS_1	GVNS_2	GVNS_3	GAP_1	GAP_2	GAP_3
rl5915.tsp	565530	695466	695602	695602	22.98	23.00	23.00
rl5934.tsp	556045	672290	672412	672412	20.91	20.93	20.93
si535.tsp	48450	48648	48803	48765	0.41	0.73	0.65
si1032.tsp	92650	92864	93285	92925	0.23	0.69	0.30
u574.tsp	36905	40248	39803	39467	9.06	7.85	6.94
u724.tsp	41910	44972	45492	44598	7.31	8.55	6.41
u1060.tsp	224094	286667	251451	240193	27.92	12.21	7.18
u1432.tsp	152970	181206	188807	164045	18.46	23.43	7.24
u1817.tsp	57201	71024	72030	63539	24.17	25.92	11.08
u2152.tsp	64253	78581	79260	79217	22.30	23.36	23.29
u2319.tsp	234256	274542	278765	266890	17.20	19.00	13.93
vm1084.tsp	239297	265725	279047	255832	11.04	16.61	6.91
vm1748.tsp	336556	407007	408102	361567	20.93	21.26	7.43
Average	**253944.13**	**301717.44**	**322626.30**	**273781.10**	**14.50**	**7.41**	**5.26**

In some cases, GVNS_3 with a time limit of 1 min produced better results than using the 2 min time limit in solving sTSP instances. This might be happened due to the use of pure random diversification method such as the shuffle operator. More precisely, by executing more times GVNS_3, the shuffle operator is also executed more times and consequently it can shift the search into not so promising areas. Thus, the search may be trapped into low quality local optima.

Table 8. Results using the Best Improvement search strategy and an execution time limit of 2 min.

Instance	zOpt	GVNS_1	GVNS_2	GVNS_3	GAP_1	GAP_2	GAP_3
a280.tsp	2579	2630	2725	2706	1.98	5.66	4.92
att48.tsp	10628	10628	10628	10820	0.00	0.00	1.81
bayg29.tsp	1610	1610	1610	1613	0.00	0.00	0.19
bays29.tsp	2020	2020	2020	2029	0.00	0.00	0.45
bier127.tsp	118282	118421	119583	119527	0.12	1.10	1.05
kroA100.tsp	21282	21282	21312	21631	0.00	0.14	1.64
burma14.tsp	3323	3323	3323	3323	0.00	0.00	0.00
ch130.tsp	6110	6137	6208	6337	0.44	1.60	3.72
ch150.tsp	6528	6564	6680	6749	0.55	2.33	3.39
d493.tsp	35002	36232	37168	36822	3.51	6.19	5.20
kroB100.tsp	22141	22168	22249	22517	0.12	0.49	1.70
kroC100.tsp	20749	20749	20804	21402	0.00	0.27	3.15
kroD100.tsp	21294	21317	21461	21935	0.11	0.78	3.01
kroE100.tsp	22068	22122	22201	22450	0.24	0.60	1.73
kroA150.tsp	26524	26656	27119	27400	0.50	2.24	3.30
kroB150.tsp	26130	26232	26605	26816	0.39	1.82	2.63
kroA200.tsp	29368	29494	30276	30135	0.43	3.09	2.61
kroB200.tsp	29437	29732	30481	30843	1.00	3.55	4.78
d198.tsp	15780	15832	16036	16036	0.33	1.62	1.62
brg180.tsp	1950	1955	2014	2166	0.26	3.28	11.08
berlin52.tsp	7542	7542	7542	7737	0.00	0.00	2.59
dantzig42.tsp	699	699	699	703	0.00	0.00	0.57
eil51.tsp	426	426	426	432	0.00	0.00	1.41
eil76.tsp	538	538	542	543	0.00	0.74	0.93
eil101.tsp	629	630	644	648	0.16	2.38	3.02
fri26.tsp	937	937	937	940	0.00	0.00	0.32
gil262.tsp	2378	2432	2505	2497	2.27	5.34	5.00
gr17.tsp	2085	2085	2085	2086	0.00	0.00	0.05
gr21.tsp	2707	2707	2707	2707	0.00	0.00	0.00
Average	**253944.1262**	**272756.94**	**277586.89**	**273027.78**	**3.55**	**4.49**	**4.24**

Table 8. *Cont.*

Instance	zOpt	GVNS_1	GVNS_2	GVNS_3	GAP_1	GAP_2	GAP_3
gr24.tsp	1272	1272	1272	1273	0.00	0.00	0.08
gr48.tsp	5046	5046	5046	5089	0.00	0.00	0.85
gr96.tsp	55209	55259	55431	56269	0.09	0.40	1.92
gr120.tsp	6942	6975	7072	7130	0.48	1.87	2.71
gr137.tsp	69853	69869	70885	71215	0.02	1.48	1.95
gr202.tsp	40160	40860	41588	41762	1.74	3.56	3.99
gr229.tsp	134602	136101	139315	138770	1.11	3.50	3.10
gr431.tsp	171414	176629	181296	178732	3.04	5.76	4.27
hk48.tsp	11461	11461	11461	11519	0.00	0.00	0.51
lin105.tsp	14379	14379	14398	14505	0.00	0.13	0.88
lin318.tsp	42029	42989	43941	43768	2.28	4.55	4.14
pcb442.tsp	50778	52381	54108	52870	3.16	6.56	4.12
pr76.tsp	108159	108159	108227	109317	0.00	0.06	1.07
pr107.tsp	44303	44384	44473	44502	0.18	0.38	0.45
pr124.tsp	59030	59030	59039	59460	0.00	0.02	0.73
pr136.tsp	96772	97202	98613	100072	0.44	1.90	3.41
pr144.tsp	58537	58537	58544	60558	0.00	0.01	3.45
pr152.tsp	73682	73808	73884	74209	0.17	0.27	0.72
pr226.tsp	80369	80411	80677	81071	0.05	0.38	0.87
pr264.tsp	49135	49324	50715	52468	0.38	3.22	6.78
pr299.tsp	48191	48906	50363	51424	1.48	4.51	6.71
pr439.tsp	107217	110910	112735	114367	3.44	5.15	6.67
rat99.tsp	1211	1211	1232	1251	0.00	1.73	3.30
rat195.tsp	2323	2349	2448	2395	1.12	5.38	3.10
rd100.tsp	7910	7912	7943	8190	0.03	0.42	3.54
rd400.tsp	15281	15684	16272	16102	2.64	6.49	5.37
si175.tsp	21407	21422	21463	21510	0.07	0.26	0.48
st70.tsp	675	676	675	690	0.15	0.00	2.22
swiss42.tsp	1273	1273	1273	1273	0.00	0.00	0.00
ts225.tsp	126643	126654	127458	128716	0.01	0.64	1.64
tsp225.tsp	3916	3976	4107	4044	1.53	4.88	3.27
u159.tsp	42080	42282	42941	44205	0.48	2.05	5.05
ulysses16.tsp	6859	6859	6859	6860	0.00	0.00	0.01
ulysses22.tsp	7013	7013	7013	7041	0.00	0.00	0.40
ali535.tsp	202339	211954	217611	217607	4.75	7.55	7.55
att532.tsp	27686	28736	29545	29247	3.79	6.71	5.64
brazil58.tsp	25395	36518	25395	25395	43.80	0.00	0.00
brg180.tsp	1950	1964	2012	2178	0.72	3.18	11.69
d657.tsp	48912	51393	52852	51827	5.07	8.06	5.96
d1291.tsp	50801	54485	57344	55645	7.25	12.88	9.54
d1655.tsp	62128	67025	70119	67295	7.88	12.86	8.32
d2103.tsp	80450	83138	86653	83452	3.34	7.71	3.73
dsj1000.tsp	18659688	20039179	20433714	20142317	7.39	9.51	7.95
fl417.tsp	11861	11987	12111	12128	1.06	2.11	2.25
fl1400.tsp	20127	21201	21064	21204	5.34	4.66	5.35
fl1577.tsp	22249	23558	24252	23746	5.88	9.00	6.73
fl3795.tsp	28772	34390	32152	30584	19.53	11.75	6.30
fnl4461.tsp	182566	199718	204450	196648	9.39	11.99	7.71
gr666.tsp	294358	312215	319791	314260	6.07	8.64	6.76
nrw1379.tsp	56638	60460	62354	60951	6.75	10.09	7.62
p654.tsp	34643	37951	35465	36421	9.55	2.37	5.13
pa561.tsp	2763	2880	2886	3067	4.23	4.45	11.00
pcb1173.tsp	56892	61479	63387	61469	8.06	11.42	8.05
pcb3038.tsp	137694	151130	155733	149676	9.76	13.10	8.70
pr1002.tsp	259045	275588	284850	279390	6.39	9.96	7.85
pr2392.tsp	378032	410246	428532	408360	8.52	13.36	8.02
rat575.tsp	6773	7133	7153	7245	5.32	5.61	6.97
rat783.tsp	8806	9344	9250	9390	6.11	5.04	6.63
rl1304.tsp	252948	278657	280329	275084	10.16	10.82	8.75
Average	**253944.1262**	**272756.94**	**277586.89**	**273027.78**	**3.55**	**4.49**	**4.24**

Table 8. *Cont.*

Instance	zOpt	GVNS_1	GVNS_2	GVNS_3	GAP_1	GAP_2	GAP_3
rl1323.tsp	270199	291236	298519	285864	7.79	10.48	5.80
rl1889.tsp	316536	343698	353960	341771	8.58	11.82	7.97
rl5915.tsp	565530	677844	654919	633460	19.86	15.81	12.01
rl5934.tsp	556045	661867	644987	604596	19.03	16.00	8.73
si535.tsp	48450	48588	48769	48812	0.28	0.66	0.75
si1032.tsp	92650	92889	93382	92896	0.26	0.79	0.27
u574.tsp	36905	41429	39874	39865	12.26	8.04	8.02
u724.tsp	41910	44377	45604	44785	5.89	8.81	6.86
u1060.tsp	224094	241290	245211	243817	7.67	9.42	8.80
u1432.tsp	152970	164667	170476	165505	7.65	11.44	8.19
u1817.tsp	57201	62417	65840	61861	9.12	15.10	8.15
u2152.tsp	64253	69701	74338	70656	8.48	15.70	9.97
u2319.tsp	234256	243207	249416	245493	3.82	6.47	4.80
vm1084.tsp	239297	254315	262020	256838	6.28	9.50	7.33
vm1748.tsp	336556	359808	373774	356880	6.91	11.06	6.04
Average	**253944.1262**	**272756.94**	**277586.89**	**273027.78**	**3.55**	**4.49**	**4.24**

Tables 9–12 contain the aggregated experimental results for the National TSP instances. They contain the benchmark name, the optimal value (zOpt), the cost of the three GVNS algorithms (GVNS_1, GVNS_2 and GVNS_3) and the solution gaps from the optimal value for each method. Table 9 depicts GVNS using the First Improvement search strategy and an execution time limit of 1 min. Table 10 shows GVNS using the Best Improvement search strategy and an execution time limit of 1 min. Tables 11 and 12 provide the results achieved by the developed GVNS algorithms with a 2 min time limit within the First and Best improvement search strategy respectively. A notable observation is that in general there are not any significant differences between different methods. However, we notice that on First Improvement for both 1 and 2 min GVNS_3 outperforms GVNS_1 and GVNS_2 implementations. Contrariwise, on Best Improvement all three methods perform better in general than on First Improvement. However, there is no significant difference between them.

Table 9. Results using the First Improvement search strategy and an execution time limit of 1 min.

Instance	zOpt	GVNS_1	GVNS_2	GVNS_3	GAP_1 (%)	GAP_2 (%)	GAP_3 (%)
ar9152.tsp	837479	1648442	1648596	1648596	96.83	96.85	96.85
gr9882.tsp	300899	388910	388944	388944	29.25	29.26	29.26
eg7146.tsp	172387	220232	220315	220315	27.75	27.80	27.80
fi10639.tsp	520527	649604	649604	649604	24.80	24.80	24.80
ho14473.tsp	177105	484571	484812	484812	173.61	173.74	173.74
ei8246.tsp	206171	258851	258889	258889	25.55	25.57	25.57
ja9847.tsp	491924	612157	612304	612304	24.44	24.47	24.47
kz9976.tsp	1061882	1358247	1358247	1358247	27.91	27.91	27.91
lu980.tsp	11340	18708	23688	12236	64.97	108.89	7.90
mo14185.tsp	427377	529729	529729	529729	23.95	23.95	23.95
nu3496.tsp	96132	220359	221920	221920	129.23	130.85	130.85
mu1979.tsp	86891	119104	120908	120908	37.07	39.15	39.15
qa194.tsp	9352	9501	9727	9706	1.59	4.01	3.79
rw1621.tsp	26051	53343	58148	58052	104.76	123.21	122.84
tz6117.tsp	394718	500936	501184	501184	26.91	26.97	26.97
uy734.tsp	79114	84131	86022	83282	6.34	8.73	5.27
wi29.tsp	27603	27603	27603	27681	0.00	0.00	0.28
ym7663.tsp	238314	308477	308747	308747	29.44	29.55	29.55
zi929.tsp	95345	101541	112775	100572	6.50	18.28	5.48
ca4663.tsp	1290319	1646429	1646889	1646889	27.60	27.63	27.63
Average	**327546.50**	**462043.75**	**463452.55**	**462130.85**	**44.43**	**48.58**	**42.70**

Table 10. Results using the Best Improvement search strategy and an execution time limit of 1 min.

Instance	zOpt	GVNS_1	GVNS_2	GVNS_3	GAP_1 (%)	GAP_2 (%)	GAP_3 (%)
ar9152.tsp	837479	1317886	1648596	1648596	57.36	96.85	96.85
gr9882.tsp	300899	368968	388944	388944	22.62	29.26	29.26
eg7146.tsp	172387	206756	220315	220315	19.94	27.80	27.80
fi10639.tsp	520527	625547	649604	649604	20.18	24.80	24.80
ho14473.tsp	177105	362827	484812	459459	104.87	173.74	159.43
ei8246.tsp	206171	246249	258889	258889	19.44	25.57	25.57
ja9847.tsp	491924	595913	612304	612304	21.14	24.47	24.47
kz9976.tsp	1061882	1296253	1358247	1358247	22.07	27.91	27.91
lu980.tsp	11340	12052	12388	12530	6.28	9.24	10.49
mo14185.tsp	427377	514995	529729	529729	20.50	23.95	23.95
nu3496.tsp	96132	108839	108639	108685	13.22	13.01	13.06
mu1979.tsp	86891	93453	95413	93433	7.55	9.81	7.53
qa194.tsp	9352	9468	9717	9791	1.24	3.90	4.69
rw1621.tsp	26051	28360	28866	29127	8.86	10.81	11.81
tz6117.tsp	394718	478074	483082	445901	21.12	22.39	12.97
uy734.tsp	79114	83595	86201	84628	5.66	8.96	6.97
wi29.tsp	27603	27603	27603	27603	0.00	0.00	0.00
ym7663.tsp	238314	291939	308747	308747	22.50	29.55	29.55
zi929.tsp	95345	100461	103557	100474	5.37	8.61	5.38
ca4663.tsp	1290319	1532650	1463565	1428402	18.78	13.43	10.70
Average	**327546.50**	**415094.40**	**443960.90**	**438770.40**	**20.93**	**29.20**	**27.66**

Table 11. Results using the First Improvement search strategy and an execution time limit of 2 min.

Instance	zOpt	GVNS_1	GVNS_2	GVNS_3	GAP_1 (%)	GAP_2 (%)	GAP_3 (%)
ar9152.tsp	837479	1648304	1648596	1648596	96.82	96.85	96.85
gr9882.tsp	300899	388916	388944	388944	29.25	29.26	29.26
eg7146.tsp	172387	220290	220315	220315	27.79	27.80	27.80
fi10639.tsp	520527	649604	649604	649604	24.80	24.80	24.80
ho14473.tsp	177105	484793	484812	484812	173.73	173.74	173.74
ei8246.tsp	206171	258867	258889	258889	25.56	25.57	25.57
ja9847.tsp	491924	612304	612304	612304	24.47	24.47	24.47
kz9976.tsp	1061882	1358246	1358247	1358247	27.91	27.91	27.91
lu980.tsp	11340	20909	23688	14382	84.38	108.89	26.83
mo14185.tsp	427377	529699	529729	529729	23.94	23.95	23.95
nu3496.tsp	96132	221466	221920	221920	130.38	130.85	130.85
mu1979.tsp	86891	119586	120908	120908	37.63	39.15	39.15
qa194.tsp	9352	9543	9811	9731	2.04	4.91	4.05
rw1621.tsp	26051	56293	58148	58148	116.09	123.21	123.21
tz6117.tsp	394718	501137	501184	501184	26.96	26.97	26.97
uy734.tsp	79114	85036	99005	83383	7.49	25.14	5.40
wi29.tsp	27603	27603	27603	27701	0.00	0.00	0.36
ym7663.tsp	238314	308713	308747	308747	29.54	29.55	29.55
zi929.tsp	95345	104572	113927	101795	9.68	19.49	6.76
ca4663.tsp	1290319	1645854	1646889	1646889	27.55	27.63	27.63
Average	**327546.50**	**462586.75**	**464163.50**	**462311.40**	**46.30**	**49.51**	**43.76**

Table 12. Results using the Best Improvement search strategy and an execution time limit of 2 min.

Instance	zOpt	GVNS_1	GVNS_2	GVNS_3	GAP_1 (%)	GAP_2 (%)	GAP_3 (%)
ar9152.tsp	837479	1372411	1648596	1648596	63.87	96.85	96.85
gr9882.tsp	300899	377898	388944	388944	25.59	29.26	29.26
eg7146.tsp	172387	209515	220315	220315	21.54	27.80	27.80
fi10639.tsp	520527	635913	649604	649604	22.17	24.80	24.80
ho14473.tsp	177105	449416	484812	474609	153.76	173.74	167.98
ei8246.tsp	206171	248441	258889	258889	20.50	25.57	25.57
ja9847.tsp	491924	602674	612304	612304	22.51	24.47	24.47
kz9976.tsp	1061882	1327647	1358247	1358247	25.03	27.91	27.91
lu980.tsp	11340	12151	12458	12559	7.15	9.86	10.75
mo14185.tsp	427377	527335	529729	529729	23.39	23.95	23.95
nu3496.tsp	96132	115162	108803	109804	19.80	13.18	14.22
mu1979.tsp	86891	94072	95874	94358	8.26	10.34	8.59
qa194.tsp	9352	9525	9757	9759	1.85	4.33	4.35
rw1621.tsp	26051	28784	29118	29131	10.49	11.77	11.82
tz6117.tsp	394718	475863	501184	501184	20.56	26.97	26.97
uy734.tsp	79114	84585	86358	84686	6.92	9.16	7.04
wi29.tsp	27603	27603	27603	27612	0.00	0.00	0.03
ym7663.tsp	238314	295414	308747	308747	23.96	29.55	29.55
zi929.tsp	95345	101161	104327	100881	6.10	9.42	5.81
ca4663.tsp	1290319	1573320	1646889	1470076	21.93	27.63	13.93
Average	**327546.50**	**428444.50**	**454127.90**	**444501.70**	**25.27**	**30.33**	**29.08**

4. Statistical Analysis on Computational Results

This section presents the statistical tests which were performed on the computational results, to evaluate the performance of the three different GVNS methods. Different statistical tests are applied to different data structures. In particular, statistical analysis methods can be divided on parametric and non-parametric tests. The first category examines normal variables whereas the other methods concern non-normal variables [20].

Initially, the application of a normality test showed that the numerical data does not follow the normal distribution. Consequently, we applied the Kruskal–Wallis test for checking the existence of a statistically significant difference between the methods. In this test receiving a p-value less than 0.05 means that there is a statistically significant difference.

At the present point, it should be mentioned that the statistical analysis was performed on median values to eliminate potential extreme deviation on the average values based on extreme values. Also, the related to the aTSP analysis is taken from our previous work [14] and it is presented here for building a more comprehensive view.

4.1. Statistical Analysis on aTSP Results

In Table 13 we can see that for all cases p-value is less than 0.05, which means that there is a statistically significant difference between the three methods in all cases. For further examination, pairwise Wilcoxon tests were performed.

Table 13. Kruskal–Wallis rank sum test.

	X^2	df	p-Value
FI_1min	6.8689	2	0.0322
FI_2mins	9.0314	2	0.0109
BI_1min	9.2739	2	0.0097
BI_2mins	9.6658	2	0.008

In accordance with the pairwise tests which are summarized in Table 14, it is clear that the GVNS_1 has significant differences with the other two schemes in all cases. Both GVNS_2 and GVNS_3 perform equivalently using the First Improvement search strategy independently of the execution time limit, while using the Best Improvement strategy they have significant reported differences with both time limits [14].

Table 14. KPairwise comparisons using Wilcoxon signed rank test.

FI_1min		
	GVNS1	GVNS2
GVNS2	0.00064	
GVNS3	0.00064	0.6701
FI_2mins		
	GVNS1	GVNS2
GVNS2	0.00064	
GVNS3	0.00064	0.4488
BI_1min		
	GVNS1	GVNS2
GVNS2	0.00109	
GVNS3	0.00064	0.00064
BI_2mins		
GVNS2	0.00109	
GVNS3	0.00064	0.00064

In respect to this Kruskal–Wallis statistical analysis, we have four box-plots illustrated in Figures 1a,b and 2a,b four box plots. Each one depicts either First Improvement or Best Improvement for one minute, as well as for two minutes runs.

Moreover, based on the corresponding of the previous statistical summary, box plots, it is confirmed that the GVNS_1 produces much better results than the other two algorithms in all cases for the aTSP, while the GVNS_2 outperform the GVNS_3 in also all cases. Also, by checking the medians at the box plots it can be seen that the GVNS_2 performs significantly better on Best Improvement, as it produces results that are "close" to the results of GVNS_1.

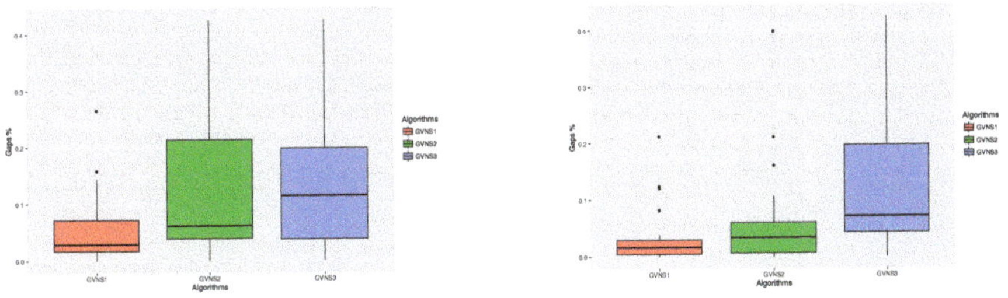

(a) (b)

Figure 1. Statistical test for aTSP (1/2). (**a**) Statistical test box plots for aTSP 1min FI; (**b**) Statistical test box plots for aTSP 1min BI.

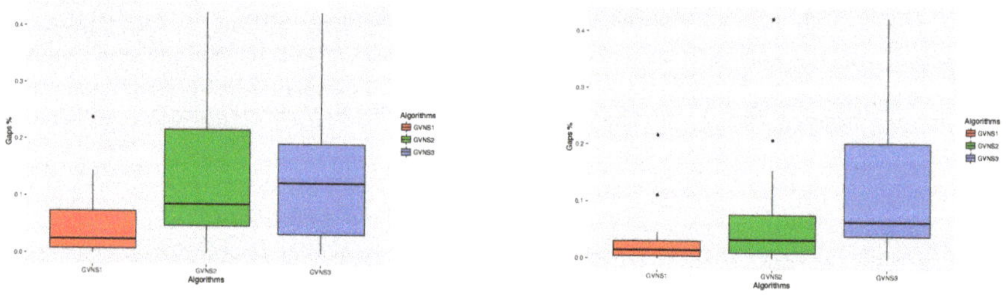

(a) (b)

Figure 2. Statistical test for aTSP (2/2). (**a**) Statistical test box plots for aTSP 2mins FI; (**b**) Statistical test box plots for aTSP 2mins BI.

4.2. Statistical Analysis on sTSP

In this subsection the statistical analysis on the results achieved by the three GVNS schemes on sTSP instances is provided.

According to the values in Table 15 we can see that only using the Best Improvement search strategy within a 2 min execution time limit, there are statistically significant differences.

Table 15. Kruskal–Wallis rank sum test.

	X^2	df	*p*-Value
FI_1min	2.4392	2	0.2954
FI_2min	4.5181	2	0.1045
BI_1min	5.5397	2	0.06267
BI_2min	11.677	2	0.002913

More specifically, the values in Table 16 highlights that there is a difference between the GVNS_1 and the other two GVNS algorithms. In particular, based on the following box plots, the GVNS_1 is slightly better than the GVNS_2 and GVNS_3, which they perform almost equivalently.

Table 16. KPairwise comparisons using Wilcoxon signed rank test.

BI_2min		
	GVNS1	GVNS2
GVNS2	0.0000000990	
GVNS3	0.0000002300	0.6800000000

Subsequently of this Kruskal–Wallis statistical analysis, we have four box-plots illustrated in Figures 3a,b and 4a,b four box plots. Each one depicts either First Improvement or Best Improvement for one minute, as well as for two minutes runs.

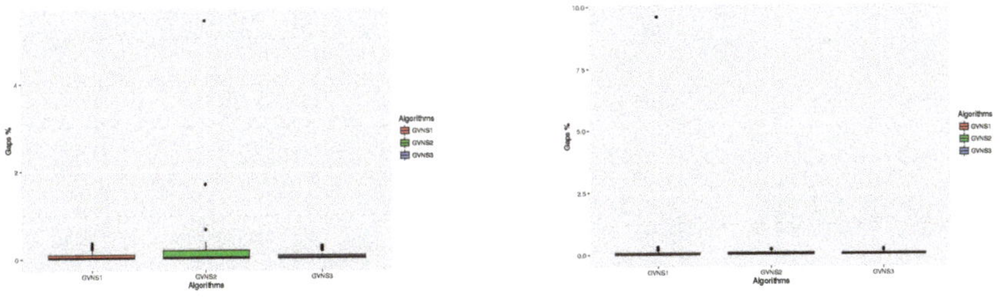

(a) (b)

Figure 3. Statistical test for sTSP (1/2). (**a**) Statistical test box plots for sTSP 1min FI; (**b**) Statistical test box plots for sTSP 1min BI.

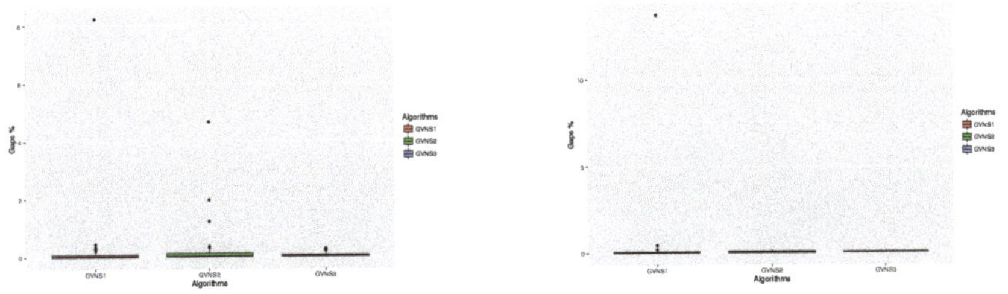

(a) (b)

Figure 4. Statistical test for sTSP (2/2). (**a**) Statistical test box plots for sTSP 2min FI; (**b**) Statistical test box plots for sTSP 2min BI.

4.3. *Statistical Analysis on nTSP*

In the case of the National TSP instances and based on the values given in Table 17 we can see that there is no significant statistical difference between the three methods.

Table 17. Kruskal–Wallis rank sum test.

	X^2	df	p-Value
FI_1min	0.22253	2	0.8947
FI_2min	0.18068	2	0.9136
BI_1min	1.825	2	0.4015
BI_2min	1.9646	2	0.3744

As a result of this Kruskal–Wallis statistical analysis, we have four box-plots illustrated in Figures 5a,b and 6a,b four box plots. Each one depicts either First Improvement or Best Improvement for one minute, as well as for two minutes runs. By checking the median value of each method, it is clear that the three algorithms perform equivalently.

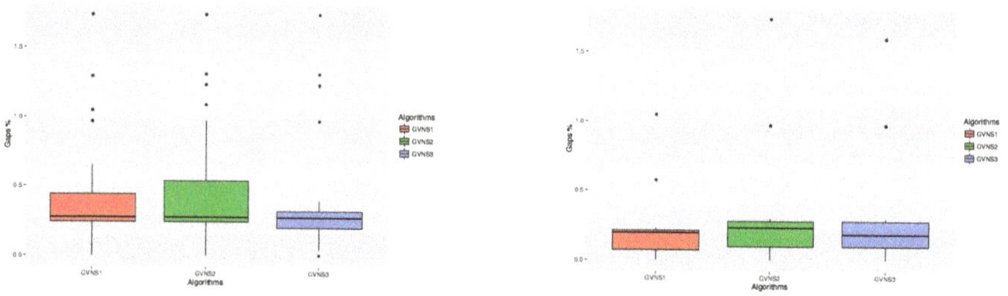

(**a**) (**b**)

Figure 5. Statistical test for nTSP (1/2). (**a**) Statistical test box plots for nTSP 1min FI; (**b**) Statistical test box plots for nTSP 1min BI.

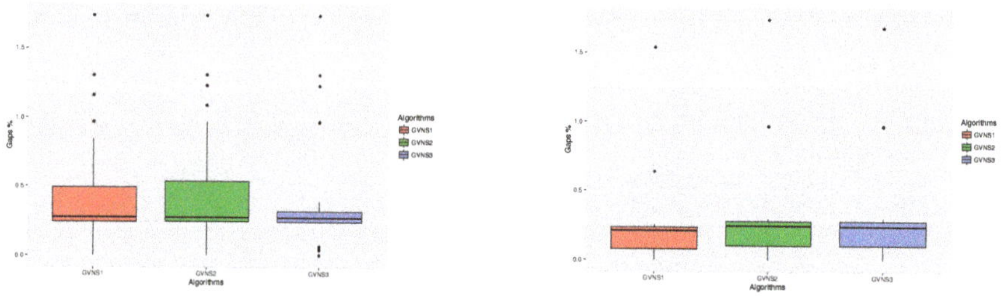

(**a**) (**b**)

Figure 6. Statistical test for nTSP (2/2). (**a**) Statistical test box plots for nTSP 2min FI; (**b**) Statistical test box plots for nTSP 2min BI

5. Comparison with Recent Similar Works

In their recent work, Halim et al. have presented an extensive analysis over the performance of different heuristic and metaheuristic algorithms on some TSPLib instances [21]. In addition, Hore et al. [22] proposed an improved hybrid VNS algorithm for solving TSP instances. In this section, a comparison between our proposed GVNS schemes (GVNS_1 and GVNS_2) and those algorithms presented on papers [21,22] is performed.

Table 18 shows the comparison between our GVNS_1 and GVNS_2, within the Best Improvement search strategy and a time limit of 120 s, with all metaheuristic solution approaches presented in the work of Halim et al. [21]. The results show that our methods produce better results than the previously mentioned metaheuristics, except the case of instance rat195.tsp in which the GA and the TS perform better than GVNS_2.

Table 18. Comparisons between GVNS_1 and GVNS_2 with BI for 2mins with the recent work of Halim et al.

Instance	OV	GVNS_1	GVNS_2	GA	SA	TS	ACO	TPO
eil51.tsp	426	426	426	454.1	439.13	439.1	467.46	437.26
berlin52.tsp	7542	7542	7542	7946.4	7960.67	7740.1	7922.32	7705.8
st70.tsp	675	676	675	700.72	696.33	690.27	756.55	697.12
kroA100.tsp	21282	21282	21312	22726.2	22277.5	22521.64	22941.68	22463.6
ch130.tsp	6110	6137	6208	6610.8	6558.7	6717.06	6913.99	6515.28
rat195.tsp	2323	2349	2448	2414.52	2537.99	2373.94	2465.11	2573.47
a280.tsp	2579	2630	2725	2789.83	2830.18	2800.79	2867.85	2790.54
rd400.tsp	15281	15684	16272	16567.29	16816.65	20723.56	19259.06	18190.84
pcb442.tsp	50778	52381	54108	55718.9	57421.04	83123.01	63436.7	60750.43

Table 19 shows the comparison between our GVNS_1 and GVNS_2, within the Best Improvement search strategy and a time limit of 120 s with the results obtained by a hybrid VNS in the second mentioned work [22]. It is clearly observed that our methods outperform the hybrid VNS. More specifically, the GVNS_1 outperforms the hybrid VNS algorithm in all tested instances, while the GNVS_2 produce better results than those achieved by Hore et al. approach in seven out of 10 problem instances.

Table 19. Comparisons between GVNS_1 and GVNS_2 with BI for 2 mins with the recent VNS work of Hore et al.

Instance	OV	GVNS_1	GVNS_2	Average	Average Time
eil51.tsp	426	426	426	428.98	454.1
berlin52.tsp	7542	7542	7542	7544.36	7946.4
st70.tsp	675	676	675	677.11	700.72
kroA100.tsp	21282	21282	21312	21695.79	22726.2
ch130.tsp	6110	6137	6208	6153.72	6610.8
rat195.tsp	2323	2349	2448	2453.81	1382.34
rd400.tsp	15281	15684	16272	16250.21	1953.49
pcb1173.tsp	56892	61479	63387	63435.95	9531.54
pcb442.tsp	50778	52381	54108	50800.24	2183.27

Table 20 shows the abbreviations regarding the metaheuristic algorithms presented in [21].

Table 20. Abbreviations.

GA	Genetic Algorithm
SA	Simulated Annealing
TS	Tabu Search
ACO	Ant Colony Optimization
TPO	Tree Physiology Optimization

6. Conclusions and Future Work

A thorough and comprehensive performance analysis on the efficiency of three GVNS algorithms has been presented in this work. The main difference lies on the perturbation strategy used. Our comparative performance analysis involves problems that are modelled as asymmetric and Symmetric TSP instances that are resolved using GVNS. The well-known TSP benchmarks from the TSPLIB are used for extensive experimental testing. We believe that the experimental results are quite conclusive, as they confirm that for asymmetric TSP instances GVNS_1 outperforms the other two methods and GVNS_2 consistently provides better solutions in all cases compared to GVNS_3. Simultaneously, the perturbation strategy does not seem to critically affect the solutions of Symmetric TSP instances.

It is also worth emphasizing that even though the improvement stage of the GVNS schemes has been given limited attention, the present paper shows that the tested approaches to the solution, can be quite promising. This is justified by the solutions produced, which are significantly better than other metaheuristic approaches in recent literature.

The investigation of alternative neighborhood structures and neighborhood change movements in VND under the GVNS framework could be a possible direction for future work. In the same vein, one might potentially study modifications or specific combinations with more than one perturbation strategy during the perturbation phase in an effort to determine whether optimal solutions can be achieved even closer, especially on large asymmetric benchmarks.

Author Contributions: All of the authors have contributed extensively to this work. C.P. and P.K. conceived the initial algorithm and worked on the first prototypes. P.K. and C.P. thoroughly analyzed the current literature gathering all the necessary material. T.A. assisted C.P. in designing the methods used in the main part. T.A. was responsible for supervising the construction of this work. C.P. was responsible for the interlinking between the theoretic model and the actual application. C.P. contributed to the appropriate typing of the formal definitions and the math used in the paper.

Funding: This research received no external funding.

Conflicts of Interest: The authors declare no conflict of interest.

Abbreviations

The following abbreviations are used in this manuscript:

TSP	Traveling Salesman Problem
VNS	Variable Neighborhood Search
GVNS	General Variable Neighborhood Search
GA	Genetic Algorithm
SA	Simulated Annealing
TS	Tabu Search
ACO	Ant Colony Optimization
TPO	Tree Physiology Optimization

References

1. Mladenovic, N.; Hansen, P. Variable neighborhood search. *Comput. Oper. Res.* **1997**, *24*, 1097–1100. [CrossRef]
2. Hansen, P.; Mladenovic, N.; Todosijevic, R.; Hanafi, S. Variable neighborhood search: Basics and variants. *EURO J. Comput. Optim.* **2017**, *5*, 423–454. [CrossRef]
3. Mladenovic, N.; Todosijevic, R.; Uroševic, D. Less is more: Basic variable neighborhood search for minimum differential dispersion problem. *Inf. Sci.* **2016**, *326*, 160–171. [CrossRef]
4. Mladenović, N.; Sifaleras, A.; Sörensen, K. Editorial to the Special Cluster on Variable Neighborhood Search, Variants and Recent Applications. *Int. Trans. Oper. Res.* **2017**, *24*, 507–508. [CrossRef]
5. Karakostas, P.; Sifaleras, A.; Georgiadis, C. Basic VNS algorithms for solving the pollution location inventory routing problem. In *Proceedings of the LNCS Proc. of the 6th International Conference on Variable Neighborhood Search (ICVNS 2018), Sithonia, Greece, 4–7 October 2018*; Sifaleras, A., Salhi, S., Brimberg, J., Eds.; Springer: Berlin, Germany, 2019; Volume 11328.
6. Karakostas, P.; Sifaleras, A.; Georgiadis, C. A general variable neighborhood search-based solution approach for the location-inventory-routing problem with distribution outsourcing. *Comput. Chem. Eng.* **2019**, *126*, 263–279. [CrossRef]
7. Papalitsas, C.; Karakostas, P.; Andronikos, T.; Sioutas, S.; Giannakis, K. Combinatorial GVNS (General Variable Neighborhood Search) Optimization for Dynamic Garbage Collection. *Algorithms* **2018**, *11*, 38. [CrossRef]
8. Sifaleras, A.; Konstantaras, I. General variable neighborhood search for the multi-product dynamic lot sizing problem in closed-loop supply chain. *Electron. Notes Discret. Math.* **2015**, *47*, 69–76. [CrossRef]
9. Sifaleras, A.; Konstantaras, I. Variable neighborhood descent heuristic for solving reverse logistics multi-item dynamic lot-sizing problems. *Comput. Oper. Res.* **2017**, *78*, 385–392. [CrossRef]
10. Silva, M.A.; Souza, S.R.; Souza, M.J.; Filho, M.F.F. Hybrid metaheuristics and multi-agent systems for solving optimization problems A review of frameworks and a comparative analysis. *Appl. Soft Comput.* **2018**, *71*, 433–459. [CrossRef]
11. Duan, Q.; Liao, T.; Yi, H. A comparative study of different local search application strategies in hybrid metaheuristics. *Appl. Soft Comput.* **2013**, *13*, 1464–1477. [CrossRef]
12. Huber, S.; Geiger, M.J. Order matters–A Variable Neighborhood Search for the Swap-Body Vehicle Routing Problem. *Eur. J. Oper. Res.* **2017**, *263*, 419–445. [CrossRef]
13. Papalitsas, C.; Giannakis, K.; Andronikos, T.; Theotokis, D.; Sifaleras, A. Initialization methods for the TSP with Time Windows using Variable Neighborhood Search. In Proceedings of the IEEE Proc. of the 6th International Conference on Information, Intelligence, Systems and Applications (IISA 2015), Corfu, Greece, 6–8 July 2015.

14. Papalitsas, C.; Karakostas, P.; Andronikos, T. Studying the impact of perturbation methods on the efficiency of GVNS for the ATSP. In *LNCS Proc. of the 6th International Conference on Variable Neighborhood Search (ICVNS 2018), Sithonia, Greece, 4–7 October 2018*; Sifaleras, A., Salhi, S., Brimberg, J., Eds.; Springer: Berlin, Germany, 2019; Volume 11328.
15. Papalitsas, C.; Karakostas, P.; Kastampolidou, K. *A Quantum Inspired GVNS: Some Preliminary Results*; Vlamos, P., Ed.; GeNeDis 2016; Springer International Publishing: Cham, Switzerland, 2017; pp. 281–289.
16. Nunes, D.C.L. *Fundamentals of Natural Computing: Basic Concepts, Algorithms, and Applications*; Chapman & Hall/CRC: Boca Raton, FL, USA, 2006.
17. Dey, S.; Bhattacharyya, S.; Maulik, U. New quantum inspired meta-heuristic techniques for multi-level colour image thresholding. *Appl. Soft Comput.* **2016**, *46*, 677–702. [CrossRef]
18. Feynman, R.P. Simulating physics with computers. *Int. J. Theor. Phys.* **1982**, *21*, 467–488. [CrossRef]
19. Feynman, R.P.; Hey, J.; Allen, R.W. *Feynman Lectures on Computation*;Longman Publishing Co., Inc.: Cambridge, MA, USA, 1998.
20. Coffin, M.; Saltzman, M.J. Statistical Analysis of Computational Tests of Algorithms and Heuristics. *INFORMS J. Comput.* **2000**, *12*, 24–44, doi:10.1287/ijoc.12.1.24.11899. [CrossRef]
21. Halim, A.H.; Ismail, I. Combinatorial optimization: Comparison of heuristic algorithms in travelling salesman problem. *Arch. Comput. Methods Eng.* **2019**, *26*, 367–380. [CrossRef]
22. Hore, S.; Chatterjee, A.; Dewanji, A. Improving variable neighborhood search to solve the traveling salesman problem. *Appl. Soft Comput.* **2018**, *68*, 83–91. [CrossRef]

applied system innovation

MDPI

Article

Adaptive Neuro-Fuzzy Inference System Based Grading of Basmati Rice Grains Using Image Processing Technique

Dipankar Mandal

Centre of Studies in Resources Engineering, Indian Institute of Technology Bombay, Mumbai, India; dipankar_mandal@iitb.ac.in; Tel.: +91-22-2576-4654

Received: 10 April 2018; Accepted: 15 June 2018; Published: 20 June 2018

Abstract: Grading of rice intents to discriminate broken and whole grain from a sample. Standard techniques for image-based rice grading using advanced statistical methods seldom take into account the domain knowledge associated with the data. In the context of a high product value basmati rice with an image based grading process, one ought to consider the physical properties of grain and the associated knowledge. In this present work, a model of quality grade testing and identification is proposed using a novel digital image processing and knowledge-based adaptive neuro-fuzzy inference system (ANFIS). The rationale behind adopting a grading system based on fuzzy rules relies on capabilities of ANFIS to simulate the behaviour of an expert in the characterization of rice grain using the physical properties of rice grains. The rice kernels are characterized with the help of morphological descriptors and geometric features which are derived from sample images of milled basmati rice. The predictive capability of the proposed technique has been tested on a sufficient number of training and test images of basmati rice grain. The proposed method outperforms with a promising result in an evaluation of rice quality with >98.5% classification accuracy for broken and whole grain as compared to standard machine learning technique viz. support vector machine (SVM) and K-nearest neighbour (KNN). The milling efficiency is also assessed using the ratio between head rice and broken rice percentage and it is 77.27% for the test sample. The overall results of the adopted methodology are promising in terms of classification accuracy and efficiency.

Keywords: ANFIS; basmati rice; image processing; grading; quality assessment; fuzzy inference system

1. Introduction

India is the leading exporter of the basmati rice (*Oryza sativa*) to the global market. The annual export of basmati rice was ~4.05 million MT to the global market during the year 2015–2016 [1]. Basmati rice is a protracted slender grain variety of aromatic rice grown in the Indian sub-continent. It has a high product value due to its flavour, delicate texture, delightful fragrance, and softness. The length of basmati rice grain is longer than the width, and it grows even longer during cooking [2].

The high-value basmati rice grain has to go through several operations (such as threshing, handling, de-husking, milling and whitening of grains) starting from harvesting of paddy to final production of rice grains by means of several mechanical systems [3]. Thereby, the grade of the produced grain exclusively depends on the adjustment of the equipment used in the various mentioned operations. In general, in a rice milling facility, the quality grade of product is being monitored by visual inspection by experienced quality control personnel at 2–3 h intervals, rather utilizing a continuous operational measurement method. This means that the operator, based on his experience and proficiency with the processing machinery, assesses the quality grade of the product by mere visual inspection of rice grain appearance and making the required adjustments which are time-consuming and subjective.

Alternatively, image-based grading approaches are nondestructive and rapid. With suitable statistical or machine learning techniques, the image-based approaches are proven to be an efficient way to achieve automatic inspection and grade evaluation efficiently [4,5]. During last decade, researchers have investigated several techniques based on machine vision and digital image processing for quality assessment of rice kernels which are fast, non-destructive, accurate, and cost-effective as compared to traditional methods [6,7]. Image-based approaches have been applied for characterizing rice grains using either one of the morphological, colour, and textural features, or a combination. However, in order to find suitable rice grain descriptor and to improve the classification accuracy, it is imperative that some key features should be selected to describe grain feature exactly.

The marketing value of rice depends on its physical qualities after processing. Major axis/minor axis ratio of the rice kernel is reported as a key feature of basmati rice which might identify the adulteration of basmati rice with other rice varieties [8]. Vaingankar and Kulkarni [8] reported the major axis/minor axis ratio of 3.92–4.09 as an indicator of pure Basmati-370 variety. In the context of grading of rice, the percentage of the head or whole grain and broken grain is a most important factor which determines the milling efficiency. Till date, several studies reported improvement in classification accuracy of rice grain using machine vision and image processing techniques [9–12].

Pazoki et al. [13] illustrated that determining grain variety using a simple mathematical function is difficult because the grain has various morphologies, colours, and textures. Alternatively, artificial neural network (ANNs) techniques have been applied for grain quality control and discrimination of grain variety. Chen et al. [14] proposed a methodology to identify five corn varieties with the accuracy of more than 90% using pattern recognition techniques and neural networks. In a comparative analysis in [11] of artificial neural networks, support vector machines, decision trees and Bayesian Networks to classify milled rice samples, it has produced highest classification accuracy with ANN. Despite promising results, there are several problems might arise with ANN's training and designing [15–18]. The assignment of the weights in ANN structure is one of the most important problems [19] which has a direct effect on its performance. Moreover, the uncertainties in ANN output is proven to be a challenging issue [20].

ANN optimization is limited in practice by a finite training sample and is accomplished through a stochastic training process which gives ANN the ability to avoid being trapped at local minima. On the contrary, this stochastic process makes ANN optimization empirical and subject to strong influence from statistical variations [20]. To overcome these issues with ANN, a hybrid approach with the fuzzy system has introduced. Fuzzy systems are quite good at handling uncertainties and can interpret the relationship between input and output by producing rules. Therefore, to increase the capability of Fuzzy and ANN, hybridization of ANN and fuzzy is usually implemented. Sabanci et al. [21] used ANFIS for wheat grain classification with 99.46% of classification accuracy. Zareiforoush et al. [22] coupled a fuzzy inference system (FIS) with image processing technique for a decision-support system for qualitative grading of milled rice. The results are reported with 89.8% agreement between the grading results obtained from the FIS system and those determined by the experts.

In the context of rice grading, some head grains are easily misclassified with broken grain due to the resemblance in single feature (e.g., eccentricity) extracted from digital images and are not deterministically separable. In such cases, fuzzy approach [23] is more convenient for discrimination of head and broken rice grains [24]. Shiddiq et al. [25] investigated the rice milling degree using colour features (RGB) with an adaptive network-based fuzzy inference model. It was reported an error of 3.55–5.62% in milling degree using this process. However, the morphological features can improve the efficiency in terms of classification. In this context, a grading system based on fuzzy rules can simulate the behaviour of an expert in the evaluation and classification of physical properties of rice grains for grading. In this present work, the predictive capacity of ANFIS is assessed for quality testing and identification of basmati rice based on morphological features. This has been motivated by the fact that well-documented knowledge regarding rice kernels are usually available [26,27]. This knowledge has been incorporated in forming the rules of the fuzzy inference system used to determine head and broken rice kernels. Moreover, the proposed ANFIS based classification method provides a rationale behind the knowledge of morphological features and their underlying dependencies with rice grains. Subsequently, the milling efficiency is estimated with broken grain and head grain ratio for test images. Furthermore, the proposed classification method is compared with standard data-driven machine learning techniques viz., support vector machine (SVM) and k-nearest neighbour (KNN) classifier.

The rest of the manuscript is organized as follows: Section 2 briefly describes the materials and methods. Section 3 explain in detail the results and finally the work is succinctly summarized and concluded in Section 4.

2. Materials and Methods

The schematic workflow of the proposed ANFIS based grading of basmati rice grains is given in this section. Subsequently, the steps involved in the technique are detailed in the following subsections.

2.1. Sample Preparation

Basmati rice grains of different grades (Pusa basmati 1121), were used in this study. This variety of basmati rice possesses extra-long slender milled grains ($\sim$9.0 mm), pleasant aroma, and an exceptionally high cooked kernel elongation ratio of $\sim$2.5 [28]. It is the most common Basmati rice variety in rice grain quality research for developing standard Basmati quality traits. The grades of the rice sample are based on the percentage of broken rice content (e.g., 5% broken rice).

The rice grain samples can be taken as heaped together or in a scattered arrangement for imaging. These arrangements are important which is likely due to the fact that the grain characterization method employs the visual attributes of grains obtained from image-processing techniques. Thus, the heaped grain images might attain certain disadvantages e.g., boundaries of grains not completely visible and distinguishable and noise appearing more prominent than the boundaries if grains are overlapping with each other [21]. Thereby, it is desirable to take samples in a scattered arrangement with a black background (it can improve the contrast of the image in the scattered configuration). Furthermore, it should be ensured that not too many rice grains are clustering in scattered configuration.

2.2. Imaging System and Image Acquisition

A schematic diagram of the image acquisition system is shown in Figure 1. Typically, a vision system consists of the illumination component to illuminate the sample under test; the camera to acquire an image; personal computer or microprocessor system to provide disk storage of images and computational capability.

In an image acquisition system, choosing the right lighting strategy remains a difficult problem because there is no specific guideline for integrating lighting and machine vision application. Despite this, some rules of thumb exist [29] which suggest that fluorescent bulbs are inherently more efficient and produce more intense illumination at specific wavelengths. Moreover, the fluorescent light provides a more even, uniform dispersion of light from the emitting surface [30]. A 25–40 kHz ring-shaped compact fluorescent light is used for illumination in this setup. Apart from the illuminant, the surface geometry is also important in the illumination design. In this present work, a diffuse illuminator is used to produce uniform lighting as shown in Figure 1. Such a setup is extremely useful for visual inspection of grains and oilseed with a success rate almost reaching 100% [31].

The system was enclosed in a dark chamber to prevent exposure to stray light. The digital camera (Canon EOS 1300D) was set in the manual mode for image acquisition with an ISO of 400 and a shutter of 1/30 s. The images were taken with a black background for basmati rice sample with different orientation and quality in a scattered arrangement for training and testing. Camera aperture and focus were adjusted to make individual grain boundaries distinguishable in the picture. A total of 40 images were acquired and saved in raw format, in which no adjustment (e.g., white balance) was applied.

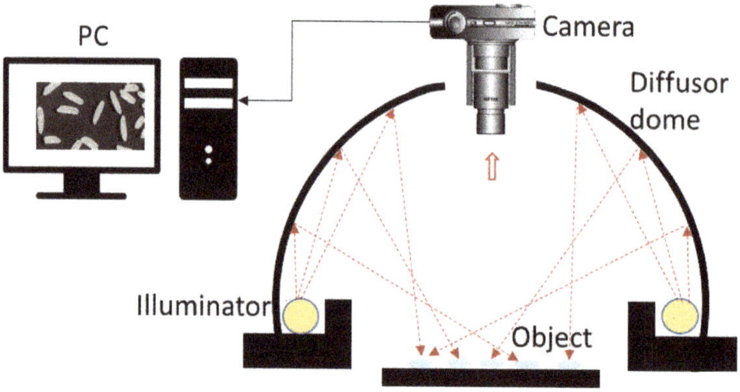

Figure 1. Schematic diagram of image acquisition system equipped with a camera, illumination source and geometry, and connected PC.

2.3. Image Processing

The image processing was carried out with MATLAB to acquire the feature data. At first, the acquired RGB image was separated in single R, G and B channel in grayscale mode. Subsequently, each channel grey image was converted to a binary image using Otsu's method [32]. This method converts the grayscale image to a binary image based on image clustering in accordance with a threshold value. This threshold value is optimally determined between 0 and 1 by Otsu's method. The grey level is normalized from 0–255 to 0–1. The method then splits the normalized image into two classes having lower or higher grey level than the threshold value. Each pixel is set to white (1) if the grey level is higher than the threshold value, otherwise, it is set to black (0). Thus, the image segmentation considers the identification of objects within an image using an edge detection algorithm which identifies the boundaries of individual object and labels the centre of each object for further processing. Eventually, each grain's position is fixed and it is tagged according to its position through a segmentation process.

The noise of each image is then eliminated using a morphological process. It is followed by morphological opening operation [33] were applied with 'disk' type structuring element using 'imopen' function followed by hole filling and clear borders. Morphological opening operations generally smooth the objects of the image. Opening operation eliminates thin protrusions of the objects. Opening operation eliminates the objects which cannot accommodate the structuring element completely. Thus it removes the noise from the image. Then, each object was labelled followed by counting the objects.

Feature extraction involves the retrieval of quantitative information from the segmented images. Here, extraction of parameters e.g., eccentricity, equivalent diameter, area, perimeter, major axis length and minor axis length have been carried out further with 'regionprops' function for differentiating head grain from broken grain. Aspect ratio i.e., major axis length/minor axis length was estimated from object feature set. The schematic processing chain is shown in Figure 2. The similar operation was conducted for other training images of basmati rice sample and for the test image too. Feature dataset for training as well as testing was created from the object properties and object class (Whole grain = 1 or broken grain = 0).

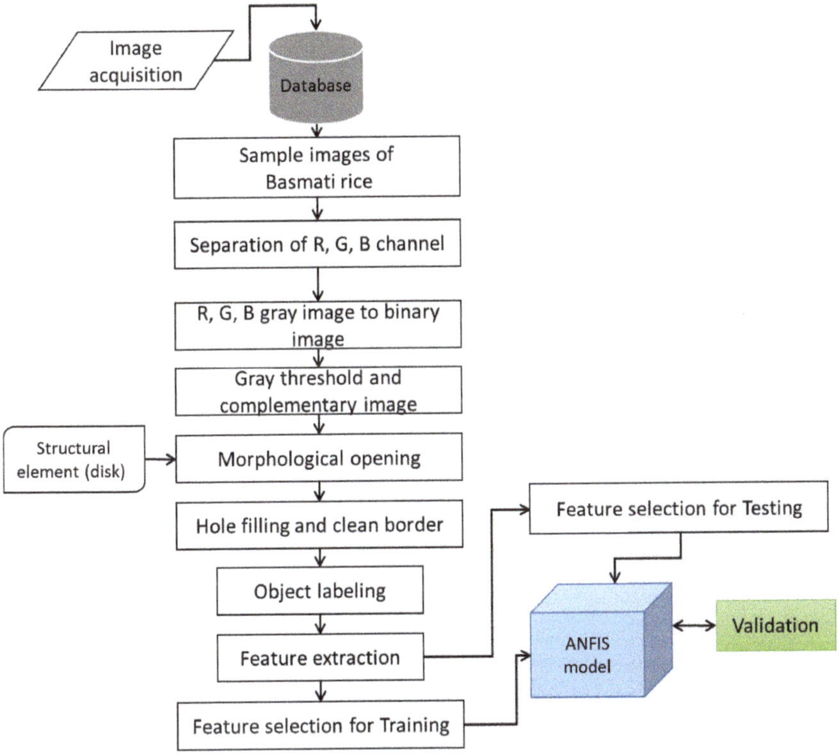

Figure 2. Schematic workflow for image processing.

2.4. Features of Basmati Rice

To physical properties of the rice kernel is characterised using the morphological features. Rice grains are generally considered as an ellipse as shown in Figure 3. Based on this assumption of the object, the following morphological features as reported by Shantaiya and Ansari [34] are considered for the present study. The devised features are also reported to be promising in [35] as optimal morphological features for rice kernel identification using standard sequential forward (SFS) algorithm. These features are:

- *Major axis length*: It is the total number of pixels between the extreme points along the major axis of the rice kernel.
- *Minor axis length*: It measures the number of pixels between the extreme points of the along the minor axis of the rice kernel.
- *Perimeter*: It is the total number of pixels along the boundary of rice grain.
- *Area*: It is the total number of pixels in rice grain object.
- *Aspect ratio (a/b)*: It is the ratio of major axis length and minor axis length of the rice grain.
- *Eccentricity*: The eccentricity is calculated by a fraction of the number of pixels between the major axis length and foci of the ellipse containing the grain. The value of eccentricity ranges in between 0 to 1.
- *Equivalent diameter*: Equivalent diameter of rice grains is calculated as, $Eqd = \sqrt{(4 * Area)/\pi}$

These parameters were extracted with image processing techniques as discussed in Section 2.3.

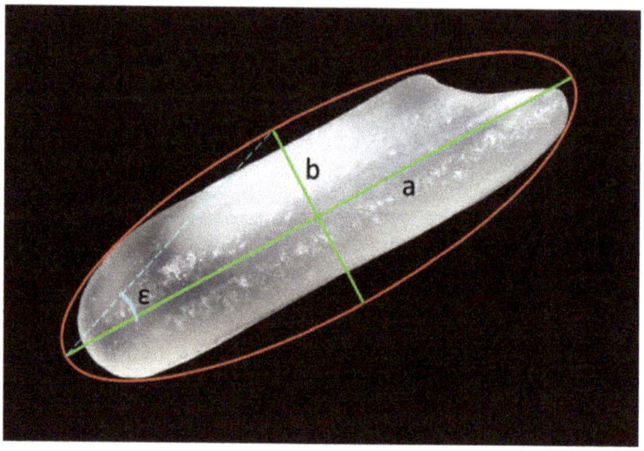

Figure 3. Rice grain properties.

2.5. Fuzzy Inference System

A Fuzzy Inference System (FIS) incorporates the knowledge of an expert, during design a model in between input and output parameters. In FIS, the input-output relations are defined by a set of fuzzy rules, e.g., IF-THEN rules [36]. Fuzzy logic-reasoning involves the assignment of membership function to the input and output parameters; and the rule base which processes the fuzzy values of the inputs to fuzzy values of the outputs. The accurate selection of these membership function and the rules is one of the most critical stages in the FIS which needs expert knowledge. FIS consists of three segments viz. fuzzification, inference engine and defuzzifier. Fuzzification converts the numeric value of the input to a linguistic variable with the help of the membership functions e.g., triangular, trapezoidal, Gaussian, etc. The inference engine evaluates the degree of the membership function of the input variables (premise) to the fuzzy consequent part using the fuzzy IF-THEN rules. The conditional statement contains a premise, the if-part, and a conclusion, the then-part [37]. The knowledge involved in a fuzzy inference system contains a group of several rules [38]. At last, the defuzzifier converts the fuzzy output into a crisp value. The fuzzy inference engine is the core of FIS which can represent the human decision-making process [36].

The Takagi-Sugeno (T-S) FIS has fuzzy inputs and a crisp output which is a linear combination of the inputs or constant. This method is computationally efficient and suitable to work with optimization and adaptive techniques [39]. The T-S method involves a systematic approach to generating fuzzy rules from a given input-output data set (Figure 4). It uses a membership function of the input variables for producing the consequent (then part). It uses the fuzzy rule: *IF x is A AND y is B THEN z is* $f(x, y)$ where x, y, and z are linguistic variables, A and B are fuzzy sets and $f(x, y)$ is a mathematical function [39]. T-S FIS uses a weighted average to generate the crisp output.

In this present work, zero order T-S was adopted for grading of basmati rice. The membership functions were taken as 'gbellmf' [40] for all inputs viz. eccentricity, equivalent diameter, perimeter and the major axis length/minor axis length (a/b); the outputs were taken as constant (for whole grain, output = 1 and broken grain output = 0). An example of fuzzy membership function is shown in Figure 5. The rules of the T-S method were taken as follows:

- **Rule 1:** If (eccentricity is high) and (equivalent diameter is high) and (perimeter is high) and (a/b is high) then (output is Whole grain).
- **Rule 2:** If (eccentricity is low) and (equivalent diameter is low) and (perimeter is low) and (a/b is low) then (output is broken grain).

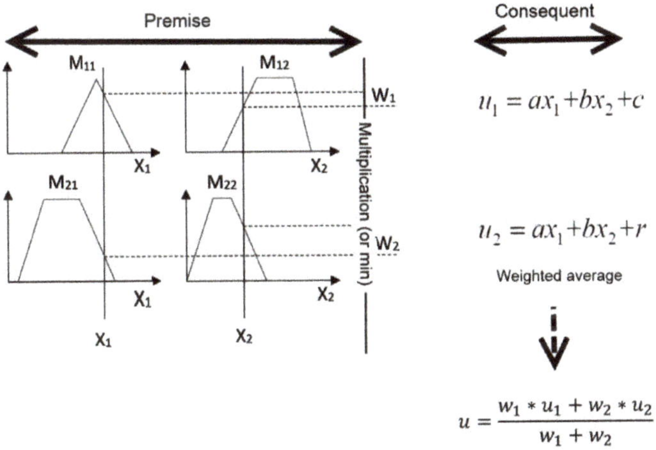

Figure 4. Takagi-Sugeno type FIS system with premise and consequent part.

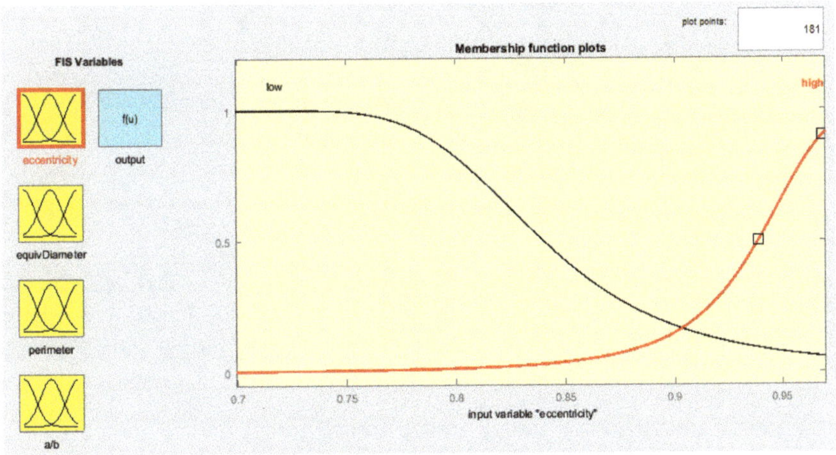

Figure 5. Membership function for Eccentricity feature.

2.6. Adaptive Neuro-Fuzzy Inference System (ANFIS)

The adaptive neural network based fuzzy inference system (ANFIS) is a hybrid system. It includes both the advantages of the self-adaptability and learning competence of the neural network and the ability of the fuzzy system to take into account the prevailing uncertainty and imprecision of real systems. The neuro-fuzzy modeling approach is concerned with model extraction from numerical data which represents the dynamic behaviour of the scheme. With ANFIS method, an initial fuzzy model is generated with the help of the rules extracted from the input-output data. Next, the neural network is used to tune the rules of the initial fuzzy model to produce the final ANFIS model. The formulations and discussion of ANFIS architecture can be found in [41,42]. Unlike ANN, it has a higher capability in the learning process to adapt to its environment. Therefore, it can be used to automatically adjust the membership function's parameters and reduce the rate of errors in the determination of rules in fuzzy logic.

The ANFIS architecture shown in Figure 6 is an adaptive network that uses supervised learning algorithm and has a function similar to the model of Takagi-Sugeno fuzzy inference system as discussed in Section 2.5. Let's assume that there are two inputs x and y, and one output f of the architecture. Two rules are used in the method of "If-Then" for Takagi–Sugeno model, as follows:

- **Rule 1:** If x is A_1 and y is B_1 Then $f_1 = p_1x + q_1y + r_1$.
- **Rule 2:** If x is A_2 and y is B_2 Then $f_2 = p_2x + q_2y + r_2$.

where A_1, A_2 and B_1, B_2 are the membership functions of each input x and y (premises), while p_1, q_1, r_1 and p_2, q_2, r_2 are linear parameters in the consequent part of Takagi–Sugeno fuzzy inference model. ANFIS architecture has five layers. The first and fourth layers contain an adaptive node, while the other layers are fixed nodes.

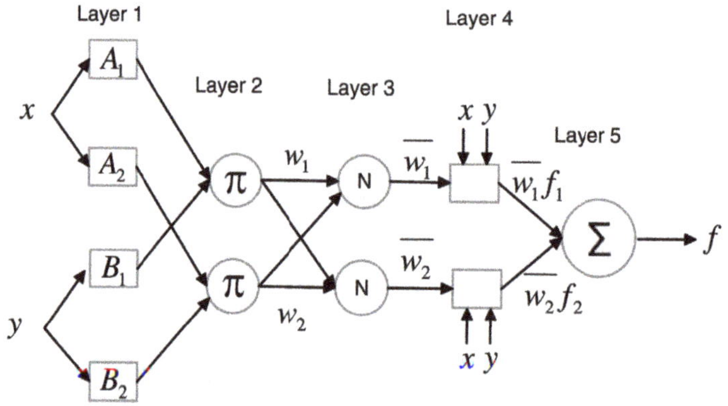

Figure 6. ANFIS architecture with input, hidden and output layer [43].

Layer 1: Each node adapts to a function parameter. The output from each node is a degree of membership value that is given by the input of the membership functions. For example, the membership function used in this study is a generalized bell membership function (c.f. Section 2.5).

$$\mu_{Ai}(x) = \frac{1}{1 + \left|\frac{x-c}{a}\right|^{2b}} \tag{1}$$

where μ_{Ai} is the degree of membership functions for the fuzzy set Ai, and $\{a, b, c\}$ are the parameters of a membership function which can change the shape of the membership function as shown in Figure 5.

Layer 2: Each node in this layer is fixed or non-adaptive and represented with a product operator Π. Each node in this layer represents the firing strength for each rule.

Layer 3: Each node in this layer is fixed or non-adaptive and labeled as N. It is normalizing the firing strength as $\bar{w}_i = w_i / \sum w_i$.

Layer 4: Each node in this layer is an adaptive node with a node function defined as $\bar{w}_i f_i = p_i x + q_i y + r_i$. The parameters in this layer are referred to as consequent parameters.

Layer 5: The single node in this layer is a fixed or non-adaptive node that computes the overall output as the summation of all incoming signals from previous nodes as $\sum \bar{w}_i f_i$.

In the ANFIS architecture, the first layer and the fourth layer contain the parameters which are tuned during the training phase. The number of training epochs, the membership functions and the number of fuzzy rules should be selected accurately while designing of ANFIS model [41], as it may lead system to overfit the data. This tuning is obtained with a hybrid algorithm combining the least-squares method and the gradient descent method with a mean square error method [44].

The training error, as well as test error, are also assessed using the rice grain sample data. A threshold was applied to the output of ANFIS to get a binary class of whole grain or broken grain.

2.7. Design of Experiment

The features (c.f. Section 2.4) were obtained form all the basmati rice sample images (in total 40 images) as discussed in aforementioned Sections. Among them, features obtained from 30 images were used to train the ANFIS and features derived from the remaining 10 images were used for testing. Subsequently, the classification accuracy was estimated with actual class of test data and ANFIS output results. Furthermore, the performance of ANFIS was compared with standard classification techniques of support vector machines (SVM) and K-nearest neighbours (KNN). The optimal margin and kernel parameters (radial basis function) for the Soft-margin SVM classifiers were determined using grid search and 10-fold cross-validation. Each image contains ~15 rice kernels (objects). However, for representation we have kept only 1 image and have shown all image processing steps involved in result section (c.f. Section 3).

Furthermore, the effectiveness of morphological features was analyzed for a segmented test image (for each object). The histograms of individual feature can provide a rationale by relating the features to the physical properties derived from an image based technique associated with the grain class. In addition to the classification accuracy assessment, the milling efficiency (the ratio of broken grain and whole grain) was assessed using the broken grain and whole grain objects derived from a test image.

3. Results and Discussion

This section explains the results of the ANFIS based rice grading technique. The proposed morphological features are generated as detailed in Section 3.1 using the image processing technique which was performed using the steps described in Section 2.3. In Section 3.2, the ANFIS classification result using the features is analyzed and subsequently compared with the standard classification method. Furthermore, a histogram analysis of features are followed by estimation of the milling efficiency in Sections 3.3 and 3.4.

3.1. Image Processing Outputs

The sample images were processed as mentioned in Section 2.3. An example of the processed images is shown in Figure 7. These images were used for feature extraction which was utilized for a fuzzy model generation. The images are consisting of both whole and broken grain rice. After segmentation and morphological operations, each object of an individual image is labelled as shown in Figure 7h–l. The features associated with each object (e.g., Object 1 in Figure 7h) are stored with the associated objectID in a tuple. These data set are further being used in training and testing of the classifier.

3.2. Classification Performance

The features extracted from processed training images were used to build the ANFIS model. Test error was found to be on training data for 20 epochs during ANFIS training (Figure 8). The error was zero for test data after thresholding (threshold at 2) on ANFIS output. The results of training and testing are quite impressive as shown in Figure 8 and it is further analyzed in Table 1.

From Table 1 it is observed that for the majority of the objects (rice grains in test image) the ANFIS output threshold class is similar to actual class (i.e., 0 or 1). Therefore, the classification accuracy of ANFIS is 100%, as an actual class and ANFIS output class for the test objects are alike.

Here it is important to note that the proposed approach categorizes basmati rice grains into two categories: (1) whole grain and (2) broken grain. Perhaps the readers can imagine what happens with imperfect grains or the imperfect whole grain? For an imperfect grain, the morphological parameters are different from a whole grain. The present study considers a binary classification ('whole grain' = 1, 'broken grains or/and others' = 0). The membership function is taken as 'High' and 'Low' (cf. Section 2.5). For an imperfect grain eccentricity will be 'low', (a/b) is 'low'. Perimeter might be 'medium'. However, in our case membership function is set to 'High' and 'Low'. Therefore, it will be classified as 'broken grain'. By virtue of these, imperfect grain anyhow is separated from 'whole grain', thus it is not affecting 'milling efficiency'. Figure 9 illustrates a similar scenario which is analyzed from a test image. It is clearly showing a dissimilarity between grain physical parameters for three-grain types.

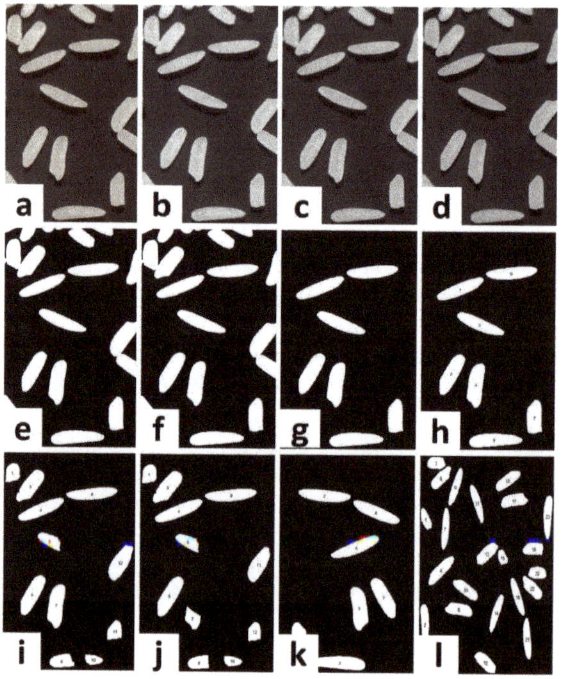

Figure 7. Image processing outputs. (**a**) Raw sample image (RGB); (**b**) Red channel image; (**c**) Green channel image; (**d**) Blue channel image; (**e**) Binary image; (**f**) Image after morphological opening; (**g**) Hole filled and border cleared image; (**h**) Labelled objects in training sample1 image; (**i**) Labelled objects in training sample 2 image; (**j**) Labelled objects in training sample 3 image; (**k**) Labelled objects in training sample 4 image; (**l**) Labelled objects in test sample image.

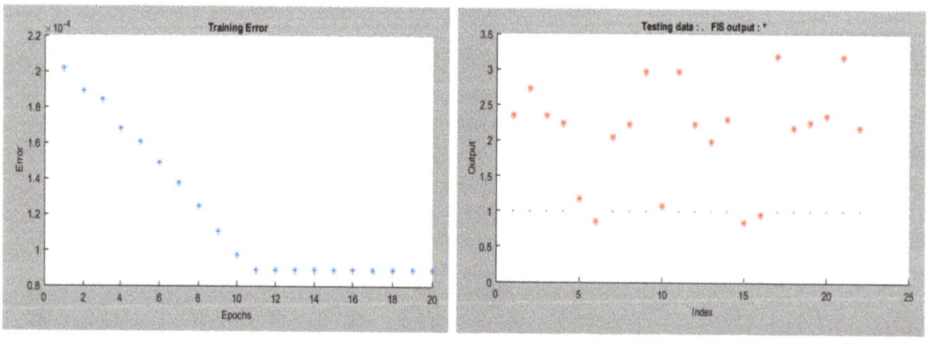

Figure 8. Training and test error for sample images. During training, after 11 epochs the error significantly reduces. In test error plot, the blue dots are actual output, and the red stars are ANFIS output corresponding to each object.

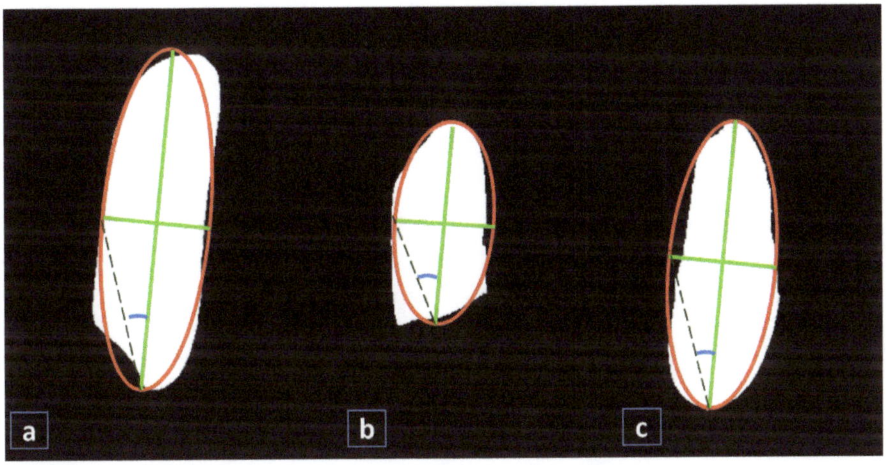

Figure 9. Variations in features in between different grains. (**a**) Whole rice grain; (**b**) Broken rice grain; (**c**) Imperfect rice grain.

Table 1. ANFIS results for test image for each rice object. Threshold is applied with rule: If (ANFIS output class > 2) Then (ANFIS output class thresholded = 0).

Object Label	Actual Class	ANFIS Output Class	ANFIS Class Output Thresholded
1	1	2.353	1
2	1	2.723	1
3	1	2.352	1
4	1	2.240	1
5	0	1.168	0
6	0	0.859	0
7	1	2.050	1
8	1	2.228	1
9	1	2.973	1
10	0	1.073	0
11	1	2.971	1
12	1	2.231	1
13	1	1.992	1
14	1	2.309	1
15	0	0.841	0
16	0	0.959	0
17	1	3.189	1
18	1	2.183	1
19	1	2.251	1
20	1	2.347	1
21	1	3.184	1
22	1	2.182	1

From the experimental results, it is observed that the ANFIS perform satisfactorily in evaluating the percentage of broken rice with an overall accuracy of >98.5%. However, the comparative analysis with the SVM and KNN are less favourable with accuracies <95% for 10 test images as shown in spider plot in Figure 10.

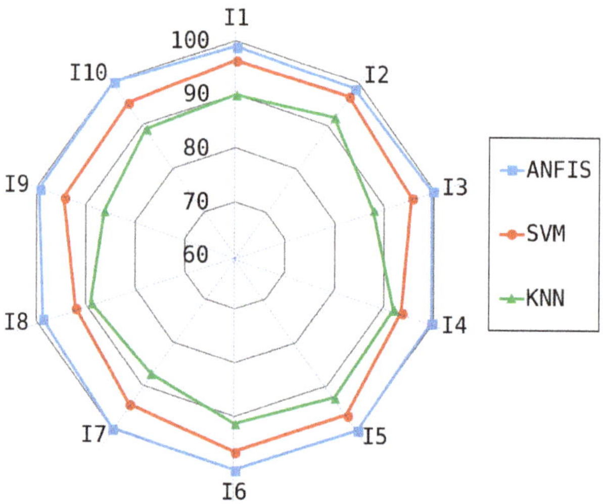

Figure 10. Classification performance of ANFIS, SVM and KNN for 10 test image samples. I1-I10 represents the test image IDs.

3.3. *Histogram of Features in Testing Images*

Histograms of feature extracted from the test image objects are shown in Figure 11. All the features are positively skewed representing a positive relation with basmati grain size (head grain). The occurrence of eccentricity >0.9 is found for more than 16 rice objects in the test image. The aspect ratio is >2.4 for more than 17 objects. Similar results are also shown in the case of area, perimeter and major axis length. The histogram images as shown in Figure 11, also exemplify the head grain and broken grain percentage in the test images.

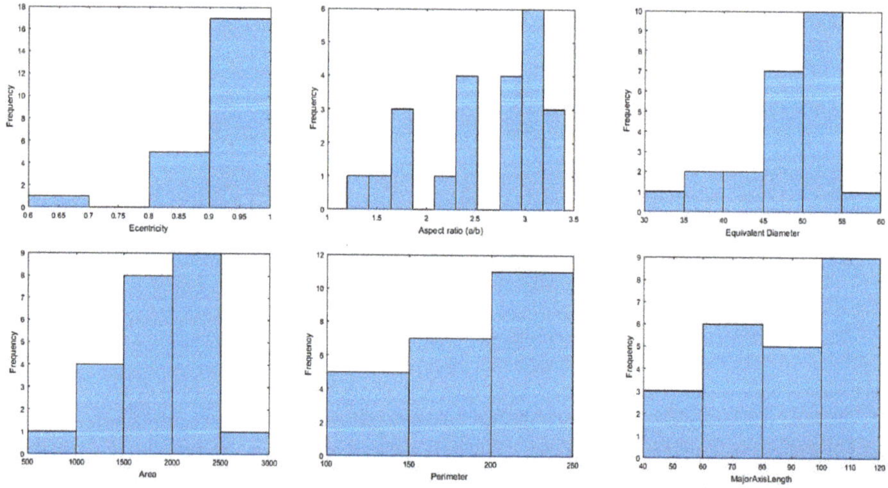

Figure 11. Histogram of eccentricity, aspect ratio, equivalent diameter, area, perimeter and major axis length of test image objects.

3.4. Milling Efficiency

From the test image output results, the number of the whole grains = 17 out of 22 objects, whereas, the broken grain objectsare = 5 out of 22 objects. Therefore, the percentage of whole grain = $(17/22) \times 100 = 77.27\%$. This milling efficiency (η) was found for a specific roller characteristic (rpm = 2000 and gap between roller ~0.5 mm) of the milling machine [45]. The milling efficiencies were evaluated in a similar way for the 10 test images. Subsequently, the average of all the 10 milling efficiencies was determined, which is ~77.3% with a standard deviation (σ) of 1.5. The milling efficiency (η_{avg}) derived using the image-based method was in accordance with the manual calculation results ($\eta = 76.87\%$).

4. Summary and Conclusions

Standard classification technique seldom incorporates domain knowledge associated with the physical properties of rice grain. Hence, a knowledge-based neuro-fuzzy classification technique was proposed in this study for grading of basmati rice grains. This technique takes into account the physical properties of grain devised from an image-based method to classify whole and broken grain.

A novel image processing technique was adopted for morphological feature extraction followed by ANFIS model building for discrimination of grains. The classification accuracy for the test images were >98.6%, which comparatively better than standard SVM and KNN classifier (<95%). Moreover, the proposed ANFIS classification results seem to be more reliable than the results obtained from SVM and KNN, since it deals with uncertainty in output. It is important to note that the standard technique does not take into account any domain knowledge associated with grain physical properties. In fact, the physical properties are essential for grain grading and characterization as analyzed in this study.

The milling efficiency was estimated in terms of percentage of whole grain or head grain and it was 77.27% for the test sample. However, the colour and texture based quality and grading was not considered during feature selection. The overall results of the adopted methodology were promising in terms of classification accuracy and efficiency. This work can be extended to discrimination of different rice varieties for determining the degree adulteration. Furthermore, the real-time image processing based grading can be addressed equally. This can be extended for optimization of milling machine parts characteristics during milling operation (roller parameter-speed, the gap between rollers) for process automation using micro-controller units.

Funding: This research received no external funding.

Acknowledgments: The author would like to thank Rice Processing Unit, Department of Postharvest Engineering, Bidhan Chandra Krishi Viswavidyalaya, India for providing sample images of basmati rice.

Conflicts of Interest: The authors declare no conflict of interest.

References

1. All India Rice Exporters Association. Export Statistics of Basmati Rice. 2015. Available online: http://www.airea.net/page/58/statistical-data/export-statistics-of-basmati-rice/ (accessed on 10 April 2018).
2. Bhattacharjee, P.; Singhal, R.S.; Kulkarni, P.R. Basmati rice: A review. *Int. J. Food Sci. Technol.* **2002**, *37*, 1–12. [CrossRef]
3. Mandal, D. *Concepts of Farm Machinery and Power*; Narendra Publishing House: Delhi, India, 2016.
4. Brosnan, T.; Sun, D.W. Improving quality inspection of food products by computer vision—A review. *J. Food Eng.* **2004**, *61*, 3–16. [CrossRef]
5. Zheng, C.; Sun, D.W.; Zheng, L. Recent developments and applications of image features for food quality evaluation and inspection—A review. *Trends Food Sci. Technol.* **2006**, *17*, 642–655. [CrossRef]
6. Cheng, F.; Ying, Y.B. Machine vision inspection of rice seed based on Hough transform. *J. Zhejiang Univ. Sci. A* **2004**, *5*, 663–667.
7. Maheshwari, C.V.; Jain, K.R. Parametric quality analysis of indian Ponia Oryza Sativa SSP Indica (Rice). *Int. J. Sci. Res. Dev.* **2013**, *1*, 114–118.

8. Vaingankar, N.M.; Kulkarni, P.R. A cooking quality parameter as an indicator of adulteration of Basmati rice. *J. Sci. Food Agric.* **1989**, *48*, 381–384. [CrossRef]
9. Verma, B. Image processing techniques for grading & classification of rice. In Proceedings of the 2010 International Conference on Computer and Communication Technology (ICCCT), Allahabad, India, 17–19 September 2010; pp. 220–223.
10. Zareiforoush, H.; Minaei, S.; Alizadeh, M.R.; Banakar, A. Potential applications of computer vision in quality inspection of rice: a review. *Food Eng. Rev.* **2015**, *7*, 321–345. [CrossRef]
11. Zareiforoush, H.; Minaei, S.; Alizadeh, M.R.; Banakar, A. Qualitative classification of milled rice grains using computer vision and metaheuristic techniques. *J. Food Sci. Technol.* **2016**, *53*, 118–131. [CrossRef] [PubMed]
12. Liu, J.; Tang, Z.; Xu, P.; Liu, W.; Zhang, J.; Zhu, J. Quality-related monitoring and grading of granulated products by weibull-distribution modeling of visual images with semi-supervised learning. *Sensors* **2016**, *16*, 998. [CrossRef] [PubMed]
13. Pazoki, A.; Farokhi, F.; Pazoki, Z. Classification of rice grain varieties using two Artificial Neural Networks (MLP and Neuro-Fuzzy). *J. Anim. Plant Sci.* **2014**, *24*, 336–343.
14. Chen, X.; Xun, Y.; Li, W.; Zhang, J. Combining discriminant analysis and neural networks for corn variety identification. *Comput. Electron. Agric.* **2010**, *71*, S48–S53. [CrossRef]
15. Pamučar, D.; Ljubojević, S.; Kostadinović, D.; Đorović, B. Cost and risk aggregation in multi-objective route planning for hazardous materials transportation—A neuro-fuzzy and artificial bee colony approach. *Expert Syst. Appl.* **2016**, *65*, 1–15. [CrossRef]
16. Lukovac, V.; Pamučar, D.; Popović, M.; Đorović, B. Portfolio model for analyzing human resources: An approach based on neuro-fuzzy modeling and the simulated annealing algorithm. *Expert Syst. Appl.* **2017**, *90*, 318–331. [CrossRef]
17. Adhikari, R. A neural network based linear ensemble framework for time series forecasting. *Neurocomputing* **2015**, *157*, 231–242. [CrossRef]
18. Ioannou, K.; Arabatzis, G.; Lefakis, P. Predicting the prices of forest energy resources with the use of Artificial Neural networks (ANNs). The case of conifer fuel wood in Greece. *J. Environ. Protect. Ecol.* **2009**, *10*, 678–694.
19. Vishwakarma, M.D.D. Genetic algorithm based weights optimization of artificial neural network. *Int. J. Adv. Res. Electr. Electron. Instrum. Eng.* **2012**, *1*, 206–211.
20. Jiang, Y. Uncertainty in the output of artificial neural networks. *IEEE Trans. Med. Imaging* **2003**, *22*, 913–921. [CrossRef] [PubMed]
21. Sabanci, K.; Toktas, A.; Kayabasi, A. Grain classifier with computer vision using adaptive neuro-fuzzy inference system. *J. Sci. Food Agric.* **2017**, *97*, 3994–4000. [CrossRef] [PubMed]
22. Zareiforoush, H.; Minaei, S.; Alizadeh, M.R.; Banakar, A. A hybrid intelligent approach based on computer vision and fuzzy logic for quality measurement of milled rice. *Measurement* **2015**, *66*, 26–34. [CrossRef]
23. Mamdani, E.H. Application of fuzzy algorithms for control of simple dynamic plant. *Proc. Inst. Electr. Eng.* **1974**, *121*, 1585–1588. [CrossRef]
24. Hosseinzadeh, B.; Esmaeili, Z.; Rostami, S.; Zareiforoush, H. Representing the Human Experts Judgment on Quality Indices of White Rice by Image Processing and Artificial Intelligence Techniques. *Agric. Eng. Int. CIGR J.* **2016**, *18*, 97–106.
25. Shiddiq, D.M.; Nazaruddin, Y.Y.; Muchtadi, F.I.; Raharja, S. Estimation of rice milling degree using image processing and adaptive network based fuzzy inference system (ANFIS). In Proceedings of the 2011 2nd International Conference on Instrumentation Control and Automation (ICA), Bandung, Indonesia, 15–17 November 2011; pp. 98–103.
26. Kamath, S.; Stephen, J.; Suresh, S.; Barai, B.; Sahoo, A.; Radhika Reddy, K.; Bhattacharya, K.R. Basmati rice: Its characteristics and identification. *J. Sci. Food Agric.* **2008**, *88*, 1821–1831. [CrossRef]
27. Singh, R.K.; Khush, G.S. *Aromatic Rices*; International Rice Research Institute: Los Baños, Philippines, 2000.
28. Singh, V.; Singh, A.K.; Mohapatra, T.; Ellur, R.K. Pusa Basmati 1121—A rice variety with exceptional kernel elongation and volume expansion after cooking. *Rice* **2018**, *11*, 19. [CrossRef] [PubMed]
29. Sun, D.W. *Computer Vision Technology for Food Quality Evaluation*; Academic Press: Amsterdam, The Netherland, 2016.
30. Abdullah, M.; Fathinul-Syahir, A.; Mohd-Azemi, B. Automated inspection system for colour and shape grading of starfruit (*Averrhoa carambola* L.) using machine vision sensor. *Trans. Inst. Meas. Control* **2005**, *27*, 65–87. [CrossRef]
31. Paulsen, M. Using machine vision to inspect oilseeds. *Int. News Fats Oils Relat. Mater.* **1990**, *1*, 50–55.

32. Otsu, N. A threshold selection method from gray-level histograms. *IEEE Trans. Syst. Man Cybern.* **1979**, *9*, 62–66. [CrossRef]
33. Soille, P. *Morphological Image Analysis: Principles and Applications*; Springer Science & Business Media: Berlin/Heidelberg, Germany, 2013.
34. Shantaiya, S.; Ansari, U. Identification of food grains and its quality using pattern classification. In Proceedings of the 12th IEEE International Conference on Communication Technology (ICCT), Nanjing, China, 11–14 November 2010; Volume 1114, p. 35.
35. Mousavirad, S.; Tab, F.A.; Mollazade, K. Design of an expert system for rice kernel identification using optimal morphological features and back propagation Neural Network. *Int. J. APpl. Inf. Syst.* **2012**, *3*, 33–37.
36. Camastra, F.; Ciaramella, A.; Giovannelli, V.; Lener, M.; Rastelli, V.; Staiano, A.; Staiano, G.; Starace, A. A fuzzy decision system for genetically modified plant environmental risk assessment using Mamdani inference. *Expert Syst. Appl.* **2015**, *42*, 1710–1716. [CrossRef]
37. Cornelissen, A.; van den Berg, J.; Koops, W.; Grossman, M.; Udo, H. Assessment of the contribution of sustainability indicators to sustainable development: A novel approach using fuzzy set theory. *Agric. Ecosyst. Environ.* **2001**, *86*, 173–185. [CrossRef]
38. Klir, G.; Yuan, B. *Fuzzy Sets and Fuzzy Logic*; Prentice Hall: Upper Saddle River, NJ, USA, 1995.
39. Takagi, T.; Sugeno, M. Fuzzy identification of systems and its applications to modeling and control. In *Readings in Fuzzy Sets for Intelligent Systems*; Morgan Kaufman: San Mateo, CA, USA, 1993; pp. 387–403.
40. Gulley, N. *Fuzzy logic toolbox for use with MATLAB*; MathWorks, Inc.: Natick, MA, USA, 1996.
41. Jang, J.S. ANFIS: Adaptive-network-based fuzzy inference system. *IEEE Trans. Syst. Man Cybern.* **1993**, *23*, 665–685. [CrossRef]
42. Walia, N.; Singh, H.; Sharma, A. ANFIS: Adaptive neuro-fuzzy inference system—A survey. *Int. J. Comput. Appl.* **2015**, *123*, 32–38. [CrossRef]
43. Al-Hmouz, A.; Shen, J.; Al-Hmouz, R.; Yan, J. Modeling and simulation of an adaptive neuro-fuzzy inference system (ANFIS) for mobile learning. *IEEE Trans. Learn. Technol.* **2012**, *5*, 226–237. [CrossRef]
44. Uçar, T.; Karahoca, A.; Karahoca, D. Tuberculosis disease diagnosis by using adaptive neuro fuzzy inference system and rough sets. *Neural Comput. Appl.* **2013**, *23*, 471–483. [CrossRef]
45. Sharma, P.; Chakkaravarthi, A.; Singh, V.; Subramanian, R. Grinding characteristics and batter quality of rice in different wet grinding systems. *J. Food Eng.* **2008**, *88*, 499–506. [CrossRef]

applied system innovation

MDPI

Article

A Fuzzy Inference System for Unsupervised Deblurring of Motion Blur in Electron Beam Calibration

Salaheddin Hosseinzadeh

Department of Engineering, Design and Physical Sciences, Brunel University London, Kingston Lane, Uxbridge, London UB8 3PH, UK; Salaheddin.Hosseinzadeh@gmail.com

Received: 18 October 2018; Accepted: 25 November 2018; Published: 4 December 2018

Abstract: This paper presents a novel method of restoring the electron beam (EB) measurements that are degraded by linear motion blur. This is based on a fuzzy inference system (FIS) and Wiener inverse filter, together providing autonomy, reliability, flexibility, and real-time execution. This system is capable of restoring highly degraded signals without requiring the exact knowledge of EB probe size. The FIS is formed of three inputs, eight fuzzy rules, and one output. The FIS is responsible for monitoring the restoration results, grading their validity, and choosing the one that yields to a better grade. These grades are produced autonomously by analyzing results of a Wiener inverse filter. To benchmark the performance of the system, ground truth signals obtained using an 18 μm wire probe were compared with the restorations. Main aims are therefore: (a) Provide unsupervised deblurring for device independent EB measurement; (b) improve the reliability of the process; and (c) apply deblurring without knowing the probe size. These further facilitate the deployment and manufacturing of EB probes as well as facilitate accurate and probe-independent EB characterization. This paper's findings also makes restoration of previously collected EB measurements easier where the probe sizes are not known nor recorded.

Keywords: fuzzy inference system; fuzzy logics; linear motion blur; fuzzy deblurring; electron beam calibration; signal and image processing

1. Introduction

The main goal of fuzzy systems is to define and control sophisticated processes by incorporating and taking advantage of human knowledge and experience. Nowadays, fuzzy logics are widely used in industry for various applications ranging from cameras to cement kilns, trains, and vacuum cleaners [1]. Furthermore, deblurring techniques have versatile applications and they are either performed in spatial [2] or frequency domains [3–5]. Hosseinzadeh [6] modeled the electron beam (EB) measurement process with a linear motion blur and evaluated three of the well-established deblurring techniques for EB restoration. In this study [6], Hosseinzadeh used a Weiner inverse filter and blind Richardson-Lucy deconvolutions to restore the EB distribution and correct the measurements through deblurring. A simple motion blur is formulated in Equation (1).

$$g(x) = \int f(x)h(x) + n(x), \tag{1}$$

where in the spatial domain, f, g, h, and n are the ground truth signal (EB distribution) of length L_f, degraded signal (measurement from probe), point spread function (PSF) of length L_h, and noise respectively. Their frequency domains are represented by uppercase letters F, G, and H. In the case of electron beam measurements, the ground truth signal is the distribution of EB and the degraded signal is the measurement acquired from the probe. The electron absorption of a slit or wire probe of size L_h is modeled with a PSF kernel [6].

Linear motion blur point spread function has two distinct characteristics of motion direction and length (L) [7]. The PSF is known for having harmonically spaced vanishing magnitudes in the frequency domain due to its limited length in the spatial domain [8]. There are several approaches to estimate L_h such as log power spectrum, cepstrum, bispectrum, and pitch detection algorithms. In image deblurring jargon, it is assumed that the frequency spectrum of F is smooth and does not contain vanishing frequencies, hence any vanishing frequencies in G are associated to H [9,10]. However, this assumption usually does not hold for EB measurements, especially where the L_f is in the same order of L_h. This similarity makes it complicated to distinguish between L_f and L_h and therefore compromises the deblurring process by an incorrect detection of null frequencies. Such an erroneous deblurring process is likely to produce an incorrect but convincing result, notably when f and h have remarkable cross-correlation. This ambiguity is likely to happen in EB measurements, because: (a) f and h are usually in the same order of magnitude and they have relatively high cross-correlation; and (b) the L_f can be inconsistent. In Reference [6], a prior knowledge of L_h is used to estimate the position of null frequency of h from the spectrum analysis of G. Hosseinzadeh limited the spectrum of G to $\pm 15\%$ of the nominal L_h by applying a window to its log-power spectrum, thereby ignoring vanishing frequencies outside of this interval. This algorithm is available in Reference [11]. This strategy relies on knowing the L_h. Therefore, it is a good approach when it is known accurately. There are a few limitations with this method due to the varying nature of L_f during the calibration and measurement process. As a result, the beam's vanishing frequency (or its harmonics) can be located within the applied window and cause a false detection. Furthermore, if the inaccuracy of L_h is more than 15%, the null frequency of h is ignored by the window resulting in an erroneous restoration. In addition, any inaccuracy of more than $\pm 15\%$ cannot be compensated.

One solution to effectively address this uncertainty is to use fuzzy systems. Fuzzy inference systems are widely used to address instrumental uncertainties. A comprehensive review and explanation of fuzzy inference systems are provided in Reference [12].

It is known that a wrong estimation of L_h can lead to drastic noise-like errors in the restorations [13]. Furthermore, utilizing deblurring techniques for industrial purposes requires real-time, reliable, and unsupervised methods. To satisfy these requirements, this article proposes a Wiener filter that is monitored by a fuzzy inference system. A Wiener filter is selected due to its simplicity, real-time execution, and superior performance in the restoration of linear motion blur [6]. The fuzzy inference system deals with the uncertainty of the deconvolution by monitoring the entire restoration process. This FIS is comprised of three crisp inputs that included the PSF length or probe size (L_h) deviation, attenuation of the vanishing frequencies, and deconvolution residue.

However, probe size deviation is an optional input, which is based on a previous rough knowledge of L_h. If L_h is roughly known, it serves as a reference point from which the PSF length deviation is calculated. Therefore, unlike Reference [6], prior knowledge of L_h does not limit the inaccuracy compensation to $\pm 15\%$. It is demonstrated in Reference [6] that the spatial domain of h has a sharper transition compared to the EB distribution (f). This is due to the semi-Gaussian distribution of f compared to h. Therefore, vanishing frequencies of h are expected to have higher attenuation or lower magnitude compared to f. Hence, the normalized magnitude of the detected null frequencies in G are the second crisp input to the fuzzy inference systems. The last input of the system is the quantified deblurring artifacts that are introduced during the restoration of f from g. The restored beam distributions are denoted as ($\hat{f}$). These residual artifacts are inevitable and they increase as the h deviates from its mathematical definition. Extraction of residues from $\hat{f}$ is explained in section II. The output of the fuzzy system (E_i) is defuzzified to represent the quality of the restorations. This output is generated based on the definition of the fuzzy rules that are explained in the next section.

The rest of this paper is arranged as follows: Section 2 illustrates the details of FIS implementation. This includes specifying the crisp inputs and fuzzifying them, defining the membership functions, and formulating the fuzzy sets. The section continues by identifying the fuzzy rules and making an inference to generate the output. Section 3 presents the practical results of the proposed method and the ability of the system to distinguish the correct deblurring results. The values of membership functions parameters are provided and a comparison is made between implementing the fuzzy system with and without the knowledge of probe size (L_h).

2. Modeling and Implementation

As mentioned, when there is similarity between L_f and L_h it is difficult to discriminate between their null frequencies just by looking at G. This introduces an uncertainty and makes it hard to decide which null frequency belongs to the probe (H) because null frequencies can belong to either beam (F) or probe (H). To address the uncertainty of unsupervised L_h detection, all the null frequencies in G are identified and only the first two nulls with lowest frequencies are extracted while avoiding the harmonics. This implies that a maximum of two null frequencies ($\omega_{i=1,2}$) are to be extracted from G. There are three possibilities based on the extracted number of null frequencies: (a) If no null frequency is detected due to $L_h \ll L_f$, then motion blur effect is negligible and deconvolution is not necessary; (b) if a single null frequency is detected as a result of $L_h \gg L_f$, then the deconvolution can progress without involving the fuzzy system as the null frequency belongs to L_h; (c) in case two null frequencies are extracted (ω_1, ω_2), two deconvolutions are performed where each of the deconvolutions are performed by adjusting their corresponding $\hat{L}_{i=1,2}$ ($\hat{L}_{i=1,2} \propto 1/\omega_{i=1,2}$). This is done because both ω_1 and ω_2 could be belonging to h of different sizes.

The FIS is defined with three merits to grade the deblurrings. Deblurrings are performed by two individual Weiner filters that use $\hat{L}_1$ and $\hat{L}_2$ resulting in $\hat{f}_1$ and $\hat{f}_2$ respectively. The fuzzy system produces a single crisp output deconvolution grade ($E_{i=1,2}$) for each restoration. The restoration process that produces a higher E_i is then chosen as the correct process with its corresponding $\hat{L}_i$ being the correct probe size ($L_h \leftarrow \hat{L}_i$). A single layer (non-hierarchal) fuzzy inference system of three inputs and a single output is designed to evaluate the overall deblurring process. These inputs are: PSF length deviation, null frequency magnitude, and residue, and the deconvolution grade is the only output. These inputs and the output are explained in detail as follows.

2.1. PSF Length Deviation

As mentioned, ω_1 and ω_2 are extracted to accurately adjust the L_h during the restoration process. By having rough prior knowledge of the probe size (L_h) and the estimated sizes ($\hat{L}_i$) from G, we can define PSF length deviation as the distance between the expected and the estimations ($|L_h - \hat{L}_i|$). This definition converges to zero if the estimation is close to the prior knowledge, whereas it increases if $\hat{L}_i$ is deviated from L_h. Two fuzzy sets (A_{far} & A_{close}) with membership functions of μ'_m and μ_m are defined to account for the probe inaccuracy and assign a degree of membership to each $\hat{L}_i$ based on its deviation from L_h. Membership functions are defined by polynomial-Z (zmf) and polynomial-S (smf). The A_{close} fuzzy set definition and its membership function is formulated in Equation (2). A thorough evaluation of fuzzy membership functions are provided in Reference [14].

$$A_{close} = \{(\hat{L}_i, \mu_{m(\hat{L}_i)}) |\, 0 < \hat{L}_i < \infty, m(\hat{L}_i) = \tfrac{2|L_h - \hat{L}_i|}{L_h}\},$$
$$\mu_m = \begin{cases} 1 & m \le a_m \\ 1 - 2\left(\frac{m-a_m}{c_m-a_m}\right)^2 & a_m < m \le \frac{a_m+c_m}{2} \\ 2\left(\frac{m-c_m}{c_m-a_m}\right)^2 & \frac{a_m+c_m}{2} < m \le c_m \\ 0 & m > c_m \end{cases}, \tag{2}$$

where a_m and c_m are the membership function parameters that are found heuristically through analysis of several measurements.

2.2. Null Frequency Magnitude

The second input of the fuzzy system is the magnitude of the extracted null frequencies. This is extracted from the normalized log-power spectrum of g and has a dynamic range of 0 to 1 dB, demonstrated in Figure 1.

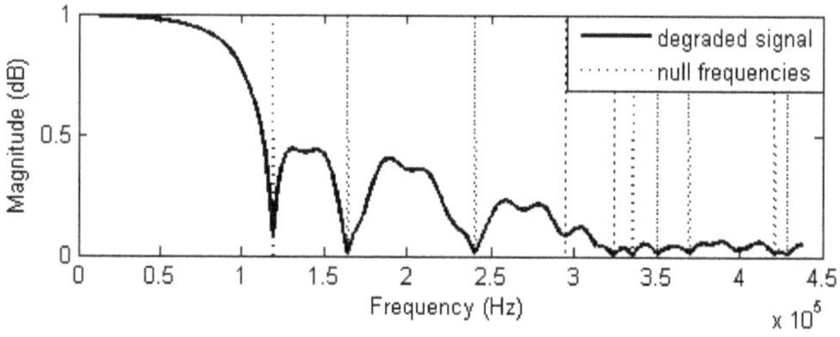

Figure 1. Normalized power spectrum of G exhibits ω_1 and ω_2 at 0.12 and 0.165 MHz frequencies with their harmonics at higher frequencies.

As explained, h is most likely to have rapid spatial transitions compared to f. This implies that H is likely to have the nulls with higher attenuation in G (nulls with lower magnitude). As a result, two fuzzy sets (B_{high} & B_{Low}) with membership functions of μ'_o and μ_o are defined to assign a higher membership value to the nulls with more attenuation (or lower magnitude), whereas a lower degree of membership is assigned to less attenuated (higher magnitude) nulls. Membership functions are defined with zmf and sfm. B_{Low} is formulated in Equation (3), where G_N is the normalized frequency spectrum of the degraded signal G and a_o and c_o are the membership function parameters. A_{Far} membership function definition is similar to B_{Low} as they are both defined by smf.

$$B_{Low} = \{(\hat{L}_i, \mu'_{o(\hat{L}_i)}) | \, 0 < \hat{L}_i < \infty, o(\hat{L}_i) = \log(|G_N(\hat{L}_i) + 1|)\},$$
$$\mu'_o = \begin{cases} 0 & o \leq a_o \\ 2\left(\frac{o-a_o}{c_o-a_o}\right)^2 & a_o < o \leq \frac{a_o+c_o}{2} \\ 1-2\left(\frac{o-c_o}{c_o-a_o}\right)^2 & \frac{a_o+c_o}{2} < o \leq c_o \\ 1 & o > c_o \end{cases}, \quad (3)$$

2.3. Deconvolution Artifact Residues

Deconvolutions are performed using the Wiener inverse filtering process in Equation (4).

$$\hat{F}_i = \frac{1}{H(\omega_i)}\left[\frac{|H(\omega_i)|^2}{|H(\omega_i)|^2 + \frac{1}{SNR(\omega)}}\right] G(\omega), \quad (4)$$

where in the frequency domain, $\hat{F}_i$ is the restored ground truth signal and SNR is the signal-to-noise ratio. After the deconvolutions,$\hat{f}_{i=1,2}$ has shorter lengths in spatial domain compared to g. We first normalized g and both of the restorations ($\hat{f}_{i=1,2}$) between $[0,-1]$, g_N is then shifted so its minimum is matched with the minimums of each $\hat{f}_i$ in the spatial domain to obtain $\hat{g}_N$. Finally, every restoration residue (r_i) is quantified as in Equation (5).

$$r_i = \frac{4}{\int g(x)dx} \cdot \int \hat{f}_i(\tau)d\tau \qquad \{\tau \in x | \hat{g}_N(\tau) > -0.05\}, \quad (5)$$

The deconvolution process using both of the extracted PSFs and their corresponding residues are showed in Figure 2. The deconvolution was performed with a Wiener inverse filter, where h is formulated in Equation (6).

$$h_{\hat{L}_i}(x) = \begin{cases} 0 & o.w \\ 1 & |x| < \frac{\hat{L}_i}{2} \end{cases}, \quad (6)$$

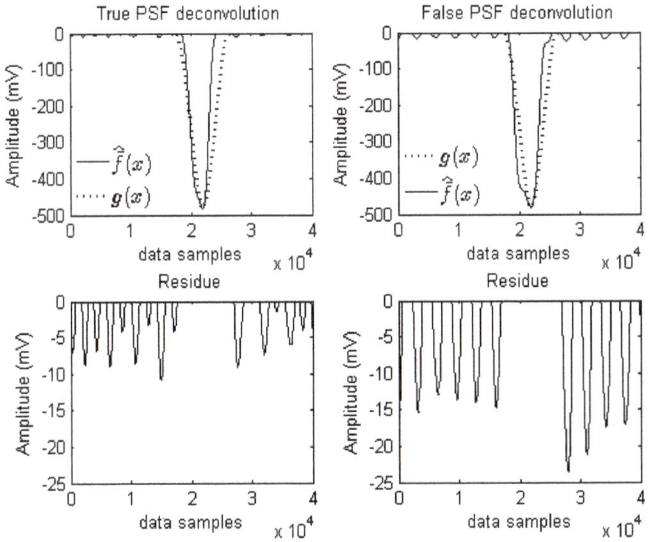

Figure 2. Deconvolution of the degraded pulse in Figure 1, using two different point spread function (PSF) lengths and demonstration of their deconvolution residues.

Two fuzzy sets (C_{low} and C_{high}) are defined with membership functions of μ_r and μ_r' using zmf and smf respectively, where the overall shape of the functions is determined by a_r and c_r. These functions are designed to assign a higher degree of membership to the $\hat{L}_i$ that produces a smaller number of residues after restoration.

2.4. Deconvolution Grade

All the combinations of the aforementioned inputs are used to form eight if–then rule statements with different weights. These statements, with their corresponding weights, are provided in Table 1. Fuzzy AND operator is then used for the implication of the fuzzy consequences.

Table 1. Rule base formation criteria.

Antecedent			Consequence	Rule Weight
PSF Dev	Attenuation	Residue	Restoration Quality	
μ_m	μ_o	μ_r	μ_g	1
μ_m	μ_o	μ_r'	μ_g	0.66
μ_m	μ_o'	μ_r	μ_g	0.66
μ_m	μ_o'	μ_r'	μ_b	0.66
μ_m'	μ_o	μ_r	μ_g	0.66
μ_m'	μ_o	μ_r'	μ_b	0.66
μ_m'	μ_o'	μ_r	μ_b	0.66
μ_m'	μ_o'	μ_r'	μ_b	1

Rule weight is added to scale the consequences and account for the certainty of the rules. The consequence is the restoration quality with two fuzzy sets (D_{good} & D_{bad}) and membership functions of μ_q and μ'_q respectively defined by smf and zmf. Aggregations of the rules are performed by using a Zadeh T-norm and defuzzifications are carried out by mean of maximum (MoM) method [15]. The resulting crisp values are the deconvolution grades ($E_{i=1,2}$). Therefore, there is a grade ($E_{i=1,2}$) for each deconvolution. In other words, for each $\hat{f}_{i=1,2}$ that is deblurred by its corresponding $h_{\hat{L}_{i=1,2}}$, there is an overall grade of restoration ($E_{i=1,2}$). According to the definition of the consequence membership functions, a greater value of E_i represents a better restoration and, on the contrary, a lower value of E_i represents a possible erroneous process, (E_i is ranging from 0 to 1). With this proposed system, if by mistake L_f is used instead of L_h in the formation of the h (Equation (6)), then the resulting E_i will be lower. Overall, E_1 and E_2 are used comparatively to determine and select the best restoration between $\hat{f}_1$ and $\hat{f}_2$ that are emerged from restoring a degraded sample (g). This proposed system and its overall restoration processes are demonstrated in Figure 3.

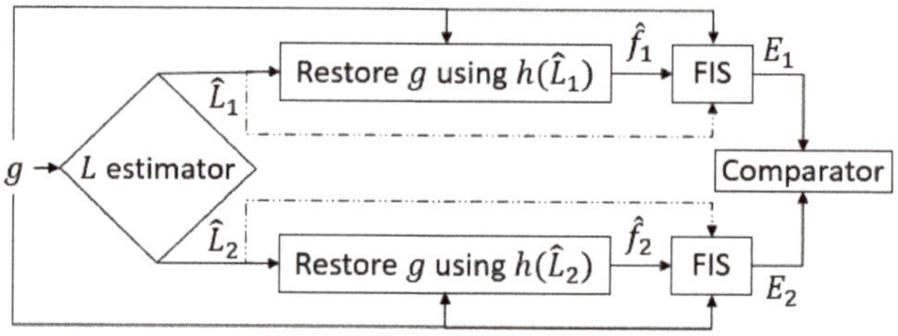

Figure 3. Process diagram, $\hat{L}_i$ connections to the fuzzy inference system (FIS) are optional.

3. Practical Result

Membership Function Parameters

Membership function parameters were investigated pragmatically by testing the explained algorithm for various degraded EB measurement samples. In all degraded measurements, h and f had approximately similar sizes as a result of which $\hat{L}_1 \cong \hat{L}_2$. The membership functions were designed with smooth transitions to provide a general solution and more flexibility, except for the attenuation. To further discriminate between E_1 and E_2, the attenuation membership function parameters were adjusted to have more emphasis between the interval of 0 to 0.3 dB. This intuitive definition was done by observing the magnitude of null frequencies in several degraded signals where the attenuation of the null frequencies was always under 0.3 dB. The membership function parameters are presented in Table 2.

Table 2. Membership function definition details.

PSF Deviation				Attenuation				Residue				Restoration Quality			
μ_m		μ'_m		μ_o		μ'_o		μ_r		μ'_r		μ_q		μ'_q	
a_m	c_m	a_m	c_m	a_o	c_o	a_o	c_o	a_r	c_r	a_r	c_r	a_q	c_q	a_q	c_q
0.02	1	0.04	1	0.02	0.3	0.05	0.3	0	1	0	1	0	1	0	1

The membership functions of attenuation (B_{high} & B_{Low}) and residue (C_{low} & C_{high}) fuzzy sets are depicted in Figure 4, according to their values in Table 2. The fuzzy sets of PSF deviation and restoration quality were also defined with the similar membership functions to that of residues.

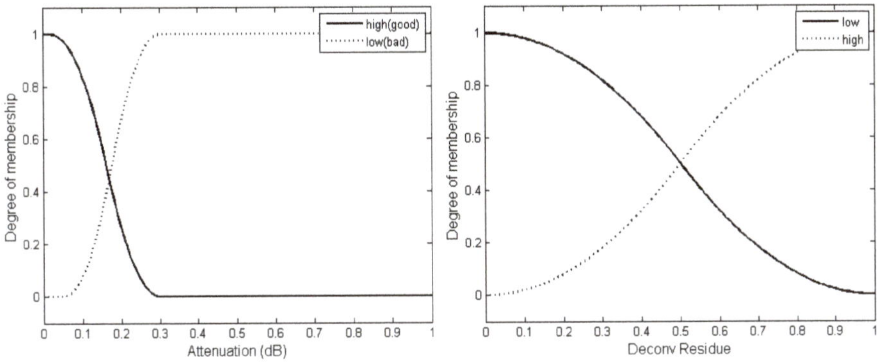

Figure 4. Attenuation and deconvolution residue membership functions.

The analysis of a few of the samples are shown in Figures 5 and 6. For a few of the EB measurements, the L_h (probe sizes) were known to be 1.00, 0.20, and 0.40 mm respectively. The crisp fuzzy inputs and deconvolution grades E_i were also provided for every sample. The restoration that resulted in the higher E_i was selected by the system as the correct solution and its corresponding $\hat{L}_i$ therefore represents the probe size ($\hat{L}_h \leftarrow \hat{L}_i$). To validate the proposed system with the ground truth signal (f) [6], both restorations ($\hat{f}_{1,2}$) were compared against their ground truth signal using cross-correlation. For the $\hat{f}_i$ with the higher E_i, the cross-correlation of $\hat{f}_i$ and f also produced greater coefficients, supporting the accuracy and reliability of the system. As another benchmark, full width at half maximum (FWHM) analysis was used, as it is a popular measure in the EB calibration jargon. The FWHM of f and the $\hat{f}_i$ that had the higher E_i produced a similar result, further confirming that the FIS had successfully identified the correct restoration process.

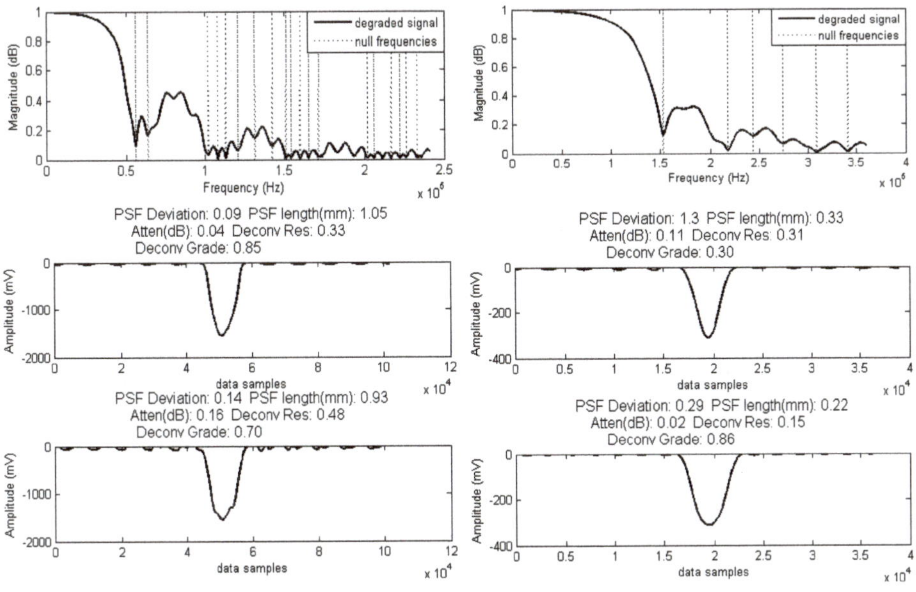

Figure 5. Null frequencies in the spectrum of the degraded pulse. Result of restoration with detected null frequencies, expected PSF length of 1 mm on the left and 0.2 mm on the right.

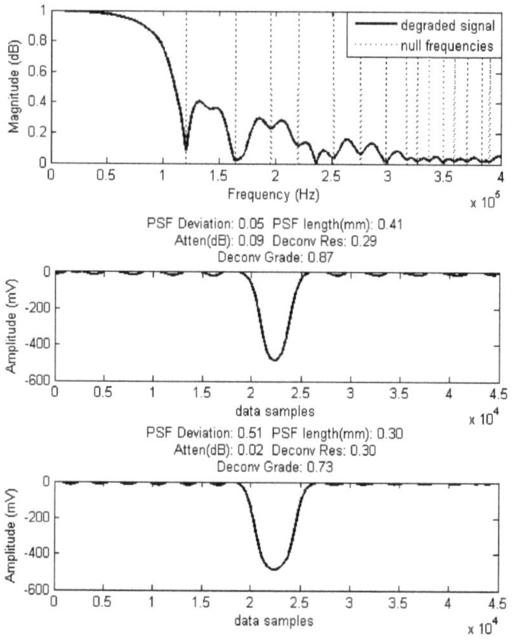

Figure 6. Null frequencies in the spectrum of the degraded pulse. Result of restoration with detected null frequencies, expected PSF length of 0.4 mm.

4. Conclusions and Discussion

The algorithm showed superior performance when a rough prior knowledge of L_h was provided for the fuzzy inference system. The $\Delta E_i = (|E_1 - E_2|)$ was greater than 0.5 thereby clearly identifying and segregating the correct deconvolution process. The algorithm was also tested without including the PSF knowledge, in which case ΔE_i was in the interval of 0.1 to 0.5, which was enough to confidently separate the correct deconvolution process.

Figure 6 depicted a special case where H had a null frequency at ω_h = 120 kHz with a normalized magnitude of 0.09 dB, whereas, F null was at ω_f = 170 kHz with a magnitude of 0.02 dB and had four times higher attenuation. Although ω_f had a magnitude that was in its favor, the PSF deviation of 0.51 was not, yet the PSF deviation outweighed its low magnitude and the correct restoration was successfully distinguished with 14% separation in the deconvolution grades ($|E_1 - E_2|$ = 0.14). This high attenuation of ω_f was most likely due to it being closer to the second harmonic of ω_h and, therefore, it experienced further attenuation. Nevertheless, owing to the FIS implementation, the correct restoration process was identified. All the possible rules were considered for the implementation of this FIS and its tuning was performed heuristically by an expert. However, clustering algorithms could be used for FIS with multiple inputs and membership functions to determine the optimum number of rules. Furthermore, adaptive FISs can be used to automate the tuning and learning process of the FIS in a more complicated and complex scenario.

Funding: This research received no external funding.

Acknowledgments: The author would like to thank A. Ferhati, A. Faghihi for their help and cooperation, C. Longman for reviewing the article and V. Jefimovs for his laboratory assistance. Many thanks to NSIRC, TWI Ltd. and Brunel University for providing the measurement facilities and research funds.

Conflicts of Interest: The author declares no conflicts of interest.

References

1. Zimmermann, H.J. Fuzzy control. In *Fuzzy Set Theory—And Its Applications*; Springer: New York, NY, USA, 1996; pp. 203–240.
2. Jansson, A.P.; da Silva, L.; Crilly, P.B.; Bernardi, A. Improving the convergence rate of Johnson's deconvolution method. *IEEE Trans. Instrum. Meas.* **2002**, *51*, 1142–1144.
3. Tiwari, S.; Shukla, V.P.; Singh, A.K.; Biradar, S.R. Review of motion blur estimation techniques. *J. Image Graph.* **2013**, *1*, 176–184. [CrossRef]
4. Bidyut, P.; Riad, S.M. Study and performance evaluation of two iterative frequency-domain deconvolution techniques. *IEEE Trans. Instrum. Meas.* **1984**, *33*, 281–287.
5. Gonzalez, R.C.; Woods, R.E.; Eddins, S.L. *Digital Image Processing Using MATLAB*; Pearson-Prentice-Hall: Upper Saddle River, NJ, USA, 2004; Volume 624.
6. Hosseinzadeh, S. Unsupervised spatial-resolution enhancement of electron beam measurement using deconvolution. *Vacuum* **2016**, *123*, 179–186. [CrossRef]
7. Tanaka, M.; Yoneji, K.; Okutomi, M. Motion blur parameter identification from a linearly blurred image. In Proceedings of the 2007 IEEE International Conference on Consumer Electronics (ICCE 2007), Las Vegas, NV, USA, 10–14 January 2007.
8. Bennia, A.; Riad, S.M. An optimization technique for iterative frequency-domain deconvolution. *IEEE Trans. Instrum. Meas.* **1990**, *39*, 358–362. [CrossRef]
9. Lo, W.Y.; Puchalski, S.M. Digital image processing. *Vet. Radiol. Ultrasound* **2008**, *49*, S42–S47. [CrossRef] [PubMed]
10. Moghaddam, M.E.; Jamzad, M. Linear motion blur parameter estimation in noisy images using fuzzy sets and power spectrum. *EURASIP J. Adv. Signal. Process.* **2006**, *2007*, 068985. [CrossRef]
11. Hosseinzadeh, S. Electron Beam Measurement Using Deblurring (Deconvolution). 2016. Available online: https://uk.mathworks.com/matlabcentral/fileexchange/60414-electron-beam-measurement-using-deblurring-deconvolution (accessed on 3 December 2018).
12. Ferrero, A.; Federici, A.; Salicone, S. Instrumental uncertainty and model uncertainty unified in a modified fuzzy inference system. *IEEE Trans. Instrum. Meas.* **2010**, *59*, 1149–1157. [CrossRef]
13. Bennia, A.; Riad, S.M. Filtering capabilities and convergence of the Van-Cittert deconvolution technique. *IEEE Trans. Instrum. Meas.* **1992**, *41*, 246–250. [CrossRef]
14. Zhao, J.; Bose, B.K. Evaluation of membership functions for fuzzy logic controlled induction motor drive. In Proceedings of the 28th Annual Conference of the IEEE Industrial Electronic Society, Sevilla, Spain, 5–8 November 2002.
15. Zadeh, L.A. Toward a theory of fuzzy information granulation and its centrality in human reasoning and fuzzy logic. *Fuzzy Sets Syst.* **1997**, *90*, 111–127. [CrossRef]

applied system innovation

MDPI

Article

Gaze-Guided Control of an Autonomous Mobile Robot Using Type-2 Fuzzy Logic

Mahmut Dirik [1,*], Oscar Castillo [2] and Adnan Fatih Kocamaz [1]

[1] Faculty of Engineering, Inonu University, 44280 Malatya, Turkey; fatih.kocamaz@inonu.edu.tr
[2] Tijuana Institute of Technology, 22414 Tijuana, Mexico; ocastillo@tectijuana.mx
* Correspondence: mahmut.dirik@inonu.edu.tr

Received: 24 March 2019; Accepted: 16 April 2019; Published: 24 April 2019

Abstract: Motion control of mobile robots in a cluttered environment with obstacles is an important problem. It is unsatisfactory to control a robot's motion using traditional control algorithms in a complex environment in real time. Gaze tracking technology has brought an important perspective to this issue. Gaze guided driving a vehicle based on eye movements supply significant features of nature task to realization. This paper presents an intelligent vision-based gaze guided robot control (GGC) platform that uses a user-computer interface based on gaze tracking enables a user to control the motion of a mobile robot using eyes gaze coordinate as inputs to the system. In this paper, an overhead camera, eyes tracking device, a differential drive mobile robot, vision and interval type-2 fuzzy inference (IT2FIS) tools are utilized. The methodology incorporates two basic behaviors; map generation and go-to-goal behavior. Go-to-goal behavior based on an IT2FIS is more soft and steady progress in data processing with uncertainties to generate better performance. The algorithms are implemented in the indoor environment with the presence of obstacles. Experiments and simulation results indicated that intelligent vision-based gaze guided robot control (GGC) system can be successfully applied and the IT2FIS can successfully make operator intention, modulate speed and direction accordingly.

Keywords: eye gaze tracking; interval type-2 fuzzy logic; vision system; mobile robots; intelligent control

1. Introduction

Mobile robot motion control problems have attracted considerable interest from researchers. The environments where the robot moves may vary from static or dynamic obstacles. In such an uncertain environment without prior information about the robot, target and obstacle, the basic aim is to safely move the robot without colliding with obstacles and reach the target point. The problem is the design of a safe and efficient algorithm that will guide robots to the target. The control system of these robots includes a set of algorithms. Recently, various methodologies have been utilized for mobile robot navigation [1–3]. These are, artificial potential fields (APF) [4–6], vision-based gaze guided control (GGC) [7–15], fuzzy logic control [12,16–18], vector field histogram (VFH) [19,20], rapidly-exploring random trees [21,22], and obstacle avoidance and path planning [23] algorithms. The fuzzy logic controller is one of the efficient techniques, which tends to be used actually for steering control in unpredictable environments [3,24,25]. Fuzzy logic can smoothly be concerted to handle various types of linear and nonlinear systems like the human model perception process in uncertainty [26]. Motion control of a mobile robot in an unstructured dynamic or static environment is usually characterized by a number of uncertainties that characterize real-world environments. To have exact and complete for prior knowledge of these environments is not possible. As the membership sets of the type-2 fuzzy systems are fuzzy, they offer a stronger solution to the type-1 fuzziness to represent and address uncertainty types. Type-1 fuzzy control cannot handle such kind of uncertainties. For this reason, the type-2 fuzzy system [24,27] is preferred to handle uncertainties and increase performance of the control navigation system that was introduced by Zadeh [28,29]. Type-2 fuzzy sets theory has been further developed by Mendel and Karnik [24,30,31]. Theoretical background computational methods of IT2FIS and its design principles were developed for type reduction [32–35].

In the paper, a gaze guided robot control system, based on eye gaze using the type-2 fuzzy controller, is presented. It aimed to determine the robot's direction based on the target location and create wheel speeds based on gaze direction and demonstrate how gaze could be used for automatic control of a mobile robot. That means to obtain a model of the robot moving towards the objects by looking directly at the object of interest in the real world. The interaction methods are expected to benefit users with mobility in their daily works. The intent of gaze direction amplitudes and the angle variables are applied for designing the motion control system's input. The input variables are image coordinates in 2D image space of the scene monitored by the overhead camera. Eye movements identified such as looking up, down in y (eye width) coordinates and left and right in x (eye height) coordinates values were mapped to the mobile robot steering commands. The approximate width (X) and height (Y) of the eye was taken into account for the inputs. Gaze combination control is a benefit from effective hands-free input parameter of remote vehicles control. Gaze-based robot control work done in this regard is relatively scarce. Combining different types of technology with eye based systems [36–40] integration can be useful for disabled, elderly and patients peoples that especially with neuro-motor disabilities. They can remotely control a moving platform [11,12,41,42] with these technologies in complex daily tasks [14,43]. A wheelchair is a good substitute, for example Reference [44]. It's orientation commands are created by a wheelchair mounted eye movement tracking system [45–47].

The proposed gaze guided robot control (GGC) system consist of an EV3 robotic platform. It is highly customizable and provides advanced LabVIEW programming features [48,49]. In this platform, two high-resolution cameras are used. The first camera is mounted on the robot moving platform for transmitting a live video to the user screen. The second camera is used to identify the user's eye movement and translate this gaze direction to the host system to calculate the robot's steering control by using soft computing technique. The goal is to process to robot's movements where the user is looking at the display screen. To make the robot move around in the vibrant environment under the visibility of an overhead camera, it is focused on the remote settings with a 2D gaze guided driving system such as forward, backward, left and right.

This paper is organized as follows. Section 2 represents the procedure of the gaze guided system including a vision system of gaze tracking, the control system of the type-2 fuzzy mechanism and its application and experimental platform. The experimental design and results of the proposed methods are given in Section 3. Section 4 includes the discussion. Finally, Section 5 concludes the paper and recommendation for future works.

2. Method of Gaze Guided System

The intention of the system's architecture is to achieve an influential integration of eye tracking technology with practical GGC system. The overall concept of gaze guided robot control system can be demonstrated in Figure 1. In this structure, it is illustrated how different physical elements can be attached and communicate. The system architecture is divided into three parts. The first part is (user side) the user inferences, which are related to the eye tracking subsystem and the eye movement translation into robotic platform commands (1), the second part (host-IT2FIS) is data processing and command execution (2), the third part (robot side) is a robot moving environment under the visibility of an overhead camera. This framework includes four parts: an eye tracking system which can track user's eyes, an overhead camera system that supplies the video feedback to the user, a wheeled mobile robot and the host computer system which is accountable for collecting the gaze data and commentate it into robot motion commands. The mobile robot used in this platform has an ARM9-based processor at 300 MHz Linux-based operating system. Detailed properties can be viewed in [50]. A well-developed programming interface based LabVIEW programming language allows us to transmit movement information, depending on the direction of the gaze, to the robotic platform via Bluetooth. The image acquisition camera [51] is able to provide live video with a resolution of 480 × 640 at 30 fps. The overhead camera height from the floor is approximation 2 m. The system software principally consists of two parts: the vision detection (eye tracking, robot tracking) and motion control algorithm. Sub-titles and details of the system are given below.

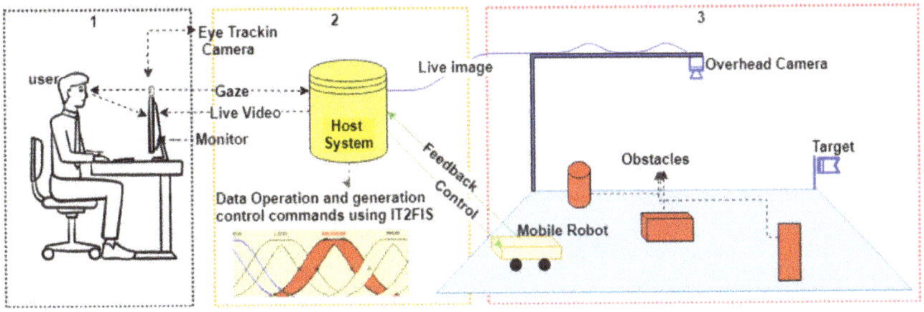

Figure 1. Interaction of the gaze-guided robot control system (1. User side, 2. Control system, 3. Robot side).

2.1. Eye-Gaze Tracking GGC System

The GGC control system architecture is shown in Figure 1. In the first block (1), a user sitting in a chair and watches the live video from an overhead camera. The movement of the robot is monitored on this screen. Wherever the robot is required to move, the user looks at that side. The visual attention on that side is extracted from gaze data. To perform the robot navigation, the visual attention is converted from image space to 2D Cartesian space and produces control commands from this coordinate system. The point of gaze as the user observes the video frames is utilized for robot motion control inputs. The direction and speed are regulated by distance from the center point of the eye as seen in Figure 2. X-axis regulates steering and y-axis regulates robot speed.

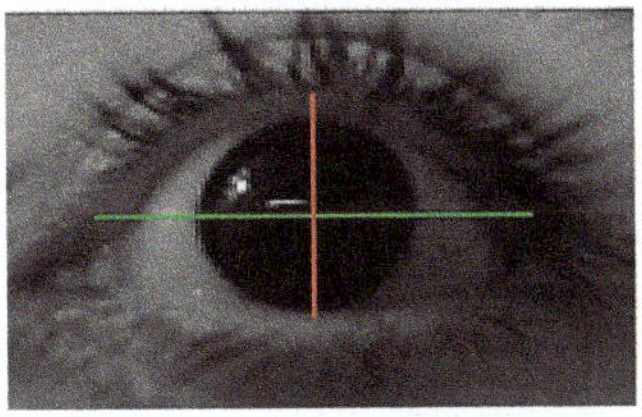

Figure 2. Robot control input parameters range (the X-axis regulate steering and the Y-axis regulates speed).

A high-resolution webcam [51] is used to track eyes where the user is looking. It is a video-based remote eye tracking system. A shape adapted mean shift algorithm [52] is utilized which is asymmetric and anisotropic kernels for object tracking which is process the image and calculates the point of gaze coordinates (see Figure 3).

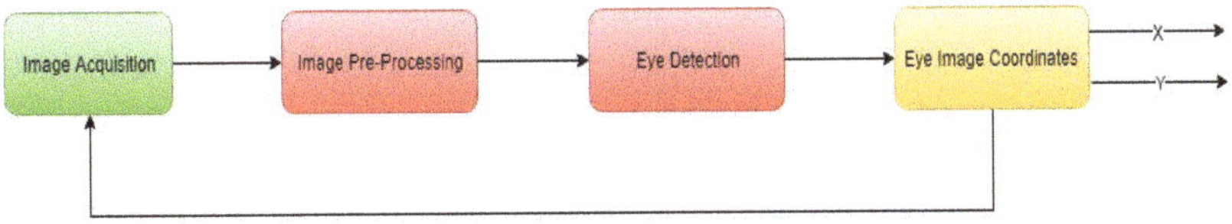

Figure 3. Structure and functionality of real-time eye tracking algorithm.

A series of raw image continuously capture in image acquisition step (see Figure 4). The image tackled in this step is important in order to get the relevant Region of interest (ROI). The algorithms have been implemented using NI Vision Builder programming tools. The images are sent to the NI Vision system for machine vision and image processing applications. Image pre-processing makes easier to extract ROI. The analysis has purposed the removal of uninterested information to reduce the processing area. The original image is transformed into a grayscale and masked image that does not need to detect an eye feature point. A morphological operation has been applied for increasing the detection accuracy. A shape adapted mean shift algorithm has been applied which provides angular as well as a linear offset in object shapes.

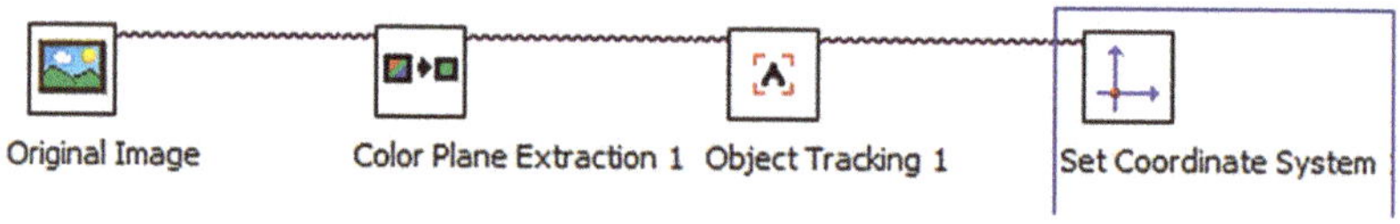

Figure 4. Image Analysis Steps.

This algorithm allows tracking eye templates with changing shapes and size and a built-in function using the NI vision assistant. The tracking template position is well-considered as the reference and its coordinates are taken into account. In the final image analysis, the eye coordinate system is obtained (see Figure 5).

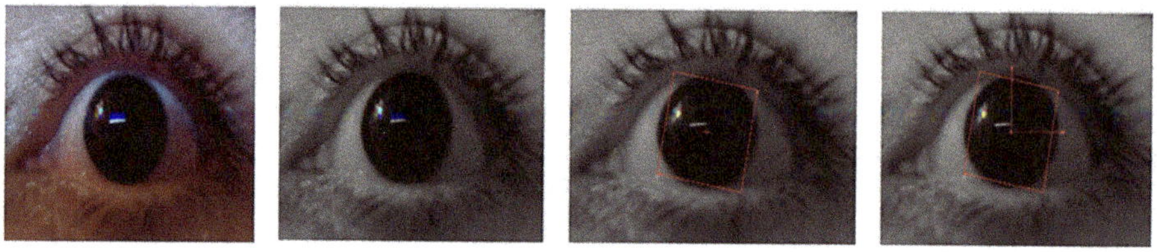

Figure 5. Real-image analysis.

2.2. Overhead Camera System-Robot Tracking

In Figure 1 (3. robot side), the graphical representation of an overhead camera based robot tracking field is shown. The environment where the robot moves is characterized by static obstacles. Gaze-based collision-free motion control of the robot in such a static environment is important. Environment model is captured by an overhead camera and sent to the NI vision system then the mobile robot on the field is tracked by a standardized vision system. The center of the moving robot area is accepted as (0,0). Horizontal (x) and vertical (y) axes represent the robot direction and wheel velocity. The information of the robot motion control environment is received by the host computer, and the robot tracking algorithm is executed. The robot's position is continuously tracked and updated considering the acquired information from the sequentially captured images. Decision strategy is improved in the host system for plans the robot motion and corresponding velocity commands. The host computer regulates the rotational velocities of wheels and sends these commands to the robot. An experiment image shown in Figure 6 has captured by the overhead camera and this image is sent to the NI vision system and user monitor.

The robot steering control command is produced by eye gaze movement as explained before. After defining the eye gaze coordinates, the output signal calculated for robot motor speeds by using type-2 fuzzy control sent via Bluetooth.

Figure 6. The graphical representation of the robot field.

2.3. Interval Type2 Fuzzy Control System

The design and theoretical basis of a type-2 fuzzy model for mobile robot motion control has been presented in this section. This is planned to supply the basic thoughts needed to explain the algorithm using gaze input variables and rule base to determine the value of the output system. Fuzzy type-2 has been verified to be a powerful tool for controlling a complex system because of its robustness for controlling nonlinear systems with characteristic and uncertainties [33,53]. The concept of the type-2 fuzzy set was proposed by Zadeh [28,54] as an extension of type-1 fuzzy logic. It is able to model uncertainties in a much better way for control application. The appearance of uncertainties in nonlinear system control using the highest and lowest value of the parameter extending type-1 (Figure 7a) fuzzy to type-2. (Figure 7b). Uncertainty is a characteristic of information, which may be incomplete, inaccurate, undefined, and inconsistent and so on. These are represented by a region called a footprint of uncertainty (FOU) that is a limited region of upper and lower type-1 membership function.

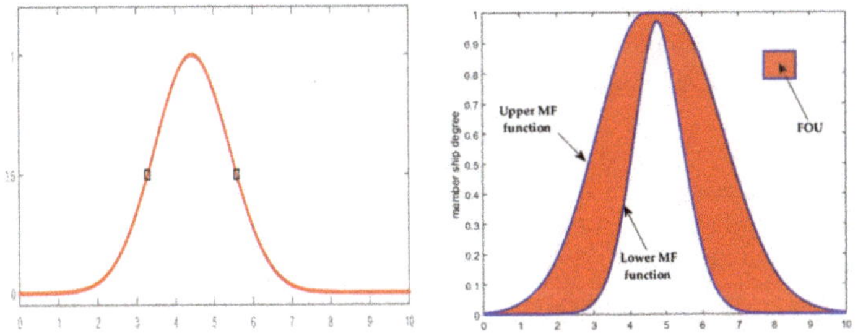

(a) **(b)**

Figure 7. (**a**) Type-1 membership function. (**b**) Type-2 membership function.

An interval type-2 fuzzy set denoted by $\widetilde{A}$, it is expressed in (1) or (2).

$$\widetilde{A} = \{(x, y),\ \mu_{\widetilde{A}}(x, y) \big| \ \forall_x \in X,\ \forall_u \in J_x \ \subseteq [0\ 1]\}. \quad (1)$$

Hence, $\mu_{\widetilde{A}}(x, u) = 1, \forall_u \in J_x \subseteq [0\ 1]$ it is considered as interval type-2 membership function as shown in Figure 8.

$$\widetilde{A} = \int_{x \in X} \int_{u \in J_x} 1/(x, u)\ J_x \subseteq [0\ 1] \quad (2)$$

where $\int \int$ donate the union of all acceptable x and u. An IT2FIS can be explained in terms of an upper membership function $\overline{\mu}_{\widetilde{A}}(x)$ and a lower membership function $\underline{\mu}_{\widetilde{A}}(x)$. J_x is just the interval $[\overline{\mu}_{\widetilde{A}}(x),\ \underline{\mu}_{\widetilde{A}}(x)]$. A type-2 FIS is characterized by IF-THEN rules, where the antecedent and consequent sets are of type-2. The fundamental block used for designing the type-2 controller is the same as used with type-1. As shown in Figure 8, A type-2 FLS includes a fuzzifier, a rule base, a fuzzy inference engine, and an output complement. The output processor includes a type-reducer and defuzzifier; it produces a type-1 fuzzy set output (from the type-reducer) or a crisp number (from the defuzzifier) [53]. Type reducer is added because of its association with the nature of the membership grade of the elements [35].

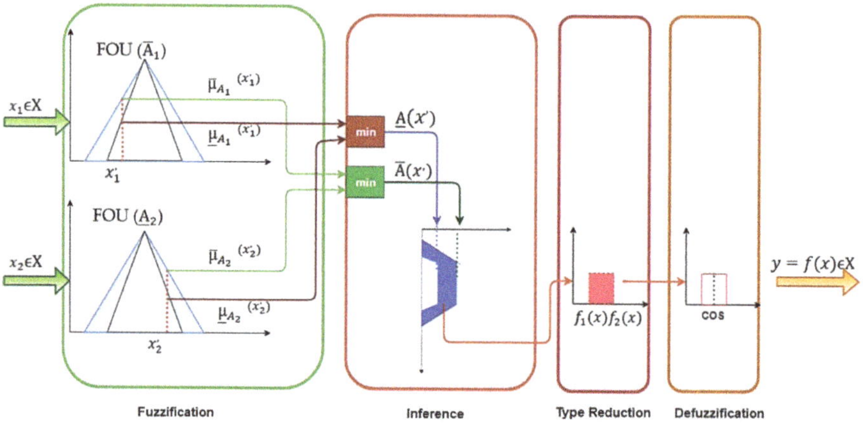

Figure 8. Structure of a type-2 fuzzy logic system.

2.3.1. Fuzzifier

In this case, the inputs of the fuzzy sets are described. There are two inputs parameter used in the proposed system. These are named robot direction and robot speed respectively. The width (X) and height (Y) of the eye was taken into account for the input membership functions' values range (see Figure 9). As mentioned before, X-axis is representing the robot direction and Y-axis is representing the robot speed which is used to determine the two crisp inputs variables. Table 1 shows the width and height information of the eye. These data are taken into account for the input function. In this table, it is illustrated that the eyes' start and end coordinate space in 2D is used for arranging the membership function. The membership functions consist of one or several type-2 fuzzy sets. A numerical vector x of fuzzifier maps into a type-2 set $\widetilde{A}$. Type-2 fuzzy singleton is considered. In a singleton fuzzification, the inputs are crisp values on nonzero membership. We contemplate three fuzzy membership functions for the robot direction with labels $\widetilde{\mathrm{L}}$, $\widetilde{\mathrm{S}}$, $\widetilde{\mathrm{R}}$. These are indicating left, straight, and right respectively as illustrated in Figure 10. We worked using Gaussian and sigmoidal membership functions. A Gaussian type-2 fuzzy set is one in which the membership grade of every domain point is a Gaussian type-1 set contained in [0,1]. These functions are unable to specify asymmetric and archive smoothness membership function which are important in certain applications. The sigmoidal membership function, which is either open left, right asymmetric closed, it is appropriate for representing concepts such as very large" or very negative". In the same method, we contemplate three membership functions for robot speed with labels $\widetilde{\mathrm{N}}$, $\widetilde{\mathrm{M}}$, $\widetilde{\mathrm{F}}$, these are indicating near, medium, far respectively as illustrated in Figure 10. The variable range of functions is not infinite (see Figure 10).

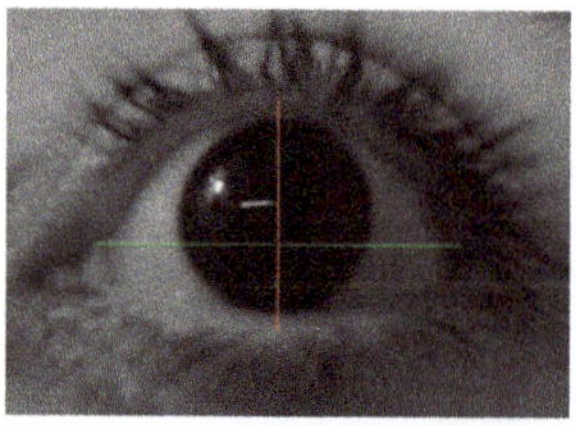

Figure 9. Eye tracking region variable.

Table 1. Eye region information.

LengthX	409 pixel	Start (X) = 104	Start (Y) = 253	End (X) = 513	End (Y) = 253	Mean = 61.1	StdDev = 45.7	Min = 9	Max = 157
LengthY	247 pixel	Start (X) = 310	Start (Y) = 98	End (X) = 310	End (Y) = 345	Mean = 30.6	StdDev = 29.4	Min = 5	Max = 131

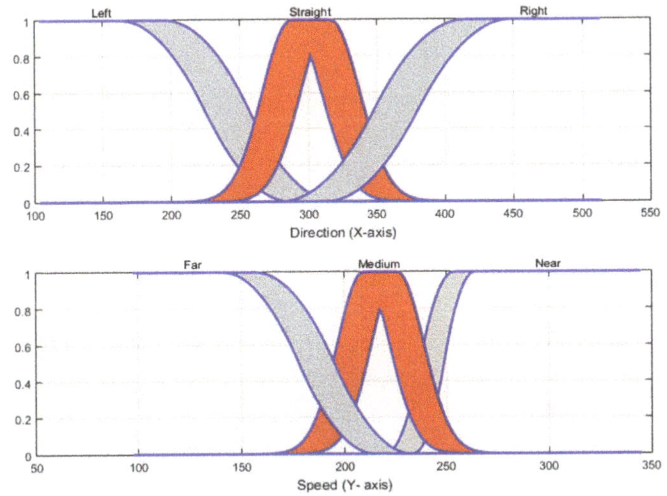

Figure 10. Membership functions of robot direction and speed.

The output fuzzy controllers are the left and right velocities of the wheel speed. The linguistic variables are implemented with tree membership function. For both right and left wheel speed, these are labeled as $\widetilde{S}$, $\widetilde{M}$, $\widetilde{F}$—slow, medium and fast respectively. It is illustrated in Figure 11.

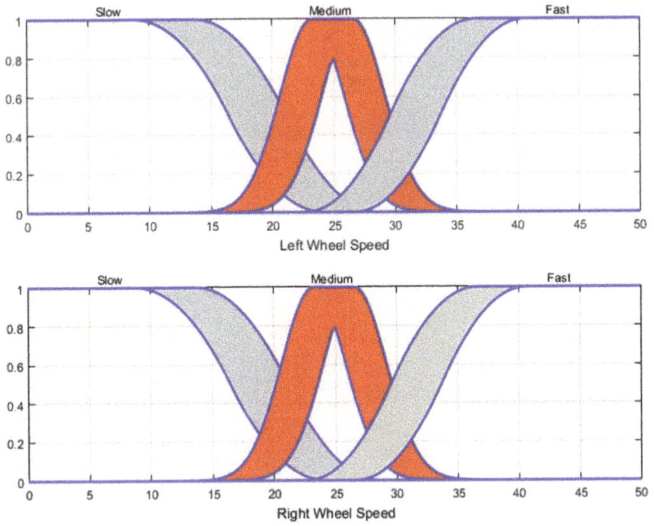

Figure 11. Membership functions of the wheels speed.

2.3.2. Fuzzy Inference Engine

The inference engine combines rules and gives an outline from input to output type-2 fuzzy sets. Figure 8 shows a graphical representation of the relationship between input and output. It is necessary to compute the intersection and union of type-2 sets and implement compositions of type-2 relations. The desired behavior is defined by a set of linguistic rules. It is necessary to set the rules adequately for the desired result. For instance, a type-2 fuzzy logic with p inputs ($x_1 \in X_1, \ldots x_p \in X_p$) and one output (y) with M rules have the following form.

$$R^{\mathcal{L}}: \text{ IF } x_1 \text{ is } \widetilde{F}_1^{\mathcal{L}} \ldots \text{and } x_p \text{ is } \widetilde{F}_p^{\mathcal{L}} \text{ THEN } y \text{ is } \widetilde{G}^{\mathcal{L}},\ \mathcal{L} = 1 \ldots M.$$

The knowledge bases related to the robot wheels speed are reported in Table 2. The approximate locations of the rules formed in the knowledge base in the coordinate plane are shown in Figure 12.

Table 2. Rule base for robot wheel speed fuzzy controller.

X (Direction) / Y (Speed)	Left ($\tilde{L}$)	Straight ($\tilde{S}$)	Right ($\tilde{R}$)
Near ($\tilde{N}$)	Lws = $\tilde{S}$ Rws = $\tilde{F}$	Lws = $\tilde{S}$ Rws = $\tilde{S}$	Lws = $\tilde{F}$ Rws = $\tilde{S}$
Medium ($\tilde{M}$)	Lws = $\tilde{S}$ Rws = $\widetilde{M}$	Lws = $\tilde{M}$ Rws = $\tilde{M}$	Lws = $\widetilde{M}$ Rws = $\tilde{S}$
Far ($\tilde{F}$)	Lws = $\tilde{M}$ Rws = $\tilde{F}$	Lws = $\tilde{F}$ Rws = $\widetilde{F}$	Lws = $\tilde{F}$ Rws = $\tilde{M}$

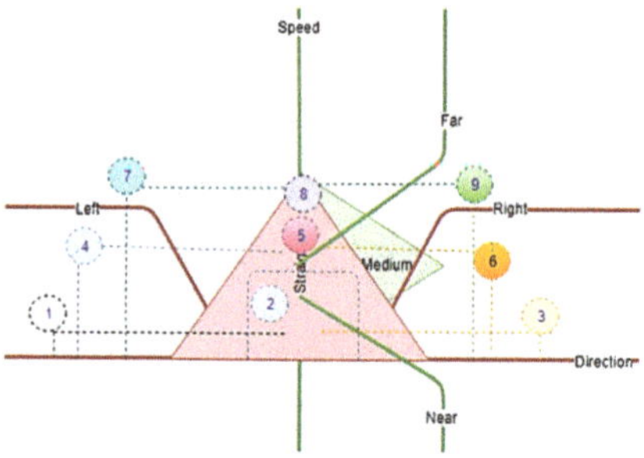

Figure 12. Graphical distribution of the rule table.

In this experiment, we used type-2 fuzzy sets and minimum t-norm operation. The rule firing strength $F^i(x)$ for crisp input vector is given by type-1 fuzzy set.

$$F^l(x') = \left[\underline{f}^l(x'), \overline{f}^l(x')\right] \equiv \left[\underline{f}^l, \overline{f}^l\right] \tag{3}$$

where

$$\underline{f}^l(x') = \underline{\mu}_{\widetilde{F}_1^l}\left(x_1'\right) * \ldots * \underline{\mu}_{\widetilde{F}_p^l}\left(x_p'\right), \tag{4}$$

$$\overline{f}^l(x') = \overline{\mu}_{\widetilde{F}_1^l}\left(x_1'\right) * \ldots * \overline{\mu}_{\widetilde{F}_1^l}\left(x_p'\right). \tag{5}$$

The graphical representation of the rules of the system is shown in Figure 12. These are composed of input and output linguistic variables. Nine inference rules are designed to determine how the mobile robot should be steered and velocity. In each rule, a logic and operation is used to deduce the output. In Table 2 we present the rule set whose format is established as follows:

Rule 1: IF Direction (X) is left and speed (Y) is medium, then left wheel speed (Lws) is small and right wheel speed is medium.

2.3.3. Type Reducer

Type reducer creates a type-1 fuzzy set output which is then transformed into a crisp output through the defuzzifier that combines the outputs sets to acquire a single output using one of the existing type reduction methods. Type reducer was proposed by Karnik and Mende [32,34,55]. In our experiments, we used the center of sets (cos) type reduction method. The expression of this method can be written in the following Equation (6).

$$Y_{cos}(x) = [y_l,\ y_r] = \int_{y^1 \in [{y_l}^1, {y_r}^1]} \cdots \int_{y^1 \in [{y_l}^M, {y_r}^M]} \int_{f^1 \in [\underline{f}^{\,1}, \overline{f}^1]} \cdots \int_{f^M \in [\underline{f}^{\,M}, \overline{f}^M]} / \frac{\sum_{i=1}^{M} f^i y^i}{\sum_{i=1}^{M} f^i}. \tag{6}$$

The consequent set of the interval type-2 determined by two endpoints $(y_l,\ y_r)$. If the values of f_i and y_i, which are associated with y_l, are donated ${f_l}^i$ and ${y_l}^i$, respectively, and the values of f_i and y_i which are associated with y_r are donated ${f_r}^i$ and ${y_r}^i$ respectively, these points are given in Equations (7) and (8).

$$y_l = \frac{\sum_{i=1}^{M} {f_l}^i {y_l}^i}{\sum_{i=1}^{M} {f_l}^i} \tag{7}$$

$$y_r = \frac{\sum_{i=1}^{M} {f_r}^i {y_r}^i}{\sum_{i=1}^{M} {f_r}^i} \tag{8}$$

where y_l and y_r are the output of IT2FIS, which can be used to verify data (training or testing) contained in the output of the fuzzy system.

2.3.4. Defuzzifier

The interval fuzzy set $Y_{\cos}(x)$ variables obtained from type reducer are defuzzified and the average of y_l and y_r are used to defuzzify the output of an interval singleton type-2 fuzzy logic system. The equation is written as

$$y(x) = \frac{y_l + y_r}{2}. \tag{9}$$

3. Experiments and Results

In this paper, we focused on intelligent vision-based gaze guided robot control systems. The evaluation and validation of this method were tested with several experiments. The experiments were performed under an overhead camera image and using type-2 fuzzy control system. The aim is to make strategic planning and implement remote control of the robot on the base of gaze coordinates where user looking for. The experiments included two stages; evaluation and determination of gaze coordination and using this information as input command effectively for robot control. In our proposed method, we have designed an interface system where the user looks at the experimental field view from the overhead camera on the computer monitor. The eye gaze tracker is calibrated based on real-world eye viewing fields. The human eye view field is an essential factor in getting the coordination system. The closed eye situation is also identified by the computer program and is used to stop the robot. Then the gaze coordinates are utilized to control the robot remotely. It aims to directly control the robot after calibration of the gaze tracker. The robot direction and speed are modulated linearly by the distance from the center of the gaze coordinate. In the eye horizontal plane, the x-axis represents the movement of the eye gaze coordinate and the vertical plane coordinate system represents the mobile robot wheel speed input variables. In order to determine robot steering, this coordinate system is considered. Commands for robot motion control are extracted and updated for every 250 ms continuously.

The simulation results show the ability of the type-2 fuzzy logic controller to simultaneously determine human intention from the combined viewpoint and eye gaze. Nevertheless, the proposed method is powerful to determine robot speed, orientation, and obstacle avoidance. In Figure 13, the simulation of the robot's initial and goal points is illustrated and robot motion data on this behavior is also shown in Table 3.

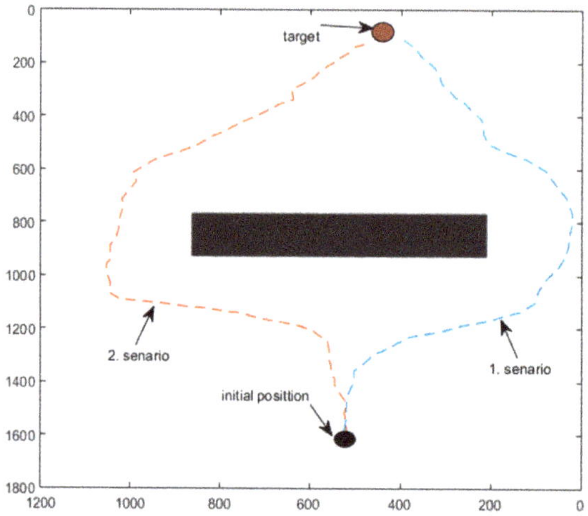

Figure 13. Robot trajectory from eye gaze based go-to-goal behavior.

Table 3. Robot trajectories data (inputs, outputs) for two scenarios from Figure 13.

Order	1. Scenario				2. Scenarios			
	Inputs (Gaze Coordinate)		Outputs		Inputs (Gaze Coordinate)		Outputs	
	X	Y	LWA	RWS	X	Y	LWS	RWS
1	269	296	12.29	12.65	269	290	14.03	14.38
5	272	204	24.88	25.04	255	272	20.46	21.2
12	509	209	25.05	8.01	20	223	7.14	25.01
19	516	183	26.73	17.52	14	157	22.69	34.56
26	501	151	36.71	30.17	36	83	24.98	42.94
33	517	53	42.08	31.27	284	66	24.99	37.8
40	493	56	42.26	31.47	187	8	26.01	37.77
47	182	58	24.99	42.37	185	15	25.12	38.4
...	...	...	...	...	...	...	...	...

The outcomes of the experiment and simulation are illustrated graphically and plotted separately to elucidate the effect of gaze based on the fuzzy type-2 rule set in Figures 14–19. In Table 3, it can be seen that robot wheel speeds are change according to the gaze coordinates. In this table, there are only a few examples. The details of these data are shown graphically below in Figures 15, 17, and 18. Various scenarios were performed to test the proposed method in a real environment. The experimental setup is shown in Figures 14 and 16. The adaptability, robustness, accuracy and efficiency of the proposed methods can be observed from these experimental results. The real experimental environment is shown in Figures 14 and 16. It includes the robot, obstacle and target token. Here are six different frames obtained from the experimental environment. The eye gaze data coordinates, robot path coordinates, and robot wheel speed are stored during the movement of the robot and this data is graphically illustrated. The affection of the relationship between robot speed and gaze point is illustrated in Figure 18. The effectiveness of the fuzzy rules and suitability of the fuzzy controller on collision-free behavior is shown in Figure 19.

In the first scenario (see Figure 14), the robot navigates in an uncomplicated environment with one obstacle. Throughout its motion to the target, the mobile robot encounters an obstacle. In this situation, the robot is moving smoothly and without colliding. The real experimentation of this scenario is illustrated in Figure 14. The eye gaze coordination for this scenario is shown in Figure 15.

Figure 14. Experimental setup; 1. Scenario—robot motion control from start (1) and goal (6) position.

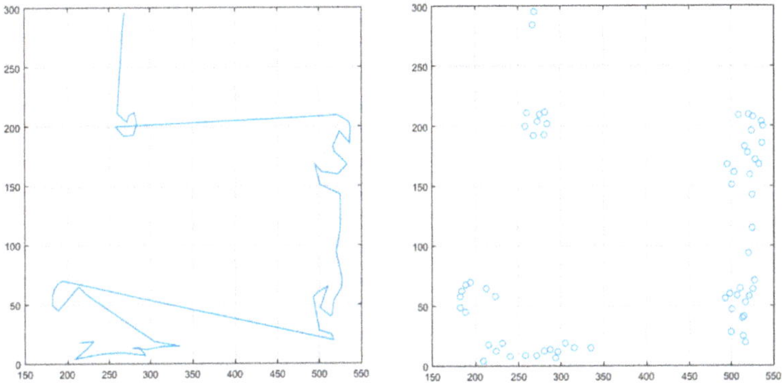

(**a**) (**b**)

Figure 15. Eye gaze coordination for 1. Scenario (Figure 13) (2D line graph (**a**), scattered (**b**)).

In the second scenario, the user has directed the robot to the left side with the eye movements. It was seen that the mobile robot behaved similarly to its first scenario. The real experimental results are shown in Figure 16. The plot of this scenario is shown in Figure 17. The objective of this driving is to confirm the efficacy of the method in different cases. As can be seen from this case, the mobile robot navigates successfully around the obstacles without collision and reach the target.

Figure 16. Experiment setup; 2. Scenario—robot motion control from start (1) and goal (6) position.

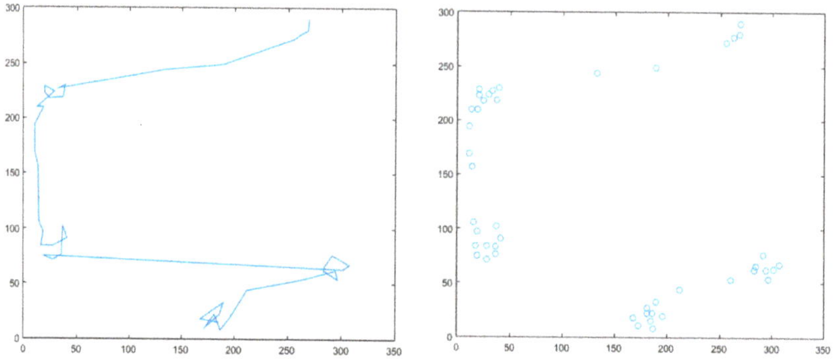

(**a**) (**b**)

Figure 17. Eye gaze coordination for 2. Scenario (Figure 12) (2D line graph (**a**), scattered (**b**)).

The robot wheel speed for the two scenarios is shown in Figure 18. And, the Robot trajectory from the experiment of Figure 14 is shown in Figure 19.

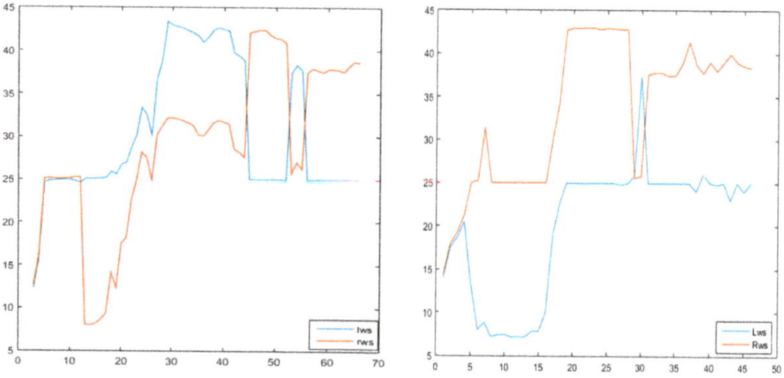

(**a**) (**b**)

Figure 18. Robot wheel speed for the two scenarios (1. Scenario (**a**), 2. Scenario (**b**)) from Figures 14–16.

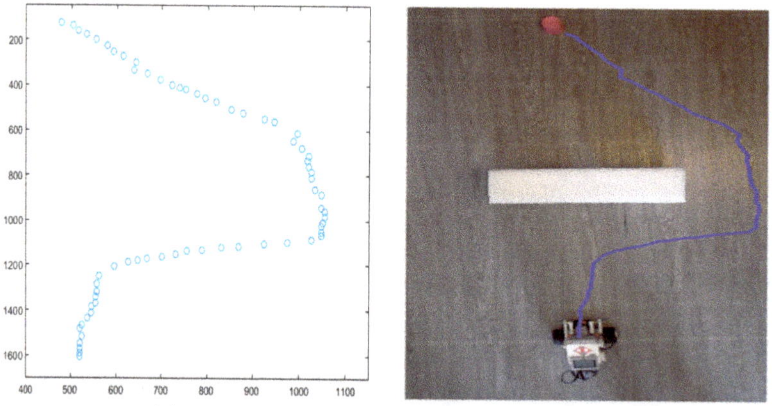

(**a**) (**b**)

Figure 19. Robot trajectory from the experiment of Figure 14 (1. Scenario). (**a**) Inputs (Gaze Coordinate); (**b**) robot path trajectory.

4. Discussion

From the experiments, the human gaze has been evidenced to be an encouraging modality for GGC. Our proposed method was expanded to explore motion control by straight-gaze input to perform hands-free control of the mobile robot. The gaze-based control system has good potential for future work. We think that this technology has a significant function in facilitating the life of a disabled person. The gaze-based control system is still in its infancy and its function and features need to be further developed in order to be implemented in more complex situations. The use of an eye-based interaction modality can reduce both physical and mental burden. The main aim will be to increase the usability of the system by increasing the learning and ease-of-use by considering the overall efficiency. A low-cost eye-tracking device used in our experiments is crucial in the stability. The user's head must be stable and the eye tracker must snapshot the eye frame in certain positions. Although this situation is difficult to set-up, it is preferred because it is inexpensive and robust in the gaze position for motion control commands. The profoundly accurate eye trackers are able to present control of vehicles that are as good in term of speed and error as mouse control. It is clear that the differentiation between our model and controlling a wheelchair-based gaze guide drive is using overhead camera images. This type of motion control could be profitable in remote control situations like hazardous, poisonous places where the hands are needed for other tasks. A signification limitation of the current work is that the surveillance system is managed from a fixed location. The correct determination of gaze width and height positions (start and end) is important for the designed controller. When the eyes are closed, the engines are driven to zero and thus the robot is stopped. For this purpose, the sub-program has been audited without choosing the rule base. This is considered as the mobile robot's go-to-goal behavior. For this reason, the behavior of robot retraction has not been taken into account. This choice was made for the purpose of feasibility in testing a novel idea for socially assistive robots. It is our expectation that a new idea is implemented for target audiences who need socially assisted robots.

The proposed method was implemented for an indoor experiment. As the work accomplished in this paper progressed, many ideas came to mind, which could be discussed in future research, such as outdoor experiments and investigations on how the gaze based control of devices and how it compares with traditional methods such as omnidirectional wheels. These wheels are troublesome individuals compared to the traditional ones. Lastly, our proposed method may sub serve as a simple, secure and affordable method for future gaze guided vehicle control prototypes.

5. Conclusions

In this paper, we proposed a vision-based gaze guide mobile robot control. Application of gaze interaction, wearable, and mobile technology has mainly been conducted on controlling the movement of robots. Users were enabled to control the robot remotely and hands-free by using their eyes to specify the target where they want the robot to move to. A central processing unit executes data communication between the user and the robotic platform. This continuously monitors the state of the robot through visual feedback and send commands to control the motion of the robot. Our experiments include an overhead camera, an eye tracking device, a differential drive mobile robot, vision and IT2FIS tools. This controller produces the required wheel velocity commands to drive the mobile robot to its destination along a smooth path. To achieve this requirement, our methodology incorporates two basic behaviors, map generation and go-to-goal behavior. Go-to-goal behavior, based on an IT2FIS, is more smooth and uniform to progress handling data in uncertainties to produce a better performance. The algorithms are implemented in an indoor environment with the presence of obstacles. Furthermore, the IT2FIS controller was strongly used to control a real-time mobile robot using exact gaze data obtained from human users using an eye tracking system. The differential drive mobile robot (EV3) was successfully commanded by the user's gaze. This method of interaction is available to most people, including those with disabilities and the elderly, who undermine motor ability. Thus, I would like to express that this technology needs to be developed in order to be able to be used in many fields. This system can also be an alternative or supplement to a conventional control interface such as a mouse or joystick, etc. The results from the proposed technique have been illustrated via simulation and experiments. It is indicated that the intelligent vision-based gaze guided robot control (GGC) system is applicable and the IT2FIS can successfully infer operator intent, modulate speed and direction accordingly. The experimental results obtained are very adequate and verify the efficacy of the proposed approach.

Author Contributions: M.D. analyzed the original Mobile Robot motion control, visual servoing system and Gaze Guide control methods and applied Type-2 fuzzy logic for dynamic parameter adaptation applied for tracking trajectories, and contributed to performing the experiments and wrote the paper; O.C. reviewed the state of the art, analyzed the data and proposed the method; A.F.K. contributed to the discussion and analysis of the results.

Funding: This research was funded by TUBITAK-BIDEB under grant 2214/A.

Conflicts of Interest: The authors declare no conflict of interest.

References

1. Choset, H.; Lynch, K.M.; Hutchinson, S.; Kantor, G.A.; Burgard, W.; Kavraki, L.E.; Thrun, S. *Principles of Robot Motion: Theory, Algorithms, and Implementation*; MIT Press: Cambridge, MA, USA, 2007.
2. Siegwart, R.; Nourbakhsh, I.R.; Scaramuzza, D.; Arkin, R.C. *Introduction to Autonomous Mobile Robots*; MIT Press: Cambridge, MA, USA, 2011.
3. Abiyev, R.H.; Erin, B.; Denker, A. Navigation of Mobile Robot Using Type-2 Fuzzy System. In *Intelligent Computing Methodologies*; Huang, D.-S., Hussain, A., Han, K., Gromiha, M.M., Eds.; Springer International Publishing: Cham, Switzerland, 2017; Volume 10363, pp. 15–26.
4. Rimon, E.; Koditschek, D.E. Exact robot navigation using artificial potential functions. *IEEE Trans. Robot. Autom.* **1992**, *8*, 501–518. [CrossRef]
5. Dönmez, E.; Kocamaz, A.F.; Dirik, M. Visual Based Path Planning with Adaptive Artificial Potential Field. In Proceedings of the 2017 25th Signal Processing and Communications Applications Conference (SIU 2017), Antalya, Turkey, 15–18 May 2017.
6. Dönmez, E.; Kocamaz, A.F.; Dirik, M. A Vision-Based Real-Time Mobile Robot Controller Design Based on Gaussian Function for Indoor Environment. *Arab. J. Sci. Eng.* **2018**, *43*, 7127–7142. [CrossRef]
7. Barea, R.; Boquete, L.; Bergasa, L.M.; López, E.; Mazo, M. Electro-Oculographic Guidance of a Wheelchair Using Eye Movements Codification. *Int. J. Robot. Res.* **2003**, *22*, 641–652. [CrossRef]
8. Yu, C.; Schermerhorn, P.; Scheutz, M. Adaptive eye gaze patterns in interactions with human and artificial agents. *ACM Trans. Interact. Intell. Syst.* **2012**, *1*, 1–25. [CrossRef]
9. Lee, J.; Hyun, C.-H.; Park, M. A Vision-Based Automated Guided Vehicle System with Marker Recognition for Indoor Use. *Sensors* **2013**, *13*, 10052–10073. [CrossRef] [PubMed]
10. Li, S.; Zhang, X.; Kim, F.J.; da Silva, R.D.; Gustafson, D.; Molina, W.R. Attention-Aware Robotic Laparoscope Based on Fuzzy Interpretation of Eye-Gaze Patterns. *J. Med. Devices* **2015**, *9*, 041007. [CrossRef]
11. Nelson, C.A.; Zhang, X.; Webb, J.; Li, S. Fuzzy Control for Gaze-Guided Personal Assistance Robots: Simulation and Experimental Application. *Int. J. Adv. Intell. Syst.* **2015**, *8*, 77–84.
12. Nelson, C.A. Fuzzy Logic Control for Gaze-Guided Personal Assistance Robots. *Int. J. Adv. Intell. Syst.* **2014**, *8*, 77–84.
13. Yu, M.; Wang, X.; Lin, Y.; Bai, X. Gaze tracking system for teleoperation. In Proceedings of the 26th Chinese Control and Decision Conference (2014 CCDC), Changsha, China, 31 May–2 June 2014.
14. Pasarica, A.; Andruseac, G.G.; Adochiei, I.; Rotariu, C.; Costin, H.; Adochiei, F. Remote Control of an Autonomous Robotic Platform Based on Eye Tracking. *Adv. Electr. Comput. Eng.* **2016**, *16*, 95–100. [CrossRef]
15. Astudillo, L.; Melin, P.; Castillo, O. *Chemical Optimization Algorithm for Fuzzy Controller*; Springer International Publishing: Cham, Switzerland, 2014.
16. Martínez, R.; Castillo, O.; Aguilar, L.T. Intelligent Control for a Perturbed Autonomous Wheeled Mobile Robot Using Type-2 Fuzzy Logic and Genetic Algorithms. *J. Autom. Mob. Robot. Intell. Syst.* **2008**, *2*, 12–22.
17. Astudillo, L.; Castillo, O.; Melin, P.; Alanis, A.; Soria, J.; Aguilar, L.T. Intelligent Control of an Autonomous Mobile Robot using Type-2 Fuzzy Logic. *Eng. Lett.* **2013**, *13*, 5.
18. Han, J.; Han, S.; Lee, J. The Tracking of a Moving Object by a Mobile Robot Following the Object's Sound. *J. Intell. Robot. Syst.* **2013**, *71*, 31–42. [CrossRef]

19. Borenstein, J.; Koren, Y. The vector field histogram-fast obstacle avoidance for mobile robots. *IEEE Trans. Robot. Autom.* **1991**, *7*, 278–288. [CrossRef]
20. Ulrich, I.; Borenstein, J. VFH+: Reliable obstacle avoidance for fast mobile robots. In Proceedings of the 1998 IEEE International Conference on Robotics and Automation (Cat. No.98CH36146), Leuven, Belgium, 20 May 1998; Volume 2, pp. 1572–1577.
21. LaValle, S.M.; Kuffner, J.J., Jr. Randomized kinodynamic planning. *Int. J. Robot. Res.* **2001**, *20*, 378–400. [CrossRef]
22. Dönmez, E.; Kocamaz, A.F.; Dirik, M. Bi-RRT path extraction and curve fitting smooth with visual based configuration space mapping. In Proceedings of the International Artificial Intelligence and Data Processing Symposium (IDAP), Malatya, Turkey, 16–17 September 2017.
23. Rashid, A.T.; Ali, A.A.; Frasca, M.; Fortuna, L. Path planning with obstacle avoidance based on visibility binary tree algorithm. *Robot. Auton. Syst.* **2013**, *61*, 1440–1449. [CrossRef]
24. Wu, H.; Mendel, J.M. Introduction to Uncertainty Bounds and Their Use in the Design of Interval Type-2 Fuzzy Logic Systems. In Proceedings of the 10th IEEE International Conference on Fuzzy Systems (Cat. No.01CH37297), Melbourne, Victoria, Australia, 2–5 December 2001.
25. Al-Mutib, K.; Abdessemed, F. Indoor Mobile Robot Navigation in Unknown Environment Using Fuzzy Logic Based Behaviors. *Adv. Sci. Technol. Eng. Syst. J.* **2017**, *2*, 327–337. [CrossRef]
26. Sepúlveda, R.; Castillo, O.; Melin, P.; Rodríguez-Díaz, A.; Montiel, O. Experimental study of intelligent controllers under uncertainty using type-1 and type-2 fuzzy logic. *Inf. Sci.* **2007**, *177*, 2023–2048. [CrossRef]
27. Castro, J.R.; Castillo, O. Interval Type-2 Fuzzy Logic for Intelligent Control Applications. In Proceedings of the 2007 Annual Meeting of the North American Fuzzy Information Processing Society (NAFIPS'07), San Diego, CA, USA, 24–27 June 2007; pp. 592–597.
28. Zadeh, L.A. A fuzzy-algorithmic approach to the definition of complex or imprecise concepts. *Int. J. Man-Mach. Stud.* **1976**, *8*, 249–291. [CrossRef]
29. Mendel, J.M. Type-2 fuzzy sets and systems. *Inf. Sci.* **2007**, *177*, 84–110. [CrossRef]
30. Mendel, J.M.; John, R.I.B. Type-2 fuzzy sets made simple. *IEEE Trans. Fuzzy Syst.* **2002**, *10*, 117–127. [CrossRef]
31. Wu, H.; Mendel, J.M. Uncertainty bounds and their use in the design of interval type-2 fuzzy logic systems. *IEEE Trans. Fuzzy Syst.* **2002**, *10*, 622–639.
32. Karnik, N.N.; Mendel, J.M.; Liang, Q. Type-2 Fuzzy Logic Systems. *IEEE Trans. Fuzzy Syst.* **1999**, *7*, 16. [CrossRef]
33. Sadeghian, A.; Mendel, J.M.; Tahayori, H. *Advances in Type-2 Fuzzy Sets and Systems: Theory and Applications*; Springer: New York, NY, USA, 2013.
34. Mendel, J.M.; John, R.I.; Liu, F. Interval Type-2 Fuzzy Logic Systems Made Simple. *IEEE Trans. Fuzzy Syst.* **2006**, *14*, 808–821. [CrossRef]
35. Castillo, O.; Melin, P. A review on interval type-2 fuzzy logic applications in intelligent control. *Inf. Sci.* **2014**, *279*, 615–631. [CrossRef]
36. Andersh, J.; Mettler, B. Modeling the human visuo-motor system to support remote-control operation. *Sensors* **2018**, *18*, 2979. [CrossRef]
37. Andersh, J.; Li, B.; Mettler, B. Modeling visuo-motor control and guidance functions in remote-control operation. In Proceedings of the 2014 IEEE/RSJ International Conference on Intelligent Robots and Systems, Chicago, IL, USA, 14–18 September 2014.
38. Tseng, K.S.; Mettler, B. Analysis of Coordination Patterns between Gaze and Control in Human Spatial Search. *IFAC-PapersOnLine* **2019**, *51*, 264–271. [CrossRef]
39. Cai, H.; Fang, Y.; Ju, Z.; Costescu, C.; David, D.; Billing, E.; Ziemke, T.; Thill., S.; Belpaeme, T.; Vanderborght, B.; et al. Sensing-enhanced therapy system for assessing children with autism spectrum disorders: A feasibility study. *IEEE Sens. J.* **2019**, *19*, 1508–1518. [CrossRef]
40. Hernández, E.; Hernández, S.; Molina, D.; Acebrón, R.; Cena, C.E.G. OSCANN: Technical characterization of a novel gaze tracking analyzer. *Sensors* **2018**, *18*, 522. [CrossRef] [PubMed]
41. Webb, J.D. *Gaze Control for Remote Robotics*; Colorado School of Mines: Golden, CO, USA, 2016; p. 87.
42. Prabhakar, G.; Biswas, P. Eye Gaze Controlled Projected Display in Automotive and Military Aviation Environments. *Multimodal Technol. Interact.* **2018**, *2*, 1. [CrossRef]

43. Li, S.; Webb, J.; Zhang, X.; Nelson, C.A. User evaluation of a novel eye-based control modality for robot-assisted object retrieval. *Adv. Robot.* **2017**, *31*, 382–393. [CrossRef]
44. Rojas, M.; Ponce, P.; Molina, A. A fuzzy logic navigation controller implemented in hardware for an electric wheelchair. *Int. J. Adv. Robot. Syst.* **2018**, *15*, 1729881418755768. [CrossRef]
45. Solea, R.; Filipescu, A.; Filipescu, A.; Minca, E.; Filipescu, S. Wheelchair control and navigation based on kinematic model and iris movement. In Proceedings of the IEEE 7th International Conference on Cybernetics and Intelligent Systems (CIS) and IEEE Conference on Robotics, Automation and Mechatronics (RAM), Siem Reap, Cambodia, 15–17 July 2015.
46. Heo, H.; Lee, J.M.; Jung, D.; Lee, J.W.; Park, K.R. Nonwearable Gaze Tracking System for Controlling Home Appliance. *Sci. World J.* **2014**, *2014*, 303670. [CrossRef] [PubMed]
47. Matsumotot, Y.; Ino, T.; Ogsawara, T. Development of intelligent wheelchair system with face and gaze based interface. In Proceedings of the 10th IEEE International Workshop on Robot and Human Interactive Communication, ROMAN 2001 (Cat. No.01TH8591), Paris, France, 18–21 September 2001; pp. 262–267.
48. Training Program for Dyslexic Children Using Educational Robotics—PDF Free Download. zapdf.com. Available online: https://zapdf.com/training-program-for-dyslexic-children-using-educational-rob.html (accessed on 1 September 2018).
49. NI LabVIEW Module for LEGO® MINDSTORMS®—National Instruments. Available online: http://sine.ni.com/nips/cds/view/p/lang/en/nid/212785 (accessed on 1 September 2018).
50. About EV3—Mindstorms. LEGO.com. Available online: https://www.lego.com/en-us/mindstorms/about-ev3 (accessed on 3 September 2018).
51. Logitech C920 HD Pro Webcam for Windows, Mac, and Chrome OS. Available online: https://www.logitech.com/en-us/product/hd-pro-webcam-c920?crid=34 (accessed on 3 September 2018).
52. Object Tracking Techniques—NI Vision 2016 for LabVIEW Help—National Instruments. Available online: https://zone.ni.com/reference/en-XX/help/370281AC-01/nivisionconcepts/object_tracking_techniques/ (accessed on 4 September 2018).
53. Castillo, O.; Melin, P.; Kacprzyk, J.; Pedrycz, W. Type-2 Fuzzy Logic: Theory and Applications. In Proceedings of the 2007 IEEE International Conference on Granular Computing (GRC 2007), Fremont, CA, USA, 2–4 November 2007; p. 145.
54. Zadeh, L.A. The concept of a linguistic variable and its application to approximate reasoning-III. *Inf. Sci.* **1975**, *9*, 43–80. [CrossRef]
55. Karnik, N.N.; Mendel, J.M. Centroid of a type-2 fuzzy set. *Inf. Sci.* **2001**, *132*, 195–220. [CrossRef]

MDPI
St. Alban-Anlage 66
4052 Basel
Switzerland
Tel. +41 61 683 77 34
Fax +41 61 302 89 18
www.mdpi.com

Applied System Innovation Editorial Office
E-mail: asi@mdpi.com
www.mdpi.com/journal/asi

www.ingramcontent.com/pod-product-compliance
Lightning Source LLC
LaVergne TN
LVHW070136100526
838202LV00015B/1835

* 9 7 8 3 0 3 6 5 4 9 2 9 3 *